An Introduction to High-Performance Scientific Computing

Scientific and Engineering Computation
Janusz Kowalik, editor

Data-Parallel Programming on MIMD Computers, Philip J. Hatcher and Michael J. Quinn, 1991.

Unstructured Scientific Computation on Scalable Multiprocessors, edited by Piyush Mehrotra, Joel Saltz, and Robert Voigt, 1992.

Parallel Computational Fluid Dynamics: Implementation and Results, edited by Horst D. Simon, 1992.

Enterprise Integration Modeling: Proceedings of the First International Conference, edited by Charles J. Petrie, Jr., 1992.

The High Performance Fortran Handbook, Charles H. Koelbel, David B. Loveman, Robert S. Schreiber, Guy L. Steele J., and Mary E. Zosel, 1994.

Using MPI: Portable Parallel Programming with the Message Passing Interface, William Gropp, Ewing Lusk, and Anthony Skjellum, 1994.

PVM: Parallel Virtual Machine—A Users' Guide and Tutorial for Networked Parallel Computing, Al Geist, Adam Beguelin, Jack Dongarra, Weicheng Jiang, Robert Mancheck, and Vaidy Sunderam, 1994.

Enabling Technologies for Petaflops Computing, Thomas Sterling, Paul Messina, and Paul H. Smith, 1995.

Practical Parallel Programming, Gregory V. Wilson, 1995.

An Introduction to High-Performance Scientific Computing, Lloyd D. Fosdick, Elizabeth R. Jessup, Carolyn J. C. Schauble, and Gitta Domik, 1995.

An Introduction to High-Performance Scientific Computing

Lloyd D. Fosdick
Elizabeth R. Jessup
Carolyn J. C. Schauble
Gitta Domik

The MIT Press
Cambridge, Massachusetts
London, England

This book was set in LaTeX by the authors and was printed and bound in the United States of America.

Library of Congress Cataloging-in-Publication Data

An introduction to high-performance scientific computing / Lloyd D. Fosdick ... [et al].
    p.   cm.—(Scientific and engineering computation)
    Includes bibliographical references and index.
    ISBN 0-262-06181-3 (hc: alk. paper)
    1. Electronic data processing. 2. Supercomputers. 3. Science—Data processing. I. Fosdick, Lloyd Dudley.
II. Series.
QA76.A594   1995
502′.85′411—dc20   95-20383
                                                                          CIP

Contents

Series Foreword

The world of modern computing potentially offers many helpful methods and tools to scientists and engineers, but the fast pace of change in computer hardware, software, and algorithms often makes practical use of the newest computing technology difficult. The Scientific and Engineering Computation series focuses on rapid advances in computing technologies and attempts to facilitate transferring these technologies to applications in science and engineering. It will include books on theories, methods, and original applications in such areas as parallelism, large-scale simulations, time-critical computing, computer-aided design and engineering, use of computers in manufacturing, visualization of scientific data, and human-machine interface technology.

The series will help scientists and engineers to understand the current world of advanced computation and to anticipate future developments that will impact their computing environments and open up new capabilities and modes of computation.

This book has been designed and developed for senior undergraduate students in engineering and science. It helps to formulate and solve numerical problems that require significant amounts of computing power, such as: molecular dynamics, meteorology and computerized tomography. It shows how to implement several numerical algorithms efficiently on a wide range of high-performance computers, including: pipelined workstations, vector processors, SIMD distributed-memory computers, and MIMD distributed-memory computers. The textbook is supplemented by a laboratory manual and can be used for a "hands-on" practical course in large-scale scientific and engineering computation. A special feature of the book is its interdisciplinary character.

Janusz S. Kowalik

Preface

This book is the outgrowth of a project we began in the Fall of 1990 to develop an undergraduate course in high-performance scientific computing. Beginning in 1985, one of us, LDF, taught a course in parallel computing for undergraduate and graduate students at the University of Colorado at Boulder using various parallel computers and emphasizing numerical computation. The parallel computers used in that course included the Cyber 205, the Myrias SPS-2, and the Encore Multimax, among others. In the Spring of 1990, the National Science Foundation announced a program to stimulate educational activities at the undergraduate level in computer and information science, computer engineering, computational science, and artificial intelligence. This seemed like an excellent opportunity to establish a teaching laboratory and course materials, building on our experience in teaching the parallel computing course just mentioned. Accordingly we submitted a proposal to this program to develop a "Curriculum in High-Performance Scientific Computing," and we were successful in obtaining a three-year grant.

The nature of this book, its level, its content, and its organization are the result of our philosophy and objectives for the course. First of all, we have designed the course for upper-division undergraduate students in engineering and science. In particular, it is not intended for computer science students only even though that is our "home" department. Thus, one of the most important aspects of the course is that it is interdisciplinary for such is the nature of scientific computation. Although most students at our university can be expected to have the right prerequisites for this course, this is not always the case, and many have forgotten what they once learned. For that reason, this book contains some chapters reviewing material we expect the students to know upon entry.

Secondly, we want students to come away from the course with a good appreciation for the kind of computing that takes place at a big scientific research laboratory such as the National Center for Atmospheric Research (NCAR) here in Boulder, Los Alamos National Laboratory, the National Center for Supercomputer Applications in Champaign-Urbana, and other facilities of this nature. Consequently, we discuss applications and problems that demand high computing power, and we discuss algorithms for solving these problems. We discuss supercomputers and how to use them effectively in those scientific and engineering contexts. Finally, we discuss scientific visualization as a technique for understanding the huge volume of data produced by a typical computation.

We decided at the outset to aim for depth rather than breadth in the applications. Therefore, the course focuses on a small number of applications that require rather different mathematical frameworks. It also focuses on a small set of computers that represent significantly different architectures. Thus, we can address the challenge of solving several kinds of problems on a variety of architectures as well as the suitability of a particular architecture to a particular kind of problem. In this book, we have considered three applications: molecular dynamics as studied in chemistry and physics, advection from the fields of meteorology and climatology, and computerized tomography as used in medical and other applications. The problems we chose from these applications require the solution of systems of differential equations in the case of molecular dynamics, the solution of a partial differential equation in the case of advection, and the use of Fourier transforms in the case of computerized tomography. We examine the mathematical foundations of these problems and the numerical techniques used to carry them out computationally. We show how to implement the numerical algorithms efficiently on pipelined workstations, vector processors, SIMD distributed-memory computers, and MIMD distributed-memory computers.

Thirdly, we believe that this should be a "hands-on" course. Consequently it has the structure of a traditional lecture/laboratory course with two lectures and one supervised laboratory (3 hours) per week. The textbook is supplemented by a laboratory manual. The laboratory manual and supporting software are available by anonymous ftp from the directory /pub/HPSC on cs.colorado.edu. We will say more about the laboratory manual shortly. The workstations and supercomputers are the equipment used for the laboratory experiments. This poses a special problem familiar to anyone who has used a supercomputer — namely, how to get the right kind of documentation for a beginning user. Even when manuals and guides are available from the center housing the particular supercomputer, they are not always appropriate for a beginner. Thus, as part of this book, we provide introductory material on the types of architectures we use. These are supplemented by guides to some particular computers also available by anonymous ftp.

Finally, we want the course material to be useful to others and to remain so for a reasonable period of time. Given rapidly changing computer technology and the diverse interests of potential users, these goals present major challenges. We have tried to meet them both by writing our materials in a modular fashion so that an instructor can select the parts meeting his or her interests and needs. Furthermore, we hope that the structure of what we have written can serve as a guide to others who might want to write similar material for other applications and and other computers. This approach also allows us to replace one machine by another of a similar architecture more easily, replacing a CM-2 by a MasPar MP-1 or MP-2, for example. Recently, a group of eight faculty members from six schools have joined together to prepare new modules for other applications and computers.

We now have three years of experience teaching this course and have presented segments of it in five two-week summer workshops for students and faculty from other schools. Parts of the course have also been taught by faculty at five schools who collaborated with us from the outset. Thus, we have been able to test our materials in a variety of settings and to revise them in response to the comments of both students and faculty. The course as we offer it is a two-semester sequence. Originally, we required students to take both semesters, but we found that many did not have enough time in their schedules even though they eagerly wanted the course. Accordingly, we rearranged the order in which we presented the material so that the first semester represents a reasonable stand-alone unit. At the conclusion of this preface, we describe the interrelationship of the various parts of this book and give some guidance about how they might be ordered in a one- or two-semester course.

A word about the laboratory is useful to indicate the kind of environment the students have and the nature of their work there. Originally, the room was equipped with four DEC 5000/200 workstations. One acted as a file server and another as a gateway to the university network and the Internet. Each of these workstations acted as host to three NCD X-terminals. Later, we added two SGI Indigos. All of the workstations have color monitors, and all of the original X-terminals had black and white displays. The choice of this particular equipment was based more on the results of bids received and the money we had than on a strong desire for a particular machine. We could equally well have used machines from any of the other major vendors of workstations. Now the DEC workstations are giving way to HP workstations, and the black and white X-terminals have been replaced by color models.

Besides the usual compilers, the laboratory machines are equipped with MATLAB, Maple, Mathematica, IDL, AVS, the NAG library, xnetlib, and LAPACK, and the present incarnation of the laboratory manual makes use of MATLAB and AVS as well as much of our own application software. The supercomputers are accessed over the Internet. The iPSC/2 is in-house, and the Cray Y-MP is at NCAR as was the CM-2 we used. On occasion, we have also used the CM-2 at NCSA, the iPSC/860 at Oak Ridge National Laboratory, the Paragon at the National Oceanic and Atmospheric Administration, a MasPar MP-1 at MasPar Corporation, and the in-house KSR-1.

The work in the laboratory is tightly structured as students complete specific examples and specific exercises from the manual. Weekly assignments from the laboratory manual are supplemented by two "miniprojects" in the first semester and a single final project in the second semester. These project assignments are also available by anonymous ftp. One of us is always present during the scheduled laboratory period to help students and to respond to emergencies. In its first years, the course has been restricted to sixteen students, making it easy to provide individual attention as needed. We have also been aided in the laboratory by an experienced undergraduate who provides systems support.

While we developed our course to run in this well-equipped laboratory, it has subsequently been taught successfully at schools with fewer and less powerful machines. Reliable access to remote or local supercomputers is the most essential ingredient.

A Suggested Course Outline

We now outline how we have presented the course materials in our one semester and year-long courses. (More detailed syllabi that include the lab manual chapters are available by anonymous ftp.) Upon entry, we expect the student to have mastery of the basics of numerical computation and IEEE arithmetic as reviewed in chapters 2 and 3 of this book. To use the laboratory effectively, they should have a good working knowledge of the basics of UNIX and ftp reviewed in chapter 4. That chapter also includes a review of the vi editor. They should also have studied the material on the UNIX Make utility presented in chapter 5. As much of the software provided for this course is written in Fortran, they may want to review the basics of this language in chapter 6.

Our first semester course then covers the material in chapters 1, 7, 9, 10, 11 12, 13, and 15. The first part of the course concerns tools to support the study of high-performance scientific computing. We begin the course with the introduction to the field of scientific computing in chapter 1. We then move on to the presentation of MATLAB and AVS in chapters 7 and 9. Both are valuable tools for the analysis and presentation of data. We also provide an overview of IDL in chapter 8 as this package can replace MATLAB in most of our computational exercises although it is not required in our course. Once the students have mastered the basics of these tools, we proceed to some important concepts of scientific visualization (the use of graphical methods to display scientific data) in chapter 10.

In the next part of the course, we study the basics of pipelined, vector, and distributed-memory MIMD architectures and performance measures for them. The fundamentals of performance measurement are covered in chapter 11. The vector and MIMD architectures are examined in chapters 12 and 13. The time constraints of a one-semester course mean that we must defer the introduction of the SIMD distributed-memory machines to the second semester.

The students spend the rest of the semester studying how to solve simple molecular dynamics problems as presented in chapter 15. They begin with an introduction to the various computational algorithms that can be used on the workstations. Mathematical work, performance measurement, and data visualization are combined to help the student learn about the most efficient and accurate serial approaches to the problem. They then extend their knowledge to solve the problems on the three architectures covered in the first semester.

Of the three applications we study, molecular dynamics uses the simplest mathematics. For this reason, we have found it to be a good starting problem, although the other applications could readily be used in its place. The laboratory exercises for the tomography application require the most extensive programming.

The second semester course covers chapters 14, 15, 16, and 17. It begins with introductions to SIMD computing in general (chapter 14) and to solving the molecular dynamics problem on SIMD machines in particular (chapter 15). We then move on to study, in turn, the advection and tomography applications as presented in chapters 16 and 17, respectively. As in the case of molecular dynamics, we begin by studying serial, pipelined implementation of the appropriate algorithms for each problem, then we move on to implementations for the vector and parallel machines. There is generally not time in the second semester to cover all architectures for both applications, so we omit some or give students their choice.

While the chapters may be used in a variety of orders, we feel that the basics of performance measurement as presented in chapter 11 (and extended to particular architectures in chapters 12, 13, and 14) are essential to any study of high-performance computing.

Color illustrations

The color illustrations shown in Plates 1 – 4 are color versions of the following black and white figures in the text:

Figure 7.14 on page 219,	Figure 7.16 on page 221,
Figure 10.3 on page 301,	Figure 10.5 on page 303,
Figure 10.17 on page 322,	Figure 10.20 on page 323,
Figure 10.22 on page 326,	Figure 10.25 on page 329.

Acknowledgments

A project as large as this is impossible to complete without the assistance of many people. Here we acknowledge those who most helped us to pull the whole thing together.

We start by thanking our collaborators, Brian Smith at the University of New Mexico, Jean Bell at the Colorado School of Mines, Xiaodong Zhang of the University of Texas at San Antonio, Jim Wixom at Fort Lewis College in Durango, and Steve Schaffer of the New Mexico Institute of Mining and Technology, and the evaluating committee, Kris Stewart, Dennis Gannon, Vance Faber, and Paul Swarztrauber.

We also thank the many people who have assisted us by writing text or software, by keeping the HPSC computer laboratory running, by telling us about their classroom experiences with our materials, by critiquing parts of the manuscript, or by otherwise sharing their expertise. These include Salim Alam, Chris Beattie, Raj Chaudhury, Dennis Colarelli, Bill Connor, Silvia Crivelli, Jack Dongarra, Ed Donley, Zlatko Drmac, Howard Foster, Steve Goldhaber, Dirk Grunwald, Chris Hall, Steve Hammond, Radka Kerpedjieva, David Kincaid, Alan Krantz, Michael Kreutner, Jim Lane, Todd Miller, Manav Misra, Evi Nemeth, Rich Neves, Paul Pinkney, Chris Redmond, Carlin Rodgers, Wolfgang Schildbach, Gene Schumacher, Anna Szczyrba, Jim Tung, Robert van de Geijn, Andre van der Hoek, John Wilson, the participants of the student and faculty workshops, and the students in the HPSC class here at CU Boulder.

A special acknowledgment goes to the participants of the first pilot class, Jeremy Asbill, Todd Englund, Chris Fischer, Soraya Ghiasi, Steve Judd, David Lloyd, and Harijono Tedjo.

We thank Digital Equipment Corporation, Hewlett-Packard Company, and Network Computing Devices, Inc. for equipment donations, the Hayden Image Processing Group of Boulder for use of their imaging software, and MasPar Computer Corporation, the National Center for Atmospheric Research (NCAR) in Boulder, and the National Center for Supercomputing Applications (NCSA) in Illinois for providing computing time on their machines.

Finally, we wish to give our thanks to our families and teachers for their support and encouragement and to the National Science Foundation for providing the funds that made this project and this book possible.

LDF, ERJ, CJCS
Boulder

GD
Paderborn

Trademarks

The following trademarks are the property of the following organizations:

- PostScript is a registered trademark of Adobe Systems, Inc.

- AVS is a trademark of Advanced Visual Systems, Inc.

- Alliant FX/8 is a trademark of Alliant Computer Systems Corporation.

- ANSI is a trademark of the American National Standards Institute, Inc.

- Ametek-2010, Ametek S/14 are trademarks of Ametek Corporation.

- Cosmic Cube, Mark III Cosmic Cube are trademarks of the California Institute of Technology.

- CDC 6600, CDC Star 100, CDC Cyber 205 are trademarks of Control Data Corporation.

- Convex, Convex C3, Convex C3880, Convex C4/XA2, Convex Exemplar are trademarks of Convex Computer Corporation.

- Cray-3, Cray-4 are trademarks of Cray Computer Corporation.

- CF77, CFT, CFT77, Cray, Cray-1, Cray-2, Cray C90, Cray Fortran, Cray T3D, Cray X-MP, Cray Y-MP, Cray Y-MP/832, Cray Y-MP8/864, Cray Y-MP/M90, UNICOS are trademarks of Cray Research, Inc.

- DEC, DEC 5000/200, DEC 5000/240, DEC 3000-500 AXP, DEC 10000-660 AXP, DEC Alpha, DEC AVS, DEC PXG 3D Accelerator, DECstation, DECstation 5000, DECstation 5000/200, DECstation 5000/240, PXG, PXG Turbo+, ULTRIX, VAX, VAX 11/780 are trademarks of Digital Equipment Corporation.

- Encore Multimax is a trademark of Encore Computer Corporation.

- ETA-10 is a trademark of the ETA Corporation.

- T-series is a trademark of Floating Point Systems.

- Goodyear MPP is a trademark of Goodyear Tire and Rubber Company, Inc.

- Hewlett-Packard, HP-PA, HP 9000/750, HP 9000/735 are trademarks of Hewlett-Packard Company.

- Intel, Intel 8086/8087, Intel 8087, Intel 80286/80287, Intel 80287, Intel 80386/80387, Intel 80387, Intel i860, iPSC, Paragon, System 310, Touchstone Delta are registered trademarks of Intel Corporation.

- IBM, IBM/360 Model 91, IBM 3090, IBM 9076 SP1, IBM 9076 SP2, IBM ES/9000, IBM Power2-990, IBM RS/6000, IBM RS/6000-350, IBM RS/6000-980, IBM SP1, IBM SP2 are trademarks of International Business Machines Corporation.

- ICL DAP is a trademark of International Computers Limited.

- KSR-1, KSR-2 are trademarks of Kendall Square Research.

- MasPar Fortran, MasPar MP-1, MasPar MP-2, MasPar Programming Environment, MPF, MPL, MPPE, X-net are trademarks of MasPar Computer Corporation.

- X-Window System is a trademark of The Massachusetts Institute of Technology.

- Handle Graphics, MATLAB are trademarks of The MathWorks, Inc.

- MIPS, MIPS R2000, MIPS R2010, MIPS R3000, MIPS R3010, MIPS R4000, MIPS R10000 are trademarks of MIPS Computer Systems, Inc.

- Motorola 68020 is a trademark of Motorola Corporation.

- Myrias SPS-2 is a trademark of Myrias Research Corporation.

- Netscape is a trademark of Netscape Communications Corporation.

- NCD is a trademark of Network Computing Devices, Inc.

- NEC SX-A is a trademark of Nippon Electric Company.

- nCUBE, nCUBE/1, nCUBE/2, nCUBE/2S, nCUBE/3, nCUBE/ten are trademarks of nCUBE Corporation.

- NAG is a registered trademark of Numerical Algorithms Group Incorporated.

- VAST-2 is a trademark of Pacific Sierra.

- IDL is a registered trademark of Research Systems, Inc.

- SGI Challenge, SGI Crimson, SGI Indigo, SGI Indigo R3000, SGI Indigo R4000 are trademarks of Silicon Graphics, Inc.

- Trinitron is a trademark of Sony Electronics, Inc.

- Sun, Sun-3, SPARC, SPARCstation, SunView are trademarks of Sun Microsystems, Inc.

- Symbolics is a trademark of Symbolics, Inc.

- Tandem NonStop is a trademark of Tandem Computers.

- TEKtronix is a trademark of Tektronix, Inc.

- Teradata DBC/1012 Data Base Computer is a trademark of Teradata Corporation.

- TI-ASC is a trademark of Texas Instruments.

- C*, CM, CM-1, CM-2, CM-5, CM Fortran, Connection Machine, DataVault, *Lisp, Paris, Slicewise are trademarks of Thinking Machines Corporation.

- Spice is a trademark of the University of California at Berkeley.

- Illiac IV, Mosaic are trademarks of the University of Illinois.

- UNIX is a trademark of UNIX Systems Laboratories, Inc.

- PV-WAVE is a trademark of Visual Numerics.

- Maple is a registered trademark of Waterloo Maple Software.

- Weitek is a trademark of Weitek Computers.

- Mathematica is a trademark of Wolfram Research, Inc.

- Ethernet is a trademark of Xerox Corporation.

An Introduction to High-Performance Scientific Computing

1 An Overview of Scientific Computing

1.1 Introduction

The computer has become at once the microscope and the telescope of science. It enables us to model molecules in exquisite detail to learn the secrets of chemical reactions, to look into the future to forecast the weather, and to look back to a distant time at a young universe. It has become a critically important filter for those tools of science like high-energy accelerators, telescopes, and CAT scanners that generate large volumes of data which must be reduced, transformed, and arranged into a picture we can understand. And it has become the key instrument for the design of new products of our technology: gas turbines, aircraft and space structures, high-energy accelerators, and computers themselves.

The story of modern scientific computing begins with the opening of the computer era in the 1940s during World War II. The demands of war provided the motivation and money for the first developments in computer technology. The Automatic Sequence Calculator built by H. H. Aiken at Harvard, the relay computers by George Stibitz at Bell Telephone Laboratories, the Eniac by John Mauchly and J. Presper Eckert at the Moore School of the University of Pennsylvania, the Edvac growing from the Eniac effort and inspired by ideas of John von Neumann were products of this time. They were used almost exclusively for numerical computations, including the production of mathematical tables, the solution of equations for the motion of projectiles, firing and bombing tables, modeling nuclear fission, and so forth. But it was not all numerical computing, the deciphering of codes by Alan Turing on the Colossus computers at Bletchley Park in England is an important example of non-numerical computing activities in this period. The machines of this early period operated at speeds ranging from about one arithmetic operation per second to about one hundred operations per second.

Immediate problems of war did not provide the only motivation for computer development in this early period. Goldstine, von Neumann, and others recognized the importance of computers for the study of very fundamental problems in mathematics and science. They pointed to the importance of computers for studying nonlinear phenomena, for providing "heuristic hints" to break a deadlock in the advance of fluid dynamics, and for attacking the problem of meteorological forecasting. And computers themselves motivated new kinds of investigations, including Turing's work on fundamental questions in logic and the solvability of problems, and von Neumann's on self-reproducing automata. The possibility of solving systems of equations far larger than had ever been done before raised new questions about the numerical accuracy of solutions which were investigated by Turing, Goldstine, von Neumann, Wilkinson and others. While great advances have been made, these questions, these problems remain. This is not a failure of the promise of the computer but a testament to the fundamental nature of these questions.

In the following sections we take a broad look at scientific computing today. The aim is to capture your interest, to stimulate you to read further, to investigate, and to bring your own talents and energy to this field.

1.2   Large-scale scientific problems

In 1987, William Graham, who was then the director of the Office of Science and Technology Policy, presented a five-year strategy for federally supported research and development on high-performance computing. Subsequently, as a part of this strategy, a detailed plan for a Federal High Performance Computing Program (HPCP) was developed.[1] It provided a list of "grand challenge" problems: fundamental problems in science and engineering with potentially broad economic, political, or scientific impact, which could be advanced by applying high-performance computing resources. The grand challenge problems are now often cited as prototypes of the kinds of problems that demand the power of a supercomputer.

Here is a slightly shortened list of the grand challenge problems as from the 1989 report on the HPCP by the Office of Science and Technology Policy.

---

[1] Approximately $800M was proposed for this program in 1993.

Prediction of weather, climate, and global change *The aim is to understand the coupled atmosphere-ocean biosphere system in enough detail to be able to make long-range predictions about its behavior. Applications include understanding carbon dioxide dynamics in the atmosphere, ozone depletion, and climatological perturbations due to man-made releases of chemicals or energy into one of the component systems.*

Challenges in materials science *High-performance computing provides invaluable assistance in improving our understanding of the atomic nature of materials. Many of these have an enormous impact on our national economy. Examples include semiconductors, such as silicon and gallium arsenide, and high-temperature superconductors, such as copper oxide ceramics.*

Semiconductor design *As intrinsically faster materials, such as gallium arsenide, are used for electronic switches, a fundamental understanding is required of how they operate and how to change their characteristics. Currently, it is possible to simulate electronic properties for simple regular systems, however, materials with defects and mixed atomic constituents are beyond present computing capabilities.*

Superconductivity *The discovery of high temperature superconductivity in 1986 has provided the potential for spectacular energy-efficient power transmission technologies, ultra sensitive instrumentation, and new devices. Massive computational power is required for a deeper understanding of high temperature superconductivity, especially of how to form, stabilize, and use the materials that support it.*

Structural biology *The aim of this work is to understand the mechanism of enzymatic catalysis, the recognition of nucleic acids by proteins, antibody/antigen binding, and other phenomena central to cell biology. Computationally intensive molecular dynamics simulations, and three-dimensional visualization of the molecular motions are essential to this work.*

Design of drugs   *Predictions of the folded conformation of proteins and of RNA molecules by computer simulation is a useful, and sometimes primary, tool for drug design.*

Human genome   *Comparison of normal and pathological molecular sequences is our most powerful method for understanding genomes and the molecular basis for disease. The combinatorial complexity posed by the exceptionally long sequence in the human genome puts such comparisons beyond the power of current computers.*

Quantum chromodynamics (QCD)   *In high energy theoretical physics, computer simulations of QCD yield computations of the properties of strongly interacting elementary particles. This has led to the prediction of a new phase of matter, and computation of properties in the cores of the largest stars. Computer simulations of grand unified "theories of everything" are beyond current computer capabilities.*

Astronomy   *The volumes of data generated by radio telescopes currently overwhelm the available computational resources. Greater computational power will significantly enhance their usefulness.*

Transportation   *Substantial contributions can be made to vehicle performance through improved computer simulations. Examples include modeling of fluid dynamical behavior for three-dimensional fluid flow about complete aircraft geometries, flow inside turbines, and flow about ship hulls.*

Turbulence   *Turbulence in fluid flows affects the stability and control, thermal characteristics, and fuel needs of virtually all aerospace vehicles. Understanding the fundamental physics of turbulence is requisite to reliably modeling flow turbulence for the performance analysis of vehicle configurations.*

Efficiency of combustion systems   *To attain significant improvements in combustion efficiencies requires understanding the interplay between the flows of the various substances involved and the quantum chemistry that causes those* *substances to react. In some complicated cases the quantum chemistry required to understand the reactions is beyond the reach of current supercomputers.*

Enhanced oil and gas recovery   *This challenge has two parts: to locate as much of the estimated 300 billion barrels of oil reserves in the US as possible and to devise economic ways of extracting as much of this oil as possible. Thus both improved seismic analysis techniques and improved understanding of fluid flow through geological structures are required.*

Computational ocean sciences   *The objective is to develop a global ocean prediction model incorporating temperature, chemical composition, circulation and coupling to the atmosphere and other oceanographic features. This will couple to models of the atmosphere in the effort on global weather as well as having specific implications for physical oceanography.*

Speech   *Speech research is aimed at providing communication with a computer based on spoken language. Automatic speech understanding by computer is a large modeling and search problem in which billions of computations are required to evaluate the many possibilities of what a person might have said.*

Vision   *The challenge is to develop human-level visual capabilities for computers and robots. Machine vision requires image signal processing, texture and color modeling, geometric processing and reasoning, as well as object modeling. A competent vision system will likely involve the integration of all of these processes.*

Thus there is no shortage of problems for today's supercomputers. At a future time, when we have petaflop computers (10^{15} floating-point operations per second), we can be sure there will be no shortage of problems for them either. The solution of old problems raises new problems. That is the nature of science, its challenge and its mystery.

Most of the grand challenge problems involve modeling a physical system in a computer and using this model to create a simulation of its behavior. Others involve reduction and analysis of experimental data on a very large scale. Even the modeling problems involve data analysis and reduction on a large scale because running the simulation generates large files of data for analysis. Frequently, data analysis requires representation of the data in the form of pictures, graphs, and movies — a fascinating and rapidly growing activity known as scientific visualization. To get a closer look at a modeling problem we focus briefly on an important problem from the atmospheric sciences.

Computer simulation of the greenhouse effect

Global warming has been the subject of growing international attention. This problem is being studied by computer simulations that help us understand how changing concentrations of carbon dioxide in the atmosphere contribute to global warming through the greenhouse effect. A study of this type requires modeling the climate over a period of time. Studies by Washington and Bettge at the National Center for Atmospheric Research provide a typical example. A climate model known as the general circulation model (GCM) is used to study the warming which would be caused by doubling the concentration of carbon dioxide over a period of 20 years. The computations they describe were done on a Cray-1, now a relatively old computer, with a peak speed of about 200 Mflops.[2] The scientists report:

> *"The model running time was 110 seconds per simulated day. For two 19-year simulations, over 400 computational hours were required to complete the experiment."*

The effects the GCM attempts to take into account are illustrated in figure 1.1.

The atmosphere is a fluid and so the partial differential equations that govern the behavior of fluids are the mathematical basis of the GCM. Computer solution of these equations is done by a "finite difference" algorithm in which derivatives with respect to spatial coordinates and time are approximated by difference formulas in space and time. Thus a three-dimensional mesh in space is created, as illustrated in figure 1.2. Solution of the problem involves starting with some set of initial conditions, for which values are assigned to the variables at each mesh point, and stepping forward in time updating these variables at the end of each time step. There are some eight or nine vari-

---

[2]Mflop is an abbreviation for a "megaflop," 10^6 floating-point operations per second.

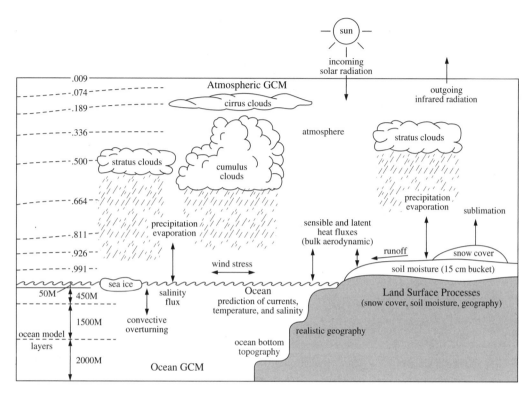

Figure 1.1: Schematic of the physical processes included in the GCM. This figure is adapted from the article by Washington and Bettge with their permission.

ables at each mesh point that must be updated, including temperature, wind velocity, CO_2 concentration, and so forth.

The mesh is the key for understanding why speed is so important. The mesh used in the computations was three dimensional with about 2000 points to cover the surface of the earth and nine layers spaced at different altitudes: altogether about 18,000 mesh points. A moment's thought will make it apparent this is an extremely coarse mesh. The surface of the earth is 2.1×10^8 sq. mi.; i.e., $100,000$ sq. mi. per mesh point. Colorado with a land area of $103,595$ sq. mi. rates one surface mesh point! Quite clearly we would like greater accuracy, and this means more mesh points. But if we double the density of points in each of the three directions we increase the number of mesh points by a factor of 8, essentially an order of magnitude increase in computational

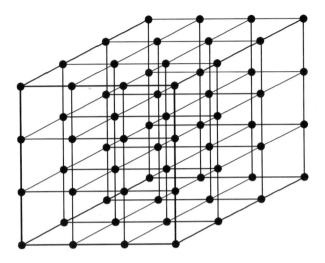

Figure 1.2: Illustration of a three-dimensional mesh.

demand. And so this computation which took 400 hours for around 18,000 mesh points, would take over 3000 hours, and we would still have only 3 or 4 surface points for Colorado.

1.3   The scientific computing environment

The scientific computing environment consists of high-performance workstations, supercomputers, networks, a wide range of software, and technical literature. In this section you will find some pointers to this material.

High-performance workstations have peak speeds of 10 to over 100 Mflops; the supercomputers have peak speeds of 500 Mflops to over 100 Gflops.[3] High-resolution color monitors with over 10^6 pixels provide excellent tools for pictorially representing data from scientific simulations. In table 1.1 there is a short list of high-performance workstations with the name of the machine, followed by the name of the manufacturer, followed by the performance in Mflops on the LINPACK Benchmark. [Dongarra 94]. This benchmark is based on the

---

[3]Gflop is an abbreviation for a *gigaflop*, 10^9 floating-point operations per second (flops).

Machine	Manufacturer	LINPACK (Mflops)
SUN SPARC10/40	SUN Microsystems	10.0
IBM RS/6000-350	IBM	19.0
IBM Power2-990	IBM	140.0
DEC 5000/240	Digital Equipment	5.3
HP 9000/735	Hewlett-Packard	41.0
SGI Indigo R4000	Silicon Graphics	12.0
SGI Crimson	Silicon Graphics	16.0

Table 1.1: A short list of high-performance workstations, and their performance on the LINPACK benchmark.

speed of solving a system of 100 simultaneous linear equations using software from LINPACK [Dongarra et al 79]. The peak performance of workstations in this table could be as much as five to ten times the LINPACK performance number.

Table 1.2 provides a short list of supercomputers, with the name of the machine series, the name of the manufacturer, and the *Theoretical Peak Performance* [Dongarra 94].[4]

As with the workstations, the machines in this list come in various models and configurations; the performance data is for the largest system listed in Dongarra's report. Note here, in contrast with the previous table, a theoretical peak performance figure is given; in an actual computation the performance could fall to one-fifth or one-tenth the performance figure given here. Roughly speaking, the speed advantage of a supercomputer over a high-performance workstation is a factor of 1000. The price difference is roughly the same or slightly higher.

The National Science Foundation (NSF) supports five supercomputer centers available to scientists and students for research and education:

---

[4]Dongarra's definition of these values: "The theoretical peak performance is determined by counting the number of floating-point additions and multiplications (in full precision) which can be performed in a period of time, usually the cycle time of the machine."

Machine	Manufacturer	Number of Processors	Theor Peak (Gflops)
CM 5	Thinking Machines	16,384	2,000
Cray Y-MP	Cray Research	16	15
Cray T3D	Cray Research	2,048	307
IBM ES/9000	IBM	6	2.7
IBM 9076 SP2	IBM	128	32
Intel iPSC/860	Intel	128	7.7
Intel Paragon	Intel	4,000	300
KSR2	Kendall Square Research	5,000	400
MP-2	MasPar	16,384	2.4
NEC SX-A	Nippon Electric Company	4	22

Table 1.2: A short list of supercomputers and their theoretical peak performances.

- Cornell Theory Center, Cornell University, Ithaca, NY.

- National Center for Atmospheric Research (NCAR), Boulder, CO.

- National Center for Supercomputer Applications (NCSA), University of Illinois, Champaign, IL.

- Pittsburgh Supercomputing Center, Carnegie Mellon University and the University of Pittsburgh, Pittsburgh, PA.

- San Diego Supercomputer Center, University of California at San Diego, San Diego, CA.

The facilities at these centers can be accessed via worldwide networks. Centers usually have facilities to accommodate on-site visitors, and they run workshops on a wide range of topics in scientific computing. In addition to these NSF centers there are other supercomputer centers at universities and national laboratories which provide network access to their facilities. These include:

- Arctic Region Supercomputing Center, University of Alaska, Fairbanks, AK.

- Army High Performance Computing Research Center, University of Minnesota, Minneapolis, MN.

- Advanced Computing Research Facility, Argonne National Laboratory, Argonne, IL.

- Center for Computational Science, Oak Ridge National Laboratory, Oak Ridge, TN.

- Lawrence Livermore National Laboratories, Livermore, CA.

- Los Alamos National Laboratory, Los Alamos, NM.

- Massively Parallel Computing Research Center, Sandia National Laboratories, Albuquerque, NM.

- Maui High Performance Computing Center, Kihei, Maui, HI.

- Research Institute for Advanced Computer Science, NASA Ames Research Center, Moffett Field, CA.

All of the supercomputers listed in table 1.2, except the NEC SX-A, are available at one of these centers.

Communication networks are a vital part of the supercomputing environment. The Internet is the worldwide system linking many smaller networks running the TCP/IP protocol. In May 1994 the Internet consisted of 31,000 networks, connecting over two million computers, and was growing at the phenomenal rate of one new network every 10 minutes [Leiner 94].

The National Science Foundation Network (NSFNET) is one of the most important components of the Internet, linking the the supercomputing centers in a network known as the backbone, which can be reached from other networks linking universities and other research organizations. Among these other networks are WESTNET in the Rocky Mountain states; NEARNET in the New England states; SURANET in the southern states; and MIDNET in the midwestern states.

The bandwidth of the NSFNET backbone has been 44.7 megabits/sec but a new backbone is under construction with a bandwidth of 155 megabits/sec. Communication networks with bandwidths in the gigabit/sec range are emerging. NSF and the Defense Advanced Research Projects Agency (DARPA) are

supporting five testbed research projects on communication networks operating at gigabit/sec rates. Included in these research projects is a study of
distributed computation on a very large scale: ocean and atmospheric climate
models will simultaneously run on separate computers exchanging data across
a network which includes the Los Alamos National Laboratory and the San
Diego Supercomputer Center.

Software to support scientific computing is available on the Internet from
the supercomputing centers and other sources. There is a particularly valuable
resource for software known as *Netlib*. It is a library of numerical software
available by e-mail or ftp from one of two centers in the U.S.A. A copy of
machine performance information as shown in the two tables above, and short
descriptions of these machines are also available from Netlib. Information
about Netlib can be obtained via e-mail to

`netlib@research.att.com` or `netlib@ornl.gov`

The body of your mail message should contain just the line

`help`

Also, you can access Netlib directly with anonymous ftp. For Europe there
is a duplicate collection of Netlib in Oslo, Norway with Internet address
`netlib@nac.no`; and for the Pacific region there is a collection at the University of Wollongong, NSW, Australia with the following Internet address
`netlib@draci.cs.uow.edu.au`.

There is a relatively new network tool called *Mosaic* now widely used for
browsing and retrieving information on the Internet. With this tool the user
can read documents located at another Internet site. These are hypertext
documents so the user can navigate through them by clicking on highlighted
keywords. High quality graphics images and animations can be included in
these documents. Menus facilitate other operations including retrieving entire
documents and programs. Mosaic was developed at NCSA and is available
from them at no charge by anonymous ftp at `ftp.ncsa.uiuc.edu`. A commercial version known as *Netscape* was recently produced.

Another important resource is represented by technical publications: journals and conference proceedings. The following lists tends to focus on computer related publications, especially those concerning parallel computing and

supercomputing, rather than applications. However an increasing number of articles concerning the application of computers to problems in physics, chemistry, and biology are finding there way into these publications. Some of the journals are:

- *ACM Transactions on Mathematical Software*

- *Computers in Physics*

- *Computer Physics Communications*

- *Computer Methods in Applied Mechanics and Engineering*

- *IEEE Transactions on Parallel and Distributed Systems*

- *International Journal of Parallel Programming*

- *International Journal of Supercomputer Applications*

- *Journal of Computational Physics*

- *The Journal of Supercomputing*

- *Methods in Computational Physics*

- *Parallel Computing*

- *SIAM Journal of Scientific Computing*

- *Supercomputing Review.*

Some of the regularly held conferences that issue proceedings are:

- Annual Symposium on Computer Architecture (IEEE Computer Society)

- Distributed Memory Computing Conference (IEEE Computer Society)

- Frontiers of Massively Parallel Computation (IEEE Computer Society)

- International Conference on Supercomputing (ACM)

- International Conference on Parallel Processing (ACM)

Workstation	Clock MHz	SPECmark	LINPACK Mflops	Theor Peak Mflops
SUN SPARC10/40	40.0	60.2	10.0	40
IBM RS6000-350	41.6	74.2	19.0	84
IBM Power2-990	71.5	260.4	140.0	286
DEC 5000/240	40.0	35.8	5.3	40
HP 9000/735	99.0	167.9	41.0	200
SGI Indigo R4000	50.0	60.3	12.0	—
SGI Crimson	50.0	63.4	16.0	32

Table 1.3: Some typical workstations and performance data. The SPECmark value is the ratio of the average speed of floating-point operations for the given machine to the average speed of floating-point operations for a VAX 11/780 on a set of benchmark programs. The LINPACK value is the speed in solving a system of 100 linear equations in double precision arithmetic. Theor Peak stands for the theoretical peak performance.

- International Parallel Processing Symposium (IEEE Computer Society)

- SIAM Conference on Parallel Processing for Scientific Computing (SIAM)

- Supercomputing (IEEE Computer Society)

- Supercomputing in Europe

- Visualization (IEEE Computer Society).

1.4   Workstations

The workstation is the desktop "supercomputer," small enough to fit on a desk but with peak speeds in the range of about 10 to 100 Mflops. Table 1.3 lists some popular workstations with performance figures. The actual speed of operation and the peak speed can differ substantially, as is illustrated by the data in this table: the speed on the LINPACK benchmark is substantially below the peak speed of these systems. Careful tuning of a program is necessary to get close to peak speed, and for some computations this is simply an unreachable goal. In some cases, best performance requires programming parts of the computation in assembly language: the compilers for high-level programming languages cannot always produce the best results.

Normally, a workstation is connected to a network giving it access to additional computing resources which include other workstations, storage devices, printers, and still more powerful computing systems. Thus it serves the scientist as a primary computing resource and as a link to a wide array of other resources. The price of a workstation ranges from about \$15K to about \$100K. The high-end machines are faster and include high quality graphics, multiple processors, and large memories.

The architecture of scientific workstations has undergone a period of rapid development since the early 1980s when they first appeared. From that time to the present, clock rates have increased to about 200 MHz; pipelining of instructions and arithmetic, and parallel functional units have been introduced; and a variety of caching mechanisms have been developed to overcome memory access delays. In a ten year period the speed of these machines has increased by about two orders of magnitude. In the following paragraphs we discuss some of the architectural features now found in popular workstations.

1.4.1 RISC architecture

Many workstations use a RISC (Reduced Instruction Set Computer) architecture [Patterson 85]. It is characterized by a relatively small set of simple instructions, pipelined instruction execution, and cache memory. The principal goal of this architecture is an execution speed of one instruction per clock cycle: with a 40 MHz clock an execution speed of 40 mips[5] is the goal. In contrast, the acronym CISC is used for architectures with larger and more complex instruction sets: the DEC VAX 11/780 with about 256 instructions is a CISC system; the DEC 5000 with 64 instructions is a typical RISC system.

The move towards RISC systems was stimulated by recognition that better performance could be achieved with a simpler and smaller instruction set.

---

[5]mips is an abbreviation for "million instructions per second," 10^6 instructions per second.

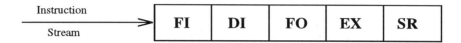

Figure 1.3: Pipelined execution of instructions. The steps are: FI, fetch instruction; DI, decode instruction; FO, fetch operand; EX execute instruction; SR store result. Since this pipeline has five segments, it can be operating on five instructions simultaneously. While the FI segment is fetching an instruction, the DI segment is decoding the previous instruction, and so on down the pipe.

Studies of programs executed on CISC machines showed that more complex instructions were not heavily used. This was attributed to the observation that the more complex instructions were too specialized and not needed for many computations, also it was difficult for compilers to recognize when they could be used effectively. By reducing and simplifying the instruction set, instruction decoding time was reduced, and space on the chip was saved. This provided more space for cache and cache management. Particularly important was the fact that simple instructions, all taking about the same amount of time to execute, made it possible to execute instructions in a *pipeline.*

Using a pipeline makes it possible to execute one instruction per clock cycle. The idea of pipelined instruction execution is easy to understand by analogy to the automobile assembly line. In the automobile assembly line the work to be done is divided into a series of steps each requiring the same amount of time, say τ seconds, so the rate of production of autos is one auto per τ seconds. Similarly, in the instruction pipeline the work of executing an instruction is divided into steps as illustrated in figure 1.3. In this pipeline, which has five steps, we expect to gain a factor of five in speed over execution without pipelining.

The speedup promised by a pipeline cannot always be attained. For example, branch instructions can cause a problem because the next instruction after the branch is not known until the branch test is executed. Thus a branch interrupts the smooth flow of instructions through the pipe. Since branches occur frequently in code they could seriously degrade performance. To deal with this problem RISC systems use a *delayed branch* that delays fetching the next instruction after the branch by a fixed number of clock cycles. This causes a bubble in the pipe that can be filled by an instruction that would be executed regardless of the direction the branch takes. Whether or not the bubble can be filled depends on the program, and if it cannot, there will be a degradation of performance.

RISC systems usually have a separate floating-point pipelined coprocessor, and some RISC systems contain multiple functional units allowing overlap of operations: the IBM RS6000 series and the Intel i860 are examples. Therefore, with these systems it is possible to achieve a performance even higher than one instruction per clock cycle, often referred to as *superscalar* performance. For example, overlap of floating-point add and multiply in the IBM RS6000 system allows evaluation of $a \times b + c$ in one clock period, giving a peak speed of 50 Mflops with a 25 MHz clock. Moreover, in one clock cycle this system is capable of executing four instructions simultaneously: a branch, a condition-register instruction, a fixed-point instruction, and a floating-point instruction. Including the possibility of overlap of floating-point add and multiply, this system can execute five instructions per clock cycle.

Operation	Single	Double
add	0.75×10^{-7}	0.75×10^{-7}
subtract	0.75×10^{-7}	0.75×10^{-7}
multiply	1.01×10^{-7}	1.26×10^{-7}
divide	3.00×10^{-7}	4.79×10^{-7}

Table 1.4: Time for arithmetic on DEC 5000/240, units are seconds.

1.4.2 The DEC 5000 workstation

One system we use in the HPSC laboratory is a DEC 5000/240 series system with a 40 MHz clock for the CPU. The memory subsystem, I/O controller, etc. operate with a 25 MHz clock [DEC 91]. It uses a MIPS[6] processor and floating-point coprocessor. Measurements of the time for elementary arithmetic operations give the results shown in table 1.4. The system can have from

---

[6]MIPS Computer Systems, Inc., Sunnyvale, CA.

8 to 480 Mbytes of DRAM memory[7] with a bandwidth of 100 Mbytes/sec. Our systems are configured with 24 or 32 Mbytes.

The processor subsystem has a 4 Gbytes virtual address space, 2 Gbytes of which are available to user processes. It has a 64 Kbyte cache memory for data and another 64 Kbyte cache memory for instructions. One word (32 bits) can be accessed from each cache in each processor cycle (25 nsec). A single memory read from a noncached address requires 690 nsec. Thus the time for a reading an operand from memory is about 28 times longer than reading it from cache.

The DEC 5000, as for most scientific workstations, can be augmented with 2D and 3D graphics options. The DEC PXG 3D Accelerator module includes an Intel i860 chip as a geometry engine and a scan converter chip to compute pixel values. The resolution of the display is 1280-by-1024 pixels. Double buffering and 24 image planes are provided with this module. The PXG system uses a 16 in. or 19 in. color Trinitron monitor. Peak speeds (for PXG Turbo+) are 436×10^3 vectors/sec and 106×10^3 polygons/sec: a "vector" is 10 pixels long; a "polygon" is a triangle, 100 pixels in area.

The MIPS R3000 and R3010 processors.

The CPU of the DEC 5000 is a MIPS R3000 processor and R3010 floating-point coprocessor. The user address space is 2 Gbytes. The coprocessor conforms to the ANSI/IEEE Standard for floating-point arithmetic.

The processor has 32 general purpose registers of 4 bytes each, and the coprocessor has 16 registers for floating-point numbers of 8 bytes each. Arithmetic instructions are register-register; e.g., add contents of register r1 to contents of register r2 and store result in register r3. Explicit move instructions move data between memory (cache) and registers in the processor or coprocessor, and between registers in the processor and coprocessor.

Times for arithmetic in the coprocessor are shown in table 1.5. Multiply and divide operations can be overlapped to some extent with other operations but not with each other; e.g., it is possible to execute a double precision

---

[7]In this section, the following abbreviations are used: Mbytes for megabytes (10^6 bytes), Gbytes for gigabytes (10^9 bytes), and nsec for nanoseconds (10^{-9} seconds).

Operation	Single	Double
add	2	2
subtract	2	2
multiply	4	5
divide	12	19

Table 1.5: Cycle time for arithmetic in MIPS R3010 coprocessor

floating-point addition and multiplication together in 5 cycles. For more information on these processors, see [Kane 88].

1.5 Supercomputers

The word "supercomputer" came into use in the late 1960s when radically new and powerful computers began to emerge from university and commercial laboratories. Since then it has symbolized the most powerful computers of the time.

Table 1.2 provides some information on the speeds of current supercomputers. The performance data given in the table provide only a rough estimate of performance. Performance depends strongly on the characteristics of the problem being solved, and hand tuning of the software can have a significant effect on performance. Because of the cost of supercomputers and the desire to solve increasingly large problems on them, the issue of squeezing the most out of these machines has received a lot of attention. New algorithms, tools, and techniques to improve their performance are constantly being developed, and research in programming languages to simplify writing effective programs is a continuing activity.

The high speed of supercomputers is the result of two factors: very fast logic elements and parallel architectures. Many people believe that the speed of logic elements is reaching the limit imposed by laws of physics and that parallel architectures are the key to supercomputers of the future. A parallel architecture allows many parts of a computation to be done simultaneously.

This feature is what really distinguishes these computers from earlier machines of the computing era. It is also the reason why supercomputers pose such an interesting challenge to the algorithm designer and why their performance depends so strongly on the problem being solved.

1.5.1   Parallel architectures

The common notion of a *parallel computer* or *multiprocessor* is a collection of processors that are connected together in a manner that lets them work together on a single problem, the idea is ten processors working together on a problem can get the job done ten times as fast as one of them working alone, and ten-thousand of them can get the job done ten-thousand times as fast, and so on. Of course, it doesn't happen quite that way.

Not all computational jobs can be divided up neatly into independent parcels in such a way as to keep all processors busy. Furthermore, one processor may need a result from another processor before it can proceed with its computation. Some jobs divide neatly while others do not; some require little interprocessor data communication while others require a lot. The following examples illustrate how hard it can be to keep the processor workloads balanced. In presenting these examples, we assume we are working with a *distributed-memory* multiprocessor. In this type of machine, a processor has direct access to its own memory only. Processors are required to send and receive messages to share data.

Evaluation of an integral

If the problem we have is to evaluate an integral, say

$$\int_a^b \int_c^d f(x,y)\, dx\, dy \quad ,$$

then an obvious way to proceed is to divide the domain of integration into n smaller rectangles, r_i, in such a way that the original problem is broken into subproblems, thus

$$\int_a^b \int_c^d f(x,y)\, dx\, dy \;=\; \int\!\!\int_{r_1} f(x,y)\, dx\, dy \quad + \qquad (1.1)$$

$$\int \int_{r_2} f(x,y)\, dx\, dy \quad + \qquad (1.2)$$

$$\ldots + \quad \int \int_{r_n} f(x,y)\, dx\, dy \qquad (1.3)$$

The computation represented by each term on the right can be done by a different processor. After all of these computations have been done, the n results must be brought together to form the sum. At this point, communication is needed. Clearly we can reduce communication by reducing the number of rectangles covering the domain of integration, but then we reduce the parallelism in the computation.

If we divide the region of integration into equal subareas it might seem that we will keep all processors equally busy. Not necessarily so! The integrand may change its value very rapidly in some regions compared with others. To maintain accuracy we will need to use smaller integration steps in some regions than others. If we know this in advance, we can try to divide the work more or less evenly. If we don't know it, we can use an adaptive scheme to adjust the integration step during the computation. In the latter case, we could find that one processor is doing most of the work and a thousand others are idle. If we want to do some sharing of the work, we will need additional communication to find out which processors can accept work and to tell them what to do.

So in this relatively simple example we see in one circumstance, a relatively smooth integrand, the job divides up nicely into parcels and there is little communication. But when the integrand is not smooth there may be difficulty in dividing the work evenly, and there may be additional communication costs.

Molecular dynamics

Imagine we have n molecules interacting with each other as in a fluid. The forces between them are such that at very short distances of separation the molecules repel each other strongly. At intermediate distances, they attract each other. Beyond a certain distance, say several molecular diameters, they hardly interact at all. We are interested in tracing the motion of the molecules of the fluid. In a problem like this n may be in the hundreds of thousands or more.

The computation we must do can be described informally as follows. Newton's laws tell us the differential equations we must solve, numerical analysis gives us a choice of algorithms to use to solve these equations. We choose an algorithm and start its execution at time 0 with each molecule at a certain point in space, moving with a certain velocity. We compute the force on each molecule from all of the others within a specified distance, i.e., its neighbors. We then compute new values for the position and velocity of each molecule at time $t + \delta t$ and repeat the process to compute new values for the intermolecular forces, position, and velocity at time $t + 2\delta t$, and so on. In this way, we generate the solution stepwise in time until the final time of interest has been reached.

An obvious idea for doing the computation in parallel is to assign each processor to one or more molecules. The processor computes the new position and velocity of its molecules at each step and communicates them to the other processors. New positions must be computed for every molecule so the new forces between the neighboring molecules can be determined. Distributing the new values for all n molecules between all of the processors involves enormous amounts of communication. In message-passing multiprocessors, the time to communicate a floating-point number is much larger than the time to perform a floating-point operation, and the algorithm we've described will not be able to make efficient use of the machine without careful balancing of computation and communication.

To reduce the communication, we could have each processor communicate only with those processors holding molecules in the neighborhood of its own molecules. But determining what processors hold molecules in the neighborhood of a given molecule is itself a major problem. As the molecules move, the population of neighbors changes. To implement this new algorithm, each processor will have to keep track of its neighbors somehow or will have to institute a search at each step. Maintaining a list of neighbors would involve redundant, non-parallel computation, and a search would require extra interprocessor communication. Minimizing this overhead presents a difficult problem.

Yet another approach would be to assign processors to specific regions of space rather than to specific sets of molecules. However, it's not immediately obvious that this organization of the problem solves the difficulties encountered with the earlier ones. As each particle moves, we must keep track of the region and, hence, processor to which it is assigned. Again, we'll either need more computation or communication for bookkeeping purposes. Whatever the approach, organizing an efficient parallel computation for this problem is going to take some careful thought. It is a computation which can cause difficult load-balancing and communication.

Types of parallel computers

We can identify two distinct types of parallel computers according to whether they obey instructions asynchronously or synchronously. We can further classify them according how they communicate information: by sharing a common memory space or by sending messages to each other. And we can classify them according to the physical nature of the interconnection network.

Parallel computers in which individual processors execute instructions asynchronously and send messages to each other are probably the easiest to understand at a logical level. These computers are labeled MIMD, which stands for *multiple instruction, multiple data* streams. The Intel computers iPSC/2, iPSC/860, and Paragon are of this type. Each processor executes its own private set of instructions. The Intel computers are also distributed-memory computers. Messages are passed between the processors by *send* and *receive* commands. There are mechanisms for causing the processor to enter a *wait state* in which it waits until receiving data from another processor. Programming these computers at a low level, that is, specifying the individual send and receive commands, is difficult and various languages have been developed attempting to simplify it: Linda, developed at Yale University by Gelernter [Carriero & Gelernter 89], and DINO, developed at the University of Colorado by Schnabel, Rosing, and Weaver [Rosing et al 91], are two examples. A group known as the High Performance Fortran Forum (HPFF) has been developing a version of Fortran, High Performance Fortran (HPF), which aims to support parallel programming (see [Koelbel et al 94]).

Parallel computers in which individual processors execute instructions asynchronously but share a common address space include machines like the Cray Y-MP and C-90, the IBM SP1 and SP2, the Silicon Graphics Challenge, and the Convex Exemplar. They are often referred to as *shared-memory computers*. These are also MIMD computers. They tend to be easier to program than the MIMD message passing computers referred to above, but they are not without their own set of difficulties. Clearly, if two processors are able to read and write into the same memory location some kind of synchronization is needed to insure the reads and writes are done in the correct sequence. Managing this synchronization adds to the complexity of programming these machines.

The other class of parallel computers in which processors operate synchronously is typified by the Maspar MP1 and MP2. In these computers there is a single sequence of instructions obeyed by all of the processors, each acting on its own data. They are labeled SIMD, standing for *single instruction, multiple data* streams. The first real parallel computer, the Illiac IV, finished in about 1970,[8] was in this class. A trivial example, matrix addition $C = A + B$, illustrates the nature of an SIMD computation. We store the matrices so that each element is on a separate processor. If we have n^2 processors connected to form a square mesh, we could map elements A_{ij} and B_{ij} to processor p_{ij} located in row i and column j of the mesh for $n \times n$ matrices A and B. With this distribution, a single add instruction performed in unison on all processors produces the sum. The result is left distributed one element per processor; in particular, processor p_{ij} holds element C_{ij}. Using this parallel algorithm, matrices of order 100 require only the time for one addition, not 10,000 additions.

SIMD machines do not share memory, rather they have a distributed memory, one memory module for each processor, and communication of data is by message passing. Like computation, message passing is generally done in parallel. For example, suppose we wish to compute the matrix product $C = AB$ on the square mesh with A_{ij} and B_{ij} initially in processor p_{ij}. As part of this computation, all processors in column j of the mesh must access all elements in column j of matrix B. Every processor in the square mesh has four neighbors (north, east, west, and south). One way to make sure all processors receive the proper elements of the matrix B is to have each processor send its element to its north neighbor. Each processor then receives an element from its south neighbor, performs the proper multiplication with its element of A, then passes the received element of B on to its north neighbor. During this data exchange, the processor at the top or northernmost position in the col-

---

[8]Now in the Computer Museum in Boston.

umn passes the elements of B to the processor at the bottom or southernmost position in the column. In this way, the elements of column j of B can cycle through column j of the processor mesh, and all n columns of processors can cycle their elements in parallel.

This partial matrix multiplication algorithm description demonstrates one way in which the processors of a SIMD multiprocessor can cycle data in lock step. It is also possible to broadcast data from one processor to all others or to exchange data by other special algorithms which take advantage of the way processors are physically interconnected in a particular machine.

In addition to the square mesh mentioned above, interconnection patterns include buses, hypercube structures, two-dimensional square meshes, switches of various kinds, rings, and combinations of these mechanisms. Bus systems, where processors are attached to one central piece of hardware or *bus* for message passing, are limited to about thirty processors at most. The others can accommodate much larger numbers of processors.

As we approach the physical limits of how fast processors can operate, machine design has begun to concentrate on increasing the number of processors in a multiprocessor. Two conflicting factors limit this number. First, the performance of a message-passing computer is determined in part by the distance a message must travel from a given processor to any other processor in the machine. This suggests that, for efficient message passing, a multiprocessor should have each of its processors connected to every other processor in the machine. In such a machine, a message would always travel from the sending processor directly to the receiving processor without being transferred through intermediate processors. As we added more processors, however, this completely-connected machine would encounter the second limiting factor— the maximum number of processors possible is limited by the number of interprocessor connections (physical wires) required. A completely-connected machine with p processors requires $p - 1$ connections per processor or a total of $p(p - 1)/2$ connecting wires.

Other processor interconnection patterns are more amenable to increased processor number. In a ring interconnection, for instance, every processor is connected to only two others. But while the number of ring-connected processors can be almost arbitrarily large, processors diametrically opposite each other in an p-processor ring would have to pass messages through about $p/2$

processors to communicate. Some multiprocessors have used an interconnection scheme known as a hypercube: with p processors each has $\log_2 p$ interconnection wires and a message need pass through no more than $\log_2 p - 1$ intermediate processors between its source and destination. This scheme has the advantage of a relatively short path length, measured in terms on the number of processor-processor connections that a message must traverse to reach its destination. However, it has the disadvantage that the number of wires connected to each processer increases with p. Some multiprocessors connect processors in a two-dimensional mesh: with p processors, assuming $p = m^2$, the mesh has m rows and m columns. In this scheme the number of wires connected to a processor is 4, assuming the edges at the top and bottom and left and right are joined, regardless of the value of p. Thus the number of wires connected to a processor does not grow with p as it does with a hypercube. On the other hand the path length grows as $m = \sqrt{p}$, thus messages in the mesh have longer paths than in a hypercube. However, the delay suggested by a longer path can be made almost insignificant by a technique called *wormhole routing*. In this technique a short header message is sent first and it sets a switch in each processor along the path, enabling the body of the message to pass through without delay.

The notion of *scalability* is important in characterizing parallel computers. As we pointed out earlier we would like the speed of a computer to increase in proportion to the number of processors; that is, if we double the number of processors then we double the speed of the computer. In practice this isn't exactly true for a number of reasons, some of which have been described above. But if the computer is designed in such a way that its speed does increase, at least approximately, in proportion to the number of processors, and its complexity in terms of the number of interconnecting wires also increases in proportion to the number of processors, we say the computer is scalable. Our preceding discussion suggests that parallel computers in which the processors are organized in two-dimensional square meshes are scalable.

1.5.2   A virtual parallel computer

Over the years systems which allow the interconnection of a set of workstations in such a way as to enable them to act as a distributed memory multiprocessor have been developed. The most successful of these is a system developed by Dongarra and his coworkers, called PVM (Parallel Virtual Machine). In addition to the fact that this system enables anyone with workstations networked together, on an Ethernet say, to have a multiprocessor, it also enables them to make a single workstation look like a multiprocessor, insofar as programming is concerned.

PVM is not restricted to workstations, versions have been built for the Intel iPSC/860 and Paragon, Thinking Machine Corporation's CM-5, Cray computers, and Convex computers. Thus PVM can serve as a common language for a number of parallel computers and in this way improve the portability of parallel programs. Furthermore, it can serve as as a platform for a heterogeneous multiprocessor consisting of, say, a Paragon, a CM-5, and a group of workstations.

The PVM software and documentation can be obtained by anonymous ftp at `netlib.att.com`, and the PVM manual is now available as a book [Geist et al 95].

From the programmers point of view, PVM consists of a library of procedures for C programs or Fortran programs to support interprocessor communication by message passing. Experience with PVM and a number of related works has led to an unofficial standard for message passing known as MPI (Message-Passing Interface), created by the Message Passing Interface Forum [Gropp et al 94b]. It specifies the syntax and semantics for message-passing procedures callable from C or Fortran programs.

1.6 Further reading

This has been a very brief treatment of the important and rapidly developing field of scientific computing. The intent was, as we said earlier, to get you interested to study it further. Here are a few suggestions for further reading. Books by Hockney and Jesshope [Hockney & Jesshope 88], Almasi and Gottlieb [Almasi & Gottlieb 89], and Leighton [Leighton 92] discuss parallel computers, their architectures, and programming. For the broad history of computing see books by Williams [Williams 85] and Augarten [Augarten 84]. For the history of scientific computing see the book by Nash [Nash 90] The work on the greenhouse effect by Washington and Bettge is described in [Washington & Bettge 90]. The federal high-performance computing program is described in [HPCC 89]. A description of High Performance Fortran is in [Koelbel et al 94]. A simple introduction to the Internet is given in the book by Krol [Krol 92]; see also the August 1994 issue of the Communications of the ACM for a number of articles on the Internet and related topics.

More complete descriptions of the architecture, performance, and applications of supercomputers can be found in the following chapters of this book. There you will find further references to the topics touched on here.

I Background

2 A Review of Selected Topics from Numerical Analysis

The course "High-Performance Scientific Computing" assumes an introductory knowledge of numerical analysis such as you might gain from any of a number of textbooks; this includes the ones by Conte and deBoor [Conte & de Boor 80], Dahlquist and Björk [Dahlquist & Björck 74], or Kahaner, Moler and Nash [Kahaner et al 89]. These are referred to as CdB, DB, and KMN in the sequel.

The purpose of this review is simply to record definitions, formulas, and concepts you are expected to know; and to define the notation we will use. It makes no serious attempt to explain or teach this material — for that you should refer to one of the texts just cited, or some similar text. The topics we cover are:

- Notation

- Error

- Taylor's series

- Elementary linear algebra

- Elementary numerical solution of ordinary differential equations

- Elementary Fourier series

2.1 Notation

$f'(x)$: This is $\frac{df}{dx}$, and similarly for the higher derivatives $f''(x)$, $f'''(x)$, etc..

$f'(x_i)$: $\frac{df}{dx}$ evaluated at the point x_i, and similarly for the higher derivatives $f''(x_i)$, $f'''(x_i)$, etc.

$\mathcal{O}(n^k)$: $f(n)$ is $\mathcal{O}(n^k)$ if

$$lim_{n \to \infty}(\frac{f(n)}{n^k}) = C,$$

where C is a nonzero constant. Alternatively, this notation is used to represent limiting behavior as a parameter goes to zero; e.g., $\mathcal{O}(h^k)$ means

$$lim_{h \to 0}(\frac{f(h)}{h^k}) = C,$$

flt(expression): the computed value of *expression*: usually different from the value of *expression* because of roundoff error.

$[a, b]$: the closed interval from a to b (i.e., includes endpoints).

(a, b): the open interval from a to b (i.e., does not include endpoints).

$[a, b)$, $(a, b]$: semiclosed intervals.

$\mathcal{T}_n(f(x_0 + \delta x))$: Taylor's polynomial of degree n for approximating $f(x)$ at $x_0 + \delta x$.

$\mathcal{R}^n$: Vector space of dimension n, real-valued elements; we use $\mathcal{C}^n$ if the elements are complex.

$\mathcal{R}^{r \times c}$: The space of matrices with r rows and c columns, real elements; we use $\mathcal{C}^{r \times c}$ if the elements are complex.

$x^{(i)}$: The ith vector of a set of vectors.

$e^{(i)}$: The ith unit vector (1 in row i, 0 elsewhere).

$A^{(i)}$: The ith matrix of a set of matrices.

x_i: The ith element of a vector.

$a_{i,j}$ **or** $(A)_{i,j}$: The element of a matrix A located in row i and column j.

$a_{i,:}$: The ith row of a matrix A.

$a_{:,j}$: The jth column of a matrix A.

$A_{i:i',j:j'}$: The rectangular submatrix of a matrix A consisting of row i to row i' and column j to column j', inclusive.

$(A)_{i,j}$: The (i,j) minor of a matrix A .

$det(A)$: Determinant of A.

x^T, A^T: Transpose of vector x, matrix A.

$\| x \|$: Norm of x; specific norms are $\| x \|_1$, $\| x \|_2$, and $\| x \|_\infty$.

$\hat{x}$: An eigenvector.

2.2   Error

Errors are caused by rounding of numbers and by formulas and procedures that give only approximate results because they are not carried to a limit (e.g., a truncated Taylor series, Simpson's formula for evaluating an integral, and Euler's method for solving a differential equation). The first kind of error we call *roundoff error*, the second *truncation error*. The total error in a computation is normally the result of these two kinds of error.

Formally, we define the *error* as the difference between the exact value of a quantity and its approximate value. Thus, the error in an approximation to x is:

$$e(x) = x_{exact} - x_{approx}.$$

This error is also referred to as the *absolute error*. The *relative error* is the ratio

$$r(x) = \frac{e(x)}{x_{exact}}.$$

From the two equations above we note that

$$x_{exact} = \frac{x_{approx}}{(1 - r(x))}.$$

In general we don't know x_{exact}, but we often can estimate an upper bound for the magnitude of the relative error, say $|r|_{ub}$, and so we can bound x_{exact} using values we know. If x_{exact} and x_{approx} are both positive then

$$\frac{x_{approx}}{(1 + |r|_{ub})} \leq x_{exact} \leq \frac{x_{approx}}{(1 - |r|_{ub})};$$

if they are both negative then

$$\frac{x_{approx}}{(1 - |r|_{ub})} \leq x_{exact} \leq \frac{x_{approx}}{(1 + |r|_{ub})}.$$

Of course, we should have $|r|_{ub} << 1$.

When two numbers nearly equal in magnitude but with opposite sign are added together, cancellation of leading digits produces a large relative error in the result. For example, suppose we have two numbers:

$$\begin{aligned} x &= +10000.2, \\ y &= -10000.0\,. \end{aligned}$$

Assume that the error in each is less than 0.05 in magnitude, then the magnitude of the relative error in each number is less than 5×10^{-6}. The sum of x and y is 0.2, with an error of at most 0.1 in magnitude, and a relative error as large as 0.5. Thus, we can have a relative error in the sum that is 10^5 larger than the relative error in the operands.

Cancellation of leading digits can lead to a result that might surprise you. A good example is the evaluation of a function at, or very near, one of its roots. For example, suppose that α is a root of the polynomial

$$x^5 + x^3 + x - 10000.19,$$

then by definition

$$\alpha^5 + \alpha^3 + \alpha - 10000.19 = 0.$$

However, if you evaluate the polynomial at α on a computer roundoff error is likely to cause a nonzero result. For example, if the computations were done to an accuracy of about seven decimal digits (corresponding to the accuracy of the type REAL in Fortran on most computers) then you could get a result as large as 0.01 and erroneously conclude that there is a mistake in the value of α.

To see why the result 0.01 for the computed value of the polynomial at α is completely reasonable, consider how the arithmetic might be done. Suppose that the polynomial is evaluated from left to right, then the last operation is

$$T - 10000.19,$$

where T represents the computed value of the expression

$$\alpha^5 + \alpha^3 + \alpha.$$

Now T should be close to 10000.19, but it could easily be off by a unit in the seventh place because of roundoff error. In that case we would get 0.01 for $T - 10000.19$. Thus, an expression which we think should be zero, or close to zero, evaluates to 0.01. The point to recognize here is that *0.01 is close to zero relative to the operands we used.* In fact, we could hardly have gotten much closer to zero: a "very small" result, much smaller than 0.01, would be impossible.

See KMN, chapter 2, for a good discussion of error.

2.3   Floating-point numbers

The form of a binary floating-point number, x, is

$$x = s \times 2^e,$$

where s is a real number, called the *significand*, and e is an integer, called the *exponent*. In a computer the value of x is usually held as an ordered triple consisting of the sign of the significand, the magnitude of the significand, and the *biased* exponent ($e_b = e + bias$). Usually the significand is normalized so that its magnitude is in $[1, 2)$, and the *bias*, a positive constant, on the exponent is chosen so that $e_b \geq 0$. The value *zero* is treated as a special case, usually with $|s| = e_b = 0$.

Many computers use the IEEE standard for floating-point arithmetic which is defined in the document *An American National Standard & IEEE Standard for Binary Floating-Point Arithmetic*. It is informally described in chapter 3 on IEEE Arithmetic. In this standard the significand for single precision has 24 bits and is normalized to lie in $\pm[1, 2)$; therefore, a change of one unit in the least significant place of the single precision significand, the *unit spacing*, is given by:

$$(\text{unit spacing})_{sp} = 2^{-23} \approx 1.2 \times 10^{-7}.$$

The single precision exponent is in $[-126, 127]$, and so the *range* for single precision is given by:

$$\begin{aligned} (\text{range})_{sp} &= \pm[2^{-126}, (2 - 2^{-23})2^{127}], \\ &\approx \pm[1.2 \times 10^{-38}, 3.4 \times 10^{38}]. \end{aligned}$$

The values for double precision are:

$$\begin{aligned} (\text{unit spacing})_{dp} &= 2^{-52} \approx 2.2 \times 10^{-16}, \\ (\text{range})_{dp} &= \pm[2^{-1022}, (2 - 2^{-52})2^{1023}], \\ &\approx \pm[2.2 \times 10^{-308}, 1.8 \times 10^{308}]. \end{aligned}$$

Normal rounding in IEEE floating-point is *round-to-nearest*, with *round-to-even* in case of a tie. This means that with normal rounding

$$\begin{aligned} |r|_{ub} &= 2^{-24} \approx 6.0 \times 10^{-8} \quad \text{(single precision)}, \\ |r|_{ub} &= 2^{-53} \approx 1.1 \times 10^{-16} \quad \text{(double precision)}. \end{aligned}$$

2.4 Taylor's series

Taylor's series provides a tool for approximating a function in terms of its value and the value of its derivatives at a single point. The series approximation is

$$f(x_0 + \delta x) \approx f(x_0) + \frac{f'(x_0)}{1!}\delta x + \frac{f''(x_0)}{2!}(\delta x)^2 + \ldots + \frac{f^{(n)}(x_0)}{n!}(\delta x)^n, \quad (2.1)$$

where

$$
\begin{aligned}
f(x) &= \text{the function to be approximated,} \\
x_0 &= \text{the point where values of } f,\ f',\ f'',\ldots \text{ are known,} \\
x_0 + \delta x &= \text{the point where } f \text{ is approximated.}
\end{aligned}
$$

The expression on the right of equation (2.1) is a polynomial in δx which is called *Taylor's polynomial* and denoted by $\mathcal{T}_n(f(x_0 + \delta x))$, thus

$$
\mathcal{T}_n(f(x_0 + \delta x)) = f(x_0) + \frac{f'(x_0)}{1!}\delta x + \frac{f''(x_0)}{2!}(\delta x)^2 + \ldots + \frac{f^{(n)}(x_0)}{n!}(\delta x)^n.
$$

The truncation error in Taylor's approximation is given by the the *remainder*, denoted by R_{n+1}:

$$
R_{n+1} = f(x_0 + \delta x) - \mathcal{T}_n(f(x_0 + \delta x)).
$$

There are two formulas for the remainder, one in terms of a derivative of $f(x)$, the other in terms of an integral of $f(x)$:

$$
\begin{aligned}
R_{n+1} &= \frac{f^{(n+1)}(x_0 + \xi)}{(n+1)!}(\delta x)^{n+1}, \quad \xi \in (0, \delta x), \\
&= \frac{1}{n!}\int_0^{\delta x}(\delta x - \zeta)^n f^{(n+1)}(x_0 + \zeta)\,d\zeta.
\end{aligned}
$$

In practice we usually cannot evaluate the remainder, but we can use it to estimate a bound on the error. For example, if we know a number F such that

$$
|f^{(n+1)}(x_0 + \xi)| < F \quad \text{if} \quad \xi \in (0, \delta x),
$$

then we know from the above formulas that

$$
|R_{n+1}| < \frac{F}{(n+1)!}|\delta x|^{n+1}.
$$

We can illustrate these ideas with a simple example. Let

$$
\begin{aligned}
f(x) &= \sin(x), \\
f(x_0 + \delta x) &\approx \mathcal{T}_4(\sin(x_0 + \delta x)).
\end{aligned}
$$

Then

$$\begin{aligned} \sin(x_0 + \delta x) \quad &\approx \quad \sin(x_0) + \frac{\cos(x_0)}{1!}\delta x - \frac{\sin(x_0)}{2!}(\delta x)^2 - \\ &\quad \frac{\cos(x_0)}{3!}(\delta x)^3 + \frac{\sin(x_0)}{4!}(\delta x)^4. \end{aligned}$$

If $x_0 = 0$, this equation becomes

$$\sin(\delta x) \approx \delta x - \frac{1}{6}(\delta x)^3.$$

The error in this approximation is R_5, which in the derivative form can be expressed as

$$R_5 = \frac{\cos(\xi)}{120}(\delta x)^5.$$

Since the magnitude of the cosine is at most 1, we have

$$|R_5| \leq \frac{1}{120}(\delta x)^5.$$

2.5 Linear algebra

Linear algebra deals with quantities we call *scalars*, *vectors*, and *matrices*. Scalars are simply numbers. Vectors and matrices are reviewed below. A good reference for this section is KMN.

2.5.1 Vectors

A vector is an ordered list of scalars often written as

$$x = \begin{pmatrix} x_1 \\ x_2 \\ \cdot \\ \cdot \\ \cdot \\ x_n \end{pmatrix}. \tag{2.2}$$

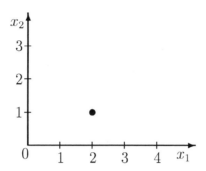

Figure 2.1: Geometric representation of $x^T = (2\ 1)$.

This is *column* form. It is also written in *row* form:

$$x^T = \left(\begin{array}{cccc} x_1 & x_2 & \ldots & x_n \end{array}\right).$$

We put the superscript T on vectors in row form unless the context makes it obvious that they are row vectors: the T stands for *transpose*. The numbers x_i are called the *elements* of x.

To identify the fact that x has n elements we write the expression $x \in \mathcal{R}^n$ if the elements of x are real numbers, the $\mathcal{R}$ standing for "real." Similarly, we write $x \in \mathcal{C}^n$ if the elements are complex numbers.

Geometrically, x represents a point in an n-dimensional space: the elements giving the coordinate values. Thus the vector

$$x = \left(\begin{array}{c} 2 \\ 1 \end{array}\right).$$

represents the point in figure 2.1.

The *length* of a vector is simply the number of elements it has: x in equation (2.2) has length n. There is a chance for confusion here because the word "length" is also used in the geometrical sense of distance. The context should make the meaning clear. Often the phrase *Euclidean length* is used when speaking in geometric terms. The Euclidean length is defined by

$$\text{Euclidean length of } x = (\sum_{i=1}^{n} x_i^2)^{1/2}. \tag{2.3}$$

Vector arithmetic

There follows a list of common arithmetic operations involving vectors and scalars. In this list a is a scalar and x and y are vectors of length n.

(scalar times vector)
$$ax \;=\; \begin{pmatrix} ax_1 \\ ax_2 \\ \vdots \\ ax_n \end{pmatrix}$$

(vector plus vector)
$$x + y \;=\; \begin{pmatrix} x_1 + y_1 \\ x_2 + y_2 \\ \vdots \\ x_n + y_n \end{pmatrix}$$

(scalar product, or dot product)
$$x^T y \;=\; \sum_{i=1}^{n} x_i y_i$$

(vector product, or cross product)
$$xy^T \;=\; \begin{pmatrix} x_1 y_1 & x_1 y_2 & \cdots & x_1 y_n \\ x_2 y_1 & x_2 y_2 & \cdots & x_2 y_n \\ \vdots & \vdots & & \vdots \\ x_n y_1 & x_n y_2 & \cdots & x_n y_n \end{pmatrix}$$

If $x^{(1)}, x^{(2)}, \ldots, x^{(k)}$ are vectors and $a_1, a_2, \ldots, a_k$ are scalars then the expression

$$a_1 x^{(1)} + a_2 x^{(2)} + \ldots + a_k x^{(k)}$$

is referred to as a *linear combination* of the vectors $x^{(1)}, x^{(2)}, \ldots, x^{(k)}$.

Special vectors: unit and zero

The unit vector, $e^{(j)}$, is defined as follows:

$$e_i^{(j)} = \left\{ \begin{array}{ll} 1 & (i = j), \\ 0 & (i \neq j). \end{array} \right.$$

We can express x in terms of unit vectors as follows:

$$x = \sum_{i=1}^{n} x_i e^{(i)}.$$

The unit vectors are also known as the *canonical vectors*.

The vector with all elements equal to zero is often written simply as "0"; e.g., in the vector equation

$$x + (-x) = 0.$$

As in this case, the meaning of 0 is usually clear from the context. Sometimes this vector is called the *null* vector and denoted ϕ.

Orthogonal vectors

A pair of vectors x and y is said to be *orthogonal* if

$$x^T y = 0.$$

Note that the unit vectors $e^{(i)}$ are orthogonal: we say they form an *orthogonal set of vectors*.

Vector norms

Informally the *norm* of a vector is a scalar that represents the size or magnitude of the vector. Formally, we say that the norm of a vector x is a scalar, represented by $\| x \|$, with the following properties:

1. $\| x \| > 0$ if $x \neq 0$;

2. $\| x \| = 0$ if and only if $x = 0$;

3. If a is any scalar, then $\parallel ax \parallel = |a| \parallel x \parallel$;

4. If y is also a vector, then $\parallel x + y \parallel \leq \parallel x \parallel + \parallel y \parallel$.

The last property is sometimes called the *triangle inequality*.

There are three commonly used norms: $\parallel x \parallel_1$, called the *1-norm*; $\parallel x \parallel_2$, called the *Euclidean norm* or *2-norm*; and $\parallel x \parallel_\infty$, called the *infinity norm*. They are defined as follows:

$$\parallel x \parallel_1 \quad = \quad \sum_{i=1}^{n} |x_i|, \tag{2.4}$$

$$\parallel x \parallel_2 \quad = \quad (\sum_{i=1}^{n} x_i^2)^{\frac{1}{2}}, \tag{2.5}$$

$$\parallel x \parallel_\infty \quad = \quad \max_{1 \leq i \leq n} |x_i|. \tag{2.6}$$

It is easy to verify that each of these norms satisfies the four conditions enumerated above. Notice that $\parallel x \parallel_2$ is the Euclidean length, equation (2.3).

Linear independence

The notion of linear independence is important because it allows us to identify sets of vectors in terms of which other vectors can be expressed. It also provides a convenient way to characterize matrices and the existence of solutions to a system of equations.

Consider the following four vectors:

$$y^{(1)} = \begin{pmatrix} 1 \\ 2 \\ 0 \end{pmatrix}, \quad y^{(2)} = \begin{pmatrix} 0 \\ 1 \\ 0 \end{pmatrix}, \quad y^{(3)} = \begin{pmatrix} 2 \\ 7 \\ 0 \end{pmatrix}, \quad y^{(4)} = \begin{pmatrix} 0 \\ 1 \\ 1 \end{pmatrix}.$$

It is obvious that

$$y^{(3)} = 2y^{(1)} + 3y^{(2)},$$

and by rearranging the terms we can equally well express $y^{(2)}$ as a linear combination of $y^{(1)}$ and $y^{(3)}$

$$y^{(2)} = -\frac{2}{3}y^{(1)} + \frac{1}{3}y^{(3)}$$

and similarly for $y^{(3)}$. The fact that any one of these three vectors can be expressed as a linear combination of the other two is expressed by saying that set of vectors $\{y^{(1)}, y^{(2)}, y^{(3)}\}$ is *linearly dependent*. Notice, however that we cannot express $y^{(4)}$ as a linear combination of $y^{(1)}$ and $y^{(2)}$: this is obvious because $y^{(1)}$ and $y^{(2)}$ have their third element equal to zero so there is no way to form a linear combination of them to produce $y^{(4)}$. We express this fact by saying that the set of vectors $\{y^{(1)}, y^{(2)}, y^{(4)}\}$ is *linearly independent*.

Formally, we say that the vectors $x^{(1)}, x^{(2)}, \ldots, x^{(k)}$ are linearly independent if the equation

$$a_1 x^{(1)} + a_2 x^{(2)} + \ldots + a_k x^{(k)} = 0$$

can only be satisfied when *all* of the coefficients a_i are zero: a set of vectors that is not linearly independent is said to be linearly dependent. The set $\{y^{(1)}, y^{(2)}, y^{(3)}, y^{(4)}\}$ must be linearly dependent because the four vectors are all of length three.

In the space $\mathcal{R}^n$, the unit vectors are linearly independent.

Any vector in $\mathcal{R}^n$ can be expressed as a linear combination of the vectors in *any* linearly independent set of vectors in $\mathcal{R}^n$. We express the notion that a linearly independent set of vectors can be used to represent any vector in $\mathcal{R}^n$ by saying that *the set spans the space*.

2.5.2 Matrices

Matrices are rectangular arrays of numbers such as

$$\begin{pmatrix} 1 & 2 & 3 \\ 1 & -1 & 0 \\ 4 & 4 & 2 \end{pmatrix}, \quad \begin{pmatrix} 1.2 & -3.4 \\ 2.5 & 1.9 \\ -0.78 & 2.9 \\ -1.0 & 0.99 \end{pmatrix}.$$

The general form for the matrix A is given in the following expression:

$$A = \begin{pmatrix} a_{1,1} & a_{1,2} & \cdots & a_{1,c} \\ a_{2,1} & a_{2,2} & \cdots & a_{2,c} \\ \vdots & \vdots & & \vdots \\ a_{r,1} & a_{r,2} & \cdots & a_{r,c} \end{pmatrix}.$$

This matrix has $r * c$ elements, arranged in r rows and c columns: we say it is an $r \times c$ matrix — the mathematical statement for this is $A \in \mathcal{R}^{r \times c}$ if the elements of A are real and $A \in \mathcal{C}^{r \times c}$ if the elements of A are complex. If $r = c$, we say that A is *square*.

The line extending from the upper left corner to the lower right corner of a square matrix is the *main diagonal*; thus the elements on the main diagonal are $a_{1,1}, a_{2,2}, \ldots, a_{n,n}$.

A matrix is said to be *order n* if it is an $n \times n$ square matrix.

A matrix is said to be *rank m* if m of its columns are linearly independent. The number of linearly independent columns of a matrix equals its number of linearly independent rows.

A vector is a special case of a matrix. The vector $x \in \mathcal{R}^n$ can also be regarded as a matrix in the space $\mathcal{R}^{n \times 1}$, and the vector x^T can be regarded as a matrix in the space $\mathcal{R}^{1 \times n}$.

It is useful to be able to refer to columns or rows of a matrix. For this we use $a_{:,j}$ to denote the jth column of A; and $a_{i,:}$ to denote the ith row of A. Similarly, we use $A_{i:i',j:j'}$ to denote the *submatrix* consisting of the rows i to i' and columns j to j', inclusive, of A.

In referring to an element of a matrix A, we use the notation $a_{i,j}$ as above: the same name but in lower case. Occasionally, we use $(A)_{i,j}$ to denote the same thing.

Matrix arithmetic

There follows a list of arithmetic operations involving matrices. In this list A and B are order n, and C is an $n \times m$ matrix; x is a column vector of n elements; p is a scalar. Notice that an expression like $A_{i,:}B_{:,j}$ is the dot product of the ith row of A and the jth column of B.

(scalar times array)

$$
pA \;=\; \begin{pmatrix} pa_{1,1} & pa_{1,2} & \ldots & pa_{1,n} \\ pa_{2,1} & pa_{2,2} & \ldots & pa_{2,n} \\ \vdots & \vdots & & \vdots \\ pa_{n,1} & pa_{n,2} & \ldots & pa_{n,n} \end{pmatrix}
$$

(matrix plus matrix)

$$A + B \;=\; \begin{pmatrix} a_{1,1} + b_{1,1} & a_{1,2} + b_{1,2} & \ldots & a_{1,n} + b_{1,n} \\ a_{2,1} + b_{2,1} & a_{2,2} + b_{2,2} & \ldots & a_{2,n} + b_{2,n} \\ \vdots & & \vdots & & \vdots \\ a_{n,1} + b_{n,1} & a_{n,2} + b_{n,2} & \ldots & a_{n,n} + b_{n,n} \end{pmatrix}$$

(array times vector)

$$Ax \;=\; \begin{pmatrix} a_{1,:}^T x \\ a_{2,:}^T x \\ \vdots \\ a_{n,:}^T x \end{pmatrix}$$

(vector times array)

$$x^T A \;=\; \begin{pmatrix} x^T a_{:,1} & x^T a_{:,2} & \ldots & x^T a_{:,n} \end{pmatrix}$$

(matrix times matrix : square)

$$AB \;=\; \begin{pmatrix} a_{1,:}^T b_{:,1} & a_{1,:}^T b_{:,2} & \ldots & a_{1,:}^T b_{:,n} \\ a_{2,:}^T b_{:,1} & a_{2,:}^T b_{:,2} & \ldots & a_{2,:}^T b_{:,n} \\ \vdots & \vdots & & \vdots \\ a_{n,:}^T b_{:,1} & a_{n,:}^T b_{:,2} & \ldots & a_{n,:}^T b_{:,n} \end{pmatrix}$$

(matrix times matrix : rectangular)

$$AC \;=\; \begin{pmatrix} a_{1,:}^T c_{:,1} & a_{1,:}^T c_{:,2} & \ldots & a_{1,:}^T c_{:,m} \\ a_{2,:}^T c_{:,1} & a_{2,:}^T c_{:,2} & \ldots & a_{2,:}^T c_{:,m} \\ \vdots & \vdots & & \vdots \\ a_{n,:}^T c_{:,1} & a_{n,:}^T c_{:,2} & \ldots & a_{n,:}^T c_{:,m} \end{pmatrix}$$

The matrix-vector and matrix-matrix products can be rewritten in a variety of other ways. For example, they may be written to show operations in terms of the columns of the matrix A instead of the rows of the matrix A. Such rearrangements are often helpful in devising efficient algorithms. For more information on alternative formulations, see, for example, [Strang 80].

Special matrices: diagonal, tridiagonal, symmetric, triangular, identity, and zero

The list of definitions follows:

(diagonal : D)

$$d_{i,j} \;=\; \begin{cases} 0 & (i \neq j) \\ \text{arbitrary} & (i = j) \end{cases}$$

(tridiagonal : T)

$$t_{i,j} \;=\; \begin{cases} 0 & (|i - j| > 1) \\ \text{arbitrary} & (|i - j| \leq 1) \end{cases}$$

(symmetric : A)

$$a_{i,j} \;=\; a_{j,i} \quad (\text{all } i, j) \;.$$

(upper triangular : U)

$$u_{i,j} \;=\; \begin{cases} 0 & (i > j) \\ \text{arbitrary} & (i \leq j) \end{cases}$$

(lower triangular : L)

$$l_{i,j} \;=\; \begin{cases} 0 & (i < j) \\ \text{arbitrary} & (i \geq j) \end{cases}$$

(identity : I)

$$(I)_{i,j} \;=\; \begin{cases} 0 & (i \neq j) \\ 1 & (i = j) \end{cases}$$

(zero : Φ)

$$\phi_{i,j} \;=\; 0 \quad (\text{all } i, j)$$

Minor, determinant, transpose, inverse, singular

Let $A^{(n)}$ be order n. The (i, j) *minor* of $A^{(n)}$ is the order $n - 1$ square obtained be deleting the ith row and j^{th} column of $A^{(n)}$. For example, the $(1, 3)$ minor of $A^{(5)}$, which we denote $(A^{(5)})_{1,3}$, is

$$(A^{(5)})_{1,3} = \begin{pmatrix} a_{2,1} & a_{2,2} & a_{2,4} & a_{2,5} \\ a_{3,1} & a_{3,2} & a_{3,4} & a_{3,5} \\ a_{4,1} & a_{4,2} & a_{4,4} & a_{4,5} \\ a_{5,1} & a_{5,2} & a_{5,4} & a_{5,5} \end{pmatrix}.$$

We define the *determinant* of $A^{(n)}$, $\det(A)$, recursively as follows:

$$\det(A^{(1)}) = a_{11}, \tag{2.7}$$

$$\det(A^{(k+1)}) = \sum_{j=1}^{k+1}(-1)^{j+1}(A^{(k+1)})_{1,j}\det((A^{(k+1)})_{1,j}). \tag{2.8}$$

When expressed in this form we say that the determinant is expanded along the first row. We could expand along any row, or column. The determinant of a product of matrices is the product of their determinants:

$$\det(AB) = \det(A)\det(B).$$

A square matrix is said to be *nonsingular* if its columns are linearly independent; if the columns are not linearly independent then it is *singular*. (If its columns are linearly independent then its rows are also linearly independent, and conversely.) The determinant of a singular matrix is zero; conversely, if the determinant of a matrix is zero then the matrix is singular.

The *transpose* of the matrix A, denoted A^T, is the matrix obtained by interchanging rows and columns of A; for example

$$\begin{pmatrix} 0 & 2 & 4 & 6 \\ 8 & 10 & 12 & 14 \\ 16 & 18 & 20 & 22 \end{pmatrix}^T = \begin{pmatrix} 0 & 8 & 16 \\ 2 & 10 & 18 \\ 4 & 12 & 20 \\ 6 & 14 & 22 \end{pmatrix}.$$

The transpose of the product of matrices is the product of their transposes in reverse order:

$$(AB)^T = B^T A^T.$$

If A is a square matrix then

$$\det(A^T) = \det(A).$$

The *inverse* of the square matrix A is a square matrix, denoted by A^{-1}, such that

$$AA^{-1} = A^{-1}A = I,$$

where I is the identity matrix. The inverse does not exist if A is singular. The inverse of a product of matrices is the product of their inverses in reverse order:

$$(AB)^{-1} = B^{-1}A^{-1}.$$

Matrix norms

The norm of a matrix A is a scalar, represented by $\| A \|$, with the following properties:

1. $\| A \| > 0$ if $A \neq$ a matrix of zeros;

2. $\| A \| = 0$ if and only if $A =$ a matrix of zeros;

3. If p is any scalar, then $\| pA \| = |p| \| A \|$;

4. If B is also a matrix, then $\| A + B \| \leq \| A \| + \| B \|$;

A matrix norm we use in this course is the infinity norm defined by

$$\| A \|_\infty \;\; = \;\; \max_{1 \leq i \leq n} \sum_{j=1}^{n} |a_{i,j}|. \tag{2.9}$$

This norm has the additional important property that

$$\|Ax\|_\infty \leq \|A\|_\infty \|x\|_\infty,$$

where x is any vector. This property is often described by saying that the matrix norm $\| \|_\infty$ defined by equation (2.9) is *consistent* with the vector norm $\| \|_\infty$ defined by equation (2.6). There are also matrix norms consistent with the two vector norms $\| \|_1$ and $\| \|_2$ defined by equations (2.4) and (2.5). For more information on matrix norms, see [Conte & de Boor 80] or [Golub & Van Loan 89].

2.5.3 Linear equations

A system of n linear equations in n unknowns ($\begin{matrix} x_1 & x_2 & \ldots & x_n \end{matrix}$)^T) is given below.

$$
\begin{aligned}
a_{1,1}x_1 + a_{1,2}x_2 + \ldots + a_{1,n}x_n &= b_1 \\
a_{2,1}x_1 + a_{2,2}x_2 + \ldots + a_{2,n}x_n &= b_2 \\
&\vdots \quad \vdots \quad \vdots \\
a_{n,1}x_1 + a_{n,2}x_2 + \ldots + a_{n,n}x_n &= b_n
\end{aligned}
\tag{2.10}
$$

This system is written in matrix form simply as

$$
Ax = b.
\tag{2.11}
$$

We call A the coefficient matrix and b the right-hand side. Notice that another way of writing this equation is

$$
a_{:,1}x_1 + a_{:,2}x_2 + \ldots + a_{:,n}x_n = b,
$$

which shows that the problem of solving a system of simultaneous linear equations can be thought of as the problem of finding a linear combination of the columns of the coefficient matrix that is equal to the right hand side. Thus if the columns of A are linearly independent we know that there is a solution because any vector in $\mathcal{R}^n$ can be expressed as a linear combination of the members of any set of linearly independent vectors in $\mathcal{R}^n$. On the other hand, even when A is singular the equations have a solution if it happens that b is expressible as a linear combination of the columns of A; or, more technically, if b lies in the space spanned by the columns of A. For example if

$$
A = \begin{pmatrix} 1 & 2 & 3 \\ 1 & 4 & 5 \\ 1 & 8 & 9 \end{pmatrix}, \quad b = \begin{pmatrix} 4 \\ 6 \\ 10 \end{pmatrix},
$$

then A is singular ($A_{:,3} = A_{:,1} + A_{:,2}$) but there is a solution:

$$
x = \begin{pmatrix} 2 \\ 1 \\ 0 \end{pmatrix}.
$$

Indeed there is an infinity of solutions. On the other hand, if

$$b = \begin{pmatrix} 1 \\ 0 \\ 0 \end{pmatrix},$$

then there is no solution.

Formally, we can express the solution by

$$x = A^{-1}b.$$

(This comes from multiplying both sides of equation (2.11) by A^{-1} and using the fact that $A^{-1}A = I$.) But without knowing A^{-1} this is useless, and computing A^{-1} is more difficult than solving the problem another way: Gaussian elimination. Before discussing Gaussian elimination we consider a special problem: solve equation (2.11) when A is triangular.

Solving a triangular system

Suppose that

$$A = \begin{pmatrix} a_{1,1} & a_{1,2} & a_{1,3} & a_{1,4} & a_{1,5} \\ 0 & a_{2,2} & a_{2,3} & a_{2,4} & a_{2,5} \\ 0 & 0 & a_{3,3} & a_{3,4} & a_{3,5} \\ 0 & 0 & 0 & a_{4,4} & a_{4,5} \\ 0 & 0 & 0 & 0 & a_{5,5} \end{pmatrix}, \tag{2.12}$$

and that none of the $a_{i,i}$ is zero, then there is an obvious procedure for solving the equations represented by equation (2.12). Starting with the last row of A and working towards the first we see that

$$
\begin{aligned}
x_5 &= b_5/a_{5,5}, \\
x_4 &= (b_4 - a_{4,5}x_5)/a_{4,4}, \\
x_3 &= (b_3 - a_{3,4}x_4 - a_{3,5}x_5)/a_{3,3}, \\
x_2 &= (b_2 - a_{2,3}x_3 - a_{2,4}x_4 - a_{2,5}x_5)/a_{2,2}, \\
x_1 &= (b_1 - a_{1,2}x_2 - a_{1,3}x_3 - a_{1,4}x_4 - a_{1,5}x_5)/a_{1,1}.
\end{aligned}
$$

This procedure is called *backsolving*. It is not difficult to verify that the computational work required for backsolving is $\mathcal{O}(n^2)$, where n is the order of the coefficient matrix.

The Gaussian elimination algorithm can be used to reduce a general system of linear equations to triangular form.

Gaussian elimination

Gaussian elimination uses *elementary row transformations* to reduce a matrix to upper triangular form. An elementary row transformation modifies a row, say row i, of a matrix according to the formula:

$$a_{i,:}^{(new)} = a_{i,:}^{(old)} + m a_{j,:}^{(old)} \quad (i \neq j), \tag{2.13}$$

where m is a scalar. Notice that if

$$m = -\frac{a_{i,k}^{(old)}}{a_{j,k}^{(old)}}, \quad \text{then} \quad a_{i,k}^{(new)} = 0.$$

Another important point is that making the same elementary row transformation on a row of the coefficient matrix and on the right-hand side vector, leaves the solution to the system of equations unchanged: if x' is a solution of equation (2.11) then

$$A^{(old)} x' = b^{(old)},$$

and

$$A^{(new)} x' = b^{(new)}.$$

Thus by a succession of properly chosen elementary row transformations it is possible to make all elements of A below the main diagonal equal to zero, producing a triangular system which has the same solution as the original system.[1]

Gaussian elimination transforms the coefficient matrix to triangular form in stages: in the first stage $n - 1$ elementary row transformations make all

---

[1] This is true only in a theoretical sense. In practice, roundoff error during Gaussian elimination perturbs the equations so that their solution is changed a little or a lot depending on circumstances. If the coefficient matrix is nearly singular, for example, then the solution may be changed a lot.

elements in the first column below the main diagonal equal to zero; in the second stage $n - 2$ elementary row transformations make all elements in the second column below the main diagonal equal to zero; and so on until the last column is reached. Thus starting with the initial coefficient matrix $A^{(0)}$, a sequence of matrices $A^{(1)}, A^{(2)}, \ldots, A^{(n-1)}$, and a corresponding sequence of right-hand side vectors is generated, with the final system

$$A^{(n-1)}x = b^{(n-1)}$$

having triangular form.

We illustrate the procedure for a system with $n = 4$.

$$A^{(0)} = \begin{pmatrix} 2.0000 & 1.0000 & -1.0000 & 4.0000 \\ 1.0000 & -0.5000 & 1.5000 & 3.0000 \\ 0.5000 & -0.2500 & 3.7500 & 2.5000 \\ 0.2500 & -0.1250 & 0.7500 & 1.8750 \end{pmatrix}, \quad b^{(0)} = \begin{pmatrix} 1.0000 \\ -0.5000 \\ 2.2500 \\ -3.8125 \end{pmatrix}. \quad (2.14)$$

After the first three elementary row transformation the first stage is completed and we have

$$A^{(1)} = \begin{pmatrix} 2.0000 & 1.0000 & -1.0000 & 4.0000 \\ 0 & -1.0000 & 2.0000 & 1.0000 \\ 0 & -0.5000 & 4.0000 & 1.5000 \\ 0 & -0.2500 & 0.8750 & 1.3750 \end{pmatrix}, \quad b^{(1)} = \begin{pmatrix} 1.0000 \\ -1.0000 \\ 2.0000 \\ -3.9375 \end{pmatrix}.$$

The elementary row transformations made here are given by the following equations:

$$A^{(1)}_{2:4,:} = A^{(0)}_{2:4,:} - \frac{a^{(0)}_{2:4,1}}{a^{(0)}_{1,1}} a^{(0)}_{1,:}, \quad (2.15)$$

$$b^{(1)}_{2:4} = b^{(0)}_{2:4} - \frac{a^{(0)}_{2:4,1}}{a^{(0)}_{1,1}} b^{(0)}_1. \quad (2.16)$$

The row $a^{(0)}_{1,:}$ in the second term on the right of equation (2.15) is called the *pivot row* and the main diagonal element in the pivot row, $a^{(0)}_{1,1}$, is called the *pivot element.* The factor multiplying the pivot row is the vector of multipliers (m in equation (2.13)):

$$-\frac{a_{2:4,1}^{(0)}}{a_{1,1}^{(0)}} = \begin{pmatrix} -1/2 \\ -1/4 \\ -1/8 \end{pmatrix}.$$

At the end of the second stage we have

$$A^{(2)} = \begin{pmatrix} 2.0000 & 1.0000 & -1.0000 & 4.0000 \\ 0 & -1.0000 & 2.0000 & 1.0000 \\ 0 & 0 & 3.0000 & 1.0000 \\ 0 & 0 & 0.3750 & 1.1250 \end{pmatrix}, \; b^{(2)} = \begin{pmatrix} 1.0000 \\ -1.0000 \\ 2.5000 \\ -3.6875 \end{pmatrix}.$$

In this stage the pivot row is the second row of $A^{(1)}$ and the pivot element is $a_{2,2}^{(1)}$. Finally, at the end of the third stage we have

$$A^{(3)} = \begin{pmatrix} 2.0000 & 1.0000 & -1.0000 & 4.0000 \\ 0 & -1.0000 & 2.0000 & 1.0000 \\ 0 & 0 & 3.0000 & 1.0000 \\ 0 & 0 & 0 & 1.0000 \end{pmatrix}, \; b^{(3)} = \begin{pmatrix} 1.0000 \\ -1.0000 \\ 2.5000 \\ -4.0000 \end{pmatrix}.$$

In this stage the pivot row is the third row of $A^{(2)}$ and the pivot element is $a_{3,3}^{(2)}$.

Now the system has been reduced to upper triangular form and the solution can be obtained by backsolving. The number of arithmetic operations required for transforming an order n matrix to triangular form by Gaussian elimination is $\mathcal{O}(n^3)$.

Pivoting

If we exchange a pair of equations in the original system this cannot change the solution: exchanging a pair of equations simply amounts to relabelling them. Thus, in the process of Gaussian elimination we are free to exchange rows of the coefficient matrix and the righthand side without affecting the solution. If the pivot element is zero, then we must exchange rows to avoid division by zero when we compute the scalar multiplier, m, for the elementary row transformation. But, aside from this, it is the practice to exchange rows during Gaussian elimination because analysis and experience shows that greater accuracy is usually obtained if the pivot element is large: the rule for exchanging rows to make the pivot element large is called *partial pivoting*.

Partial pivoting strategy can be described as follows, assuming that stage $k - 1$ has just been completed.

- Find the element of maximum magnitude in $a_{k:n,k}^{(k)}$, call it $a_{r,k}^{(k)}$, $k \leq r \leq n$.

- Exchange row r and row k.

We illustrate with a simple example. Suppose that in the last example we started out with the same set of equations but had them in a different order, say with the second and fourth equations interchanged so that the system is:

$$A^{(0)} = \begin{pmatrix} 2.0000 & 1.0000 & -1.0000 & 4.0000 \\ 0.2500 & -0.1250 & 0.7500 & 1.8750 \\ 0.5000 & -0.2500 & 3.7500 & 2.5000 \\ 1.0000 & -0.5000 & 1.5000 & 3.0000 \end{pmatrix}, \quad b^{(0)} = \begin{pmatrix} 1.0000 \\ -3.8125 \\ 2.2500 \\ -0.5000 \end{pmatrix}.$$

Stage 0 has just been "completed" so $k = 1$. The element of maximum magnitude in column 1 is 2.0. Since this element is already in row 1 no exchange is necessary and we proceed with stage 1. Then, at the end of the first stage, we have

$$A^{(1)} = \begin{pmatrix} 2.0000 & 1.0000 & -1.0000 & 4.0000 \\ 0 & -0.2500 & 0.8750 & 1.3750 \\ 0 & -0.5000 & 4.0000 & 1.5000 \\ 0 & -1.0000 & 2.0000 & 1.0000 \end{pmatrix}, \quad b^{(1)} = \begin{pmatrix} 1.0000 \\ -3.9375 \\ 2.0000 \\ -1.0000 \end{pmatrix}.$$

Now $k = 2$ and the element of maximum magnitude in $a_{2:4,2}^{(1)}$ is -1.0 in row 4; therefore, we exchange rows 2 and 4. Then we proceed with stage 2, and so forth.

The least squares problem

When the matrix A is $n \times n$, the linear system $Ax = b$ has the unique solution $x = A^{-1}b$ as long as the rank of A is n. On the other hand, when the matrix

A is $m \times n$ with $m > n$, there are more equations than unknowns to satisfy them, and the system may have no exact solutions. Such a system is termed *overdetermined*.

When the system is overdetermined, we often seek the solution $\hat{x}$ that minimizes the 2-norm of the residual error:

$$e = \| A\hat{x} - b \|_2.$$

(Note that if $A\hat{x} = b$ exactly, the residual error e is zero.)

The solution $\hat{x}$ that minimizes e is said to satisfy the system *in the least squares sense*. It is the solution that minimizes the Euclidean distance between the exact righthand side b and the approximate one $A\hat{x}$. There are several numerical methods for finding $\hat{x}$ and so solving the *least squares problem*, but we do not review them here. For more details, see CdB, DB, KMN, or [Golub & Van Loan 89].

A common special case of the least squares problem arises when we want to fit a set of data points with a polynomial. Suppose, for example, that we are studying a physical phenomenon known to be modelled by a quadratic. (An example of this is the parabolic trajectory followed by a projectile under the influence of gravity.) In this case, we expect the measured data points (x_1, y_1), (x_2, y_2), (x_3, y_3), and (x_4, y_4) to satisfy, at least approximately, the quadratic $a_1 x_i^2 + a_2 x_i + a_3 = y_i$. That is, the coefficients of the polynomial should satisfy the linear system

$$\begin{pmatrix} x_1^2 & x_1 & 1 \\ x_2^2 & x_2 & 1 \\ x_3^2 & x_3 & 1 \\ x_4^2 & x_4 & 1 \end{pmatrix} \begin{pmatrix} a_1 \\ a_2 \\ a_3 \end{pmatrix} = \begin{pmatrix} y_1 \\ y_2 \\ y_3 \\ y_4 \end{pmatrix}.$$

This system, however, is overdetermined, and we can at best expect to find a least squares approximation to $a = \begin{pmatrix} a_1 & a_2 & a_3 & a_4 \end{pmatrix}^T$. This approximation can often be a very good one, meaning that the residual error is very small.

2.5.4 Eigenvalues and eigenvectors

For a given matrix, A, there are certain vectors, $\hat{x}$ that have a special property illustrated by the formula

$$A\hat{x} = \lambda \hat{x}. \tag{2.17}$$

This equation says that the effect of multiplying $\hat{x}$ by A is the same as multiplying $\hat{x}$ by the scalar λ. Here is an example:

$$\begin{pmatrix} 2 & 1 \\ 1 & 2 \end{pmatrix} \begin{pmatrix} 1 \\ 1 \end{pmatrix} = 3 \begin{pmatrix} 1 \\ 1 \end{pmatrix}.$$

The vector $\hat{x}$ is called an eigenvector of A, and the scalar λ is called an eigenvalue of A. Eigenvectors are determined up to a scalar multiplier. This is obvious from equation (2.17): we can multiply both sides by a scalar and this is the same as multiplying $\hat{x}$ by a scalar. Generally eigenvectors are *normalized* in some way: typically by requiring their norm to have a certain value — e.g., $\| \hat{x} \|_2 = 1$.

An order n matrix has n eigenvalues. Two or more eigenvalues may be equal, and eigenvalues may be complex. In most cases, a computation is necessary to determine the eigenvalues and eigenvectors of a matrix, but there are some simple cases where the eigenvalues are obvious. The eigenvalues of a diagonal matrix are the numbers on the main diagonal, and the eigenvectors are the unit vectors. The eigenvalues of a triangular matrix are also the numbers on the main diagonal, but its eigenvectors are not the unit vectors.

An order n matrix may have up to n linearly independent eigenvectors. A symmetric matrix of order n has n linearly independent eigenvectors. In fact, the eigenvectors of a symmetric matrix are orthogonal to one another.

Eigenvalues and eigenvectors are important for studying the motion of a physical system. For example, studies of the motion of atoms in a solid lead to a matrix whose eigenvalues represent frequencies of oscillation of the atoms. The eigenvectors represent sets of amplitudes for the oscillatory motion.

Eigenvectors and eigenvalues are also important in mathematical analysis; for example, in analyzing the effect of multiplying a vector by a given matrix many times. Suppose that A is a real symmetric matrix of order n with eigenvectors $\hat{x}^{(1)}, \hat{x}^{(2)}, \ldots, \hat{x}^{(n)}$ which are real and linearly independent since A is real and symmetric. Now let y be an arbitrary vector in $\mathcal{R}^n$. Since A has n linearly independent eigenvectors in $\mathcal{R}^n$, we can express any vector, in particular y, in terms of them. Thus,

$$y = \sum_{i=1}^{n} c_i \hat{x}^{(i)},$$

where the c_i's are certain scalars. From this, and equation (2.17), it follows that

$$A^n y = \sum_{i=1}^{n} c_i \lambda^n \hat{x}^{(i)}.$$

The problem of analyzing the behavior of $A^n y$ as a function of n and y comes up in studying numerical algorithms for solving differential equations.

2.6   Differential equations

The general form of a *first order* differential equation is

$$y' = f(x, y). \tag{2.18}$$

This equation is called "first order" because the highest order derivative it contains is the the first derivative: the form of a second order differential equation is

$$y'' = f(x, y, y').$$

The problem of solving these equations amounts to determining y as a function of x on some interval, say $x \in [a, b]$: in what follows, $[a, b] = [0, 1]$.

For a first order equation a condition is necessary to get a particular solution; e.g., specification of the value of y at $x = 0$. (There is a whole family of solutions to the differential equation: the condition determines one member of the family.) For example,

$$y' = y \quad \text{has the solution } y = ce^x, \tag{2.19}$$

where c is an arbitrary constant. The family of solutions is generated by choosing different values of c. If we require that $y(0) = 2$, then it follows that $c = 2$ and we have a particular solution $2e^x$ for this differential equation.

With a second order differential equation two conditions are required to determine a particular solution. For example, we might require $y(0) = 1$ and $y'(0) = 0$, or we might require $y(0) = 1$ and $y(1) = 0$: in the first case we have an *initial value* problem, in the second a *boundary value* problem.

In practice most differential equations must be solved numerically. A numerical solution consists of a particular solution at a set of points; i.e., $y_1 = y(x_1), y_2 = y(x_2), \ldots$. The simplest way to generate such a solution is by using the first terms of Taylor's series.

2.6.1   Euler's method

Euler's method uses $\mathcal{T}_1(y(x))$ to generate the solution of a first order system with an initial condition. Thus, to solve equation (2.18) numerically on the interval $[0, 1]$ with the initial condition $y(0) = y_0$ we proceed as follows. Let $x_0, x_1, \ldots, x_n$ be a set of equally spaced points on $[0, 1]$ with

$$x_0 = 0, \ x_n = 1, \ x_{i+1} - x_i = h = 1/n, \ (0 \le i < n).$$

Then we compute the solution on these points according to the formula:

$$y_{i+1} = y_i + f(x_i, y_i)h. \tag{2.20}$$

The expression on the right is $\mathcal{T}_1(y(x_i + h))$: notice that $f(x_i, y_i) = y'(x_i)$. The initial condition defines y_0:

$$y(x_0) = \alpha \Rightarrow y_0 = \alpha.$$

Knowing y_0 we can evaluate the right side of equation (2.20) to obtain y_1, then we can compute y_2 in a similar manner, and so on. This is Euler's method.

In our notation, we make a distinction between the exact solution at x_i and the solution given by Euler's method: $y(x_i)$ is the exact solution; y_i is the solution given by Euler's method: when we want to be explicit about the value of h used we write $y_{i;h}$ in place of y_i.

To illustrate Euler's method we return to the differential equation in equation (2.19). In this case $f(x_i, y_i) = y_i$ so equation (2.20) becomes

$$y_{i+1} = y_i + y_i h. \tag{2.21}$$

If the initial condition is $y(x_0) = 1$ and we use $h = 0.1$, then $y_1 = 1.1$, $y_2 = 1.21$, $y_3 = 1.331$, and so on.

At each step, truncation error is introduced into the solution because equation (2.20) does not give the exact solution but only an approximation: indeed we know from the remainder for Taylor's approximation that after the first step

$$y(x_1) - y_{1;h} = \frac{y''(x_0 + \xi)}{2} h^2, \quad (\xi \in (0, h))$$
$$= \frac{y(x_0 + \xi)}{2} h^2. \tag{2.22}$$

The truncation error introduced at each step is called the *local truncation error*. We see from equation (2.22) that the local truncation error in Euler's method is described as $\mathcal{O}(h^2)$.

The total truncation error after k steps, due to the accumulation of local truncation errors, is called the *global truncation error*. The expression

$$y(x_k) - y_{k;h}$$

gives the global truncation error after k steps. A subtle but important point needs to be made here. The local truncation error introduced at the kth step is the difference between the computed solution at x_k and an exact solution at x_k— *an exact solution that was equal to y_{k-1} at x_{k-1}.* The idea is illustrated in figure 2.2.

In general the global error is difficult to estimate. However, it is easily computed for the problem above in which the solution was generated by equation (2.21). We see from this equation that

$$y_n = (1 + h)^n y_0.$$

If $y_0 = 1$ the exact solution for the problem is e^x so the global error is given by

$$E_{global} = e^{nh} - (1 + h)^n.$$

Expanding the expression on the right to terms involving h^2 we see that

$$E_{global} = (1 + nh + \frac{(nh)^2}{2} + \ldots) - (1 + nh + \frac{n(n-1)h^2}{2} + \ldots) \tag{2.23}$$
$$= \frac{x_n}{2} h + \ldots \tag{2.24}$$

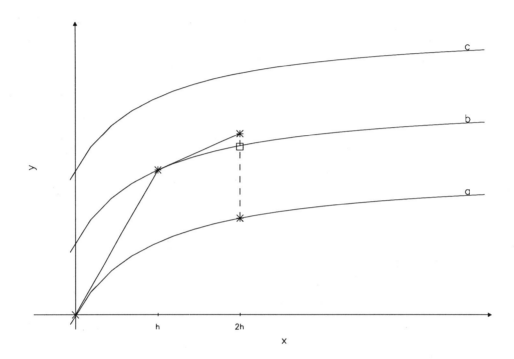

Figure 2.2: Illustration of the difference between local error and global error. Curves
a, b, c are solution curves for the differential equation. The particular solution that
satisfies the initial conditions is curve a. The straight-line segments represent the
numerical solution. The vertical dashed line extending from the numerical solution
down to curve a represents the global error — the accumulated error after two steps;
the portion of this dashed line extending just down to curve b, at the box, represents
the local error — the error made in the last step of the numerical solution.

Thus the global error at x_n is $\mathcal{O}(h)$. While this is a special case it can be shown for the general problem, equation (2.18), that the global error is $\mathcal{O}(h)$ provided that $y''(x)$ and $\partial f(x, y)/\partial y$ are bounded in the interval of interest; for details see CdB.

Next consider solving the second order initial value problem

$$y'' = f(x, y, y') \quad \text{(initial values}: y(0) = \alpha,\ y'(0) = \beta)$$

by Euler's method. We define $z = y'$ and express the problem in terms of a pair of first-order equations in the variables $x, y,\ z$:

$$
\begin{aligned}
y' &= z, \\
z' &= f(x, y, z),
\end{aligned}
$$

with initial conditions

$$y(0) = \alpha, \quad z(0) = \beta.$$

Euler's method in this case is described by the formulas:

$$
\begin{aligned}
y_{i+1} &= y_i + z_i h, \\
z_{i+1} &= z_i + f(x_i, y_i, z_i)h.
\end{aligned}
$$

The initial conditions allow computation of y_1, z_1. Substituting these values on the right side then gives us y_2, z_2, and so forth. In a similar way, a third order differential equation can be expressed as three first order differential equations which can be solved in a similar manner, and so on for any order differential equation.

2.6.2   Finite difference methods

These methods are based on *finite difference* approximations of derivatives. Simple finite difference approximations for the first derivative are

$$y'(x) \approx \frac{y(x + \delta x) - y(x)}{\delta x}, \quad y'(x) \approx \frac{y(x) - y(x - \delta x)}{\delta x}. \tag{2.25}$$

The first is called a *forward difference*, the second a *backward difference*. Using Taylor's series it is easy to verify that the error in each of these approximations is $\mathcal{O}(\delta x)$.

The *central difference* is more accurate: it is

$$y'(x) \approx \frac{y(x + \delta x) - y(x - \delta x)}{2\delta x}.$$

The error in this approximation is $\mathcal{O}(\delta x^2)$. There is a similar, $\mathcal{O}(\delta x^2)$, approximation for the second derivative:[2]

$$y''(x) \approx \frac{y(x + \delta x) - 2y(x) + y(x - \delta x)}{\delta x^2}.$$

Such approximations can be substituted for derivatives to obtain a finite difference method for solving a differential equation. We illustrate with a simple example.

Consider the second order initial value problem

$$y'' = y \quad (\text{initial values}: y(0) = \alpha, y'(0) = \beta).$$

Replacing y'' by its difference approximation leads to the formula:

$$y_{i+2} = (2 - \delta x^2)y_{i+1} - y_i.$$

Thus, if we know y_0, and y_1, we can compute y_2, and then y_3, and so on. We obtain y_1 from the formula

$$y_1 = y_0 + y_0' + \frac{y_0 \delta x^2}{2}.$$

This method is called the *explicit central difference method* in DB (see DB, p. 352-354 for a discussion of this method).

2.7   Fourier series

This subject is concerned with the representation of functions as linear combinations of sines and cosines. The functions may be continuous or be a finite set of discrete values. These two cases are discussed below. Both CdB and KMN are good references.

---

[2]For brevity of notation we use δx^2 for $(\delta x)^2$.

2.7.1 The continuous case

A Fourier series is a sum of sine and cosine terms, often written in the following form:

$$f(x) = \frac{a_0}{2} + \sum_{k=1}^{\infty}(a_k \cos(kx) + b_k \sin(kx)). \qquad (2.26)$$

The coefficients a_k and b_k are given by

$$a_k = \frac{1}{\pi}\int_{-\pi}^{+\pi} f(x)\cos(kx)\,dx, \quad b_k = \frac{1}{\pi}\int_{-\pi}^{+\pi} f(x)\sin(kx)\,dx. \qquad (2.27)$$

Typically, a Fourier series is used to represent a given function $f(x)$. The series may be truncated to provide an approximation of $f(x)$: we use $T(x; K)$ to denote the truncated series so that

$$f(x) \approx T(x; K) = \frac{a_0}{2} + \sum_{k=1}^{K}(a_k \cos(kx) + b_k \sin(kx)). \qquad (2.28)$$

A function $f(x)$ can be represented as a Fourier series if it satisfies the following conditions:

1. $f(x)$ is defined at every point in the interval $[-\pi, +\pi]$;

2. $f(x)$ is single-valued, finite, and piecewise continuous with piecewise continuous first-derivatives;

3. At a point of discontinuity x_d of $f(x)$ it is assumed that the value of $f(x_d)$ is the arithmetic mean of the limits $f(x_d - \epsilon)$ and $f(x_d + \epsilon)$, $\epsilon \to 0$.

While $f(x)$ has been specified only on the interval $[-\pi, +\pi]$, it is evident that equation (2.26) describes a periodic function, with period 2π, on the entire x−axis. A function that is not periodic may also be represented by a Fourier series; this broader application of Fourier series, the Fourier integral, is briefly discussed later.

Here is a simple example of a Fourier series. Consider the function

$$f(x) = \begin{cases} 0 & \text{if } -\pi \le x \le -\frac{\pi}{2} \\ \frac{2}{\pi}x + 1 & \text{if } -\frac{\pi}{2} < x \le 0 \\ 1 - \frac{2}{\pi}x & \text{if } 0 < x \le \frac{\pi}{2} \\ 0 & \text{if } \frac{\pi}{2} < x \le \pi \end{cases} \qquad (2.29)$$

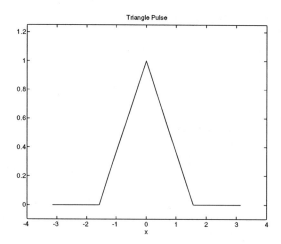

Figure 2.3: The triangular pulse.

It has a triangular shape in the middle of the interval, so we call it the *triangular pulse*. Figure 2.3 shows this function. It is continuous and has three discontinuities in its first derivative, thus it satisfies the conditions enumerated above.

It is a straightforward matter to compute the Fourier coefficients given by the integrals in equation (2.27), thus we find

$$a_k = \left\{ \begin{array}{ll} \frac{1}{2} & \text{if } k = 0 \\ \frac{4}{(k\pi)^2} & \text{if } k \text{ is odd} \\ \frac{8}{(k\pi)^2} & \text{if } k/2 \text{ is odd} \\ 0 & \text{if } k/2 \text{ is even} \end{array} \right\}, \quad b_k = 0, \text{ for all } k.$$

Thus, the Fourier series for the triangular pulse is

$$f(x) = \frac{1}{4} + \frac{4}{\pi^2}\cos(x) + \frac{8}{(2\pi)^2}\cos(2x) + \frac{4}{(3\pi)^2}\cos(3x) + \frac{4}{(5\pi)^2}\cos(5x) + \dots.$$

Note that the $\cos(4x)$ term is missing: $4/2$ is even, hence $a_4 = 0$. This Fourier series contains cosine terms only. We might have anticipated this at the outset because the triangular pulse is an even function, $f(-x) = f(x)$,

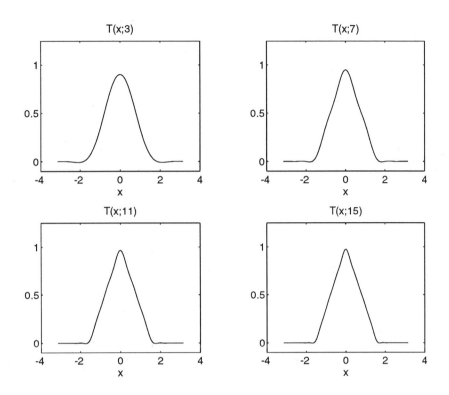

Figure 2.4: Four approximations for the triangular pulse

and the sine function is odd. Similarly, the Fourier series for an odd function, $f(-x) = -f(x)$, contains only sine terms.

Often a good approximation of $f(x)$ can be obtained from a truncated Fourier series with a small number of terms. If $K = 3$ in equation (2.28) we obtain the following approximation for the triangular pulse:

$$T(x; 3) = \frac{1}{2} + \frac{4}{\pi^2} \cos(x) + \frac{8}{(2\pi)^2} \cos(2x) + \frac{4}{(3\pi)^2} \cos(3x).$$

This approximation is illustrated in figure 2.4 along with $T(x; 7)$, $T(x; 11)$, and $T(x; 15)$. The illustration shows the improvement in the accuracy of the approximation as K increases. A bound on the error for any value of K is easily found by considering the truncated part of the series:

$$|f(x) - T(x; K)| < \frac{8}{\pi^2} \sum_{k=K+1}^{\infty} \frac{1}{k^2} < \frac{8}{\pi^2} \int_{K+1}^{\infty} \frac{1}{(t-1)^2} dt = \frac{8}{3\pi^2 K^3}.$$

Thus, we expect the error to decrease at least as fast as $1/K^3$.

We are not restricted to the interval $[-\pi, +\pi]$. With a change of variables we can move the interval of definition to any other finite interval in the usual way. If $f(x)$ is defined on $[0, 1]$, the Fourier series takes the form

$$f(x) = \frac{a_0}{2} + \sum_{k=1}^{\infty} (a_k \cos(2\pi kx) + b_k \sin(2\pi kx)), \tag{2.30}$$

where

$$a_k = 2 \int_0^1 f(x) \cos(2\pi kx) \, dx, \quad b_k = 2 \int_0^1 f(x) \sin(2\pi kx) \, dx.$$

Another form of the Fourier series uses the functions e^{ikx}, $i = \sqrt{-1}$, and $k = 0, \pm 1, \pm 2, \ldots$, instead of sines and cosines. Therefore, we have the exponential form of the Fourier series:

$$f(x) = \sum_{k=-\infty}^{+\infty} c_k e^{ikx}, \tag{2.31}$$

where

$$c_k = \frac{1}{2\pi} \int_{-\pi}^{+\pi} f(x) e^{-ikx} \, dx.$$

This form is easily derived from equation (2.26) by making use of Euler's formula

$$e^{ikx} = \cos(kx) + i \sin(kx).$$

The exponential form of the Fourier series might seem more difficult to use because it employs complex variables. For numerical computations this may be true, however it is frequently more convenient than the sine-cosine form for algebraic computations. You can appreciate this if you try to simplify the product of two Fourier series.

Notice that if we define

$$z = e^{ix},$$

then the right side of equation (2.31) becomes

$$\sum_{k=-\infty}^{+\infty} c_k z^k.$$

Thus, we see that the Fourier series has the appearance of a power series; similarly, the truncated Fourier series has the appearance of a polynomial. For this reason a truncated Fourier series is sometimes called a *trigonometric polynomial* or a *Fourier polynomial*.

Fourier series can also be used to represent functions of more than one variable. Suppose that $f(x, y)$ is defined on the square $[-\pi, +\pi] \times [-\pi, +\pi]$. Then the the Fourier series representation for this function is

$$f(x, y) = \sum_{l=-\infty}^{+\infty} \sum_{k=-\infty}^{+\infty} c_{k,l} e^{i(kx+ly)},$$

where

$$c_{k,l} = \frac{1}{4\pi^2} \int_{-\pi}^{+\pi} \int_{-\pi}^{+\pi} f(x, y) e^{-i(kx+ly)} \, dx \, dy.$$

As you might expect, the region of definition can be changed and approximations can be obtained by truncation, just as in the one-variable case.

2.7.2 The Fourier integral

A function defined on the interval $(-\infty, +\infty)$, and not periodic, can be represented as a Fourier integral, a kind of limiting form of a Fourier series in which the sum becomes an integral, the *Fourier integral*:

$$f(x) = \frac{1}{\sqrt{2\pi}} \int_{-\infty}^{+\infty} c(u) e^{iux} du,$$

where

$$c(u) = \frac{1}{\sqrt{2\pi}} \int_{-\infty}^{+\infty} f(x)e^{-iux}dx.$$

Here you can recognize that the index k has become a continuous variable u, and the sum over k has become an integral over u. It is customary to express the Fourier integral in a slightly different way:

$$f(x) = \int_{-\infty}^{+\infty} F(u)e^{i2\pi ux}du,$$

where

$$F(u) = \int_{-\infty}^{+\infty} f(x)e^{-i2\pi ux}dx.$$

This form can be obtained from the first form by a change of variables: $x \to x/\sqrt{2\pi}$ and $u \to u/\sqrt{2\pi}$. The function $F(u)$ is called the *Fourier transform of* $f(x)$; and $f(x)$ is called the *inverse* of $F(u)$.

For this relationship between $f(x)$ and $F(u)$ certain assumptions are made about $f(x)$:

1. $f(x)$ is defined on $(-\infty, +\infty)$;

2. $f(x)$ and its first derivative are piecewise continuous;

3. The integral $\int_{-\infty}^{+\infty} |f(x)|\, dx$ exists and is bounded;

4. At a point of discontinuity x_d of $f(x)$ it is assumed that the value of $f(x_d)$ is the arithmetic mean of the limits $f(x_d - \epsilon)$ and $f(x_d + \epsilon)$, $\epsilon \to 0$.

As with Fourier series there is a Fourier integral for functions of more than one variable. Thus for two variables we have the Fourier integral

$$f(x, y) = \int_{-\infty}^{+\infty} \int_{-\infty}^{+\infty} F(u, v)e^{i2\pi(ux+vy)}du\, dv,$$

and its Fourier transform

$$F(u, v) = \int_{-\infty}^{+\infty} \int_{-\infty}^{+\infty} f(x, y)e^{-i2\pi(ux+vy)}dx\, dy.$$

2.7.3   The Convolution Theorem

The notion of a convolution arises when we want to smooth a function by taking a weighted local average of its values; that is, replace the value of $f(x)$ at x_a by a weighted average of its values taken over an interval centered on x_a for all x_a. Specifically, a *convolution* of $f(x)$ and $g(x)$, the weighting function, is denoted $f * g$ and defined by

$$f * g = \int_{-\infty}^{+\infty} f(t)g(x-t)\,dt. \tag{2.32}$$

Note that the convolution is a function of x.

The Convolution Theorem states that the Fourier transform of the convolution of $f(x)$ and $g(x)$ is the product of the Fourier transforms of $f(x)$ and $g(x)$, that is

$$\int_{-\infty}^{+\infty} f * g e^{-i2\pi ux}dx = F(u)G(u).$$

This result has the important consequence of greatly simplifying the computation of a convolution in many cases. If $F(u)$ and $G(u)$ are known then

$$\int_{-\infty}^{+\infty} F(u)G(u)e^{i2\pi ux}du \tag{2.33}$$

gives the convolution $f * g$. In practice it may be easier to compute the convolution by first computing the Fourier transforms $F(u)$ and $G(u)$, then evaluating the integral equation (2.33), rather than compute it directly from equation (2.32).

2.7.4   The discrete case

We now consider the case in which f denotes a discrete and finite set of values, rather than the infinite set which we denoted with $f(x)$. In particular, let f be the set $\{f_0, f_1, \ldots, f_{K-1}\}$. In applications this set usually consists of measured values of a physical quantity. We will assume that K is even, only to simplify the discussion a little. The *discrete Fourier series representation* of f is a trigonometric polynomial:

$$f_j = \frac{a_0}{K} + \frac{2}{K}\sum_{k=1}^{K/2-1}\left(a_k\cos(2\pi k\frac{j}{K}) + b_k\sin(2\pi k\frac{j}{K})\right) + \frac{a_{K/2}}{K}, \quad j = 0, 1, \ldots, K{-}1,$$

where

$$a_k = \sum_{j=0}^{K-1} f_j\cos(2\pi k\frac{j}{K}), \quad b_k = \sum_{j=0}^{K-1} f_j\sin(2\pi k\frac{j}{K}).$$

Comparison with the Fourier series for the continuous case, equation (2.30), shows that this corresponds to using the set of points x_0, x_1, ..., x_{K-1}, where $x_j = j/K$, as the points at which $f(x)$ is evaluated.

The exponential form is

$$f_j = \frac{1}{K}\sum_{k=-K/2}^{K/2-1} F_k e^{i2\pi k\frac{j}{K}}, \quad j = 0, 1, \ldots, K-1,$$

where

$$F_k = \sum_{j=0}^{K-1} f_j e^{-i2\pi k\frac{j}{K}}.$$

It is convenient to think of f and F as vectors of length K, with elements as defined above. With this understanding, f and F can be represented in matrix equations:

$$f = WF, \quad F = W^{-1}f,$$

where W is a square matrix of order K. The matrix W has a very simple structure, easily recognized with the help of a small example.

Suppose $K = 8$, and let

$$w = e^{i\frac{2\pi}{K}},$$

then

$$W = \frac{1}{K} \begin{pmatrix} 1 & 1 & 1 & 1 & 1 & 1 & 1 & 1 \\ w^{-4} & w^{-3} & w^{-2} & w^{-1} & 1 & w^1 & w^2 & w^3 \\ w^{-8} & w^{-6} & w^{-4} & w^{-2} & 1 & w^2 & w^4 & w^6 \\ w^{-12} & w^{-9} & w^{-6} & w^{-3} & 1 & w^3 & w^6 & w^9 \\ w^{-16} & w^{-12} & w^{-8} & w^{-4} & 1 & w^4 & w^8 & w^{12} \\ w^{-20} & w^{-15} & w^{-10} & w^{-5} & 1 & w^5 & w^{10} & w^{15} \\ w^{-24} & w^{-18} & w^{-12} & w^{-6} & 1 & w^6 & w^{12} & w^{18} \\ w^{-28} & w^{-21} & w^{-14} & w^{-7} & 1 & w^7 & w^{14} & w^{21} \end{pmatrix}.$$

Thus computation of the Fourier transform of f when $K = 8$ consists in multiplying f by this matrix. The amount of arithmetic required is 64 multiplications and about the same number of additions, in general about $2K^2$ arithmetic operations. This number of operations can be reduced to about $2K \log_2(K)$ by employing a clever idea known as the *Fast Fourier Transform (FFT)*. We do not discuss the FFT here, but it is worth pointing out that it takes advantage of the special structure of W. The elements of W^{-1} should be evident from this example. For more information on the FFT, see chapter 17 on Tomography.

F is called the *discrete Fourier transform* of f, and f is called the *inverse Fourier transform* of F. Also, F is sometimes called the representation of f in *frequency space*: deriving from the fact that each element in F corresponds to a distinct value of k, which appears as a frequency in the above equations.

The equations for the discrete Fourier transform are written in other ways. Sometimes they are written

$$f_j = \sum_{k=-K/2}^{K/2-1} F_k e^{i2\pi k \frac{j}{K}}, \quad j = 0, 1, \ldots, K-1,$$

where

$$F_k = \frac{1}{K} \sum_{j=0}^{K-1} f_j e^{-i2\pi k \frac{j}{K}},$$

or they are written

$$f_j = \frac{1}{\sqrt{K}} \sum_{k=-K/2}^{K/2-1} F_k e^{i2\pi k \frac{j}{K}}, \quad j = 0, 1, \ldots, K-1,$$

where

$$F_k = \frac{1}{\sqrt{K}} \sum_{j=0}^{K-1} f_j e^{-i2\pi k \frac{j}{K}}.$$

3 IEEE Arithmetic Short Reference

Good science depends on reproducibility of results. This principle applies to computations as much as it does to laboratory experiments. In particular, we would like to have a computation done on one machine yield the same results as that computation done on another machine. When the results are not the same their reliability is questionable, and time is wasted in explaining the discrepancies. In scientific computation a lot of work goes into verifying the correctness of the program and its numerical accuracy. If the premises of this work no longer hold, as might be the case if the computation is done on a machine having different numerical characteristics, then this work is no longer valid. Thus good science and economy of effort require that these characteristics, the rules by which computers represent numbers and perform arithmetic, are the same from one computer to another.

Until recently there was no standard for floating-point arithmetic and so machines produced by different manufacturers had different numerical characteristics: different numerical precision, different number ranges, and different rules for rounding numbers. In 1987, after much work by many people, especially Prof. W. Kahan of the University of California at Berkeley, a standard for floating-point arithmetic was approved by the Institute of Electrical and Electronics Engineers (IEEE)and by the American National Standards Institute (ANSI). This standard is now followed by major chip manufacturers, including Intel (8087, 80287, 80387 processors) and MIPS (R2010 and R3010 processors). Some computers used in large-scale scientific computation do not follow the standard; for example, the Cray series of computers, the Cyber 205, and the IBM 3090.

Caution: When a manufacturer claims to have IEEE arithmetic in a chip this generally means that the rules for the number of bits used to represent the significand and exponent for single and double precision numbers have been followed, and that "round to nearest" (explained below) is done, but other aspects of IEEE arithmetic may be absent.

3.1   Single precision

3.1.1   Format

Thirty-two bits are used to represent a single precision number: one bit is used to represent the sign; eight bits are used to represent the biased exponent (e); and twenty-three bits are used to represent the magnitude of the fractional part of the significand (f). The format is shown in figure 3.1: bit (position) 0 is the sign bit, bits 1 through 8 are e, and bits 9 through 31 are f.

Figure 3.1: IEEE single precision format: s (bit 0) is the sign bit; e (bits 1 through 8) is the biased exponent; f (bits 9 through 31) is the magnitude of the fractional part of the significand.

The number represented has a sign (+ or −) according to the value of s (0 or 1). The bias on the exponent is 127, that is

$$e = p + 127,$$

where p is the exponent. The fraction has an implied binary point immediately preceding the string of bits represented by f, and the unsigned significand of the number is $1 + f$, as if a 1 were placed before the implied binary point: we sometimes express the unsigned significand as 1.f.

Thus, if the 32 bits in this format are as shown in figure 3.2, then the significand is

$$1.10100000000000000000000,$$

and the exponent is

$$-01110001.$$

Thus the value of the number represented is

$$1.10100000000000000000000 \times 2^{-0111\ 0001}$$

The result for the significand should be obvious. The result for the exponent may be less obvious. Notice that 127 (decimal) = 0111 1111 (binary), the biased exponent is 0111 0001, so the exponent, p, is given by

$$p = (00001110) - (01111111) = -(01110001).$$

The same value expressed in decimal form is

$$1.625 \times 2^{-113}.$$

Question: Suppose

```
f =   0101 0000 0000 0000 0000 0000
    e =   1000 1111
```

then what are the values of the significand and exponent; what is the number (binary) represented; what is the number (decimal) represented?

3.1.2 Special cases: $+0$, -0, $+\infty$, $-\infty$, NaN

These special cases are represented with particular values 0 and 255 of e. (*Do you see that these are the minimum and maximum values of* e*?*)
The value

```
e = 0000 0000
```

is used for representing $+0$ or -0 as illustrated in figures 3.3 and 3.4.
The number -0, which may seem a little strange, is explained later.
The value

```
e = 1111 1111
```

is used for representing $+\infty$ or $-\infty$ as illustrated in figures 3.5 and 3.6.

| 0 | 0000 1110 | 1010 0000 0000 0000 0000 000 |

Figure 3.2: Example of single precision number.

| 0 | 0000 0000 | 0000 0000 0000 0000 0000 000 |

Figure 3.3: IEEE +0 in single precision.

| 1 | 0000 0000 | 0000 0000 0000 0000 0000 000 |

Figure 3.4: IEEE -0 in single precision.

| 0 | 1111 1111 | 0000 0000 0000 0000 0000 000 |

Figure 3.5: IEEE $+\infty$ in single precision.

| 1 | 1111 1111 | 0000 0000 0000 0000 0000 000 |

Figure 3.6: IEEE $-\infty$ in single precision.

Patterns with $e = 255$ and $f \neq 0$ are used to represent cases that are "Not a Number," NaN. (The value NaN results, for example, from trying to evaluate $0/0$.)

The meaning and effect of these special values is discussed shortly.

3.1.3 Number range

Since the values 0 and 255 are reserved for special cases, the number range for nonzero, single precision numbers is constrained by the requirement that

$$0 < e < 255 \ ,$$

hence

$$p \epsilon [-126, +127].$$

(*Do you see why?*).

Therefore the smallest and largest single precision, positive numbers are:

$$1.0000\ 0000\ 0000\ 0000\ 0000\ 000 \times 2^{-0111\ 1110} = 2^{-126} \approx 1.2 \times 10^{-38}$$

and

$$1.1111\ 1111\ 1111\ 1111\ 1111\ 111 \times 2^{11111110} = \left(2 - 2^{-23}\right) \times 2^{+127} \approx 3.4 \times 10^{+38}$$

Corresponding limits hold for the negative numbers.

When a number is somewhat smaller than the smallest value, 2^{-126}, many systems set its value to zero. However, IEEE arithmetic allows for (but does not require) the representation of this smaller number in a form known as *denormalized*. In this form the unsigned significand is $0.f$. Thus, when denormalized numbers are allowed, the smallest nonzero, single precision number is $2^{-23} \times 2^{-126}$. The use of denormalized numbers is sometimes called *graceful underflow* or *gradual underflow*.

3.1.4 Numerical precision

Numerical precision is determined by the fact that the rightmost bit in f has weight 2^{-23}; i.e., a change of one unit in the least significant place of f results in a change in the value of f equal to 2^{-23}. To understand the implications of this fact consider the interval defined by the value

$$x1 = 1.00000000000000000000000 \times 2^{-3}$$

on the left end and the value

$$x2 = 1.00000000000000000000001 \times 2^{-3}$$

on the right end. In the usual mathematical notation this interval is $[x1, x2]$. Note that $x1$ and $x2$ can be represented exactly in IEEE single precision, but no value between $x1$ and $x2$ can be represented exactly (*Do you see why?*). Therefore, we must represent a value between $x1$ and $x2$ by either $x1$ or $x2$ depending on how we choose to round (see below). No matter which choice we make, there is an error in the representation of the number and this error does not exceed 2^{-23} in the value of f, and 2^{-26} in the number itself. Obviously if more than 23 bits were used for f we could represent numbers with higher precision.

Discuss the precision if 27 bits are used to represent f.

We adopt the convention that numbers that are exactly representable in the IEEE format are called *machine numbers*.

If x is the machine number

$$x = m \times 2^p,$$

where m is the significand, then the next larger machine number is

$$x_+ = (m + 2^{-23}) \times 2^p,$$

and the next smaller machine number is

$$x_- = (m - 2^{-23}) \times 2^p,$$

The spacing of these numbers is

$$x_+ - x = x - x_- = 2^{-23} \times 2^p = 2^{-23+p}.$$

The relative spacing of x and x_+ is defined by the ratio:

$$(x_+ - x)/x \approx 2^{-23} \approx 1.2 \times 10^{-7}$$

Notice that the relative spacing of a machine number and its successor is a constant (approximately), equal to about 2^{-23}. The relative spacing is a measure of the precision of the representation: the smaller the relative spacing, the higher the precision.

Question: Why do we say "approximately" above?

Question: Suppose we are given a single precision representation of 10^{30}, which we denote by $flt(10^{30})$, and we add a small machine number, ϵ, to it. What is the smallest value of ϵ such that

$$flt(10^{30}) + \epsilon \neq flt(10^{30})?$$

3.2   Double precision

3.2.1   Format

Sixty-four bits are used to represent a double precision number: 52 bits are used to represent f; 11 bits to represent the biased exponent; and one bit to represent the sign. The format, analogous to that for single precision, is shown in figure (3.7).

Figure 3.7: IEEE double precision format: s (bit 0) is the sign bit; e (bits 1 through 11) is the biased exponent; f (bits 12 through 63) is the magnitude of the fractional part of the significand.

The sign bit is interpreted as in single precision. The bias on the exponent is 1023, that is

$$e = p + 1023,$$

where p is the exponent.

You should be able to determine the range and relative spacing of the double precision numbers from the format and the previous discussion. The remarks made earlier about denormalized numbers also apply to double precision numbers *mutatis mutandis*.

Question: What are the largest and smallest double precision numbers?

Question: What is the relative spacing of the double precision numbers?

Question: Suppose we are given a double precision representation of 10^{30} which we denote by $dflt(10^{30})$ and we add a small number ϵ to it. What is the smallest value of ϵ such that

$$dflt(10^{30}) + \epsilon \neq dflt(10^{30})?$$

Values $+0$, -0, $+\infty$, $-\infty$, and NaN are represented in a fashion analogous to that for single precision.

3.3   Rounding

Arithmetic operations are performed as if they were correct to infinite precision. Rounding occurs when a number in a register of the floating-point unit of the computer is copied into memory, and so must be put into single precision or double precision format.

The default rounding mode is *round to nearest*, and *round to even* when there is a tie. There are of course only two possible choices for the rounded value of a number: the machine number to its left or the machine number to its right. In round to nearest we choose the closest of these machine numbers. If the exact value falls in the dead center of the interval then we choose the machine number that is even (i.e., has a 0 in the least significant place of f). Note that with this rounding convention the maximum error is just half a unit in the last significant place of the number.

Question: Explain the last statement "... error is just half a unit in the last significant place ...."

There are other rounding modes, called *directed rounding*. These are round toward $+\infty$, round toward $-\infty$, and round toward 0 (truncation).

Question: What is the difference in the error when using directed rounding as compared with round to nearest?

3.4   Infinity, NaN, and zero

3.4.1   Infinity

Infinity gets treated like an ordinary, but very large, number whenever it makes sense to do so. For example, the arithmetic operations below produce the results indicated on the right when x is a positive floating-point number:

$$
\begin{aligned}
+\infty + x &\rightarrow +\infty \\
+\infty \times x &\rightarrow +\infty \\
x/+\infty &\rightarrow +0 \\
+\infty/x &\rightarrow +\infty
\end{aligned}
$$

Corresponding results apply for the case x negative and $-\infty$.

3.4.2   NaN

This value is used for the results of operations that are indeterminate. Here are some examples:

$$
\begin{aligned}
0/0 &\rightarrow \text{NaN} \\
(+\infty) - (+\infty) &\rightarrow \text{NaN} \\
x + \text{NaN} &\rightarrow \text{NaN}
\end{aligned}
$$

3.4.3   Zero

As we have seen, there are two forms of zero, $+0$ and -0. Most arithmetic that would produce a zero result gives the result $+0$. The intent of -0 is to provide a mechanism for recognizing that a number that is zero to machine precision is, in fact, a very tiny negative number.

3.5   Of things not said

There is much more to the IEEE standard than what has been said here, but we have touched on the main points. The standard also includes binary-to-decimal conversion, square root, and other operations. It allows for graceful underflow, a mechanism that allows for very small numbers to be represented even when they are outside the range as described above. It also specifies the detection of various types of exceptions, such as overflow and underflow. One of the more interesting exceptions is *inexact*, which means that a numerical result is not exact because of rounding.

3.6   Further reading

The formal document defining the standard is the document *An American National Standard & IEEE Standard for Binary Floating-Point Arithmetic*.[1] You can find help for understanding the standard in a series of articles that appeared in the journal *IEEE Computer*, volume 14, number 3, 1981.

---

[1] Available from The Institute of Electrical and Electronics Engineers, Inc., 345 E. 47th St., New York, NY 10017.

4 UNIX, vi, and ftp: A Quick Review

This chapter includes three sections to help you review the common commands used in UNIX, ftp, and vi. You may wish to duplicate these pages and keep them with you when you are at your terminal or workstation.

UNIX is the name of an operating system used by most supercomputers and workstations today. This system was originally developed by AT&T and was further expanded at Berkeley. Several versions exist today with the same basic core of commands. For instance, ULTRIX is the version used by Digital Equipment, Inc, and UNICOS runs on the Cray Y-MP.

The vi (visual) screen editor is included in most UNIX systems, and the ftp utility program provides a method for communicating files between local and remote sites within UNIX.

4.1 UNIX short reference

Beginning UNIX users are often overwhelmed by the number of commands they must learn quickly to perform simple tasks. To assist such users, this section contains a sampling of commonly-used UNIX commands.

Informative Commands: These commands provide information about your login, environment, terminal, machine, and system. They also allow you to make some changes to these states.

date – to display the current date and time

kill – to kill the process with a given pid as argument

man – to get information on a UNIX command

- man f77 – shows the pages of the UNIX manual referring to f77 on the screen.

nslookup – to find the address of a given machine
- nslookup yourmach – returns the name and address of the machine yourmach, along with the name and address of its server.

passwd – to change your current password

printenv – to show the current environment setting

ps – to list your current processes with their pid

setenv – to change an environment setting
- setenv DISPLAY yourmach:0 – tells the Xserver that the Xterminal named yourmach is where any windows created are to be displayed.

- setenv PRINTER xxx – makes xxx be the default printer for any lpr commands.

time – to time the execution of a given command
- time anyunixcommand – executes anyunixcommand and returns the user, system, and total time taken for the execution

who – to list users currently logged in to given machine

whoami – to display login of user currently logged onto given terminal

File Manipulation: Remember that UNIX uses a tree structure for storing files. The main root of this tree (for the whole machine or system) is named /. Your main (or *home*) directory can be addressed as ˜ or ˜yourloginname. The home directory of another user with login guy would be ˜guy. The current directory (the one you are working in at the current time) is referred to by a single dot (.). The parent directory (the next one up the tree) is represented by two dots (..).

All file names in the following examples use capital letters, e.g., ABC. Lowercase letters and digits could have been used as well; for example, 9dec91 and afile are legal filenames. Remember that UNIX distinguishes between upper and lower case; XYZ and xyz refer to different files.

`cd` – to change directory

- `cd ABC` – moves to a directory named `ABC` located in your current directory.

- `cd ..` – moves to the parent directory of your current directory.

- `cd` – moves to your home directory.

- `cd ../ADIR` – moves to a directory named `ADIR` located in the parent directory of your current directory.

`cp` – to copy one file to another

- `cp ABC DEF` – copies file `ABC` to (or on top of) a file named `DEF`.

- `cp -i ABC ADIR/DEF` – copies file `ABC` to (or on top of) a file named `DEF` in the directory `ADIR`. Requests approval for overwriting the file if the file `ADIR/DEF` already exists. (`-i` means interactive.)

- `cp -r ADIR BDIR` – copies the entire contents of the directory `ADIR` to a new (or on top of the old) directory `BDIR`. (`-r` means recursive.)

`ls` – to display file information

- `ls` – lists all files in your current directory.

- `ls -a` – lists all files in your current directory, including any *dot* (.) files (e.g., `.login`).

- `ls -F` – lists files in your current directory, putting a slash (/) after those that are directories and an asterisk (*) after those that are executables.

- `ls -l` – lists all files in your current directory, showing protection codes, date of creation, and size.

mkdir – to make a new directory within current directory
- mkdir BDIR – creates a new subdirectory named BDIR within the current working directory.

mv – to rename (or move) a file
- mv -i ABC DEF – renames ABC to DEF; can also be thought of as moving the file ABC on top of file DEF, asking permission if the file DEF already exists. (-i for interactive.)

pwd – to display full pathname of current working directory

rm – to remove (delete) a file
- rm ABC DEF – deletes both ABC and DEF.

- rm -i ABC DEF – first asks you if you really want to delete these files; then deletes the ones for which you respond yes (y). (-i for interactive.)

rmdir – to remove (delete) a directory
- rmdir MNO – deletes the directory named MNO.

Language Commands: These commands help you to compile and debug programs.

cc – to compile a C program
- cc -O acprog.c -o acprog -lm – compiles with optimization (-O) the C program named acprog.c into the executable file named acprog, allowing the compilation to access the math library (-lm).

dbx – to debug a program
- dbx aprog – runs the executable program named aprog that was compiled with a -g option in a debugging environment.

lint – to check the syntax of a C program

f77 – to compile a Fortran program

- f77 -c fprog.f ftn1.f ftn2.f – compiles, without generating an executable file (-c), the Fortran program named fprog.f with the additional Fortran modules, ftn1.f and ftn2.f.

- f77 -g -o debug anfprog.f – compiles the Fortran program called anfprog.f with a symbol table (-g) so that the executable file named debug can be used with the dbx command.

Displaying Files: These commands allow you to see the contents of a file.

cat – to display a text file or concatenate files

- cat file1 – displays file1 on screen.

- cat file1 file2 – displays file1 followed by file2 on screen.

- cat file1 file2 > file3 – creates file3 containing file1 followed by file2.

diff – to show the differences between two files

- diff ABC DEF – displays any lines in ABC or DEF that differ from each other.

lpr – to print a file

- lpr -Pxxx ABC DEF – prints out files ABC and DEF on printer xxx.

more – to display a file, screen by screen

- more ABC DEF – displays the two files ABC and DEF sequentially on the screen. Hitting the space bar gets next screen; the return key gets the next line.

`pr` – to paginate a file before printing it

- `pr ABC DEF` – breaks both `ABC` and `DEF` into pages, puts a heading on the top with the name of the file, the date and time, and a page number. The two files are numbered independently. The result goes to the screen.

- `pr ABC | lpr -Pxxx` – paginates the file `ABC` and sends the resultant file to be printed on `xxx`. This is an example of a UNIX command that uses a *pipe* ("|"); that is, the standard output of the first part of the command (before the pipe "|") is `piped` to (is treated as the standard input for) the second part.

4.2   `vi` short reference

To Enter and Get Out of `vi`:

- `vi filename` – edit `filename` starting at line 1

- `vi -r filename` – recover `filename` that was being edited when system crashed

- `:q` – quit (exit) vi

- `:wq` – quit vi, writing out modified file to file named in original

- `:q!` – quit vi even though latest changes have not been saved for this vi call

Moving the Cursor:

- ⊔ (`space`), `l`, → – move one space to right

- `h`, ← – move one space to left

- `j`, ↓ – move down one space

- `k`, ↑ – move up one space

- `0` (`zero`) – move to start of current line (the one with the cursor)

- `$` – move to end of current line

- `^f` – move forward one screen

- `^b` – move backward one screen

- `:n` – move to line n

- `G` – move to last line in file

- `/string` – search forward for occurrence of string

- `n` – move to next occurrence of string

Adding, Changing, and Deleting Text:

- `u` – UNDO WHATEVER YOU JUST DID

- `i` – insert text before cursor, until `Esc` hit

- `a` – append text after cursor, until `Esc` hit

- `r` – replace single character under cursor (no `Esc` needed)

- `x` – delete single character under cursor

- `R` – replace characters, starting with current cursor position, until `Esc` hit

- o – open and put text in a new line below current line, until Esc hit

- O – open and put text in a new line above current line, until Esc hit

- dd – delete the current line

- Ndd – delete N lines, beginning with the current line; e.g., 5dd deletes 5 lines

- cw – change the current word with new text, starting with the character under the cursor, until Esc hit

- dw – delete the current word, starting with the character under the cursor

- cNw – change N words, N integer (works like cw); e.g., c5w changes 5 words

- dNw – delete N words, N integer (works like dw); e.g., d5w deletes 5 words

Saving and Reading Files:

- :r filename – read file named filename and insert after line with cursor

- :w – write current contents to file named in original vi call

- :w newfile – write current contents to file named newfile

- :w! prevfile – write current contents over a pre-existing file named prevfile

4.3   ftp short reference

The ftp (file transfer program) utility is commonly used for copying files to and from other computers. These computers may be at the same site or at different sites thousands of miles apart. For the purposes of this document, the *local* machine refers to the machine you are initially logged into, the one on which you type the ftp command. The *remote* machine is the other one, the one that is the argument of the ftp command.

A user interface for the standard File Transfer Protocol for ARPANET, ftp acts as an interpreter on the remote machine. The user may type a number of UNIX-like commands under this interpreter to perform desired actions on the remote machine.

Getting Started: To connect your local machine to the remote machine, type

> ftp *machinename*

where *machinename* is the full machine name of the remote machine, e.g., cs.colorado.edu. If the name of the machine is unknown, you may type

> ftp *machinenumber*

where *machinenumber* is the net address of the remote machine, e.g., 128.138.243.151. In either case, this command is similar to logging onto the remote machine. If the remote machine has been reached successfully, ftp responds by asking for a *loginname* and *password*.

When you enter your own *loginname* and *password* for the remote machine, it returns the prompt

> ftp>

and permits you access to your own home directory on the remote machine. You should be able to move around in your own directory and to copy files to and from your local machine using the ftp interface commands given on the following page.

Anonymous ftp: . At times you may wish to copy files from a remote machine on which you do not have a *loginname*. This can be done using *anonymous* ftp.

When the remote machine asks for your *loginname*, you should type in the word anonymous. Instead of a *password*, you should enter your own electronic mail address. This allows the remote site to keep records of the anonymous ftp requests.

Once you have been logged in, you are in the anonymous directory for the remote machine. This usually contains a number of public files and directories. Again you should be able to move around in these directories. However, you are only able to copy the files from the remote machine to your own local machine; you are not able to write on the remote machine or to delete any files there.

Common ftp Commands:

? – to request help or information about the ftp commands

ascii – to set the mode of file transfer to ASCII (this is the default and transmits seven bits per character)

binary – to set the mode of file transfer to binary (the binary mode transmits all eight bits per byte and thus provides less chance of a transmission error and must be used to transmit files other than ASCII files)

bye – to exit the ftp environment (same as quit)

cd – to change directory on the remote machine

delete – to delete (remove) a file in the current remote directory (same as rm in UNIX)

get – to copy one file from the remote machine to the local machine

- get ABC DEF – copies file ABC in the current remote directory to (or on top of) a file named DEF in your current local directory.

- get ABC – copies file ABC in the current remote directory to (or on top of) a file with the same name, ABC, in your current local directory.

lcd – to change directory on your local machine (same as UNIX cd)

ls – to list the names of the files in the current remote directory

mkdir – to make a new directory within the current remote directory `mget` – to copy multiple files from the remote machine to the local machine

- `mget *` – copies all the files in the current remote directory to your current local directory, using the same filenames. Notice the use of the wild card character, *.

`mput` – to copy multiple files from the local machine to the remote machine

`put` – to copy one file from the local machine to the remote machine

`pwd` – to find out the pathname of the current directory on the remote machine

`quit` – to exit the ftp environment (same as `bye`)

`rmdir` – to remove (delete) a directory in the current remote directory

Further Information: Many other interface commands are available. Also ftp can be run with different options. Please refer to the `man` page on ftp for more information.

Examples of two ftp sessions are given on the next two pages. These show the type of interaction you may expect when using the ftp utility.

```
% ftp cs.colorado.edu
Connected to cs.colorado.edu.
220 bruno FTP server (SunOS 4.1) ready.
Name (cs.colorado.edu:schauble):  anonymous
331 Guest login ok, send ident as password.
Password:
230-This server is courtesy of Sun Microsystems, Inc.
230-
230-The data on this FTP server can be searched and accessed via WAIS, using
230-our Essence semantic indexing system.  Users can pick up a copy of the
230-WAIS ".src" file for accessing this service by anonymous FTP from
230-ftp.cs.colorado.edu, in pub/cs/distribs/essence/aftp-cs-colorado-edu.src
230-This file also describes where to get the prototype source code and a
230-paper about this system.
230-
230-
230 Guest login ok, access restrictions apply.
ftp> cd /pub/HPSC
250 CWD command successful.
ftp> ls
200 PORT command successful.
150 ASCII data connection for /bin/ls (128.138.242.10,3133) (0 bytes).
ElementsofAVS.ps.Z
    .
    .
    .
execsumm_tr.ps.Z
viShortRef.ps.Z
226 ASCII Transfer complete.
418 bytes received in 0.043 seconds (9.5 Kbytes/s)
ftp> get README
200 PORT command successful.
150 ASCII data connection for README (128.138.242.10,3134) (2881 bytes).
226 ASCII Transfer complete.
local: README remote: README
2939 bytes received in 0.066 seconds (43 Kbytes/s)
ftp> bye
221 Goodbye.
% ls
....
README
....
```

Figure 4.1: An ftp session to obtain the HPSC README file from the cs.colorado.edu anonymous ftp directory using a *loginname* of anonymous and a *password* of one's own electronic mail address.

```
% ftp nordsieck.cs.colorado.edu
Connected to nordsieck.cs.colorado.edu.
220 nordsieck FTP server (Version 5.53 Tue Aug 25 10:46:12 MDT 1992) ready.
Name (nordsieck.cs.colorado.edu:schauble):   schauble
331 Password required for schauble.
Password:
230 User schauble logged in.
ftp> cd HPSC/exercises
250 CWD command successful.
ftp> ls
200 PORT command successful.
550 No files found.
ftp> put tmul.out
200 PORT command successful.
150 Opening ASCII mode data connection for tmul.out.
226 Transfer complete.
local: tmul.out remote:  tmul.out
1882 bytes sent in 0.0095 seconds (1.9e+02 Kbytes/s)
ftp> ls
200 PORT command successful.
150 Opening ASCII mode data connection for file list.
tmul.out
226 Transfer complete.
9 bytes received in 0.0021 seconds (4.3 Kbytes/s)
ftp> mput *
mput Makefile?  y
200 PORT command successful.
150 Opening ASCII mode data connection for Makefile.
226 Transfer complete.
local:  Makefile remote:  Makefile
1020 bytes sent in 0.0062 seconds (1.6e+02 Kbytes/s)
mput tmul.out?  n
ftp> quit
221 Goodbye.
% ls
....
Makefile
tmul.out
....
```

Figure 4.2: An ftp session to copy files from a remote machine back to
nordsieck.cs.colorado.edu using one's own login and password.

5 Elements of UNIX Make

5.1 Introduction

This chapter is designed to be used in a "hands-on" fashion. It is intended to introduce you to make, one of the most commonly used UNIX utilities. You should go over this material while logged on to your workstation or terminal so that you can try each example as you encounter it.

The purpose of make is to help maintain files and keep them current. It can also help to reduce keystrokes and to eliminate redundant or repetitious commands. Originally developed at Bell Labs by Stuart Feldman, make has been extended into several versions by organizations such as AT&T, Borland, Microsoft, and Berkeley. Some of these versions run on operating systems other than UNIX.

In this chapter, you first learn how to use make for a simple Fortran compilation. Later, a method for performing more complex modular compilation with make is shown. Similar techniques can be applied for compiling C programs. Additional uses of make are also discussed.

For the examples shown in this chapter, the commands that you (the reader) enter are displayed in *this font*, e.g.,

make myprog

while the computer responses are displayed in this font, e.g.,

```
f77 -O myprog.f -o myprog
```

These examples were run on a DECstation 5000/200 using ULTRIX. The given responses to the commands may differ slightly on other systems; some of the defaults may be different as well. However, the basic concepts are the same.

5.2   An example of using make

Suppose you have created a Fortran program named `myprog.f`. Also suppose that `myprog.f` is the only file in your current directory.

Naturally, the first thing you want to do is to compile your new program to check for errors. Normally, you just type something like

> *f77 -o myprog myprog.f*

and the linked object (or executable) file for the compiled program is put in `myprog`, if it compiles successfully. If you use `make` instead, you type

> *make myprog*

to accomplish the same thing. The full dialogue with `make` as displayed on the screen is as follows:

```
        ⋮
% make myprog
f77 -O myprog.f -o myprog
%
```

In other words if you ask `make` to make a file named `myprog`, `make` is clever enough to know that it has to compile `myprog.f` to do it.

Why? What happens? By what magic does `make` know to compile your program?

5.2.1   Dependencies

The `make` utility is basically used for file maintenance; it is mainly concerned with the dependencies of one file on another. By one file being *dependent* upon another, we mean that the first file is generated from the other. For instance, a file with a `.o` extension is the result of a compilation. This means it depends on a file with the same rootname and an extension that refers to a computer language, such as, `.f` for Fortran, `.c` for C, `.p` for Pascal, or even `.s` for assembly languages. In our example, the linked executable file `myprog` is dependent upon the Fortran source file `myprog.f`.

From these known dependencies, `make` generates a series of commands to form a given file, called the *target* file. These commands use the *components* of the target file, also called the *dependencies*, i.e., those files on which the *target* is dependent. This sequence of commands may be known implicitly by `make`. As mentioned above, `make` is aware that a file with a `.o` extension is the result of a compilation. Or the commands may be defined in a *description file*, called a makefile, that specifies the dependencies of a target and the commands or *rules* by which to form the target. Such a description file is usually named `Makefile` or `makefile`, and may describe rules that are unique to the files in your current directory.

5.2.2   How the `make` request works

The command

   make myprog

calls the `make` utility with the argument `myprog`. In this case, `myprog` is called the target for the given make command.

As the first step in the execution of this `make` request, your current directory is searched for the file named `myprog`. If the file were there, it might not need to be remade; the creation date would first be checked to see whether or not the file is out-of-date compared to any files from which it may have been created. However in this example, no file of this name exists.

Next `make` looks for a special file named `Makefile` or `makefile` in the current directory. This might contain specific instructions for creating `myprog`.

Figure 5.1: Derivation of `myprog` from `myprog.f`.

Since there is no makefile in the current directory assumed for this example, `make` must use the implicit rules of dependency mentioned above. Thus, it looks for a file that can be used to create the desired file, starting with `myprog.o`. If there was a `myprog.o` file, then `make` would only need to link that file to create the new executable file, `myprog`. Again no such file exists.

However, `make` finds the file named `myprog.f`. It knows that this file can be compiled to create the file `myprog`, using the Fortran compiler. And this is what it does.

Hence in trying to make the target file named `myprog`, `make` finds it must compile the only file in the directory with the correct rootname `myprog.f`. From this compilation, it can *derive* the desired file `myprog`. A picture of this *derivation* is shown in figure 5.1.

5.2.3   When make fails

If no files in the directory had the correct rootname `myprog`, this `make` command would fail. For instance, suppose you had erroneously stored the file as `myprg.f`. In response to the command

> *make myprog*

the `make` utility would produce the error message:

> `Make:  Don't know how to make myprog.  Stop.`

This is because it was unable to find anything it could logically convert to `myprog`; it was unable to find any files on which `myprog` could be dependent.

As another example, suppose there was a compilation error. Then again, `make` would not be able to complete its assigned task. This time the error message would occur in a dialogue much like the following:

```
                ⋮
% make myprog
f77 -O myprog.f -o myprog
Error on line 1 of myprog.f:  illegal ....

Stop
%
```

Any time that `make` is unable to create the final desired target file, it prints an error message to that effect. Note that some intermediate files may be created in the process and may be present in your directory when `make` is done.

5.3   Some advantages of `make`

Why is the command

> *make myprog*

better than the command

> *f77 myprog.f*

Both commands contain only two words and use about the same number of keystrokes. At this point, there does not seem to be much to recommend switching to `make`.

5.3.1   Implicit rules

One interesting point is that the command

> *make myprog*

would have worked just as well if the program in your directory were a C program (`myprog.c`), a Pascal program (`myprog.p`), or an assembly program (`myprog.s`). The `make` utility would have found a compilable program file from which to create `myprog`. Similarly, the command

> *make myprog.o*

would create an object file named `myprog.o` from `myprog.f`, `myprog.c`, `myprog.p`, or `myprog.s`. The dialogue with `make` for this command would be similar to that below.

```
         ⋮
% make myprog.o
f77 -O -c myprog.f
%
```

5.3.2   Avoiding redundant work

Now suppose that you had not worked with `myprog.f` for some time and were not certain if the copy of the executable file `myprog` in the directory included the last changes to the source program `myprog.f`. One way to find out would be to check the creation dates yourself.

```
            ⋮
% ls -l
total 80
-rwxrwxr-x 1 schauble  80516 Jul 2 11:42 myprog*
-rw-r--r-- 1 schauble     19 Jul 2 11:41 myprog.f
%
```

In this case, it appears that `myprog` was created after the latest changes to `myprog.f`. If this were not so, you would probably want to recompile the program before reusing it with the command:

f77 -o myprog myprog.f

Checking on the validity of `myprog` could also be done by using `make` instead. If you type

```
            ⋮
% make myprog
'myprog' is up to date.
%
```

the effect would be the same as examining the `ls -l` output; you know that `myprog` was more recently created than `myprog.f` without needing to look at the creation dates yourself. Further the recompilation is done automatically if the executable file was not up-to-date. In other words, `make` only recompiles the program if it is necessary.

This is another advantage of `make`. It not only checks to see if the target file already exists, but also that it is a valid version of that file. In other words, `make` assures that the target file is current with respect to the file(s) on which it is dependent.

5.4   The makefile

Using the implicit version of make has its limitations. For instance, the make
Fortran compilations in the examples above use the -O optimization flag.[1] You
might prefer a different optimization level. How can you override the implicit
rules of make? By forming your own rules and putting them in a file named
Makefile or makefile.

5.4.1   Format of a sample makefile

A simple description file or makefile for the derivation shown in figure 5.1
might look like the following:

```
#
# Simple Makefile
#     make myprog:  to compile myprog.f
#
myprog:  myprog.f
          f77 -O1 -o myprog myprog.f
```

This consists of some comments and a rule for making the target, myprog.
Let's look at the parts of this makefile in more detail.

The first few lines of the makefile are comments. It is useful to include in
these comments all the functions a given makefile performs.

```
#
# Simple Makefile
#     make myprog:  to compile myprog.f
#
```

Notice that these lines all begin with the special character #. Any characters
following an # on any line in the makefile are assumed to be part of a comment.
Blank lines are also ignored by make.

The remaining lines of this sample makefile define the rule that this makefile
is providing.

---

[1]The default compilation flags for make differ from system to system.

```
myprog:  myprog.f
         f77 -O1 -o myprog myprog.f
```

The rules of a makefile have two parts: (i) a *dependency line* that describes the target file with its dependencies and (ii) one or more *command lines* that describe how to form the target file from the dependencies. All commands forming rules for `make` are executed in the Bourne shell (`/bin/sh`).

1. The first of the lines above

   ```
   myprog:  myprog.f
   ```

   is the dependency line and defines `myprog` as being the target file and `myprog.f` as the file on which `myprog` is dependent. This means that when the command

   > *make myprog*

   is given, `make` should create the desired file `myprog` according to the rule set in this makefile provided that the file `myprog.f` exists. If there is no file named `myprog.f` in the current directory and the makefile has no rule for creating `myprog.f`, `make` complains appropriately.

   > ⋮

   ```
   % make myprog
   Make:  Don't know how to make myprog.f.  Stop.
   %
   ```

   Notice how this differs from the earlier example when `make` was unable to find a file with the correct rootname. Then it was just looking for any file that could be used to derive `myprog`. When it couldn't find one, it simply responded that it didn't know how to make `myprog`. In the current example, `make` is told that the particular file `myprog.f` is necessary for making the target file. So, when it can't find `myprog.f`, it complains that that particular file, `myprog.f`, is missing.

2. The last line of this makefile provides the actual commands or *explicit rules* (in this case a single command) that `make` is to follow in creating the target file, `myprog`.

```
f77 -O1 -o myprog myprog.f
```

This Bourne shell command line tells `make` to use the Fortran `f77` compiler on the dependency file `myprog.f` with the `-O1` optimization flag and to store the resultant executable file as the target file.

> *Caution:* One of the odd rules of `make` is that the lines containing the commands for `make` rules, such as the line discussed above, must be indented by a Tab character; spaces do not work. Using spaces instead of the Tab is a common error in makefiles.

5.4.2   Using a makefile

If we now try the command

make myprog

with only the files `makefile` and `myprog.f` in the current directory, we get the following response:

```
            ⋮
% make myprog
f77 -O1 -o myprog myprog.f
%
```

Clearly, the `-O1` optimization flag is used, as specified in our makefile rule, in contrast to the earlier examples that used the implicit `make` rules. The new file `myprog` is now in the current directory.

5.4.3   Another sample makefile

An alternate version of the description file would separate the compile step from the link step, as in the following makefile

```
# Simple 2-Step Makefile

myprog:  myprog.o
        f77 -o myprog myprog.o
```

Figure 5.2: Two-Step Derivation of `myprog` from `myprog.f`.

```
myprog.o:  myprog.f
           f77 -O1 -c myprog.f
```

This makefile has two rules replacing the one of the previous makefile. The derivation tree for this description file is shown in figure 5.2.

Given the command

make myprog

`make` sees that `myprog` is dependent on `myprog.o`, and it discovers that `myprog` is not in the directory. But it finds that there is a rule for making `myprog.o`. This rule is dependent on `myprog.f`, a file existing in the current directory. Hence, `make` uses that second rule to generate the file `myprog.o`, using the `-c` option of the Fortran `f77` compiler. Once that is accomplished, the target file, `myprog`, is created using the first rule.

⋮

```
% make myprog
f77 -O1 -c myprog.f
f77 -o myprog myprog.o
%
```

Notice that each rule is printed out as it is executed by `make`. After the successful completion of this invocation of `make`, the contents of the current working directory includes `myprog.f`, `myprog.o`, `myprog`, and `makefile`.

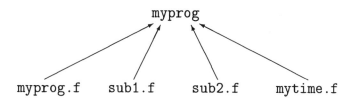

Figure 5.3: Derivation of Multi-Modular `myprog`.

Observe the order of the two rules in this makefile. The dependency for the first target is the second target. On a few systems, make appears to require that the target rules defining dependencies occur in the makefile ahead of the targets using those dependencies. On some other systems, this order is reversed. However, for most systems, the order is unimportant.

5.5   Further examples

Consider a more complex program consisting of several modules: `myprog.f`, `sub1.f`, `sub2.f`, and `mytime.f`. Here the convenience of make becomes more apparent. The makefile for the directory containing these modules might contain the following information:

```
# Simple Makefile for Several Modules

myprog:  myprog.f sub1.f sub2.f mytime.f
         f77 -O1 -o myprog myprog.f sub1.f sub2.f mytime.f
```

Notice that all four Fortran modules are components of the target, `myprog`, and are listed as dependencies. This is perfectly legal; the executable file indeed depends upon all of them as shown in figure 5.3. However, this makefile requires that all four Fortran files be recompiled whenever `myprog` is to be made, since the `*.f` version of each module is listed as part of the `f77` command line.

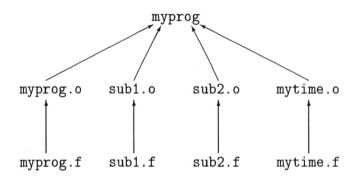

Figure 5.4: Alternate Derivation of Multi-Modular myprog.

A version taking better advantage of make separates the compile steps and has the myprog target dependent on the object versions (*.o files) for the given modules.

```
# Better Makefile for Several Modules
myprog:  myprog.o sub1.o sub2.o mytime.o
         f77 -o myprog myprog.o sub1.o sub2.o mytime.o
myprog.o:  myprog.f
         f77 -O1 -c myprog.f
sub1.o:  sub1.f
         f77 -O1 -c sub1.f
sub2.o:  sub2.f
         f77 -O1 -c sub2.f
mytime.o:  mytime.f
         f77 -O1 -c mytime.f
```

In this way, not all the Fortran files are recompiled to make myprog. Instead only those Fortran modules that have been modified since the last make are recompiled. The derivation for this description file is shown in figure 5.4.

Suppose you had these four Fortran modules in your current directory along with the makefile.

```
       ⋮
% ls -l
total 4
-rw-rw-r-- 1 schauble      293 Jul 5 09:16 makefile
-rw-r--r-- 1 schauble       19 Jul 2 11:41 myprog.f
-rw-rw-r-- 1 schauble       31 Jul 5 09:17 mytime.f
-rw-rw-r-- 1 schauble       29 Jul 5 09:17 sub1.f
-rw-rw-r-- 1 schauble       29 Jul 5 09:17 sub2.f
%
```

The first time you use make to compile these modules together you might get
the following response.

```
       ⋮
% make myprog
f77 -O1 -c myprog.f
f77 -O1 -c sub1.f
f77 -O1 -c sub2.f
f77 -O1 -c mytime.f
f77 -o myprog myprog.o sub1.o sub2.o mytime.o
%
```

Again, each rule is printed by make as it is executed. After the command
has completed, the directory contains the *.o files as well as the final target,
myprog.

```
       ⋮
% ls -l
total 94
-rw-rw-r-- 1 schauble      293 Jul 5 09:16 makefile
-rwxrwxr-x 1 schauble    81188 Jul 5 09:20 myprog*
-rw-r--r-- 1 schauble       19 Jul 2 11:41 myprog.f
-rw-rw-r-- 1 schauble      648 Jul 5 09:19 myprog.o
-rw-rw-r-- 1 schauble       31 Jul 5 09:17 mytime.f
-rw-rw-r-- 1 schauble      456 Jul 5 09:20 mytime.o
-rw-rw-r-- 1 schauble       29 Jul 5 09:17 sub1.f
```

```
-rw-rw-r-- 1 schauble     452 Jul 5 09:19 sub1.o
-rw-rw-r-- 1 schauble      29 Jul 5 09:17 sub2.f
-rw-rw-r-- 1 schauble     452 Jul 5 09:19 sub2.o
%
```

If changes were now made to one module, say `sub2.f`, then the command

> *make myprog*

merely requires that `sub2.f` be recompiled into `sub2.o`. Then all the `*.o` files, including those unchanged since the last make, are linked together to form the new `myprog`.

$$\vdots$$

```
% make myprog
f77 -O1 -c sub2.f
f77 -o myprog myprog.o sub1.o sub2.o mytime.o
% ls -l
total 94
-rw-rw-r-- 1 schauble     293 Jul 5 09:16 makefile
-rwxrwxr-x 1 schauble   81188 Jul 5 09:20 myprog*
-rw-r--r-- 1 schauble      19 Jul 2 11:41 myprog.f
-rw-rw-r-- 1 schauble     648 Jul 5 09:19 myprog.o
-rw-rw-r-- 1 schauble      31 Jul 5 09:17 mytime.f
-rw-rw-r-- 1 schauble     456 Jul 5 09:20 mytime.o
-rw-rw-r-- 1 schauble      29 Jul 5 09:17 sub1.f
-rw-rw-r-- 1 schauble     452 Jul 5 09:19 sub1.o
-rw-rw-r-- 1 schauble      30 Jul 5 09:20 sub2.f
-rw-rw-r-- 1 schauble     452 Jul 5 09:20 sub2.o
%
```

This is because `make` checks the creation dates of the `*.o` files against their corresponding `*.f` files (the dependencies), only recreating the out-of-date files (in this case, `sub2.o`).

In this last example, it becomes more obvious how `make` can save work. It knows exactly which modules need to be recompiled, and the simple command *make myprog*

performs only those compilations necessary. Observe that entering a single command (only two words in length) is all that is needed to do the job, while typing in the corresponding f77 commands is be considerably longer (and more prone to typing errors). Of course, the makefile is not short, but it can be used over and over again. And by using make, you are always certain that your files are up-to-date.

The makefile can also be used as a record of the compilation. By looking at the rules and macro settings in the makefile, you can tell how the executable files in the directory were compiled. For this reason, it is best to use different target names for executable files compiled under different flags. For instance, debug can be used for an executable file compiled under the -g option to be used by the dbx utility, as shown in section 5.9.2.

5.6   Dynamic macros

The main problem with the makefile used in this particular example is that it is long and repetitious. It seems that there must be a simpler way to accomplish this. For this reason, make provides the use of *macros* or *macro symbols*. These macros can take on the values of targets or dependencies and act much like variables in the make rules. *Dynamic macros* are defined by make to have certain properties, but the actual filenames they represent change with each rule. *User-defined macros* are covered in the next section.

5.6.1 $@ and $<

Consider the following version of our makefile:

```
# Shorter Makefile for Several Modules using Macros

myprog:  myprog.o sub1.o sub2.o mytime.o
         f77 -o $@ myprog.o sub1.o sub2.o mytime.o

.f.o :
         f77 -O1 -c $<
```

There are a few things to notice.

1. First of all, the rule for making `myprog` now contains the characters `$@`, instead of the filename `myprog`. For the `make` utility, these characters (`$@`) are interpreted as a macro that is equivalent to the name of the target. This provides a shorthand method to repeat the target name. In fact, it would be correct to modify the rule above to read as follows:

    ```
    myprog:  $$@.o sub1.o sub2.o mytime.o
             f77 -o $@ $@.o sub1.o sub2.o mytime.o
    ```

    > *Caution:* All `make` macros begin with a dollar sign (`$`). When a dynamic macro, such as "`$@`", is used in the dependency list, it must be preceded by an extra dollar sign.

2. All macros are set to the appropriate values when `make` first begins to work on a given rule and target.

3. The biggest difference in the new and shorter makefile above is that all the `.f-to-.o` rules are combined into a single rule. This rule produces a `.o` file from any `.f` file existing in the working directory. This is what the `.f.o` target means; namely, if a `.o` file is desired, create it from the corresponding `.f` file (the file with the same rootname) according to the given rule. This is called a `make` *implicit rule format* and overrides the default rule.

4. The rule itself contains the symbol, "`$<`", where normally the Fortran filename would be. This `make` *macro*, "`$<`", should only be used in implicit rules. It refers to the file that invoked the current rule, the one with the same file extension as the first one listed as part of the target. In other words, if `make` is looking for the file `mytime.o` and finds instead the file `mytime.f`, `make` interprets "`$<`" as `mytime.f` and uses that to make `mytime.o`.

Suppose we have modified the `sub1.f` and `mytime.f` files. Then typing the command

make myprog

with this new makefile shows the following response:

```
              ⋮
% make myprog
f77 -O1 -c sub1.f
f77 -O1 -c mytime.f
f77 -o myprog myprog.o sub1.o sub2.o mytime.o
%
```

If our makefile had not redefined the implicit rule, the command above would invoke the default compilation rule instead. Then the makefile would contain just one rule, much like the following:

```
# Makefile for Several Modules using Implicit Compilation

myprog:  myprog.o sub1.o sub2.o mytime.o
          f77 -o $@ myprog.o sub1.o sub2.o mytime.o
```

And the previous dialogue with make would change to show that the default -O optimization flag was used in compiling the individual modules.

```
              ⋮
% make myprog
f77 -O -c sub1.f
f77 -O -c mytime.f
f77 -o myprog myprog.o sub1.o sub2.o mytime.o
%
```

Replacing the default implicit rule allows us to define things to be done the way we want.

5.6.2 $? and $*

Two other special dynamic macros of make are the characters $? and $*. The first, $?, refers to the list of all filenames that make needs recursively to make the target. The other, $*, is only used in implicit rules; it refers to the common rootname of the two target filenames.

To illustrate this, consider the following modified version of our makefile. We explain the use of $(OBJECTS) and $(CFLAGS) in section 5.7.

```
# Makefile for Several Modules echoing Macros
OBJECTS = myprog.o sub1.o sub2.o mytime.o
CFLAGS = -O1
myprog:  $(OBJECTS)
         f77 $(CFLAGS) -o $@ $(OBJECTS)
         echo "The target file is " $@
         echo "The dependencies are " $?

.f.o :
         f77 $(CFLAGS) -c $<
         echo "The invoking file is " $<
         echo "The common rootname is " $*
         echo " "
```

Executing this version of the makefile causes the values of all four dynamic macros, including "$?" and "$*", to be written to the screen:

⋮

```
% make -s myprog
The invoking file is myprog.f
The common rootname is myprog

The invoking file is sub1.f
The common rootname is sub1

The invoking file is sub2.f
The common rootname is sub2

The invoking file is mytime.f
The common rootname is mytime

The target file is myprog
The dependencies are myprog.o sub1.o sub2.o mytime.o
%
```

Observe that the `-s` option flag has been used in this command. Normally, `make` echos each command of a makefile rule as it executes it; this flag turns the echo off. Otherwise, many more lines would have been printed out, even the "echo" commands. For more information on this and other options, see section 5.8.

5.7   User-defined macros

It is often convenient for the programmer to create his own macros in a make-file. These are called *user-defined macros* and were used in the makefile in the previous example.

5.7.1   A makefile with user-defined macros

Consider the following modifications to our makefile.

```
# Makefile for Several Modules using Macros
OBJECTS = myprog.o sub1.o sub2.o mytime.o
CFLAGS = -O1 -c
myprog:  $(OBJECTS)
         f77 -o $@ $(OBJECTS)
.f.o :
         f77 $(CFLAGS) $<
```

This performs the same as the earlier makefile using the -O1 optimization. However, two user-defined macros, $(OBJECTS) and $(CFLAGS), have been added.

1. The first of these is just a list of all the object files (*.o files) upon which the target is dependent. This macro provides a handle for this list of files, which is used twice in this makefile. We need only type the list once in the definition of the new macro, so we may prevent typing errors. This also makes it easier to add another object file to the list.

2. The second macro is $(CFLAGS). This provides the compile flags we wish to use. If we wish to change optimization levels at some time, we could change the definition of $(CFLAGS) in the makefile

   ```
   CFLAGS = -O2 -c
   ```

   As an alternative, we could call make with the specific parameter:

   make myprog "CFLAGS=-O2 -c"

`Make` allows user-defined macros to be redefined in this manner. The macros are merely given new values by using keyword parameters when the `make` command is given.

Using the second method gives more flexibility. The dialogue with `make` would be as follows, provided all the old `*.o` files have been deleted first:

> ⋮

% make myprog "CFLAGS=-O2 -c"
```
f77 -O2 -c myprog.f
f77 -O2 -c sub1.f
f77 -O2 -c sub2.f
f77 -O2 -c mytime.f
f77 -o myprog myprog.o sub1.o sub2.o mytime.o
%
```

By using different parameters for the macro `CFLAGS` with the same makefile, we can compile the program under different optimization levels. The value of any user-defined macro in a makefile may be changed in this manner for a specific run.

The main problem with this method is that we have lost the record of how the target file was created. If you were to look at the directory after some weeks, you probably would not remember whether or not `myprog` was made with the flags given in the makefile or with some parameter change. It might be wise to rename the executable file in this case, `myprog2`, to indicate that it is not the standard version of the target file. Or better yet, include a `make` rule for `myprog2` to use the `-O2` flag.

5.7.2   Another makefile using user-defined macros

As another example, we can consider using a different version of the Fortran compiler. First, we need to create a macro to define the compiler and add it to the rules, replacing all the occurrences of `f77`:

> ⋮

```
FCOMPLR = f77
```
> ⋮

```
myprog:   $(OBJECTS)
          $(FCOMPLR) -o $@ $(OBJECTS)

.f.o  :
          $(FCOMPLR) $(CFLAGS) $<
```

Now if we remove the old *.o files and remake myprog with a parameter specifying a different compiler, the dialogue looks like this:

```
                  ⋮
% make myprog "FCOMPLR=f772.1"
f772.1 -O1 -c myprog.f
f772.1 -O1 -c sub1.f
f772.1 -O1 -c sub2.f
f772.1 -O1 -c mytime.f
f772.1 -o myprog myprog.o sub1.o sub2.o mytime.o
%
```

Again, it might be best to use different target names for executable files that are compiled under different compilers. In this way, the makefile would provide a record of how each executable file was compiled.

5.7.3   Using user-defined macros

There are a few points to be observed concerning the macros.

1. Macro names are strings consisting of letters and digits and are similar to C identifiers. They can be of almost any length, subject to the system default. By convention, the letters used in macro names are usually capitalized.

2. User-defined macros should be given a default definition at the beginning of the makefile. This is to assure that they are defined before they are used.

3. Macros can be redefined by calling make with parameters. More than one parameter can be included in the make command. For instance, the command

make myprog "FCOMPLR=f772.1" "CFLAGS=-O2 -c"

is perfectly legal and redefines both the `FCOMPLR` and `CFLAGS` macros.

4. When used with a `make` rule, the macro names must be contained within parentheses and preceded by a dollar sign, e.g., `$(CFLAGS)`. The exception to this rule is that macro names of a single letter need not be put inside parentheses, e.g., `$A`. Any single character or parenthesized string that is preceded by a dollar sign is assumed to be a `make` macro.

5.8 Additional features

We have described above only a few of the features of `make` in order to get you started. Many more features are available. There are several option flags that can be used with the `make` command. And there are some special (fake) target names and command prefixes to add flexibility. More of of these features are discussed below; check the `man` pages to get a full listing.

5.8.1 Silent running

The `make` utility usually prints out each command (or rule) before it executes. This allows the user to see how the process is proceeding. However, it is possible that you might not wish to have all this output.

There are three ways to reduce this output.

1. The `-s` option flag requests a silent interaction with `make`. The commands (or rules) are not be printed out as they are executed. The dialogue with `make` simply becomes as follows:

> ⋮
> *% make -s myprog*
> *%*

Only the command line for `make` appears on the screen, as you typed it in. An example of using this option is shown in section 5.6.2.

2. If you just want certain rules not printed by make, insert the character @ at the beginning of the line. Rules beginning with an "@" are not be printed. For instance, consider the following makefile rule for a target named help that describes the functions of the given makefile:

```
help :
        @echo "This makefile supports the following:"
        @echo "make run - runs the program"
        @echo "make myprog - creates executable program"
```

Typing the command to make help gives the following response:

```
    .
    .
    .
% make help
This makefile supports the following:
make run - runs the program
make myprog - creates executable program
%
```

If the command lines do not begin with the character @, the response is as follows:

```
    .
    .
    .
% make help
echo "This makefile supports the following:"
This makefile supports the following:
echo "make run - runs the program"
make run - runs the program
echo "make myprog - creates executable program"
make myprog - creates executable program
%
```

With each echo command printed as it is executed, the output becomes confused and is not as helpful as the target name suggests.

3. If you always want to run in a silent mode, you should add the special (or fake) target name, .SILENT, to your makefile. For example, consider the following makefile:

```
# Makefile with silent responses
OBJECTS = myprog.o sub1.o sub2.o mytime.o
CFLAGS = -O1
.SILENT:
myprog:  $(OBJECTS)
         f77 $(CFLAGS) -o $@ $(OBJECTS)
.f.o :
         f77 $(CFLAGS) -c $<
```

Notice that .SILENT is given as a target with no dependencies and no
rules. If we use this makefile to produce the executable file myprog,
the dialogue does not show the actions of make while they are being
performed. The contents of the directory are shown before and after
the make command; otherwise, you might wonder if anything was indeed
done.

```
          ⋮
% ls
makefile      mytime.f      sub1.f
myprog.f      specprog.f    sub2.f
%
% make myprog
%
% ls
makefile      myprog.f      mytime.f
myprog*       myprog.o      mytime.o
specprog.f    sub1.o        sub2.o
sub1.f        sub2.f
%
```

5.8.2   Error handling

If any rule results in an error while make is executing, an error message prints
and the make command stops. Sometimes, this is not appropriate for the given
makefile. Consider, for instance, a rule to remove some files from the working

directory to clean up after running some process. If one of more of the files to be removed does not exist, the rule returns an error and make stops.

There are three ways to avoid this.

1. The first is the use of the special character – at the beginning of a rule. If an error is encountered while executing such a rule, the error is ignored and make continues.

2. The second method is to include the fake target .IGNORE as part of the makefile. This is similar to using the .SILENT fake target in section 5.8.2 and instructs make to ignore all the errors found during its execution.

3. The final method is to call make with the -i option flag, telling make to ignore all error codes for this particular invocation.

5.8.3   -n Option flag

Suppose you want to make myprog, but first you want to know which files need to be made and which are already up-to-date as well as the commands that will be executed. Of course, you could study the makefile and look at the creation dates of all the files you believe will be used. But an easier way would be to use the -n option flag for make, as follows:

```
        ⋮
% make -n myprog
f77 -O1 -c myprog.f
f77 -O1 -c sub2.f
f77 -o myprog myprog.o sub1.o sub2.o mytime.o
%
```

This appears to act just like the make command without the -n option flag. However, the n stands for no-execute-mode; so the commands that normally would be executed are merely listed without any of them being done. The example above shows that the object files, myprog.o and sub2.o, must be missing from the directory or out-of-date and so need to be created before the whole program can be linked together. In a like fashion, we can see that the object files, sub1.o and mytime.o, must be present and up-to-date. We can also see which options the Fortran compiler is using.

5.8.4   -t Option flag

The -t option flag requests that certain files be touched. When a file is touched, its creation date is changed to the current date and time. This option may be used to avoid remaking files dependent upon other files when those other files have have been changed in a manner that should not affect the dependent files.

For example, suppose we make a minor change to the comments in the file, sub2.f. This causes the current version of the object file sub2.o to be out-of-date. Consequently, the executable file for the whole program myprog is also out-of date. Touching those two files updates their creation dates. Meanwhile, the make utility knows that the other object files are still current.

```
              ⋮
% ls -l
total 89
-rw-rw-r-- 1 schauble    192 Jul  5  17:03 makefile
-rwxrwxr-x 1 schauble  81188 Jul 31  16:38 myprog*
-rw-r--r-- 1 schauble     19 Jul  2  11:41 myprog.f
-rw-rw-r-- 1 schauble    648 Jul 31  16:38 myprog.o
-rw-rw-r-- 1 schauble     31 Jul  5  09:17 mytime.f
-rw-rw-r-- 1 schauble    456 Jul 31  15:40 mytime.o
-rw-rw-r-- 1 schauble     29 Jul  5  09:17 sub1.f
-rw-rw-r-- 1 schauble    452 Jul 31  15:40 sub1.o
-rw-rw-r-- 1 schauble     30 Jul 31  16:46 sub2.f
-rw-rw-r-- 1 schauble    452 Jul 31  16:38 sub2.o
% make -t mytime.o
'mytime.o' is up to date.
% make -t sub1.o
'sub1.o' is up to date.
% make -t myprog
touch(sub2.o)
touch(myprog)
% ls -l
total 89
-rw-rw-r-- 1 schauble    192 Jul  5  17:03 makefile
```

```
-rwxrwxr-x 1 schauble 81188 Jul 31 16:47 myprog*
-rw-r--r-- 1 schauble    19 Jul 2  11:41 myprog.f
-rw-rw-r-- 1 schauble   648 Jul 31 16:38 myprog.o
-rw-rw-r-- 1 schauble    31 Jul 5  09:17 mytime.f
-rw-rw-r-- 1 schauble   456 Jul 31 15:40 mytime.o
-rw-rw-r-- 1 schauble    29 Jul 5  09:17 sub1.f
-rw-rw-r-- 1 schauble   452 Jul 31 15:40 sub1.o
-rw-rw-r-- 1 schauble    30 Jul 31 16:46 sub2.f
-rw-rw-r-- 1 schauble   452 Jul 31 16:47 sub2.o
%
```

Notice that touching the final executable file, myprog, causes the out-of-date object file, sub2.o, to be touched as well. This is because sub2.o is one of the files on which myprog is dependent.

5.8.5   -d Option flag

The −d option flag permits debugging of the makefile by generating more detailed information on each command (or rule) that is executed. For instance, typing the command

> % *make -d myprog*

results in a great deal of output, only some of which is included here.

```
           ⋮
% make -d myprog
setvar:  = noreset = 0 envflg = 0 Mflags = 040101
Reading "=" type args on command line.
Reading internal rules.
setvar:  MACHINE = mips noreset = 0 envflg = 0
    Mflags = 040101
           ⋮
Reading makefile
setvar:  OBJECTS = myprog.o sub1.o sub2.o mytime.o
    noreset = 0 envflg = 0 Mflags = 040001
```

```
setvar:  CFLAGS = -O1 -c noreset = 0 envflg = 0
     Mflags = 040001
Warning:  CFLAGS changed after being used
doname(myprog,0)
TIME(myprog)=681000472
        ⋮
TIME(mytime.o)=680996426
look for implicit rules.  0
'myprog' is up to date.
        ⋮
OBJECTS = myprog.o sub1.o sub2.o mytime.o
        ⋮
MACHINE = mips
$ = $
MAKEFLAGS = bd
mytime.f done=2
sub2.f done=2
sub1.f done=2
myprog.f done=2
mytime.o done=2
sub2.o done=2
sub1.o done=2
myprog.o done=2
done=2 (MAIN NAME)
depends on:myprog:  myprog.o sub1.o sub2.o mytime.o
commands:
$(FCOMPLR) -o $@ $(OBJECTS)
markfile done=0
        ⋮
%
```

As you can see, this option provides a lot of information. It lists all the environment settings used by make, the steps to determine the date of each target, and the rules, implicit or explicit, to make each target.

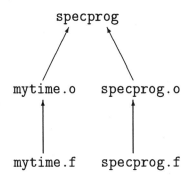

Figure 5.5: Derivation of specprog.

5.8.6   -f Option flag

The -f option flag for the make utility allows a file named something other than makefile or Makefile to act as a makefile.

Suppose we have an additional program in our current directory named specprog.f. And assume this program also uses one of the subroutines in the directory named mytime.f. The derivation for specprog is shown in figure 5.5. We can create an additional makefile that makes the executable file specprog as follows:

```
# A Special Makefile

OBJECTS = specprog.o mytime.o
CFLAGS = -O1 -c
FCOMPLR = f77

specprog:  $(OBJECTS)
        $(FCOMPLR) -o $@ $(OBJECTS)

.f.o :
        $(FCOMPLR) $(CFLAGS) $<
```

We can name this new makefile any legal filename; in this case, we call it special. To make the executable file, specprog, type the following:

make -f special specprog

and the dialogue with `make` appears as below:

       ⋮

% make -f special specprog
```
f77 -O1 -c specprog.f
f77 -o specprog specprog.o mytime.o
%
```

The `-f special` portion of the command tells `make` to use the file named `special` to find the rules for the target `specprog`. Thus it is possible to keep more than one makefile in your directory. Notice that the object file `mytime.o` had been compiled previously and was up-to-date (possibly from being used with the other program `myprog`) and so did not need to be recreated.

Of course, we could have included all the rules for both targets in a single makefile,

```
# Combined Makefile

OBJECTS = myprog.o sub1.o sub2.o mytime.o
SOBJECTS = specprog.o mytime.o
CFLAGS = -O1 -c
FCOMPLR = f77

myprog:  $(OBJECTS)
         $(FCOMPLR) -o $@ $(OBJECTS)

specprog:  $(SOBJECTS)
         $(FCOMPLR) -o $@ $(SOBJECTS)

.f.o :
         $(FCOMPLR) $(CFLAGS) $<
```

simply by renaming the `OBJECTS` macro used by one of the target rules. The combined derivation graph for this more complex makefile is shown in figure 5.6. Observe that the module, `mytime.o`, is used by both main targets.

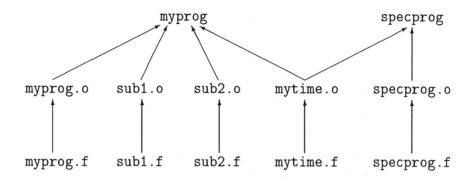

Figure 5.6: Alternate Derivation of `specprog`.

5.9 Other examples

The examples we have used above all relate to compiling modular Fortran programs. But there are a number of other things can be done with `make` as well. It is not restricted to program compilation.

5.9.1 Program execution

It is easy to include a `run` target in your makefile. This executes a program, compiling and linking modules, if necessary.

```
run:  myprog
        myprog > myprog.output
```

In this case, typing the command

make run

executes the program, `myprog`, causing the output to be redirected to the file named `myprog.output`. Observe two points about this makefile rule:

1. The target, `run`, is not a file; that is, the execution of this rule does not create a file named `run`. Since the target `run` is never created, a file named `run` is never found in the currently directory. As a result, the command

make run

always executes. This is legal and at times desirable.

It would have been quite logical to have the target be `myprog.output` (the name of the output file for the program) instead of `run`. However, the `myprog.output` output file would need to be removed or renamed before each execution.

2. The target is dependent on the executable file `myprog`. If `myprog` is unavailable or out-of-date with respect to `myprog.f` or `myprog.o`, `make` performs whatever steps are necessary to create a current version of `myprog` before executing it.

5.9.2   Compiling for `dbx`

If you are having trouble getting your program to run, you may wish to re-compile it with the `-g` option flag; this provides an executable file that can be used with the `dbx` debugging utility. To do so, you should include the following statements in your makefile:

```
        ⋮
SOURCES = myprog.f sub1.f sub2.f mytime.f
        ⋮
debug:  $(SOURCES)
        $(FCOMPLR) -g -o debug $(SOURCES)
```

Here we have added a macro named `SOURCES` to list all the source files involved with the program `myprog.f`. We cannot use the macro `OBJECTS` because that contains the `*.o` files that compile without the `-g` option. This new rule re-compiles all the modules together with the `-g` option and creates an executable file named `debug`, (so named to avoid confusion with the regular executable file, `myprog`). Since the `-g` option turns off all optimizations, there is no need to include `$(CFLAGS)` in this rule.

5.9.3 Printing files

The makefile can include rules to print out the source files as well as the output.

```
prints:  $(SOURCES)
         lpr $(SOURCES)

printo:  myprog.output
         lpr myprog.output
```

By referring to the macro SOURCES, all the modules related to myprog.f are printed. Again, no target file is created when either of these rules is executed.

5.9.4 Cleaning up

The makefile might include a rule to clean up the directory, deleting any core or *.o files.

```
clean:
         - rm core debug *.o
```

This version also removes the executable file, debug, which was created for use with dbx and is probably no longer needed. You might also wish to remove the file myprog, as that can be generated whenever needed by the makefile.

Notice that the rule begins with the special character -. If no core file happened to exist in the directory when the command make clean was done, the rule would stop when it couldn't find a file named core. But because of the special ignore-errors character - at the front of the rule, make continues onto the next file after printing out an error message.

```
          ⋮
% make clean
rm core
rm:  core nonexistent
** Error code 1 (ignored)
rm debug
rm:  debug nonexistent
** Error code 1 (ignored)
```

```
rm *.o
%
```

Here you can see that the files listed on the `rm` statement are handled individually. The error messages for non-existent files are printed but ignored. Also the commands for a rule can include wild card characters, as in `*.o`.

5.9.5   A complete makefile

A full makefile containing all the rules in this section is given in section 5.13. You may wish to use it as a model. Eventually, you will develop your own style. For instance, you might use `myprog.x` as the target name of the executable file in place of the filename `myprog`. Also, `myprog.out` might have replaced the target `run`.

5.10   A makefile for C

Designing a makefile for program maintenance in other languages should be straightforward, based on the previous example. For instance with C programs, you may want to add a list of the header (`*.h`) files to the dependencies. An example of such a makefile is given in section 5.14.

There are a few things to notice in this example:

1. The macro `LIBS` is used to include the `math.h` library in the compile-and-link command. This library is commonly used by C programs. Other library names can be added or substituted here.

2. The command

   make help

   results in a listing of all the functions of this makefile. This is a very useful and recommended procedure as one often forgets what one has done some months ago.

   This is the very first target in the make file. So if you should just type *make*

the list still appears. In other words, the `make` command with no target attempts to make the first target in the makefile. If this is the `help` target, the response is a listing of what the makefile can do.

3. The `echo` commands under the target `help` are preceded by the character `@`. This is to prevent the echo command from being printed out in addition to the string it is printing.

4. References to header `*.h` files have been added.

5. Targets entitled by macro names may substitute any of the files defined by that macro as the target file. For instance,

   make mainprog

   activates the `$(EXECFIL)` rule.

6. Target need have no commands in the rule. The target

   ```
   $(FILES) : $(HEADERS)
   ```

   merely declares that `$(HEADERS)` are dependencies for `$(FILES)`. This rule has no other purpose.

5.11 Creating your own makefile

Try creating a simple makefile in one of your directories. As you begin to use `make`, you will find more and more things that you want to do with it. As a result, your makefile begins to grow.

You may need a separate makefile in each directory, but this means that each one can be individualized to meet the requirements of its own directory.

Some points to remember when creating your own makefiles:

1. Remember that the rules in a makefile are interpreted as Bourne shell commands. Among other things, this means that it is best to use absolute pathnames when referring to other directories; the value of the tilde character ~, is not recognized by the Bourne shell.

2. The environment variables, such as $HOME and $HOST, are known by make as it executes and are treated as macros of the same names. Similarly, make assumes a macro named $SHELL that is equivalent to /bin/sh by default. On some of the newer UNIX operating systems, it is possible to change this, by redefining the macro variable:

```
SHELL = /bin/csh
```

5.12   Further information

For further information on make, you might refer to [Feldman 86] or the man pages. This utility is discussed in some C and UNIX textbooks, e.g., [Sobell 84] and [Kay & Kummerfeld 88]. Books, such as *Managing Projects with Make* [Oram & Talbott 91] and *Mastering Make* [Tondo et al 92], provide a more thorough discussion of the use of make with examples in other areas.

5.13   A makefile for Fortran modules

```
# Final Complete Makefile
#           to initialize user-defined macros
SOURCES = myprog.f sub1.f sub2.f mytime.f
OBJECTS = myprog.o sub1.o sub2.o mytime.o
CFLAGS = -O1 -c
FCOMPLR = f77

#           to execute myprog
run:  myprog
        myprog > myprog.output

#           to link and compile myprog
myprog:  $(OBJECTS)
        $(FCOMPLR) -o $@ $(OBJECTS)

#           to compile Fortran modules
.f.o :
        $(FCOMPLR) $(CFLAGS) $<

#           to compile myprog for dbx
debug:  $(SOURCES)
        $(FCOMPLR) -g -o debug $(SOURCES)

#           to print myprog source files
prints:  $(SOURCES)
        lpr $(SOURCES)

#           to print myprog output file
printo:  myprog.output
        lpr myprog.output

#           to remove unnecessary files
clean:
        - rm core debug *.o
```

5.14   A makefile for C modules

```
#
# Makefile for C modules
#
FILES = mainprog.c submod.c
HEADERS = mainprog.h submod.h
OBJECTS = mainprog.o submod.o
EXECFIL = mainprog
LIBS = -lm
FLAGS =

help :
        @echo "This makefile supports the following:"
        @echo "make run - runs the program"
        @echo "make mainprog - creates executable program"
        @echo "make lint - run lint on the program"
        @echo "make debug - provides dbx executable"
        @echo "make clean - deletes *.o, debug, core files"

run :   $(EXECFIL)
        $(EXECFIL)

$(EXECFIL): $(OBJECTS)
        cc $(FLAGS) $(OBJECTS) -o $(EXECFIL) $(LIBS)

.c.o :  $$@.c $$@.h
        cc $(FLAGS) -c $<

lint :  $(FILES)
        lint $(FILES)

debug :  $(FILES)
        cc -g $(FLAGS) -o debug $(FILES)

$(FILES): $(HEADERS)

clean :
        - rm *.o core debug
```

6 Elements of Fortran

6.1 Introduction

For many years Fortran has been the language of choice in scientific computing, and, even though C has become increasingly popular, Fortran remains an important language in scientific computing. Indeed, there are Fortran compilers for every supercomputer, and new versions of it exist for vector computers and for parallel computers. For this reason, we use Fortran in most of the laboratory exercises and examples. However, we have found that many students enrolling in our course are unfamiliar with Fortran. For this reason we have found a review of Fortran to be quite useful and therefore have included it in this book.

The brief and elementary review in this chapter describes enough of Fortran to enable you to read and understand the programs used in our laboratory exercises. Nevertheless, you will probably find a need for a more thorough description of Fortran. Two texts that you may find useful are *Fortran 77 for Humans* [Page 83] and *Effective Fortran 77* [Metcalf 85]. Besides these texts, you may find it useful to refer to the book which defines standard Fortran 77: *X3.9-1977 Programming Language Fortran* which is available from the American National Standards Institute, Inc., 1430 Broadway, New York, NY 10018.

Fortran 90 and HPF (**H**igh **P**erformance **F**ortran) are more recent versions of Fortran that include vector operations and other features for high performance computing. Descriptions of Fortran 90 can be found in *Programmers Guide to Fortran 90* [Brainerd et al 90] and in the more complete *Fortran 90 Handbook* [Adams et al 92]. A reference for HPF is *The High Performance Fortran Handbook* [Koelbel et al 94].

This chapter consists of four parts. Section 6.2 is a brief overview of Fortran. Section 6.3 contains basic definitions. Section 6.4 is a description of Fortran statements, organized alphabetically. Section 6.5 is a short description of the use of the `READ` and `WRITE` statements. Section 6.6 presents two sample programs.

6.2 Overview

6.2.1 Program structure

A Fortran program is composed of statements. Typically these statements are grouped into subprograms, also called procedures. One of the subprograms is the main program and its statements are executed first. Other subprograms are executed by procedure calls, as will be explained later. Small programs may consist of just one subprogram, the main program. An example follows.

```
PROGRAM HELLO
WRITE(*,*) 'HELLO WORLD'
END
```

6.2.2 Statements

There are two categories of statements: *executable* and *non-executable*. Executable statements specify operations that the computer must perform, or execute. An example of an executable statement is:

```
WRITE(*,*) 'HELLO WORLD'
```

When this statement is executed, the computer writes `HELLO WORLD` on the standard output device, normally your CRT display.

An example of a non-executable statement is:

```
PROGRAM HELLO
```

This statement declares the statements following it, up to and including the `END` statement to be a main program having the name `HELLO`.

Usually a statement ends at the end of the line, but provision is made for long statements to continue onto additional lines. There is no special mark, such as a semicolon as in C, to denote the end of a statement.

6.2.3 Control statements

Control statements make it possible to have *loops*, sequences of statements that are executed over and over again, and *branches*, alternate sequences of statements to execute.

The control statements are the DO statement, the IF statement, and the GOTO statement, all of which are described in section 6.4.

6.2.4 Expressions

A fundamental component of most executable statements is the *expression*, for example

        X + 1.0

The meaning of this expression is exactly what you expect: it means the value of X plus 1.0. Here X denotes a variable that has some value determined elsewhere, + denotes the arithmetic operator for addition, and 1.0 denotes itself, the numerical value one. As this small example illustrates, an expression is composed of operators and operands, the latter being either variables or constants. Besides arithmetic expressions the language also permits logical expressions and character expressions. Different kinds of expressions have different kinds, or *types*, of values. Arithmetic expressions have values that are numbers; logical expressions have only the values *true* or *false*; character expressions have values that are *characters* (i.e., the characters that you type on your keyboard as well as some you cannot) and sequences of characters, or *strings*.

The most common use of an expression is in an assignment statement; for example

        Y = X + 1.0,

where the variable Y is given the value of the expression X + 1.0. Note that in Fortran the assignment operator is =, as it is in C.

6.2.5 Types

The word *type* refers to the kind of value a variable, constant, or expression has, or is allowed to have. Each variable used in a program has a fixed type, normally defined in a non-executable statement at the beginning of the program. For example, the statement

```
INTEGER A, B
```

declares the variables A and B to have the type INTEGER. This means that the only values A and B are allowed to have are integer values. The rules for evaluating expressions depend on the types of the operands and the kinds of operators used.

6.2.6 Compiling

The UNIX command for compiling a Fortran 77 program is f77. Three sample uses of this command follow:

```
(1)    f77  hello.f
(2)    f77  hello.f  -o hello
(3)    f77  -c  hello.f
```

The first example merely compiles the program in the file hello.f, generating the executable file named a.out. The second command also compiles hello.f but gives the executable file the name hello. The third example compiles the program, but produces only an object file, hello.o; it does not produce an executable file. The standard UNIX f77 command does not produce a program listing. However, many vendors' Fortran compilers do, with a command option such as -l or -list.

6.3   Definitions and basic rules

6.3.1   Character set

The character set consists of the letters of the alphabet

```
A B C D E F G H I J K L M N O P Q R S T U V W X Y Z
```

and the digits

```
0 1 2 3 4 5 6 7 8 9
```

and the following special characters

```
= + · - * / ( ) , . $ ' :
```

There are two important exceptions. `CHARACTER` constants are allowed to contain any character representable on your computer system; the same is true for comment lines.

The following type declarations are equivalent:

```
Integer a
```

and

```
INTEGER A
```

Note, in particular, that there is no distinction between the names `A` and `a` – they stand for the same variable. Fortran is not case sensitive, but, in chapter, this we write Fortran statements using capital letters in order to make them stand out. (In the past, Fortran allowed only upper case letters, but few, if any, compilers have this restriction now.)

6.3.2   Names

Names are used to identify objects such as variables, subprograms, and constants. A name consists of one to six characters, the first of which must be a letter, the rest must be letters or digits. Examples are:

```
A   A1   A123   AB4XYZ
```

This atavism on the length of a name harks back to the days when the word length in most computers was 36 bits, and 6 bits represented a character. (A character is now a one-byte construct.) It is fast disappearing, and most Fortran compilers do not have this length restriction.

6.3.3   List of types

The types in Fortran are:

```
          INTEGER,
          REAL,
          DOUBLE PRECISION,
          LOGICAL,
          CHARACTER,
    and   CHARACTER*n
```

where n is an integer in the range (1, 2, ..., 127).

6.3.4   Labels

Labels are used to identify particular statements. They consist of one to five digits, and they must be in columns[1] 1-5 of the statement line. The leading digit may begin in any of these columns, but the last digit must not extend beyond column 5. An example of a labelled statement is:

```
    99   RETURN
```

Many Fortran compilers do not have this restriction on label position.

6.3.5   Keywords

These are words that have a special meaning in a program. Examples are

```
          DO
          REAL
          INTEGER
          COMMON
          FORMAT
          SUBROUTINE
```

---

[1]By Fortran tradition, the position of a character in a line of input or output is called the *column* where that character is located; this terminology goes back to the time when most input files and programs were kept on punched cards.

and so forth. Although Fortran does not prohibit it, you should not give names to variables that are the same as keywords, otherwise you make the program hard to read.

6.3.6   Variables

A variable that has not been assigned a value is said to be *undefined*. While most computing systems assign the value zero to all variables before execution of a program begins, it is unwise to assume that this is done. Once a variable is assigned a value it retains that value until a new value is assigned to it, usually by execution of an assignment statement.

We distinguish three kinds of variables: simple variables, array elements, and character strings. Simple variables are denoted with a name, for example

```
V1
SOLN
ROOT
```

Array elements are denoted with a name followed by *subscripts*, for example

```
A1(2,3)
MAP(J,K,L)
SCORE(100)
```

An array element denotes a value in an array of values, thus `A1(2,3)` denotes the value in row 2 and column 3 of the array `A1`. Character strings are denoted with a name, just like simple variables, or by a name followed by a pair of values inside parentheses, and separated by a colon, denoting the location of the first character and the location of the last character in a string. Thus `NAME(4:12)` denotes the string of characters starting at character position 4 and extending to position 12 (including position 12) within the character string `NAME`.

6.3.7   Arrays

An array is an n-dimensional object that in one-dimensional form is analogous to a vector and in two-dimensional form is analogous to a matrix. Thus a one-dimensional array is thought of as a sequence of values like:

```
3.45
1.0095
2.2299
```

and a two-dimensional array is thought of a set of values arranged in rows and columns like:

```
 3.45    1.004    0.998
-2.11    0.888   -0.333
```

The number of elements in each dimension is arbitrary, and up to seven dimensions are allowed.

An array has a name and a type. An element in an array is identified with *subscripts*. Thus if `ARR` is the name of the two-dimensional array above, the element `ARR(2,1)` has the value `-2.11` and the element `ARR(1,2)` has the value `1.004`. All of the elements in an array must have the same type.

The subscripts are, in general, `INTEGER` expressions. Thus we can write `ARR(I, J)`, `ARR(I-1, J+1)` and so forth, assuming that the types of `I` and `J` are `INTEGER`.

An *array declarator* is used to declare the name and size of an array. It appears in type statements. Thus in the type statement

```
REAL A(10,20)
```

`A(10,20)` is an array declarator. It declares an array consisting of 10 rows and 20 columns that has the name `A`. As illustrated here an array declarator consists of a name (the name of the declared array) followed by a sequence of values separated by commas, and enclosed in parentheses, that specify the number of elements in each dimension. The numbering of elements in each dimension begins at 1, unlike C where it begins at 0. However, Fortran also allows arbitrary array bounds; for example

```
REAL A(-1:8,0:19)
```

declares a 10×20 array with row subscripts running $-1, 0, \ldots 8$ and column subscripts running $0, 1, \ldots 19$.

An array declarator like `A(N, M)` is allowed only if `N` and `M` are symbolic constants or if the declarator appears in a `SUBROUTINE` or `FUNCTION` and `A`, `N`, and `M` are formal parameters.

6.3.8 Constants

There are two forms of constants, *symbolic* and *literal*. Symbolic constants have names and are defined with a `PARAMETER` statement. Literal constants represent themselves; e.g., `1.2`, `-6.9856`, `0.004`, etc.

A constant has a type: `INTEGER`, `REAL`, `DOUBLE PRECISION`, `CHARACTER`, or `LOGICAL`.

- A literal `INTEGER` constant is written without a decimal point, thus `35`, `1999`, and `-456` are examples of literal `INTEGER` constants. A comma is not allowed: thus `1999` not `1,999`.

- A literal `REAL` constant may be expressed in ordinary decimal form, or in *floating-point* form. Examples of the first form are `3.1415927`, `-0.004475`, and `136.0084`. Examples of the second form (floating-point) are `0.31415927e1`, `-4.475e-3`, `0.1360084e3`. In floating-point form, the integer following the `e` denotes a power of 10 that is to multiply the number standing before the `e`. Thus `0.35e6` means `0.35` times 10 raised to the power 6, so `0.35e6 = 350000.0`, and `0.35e-6 = 0.00000035`.

- A literal `DOUBLE PRECISION` constant is written like a `REAL` floating-point constant except that `d` is used in place of `e`. Thus
  $$0.3141592653589793d1$$
  is a `DOUBLE PRECISION` constant representing the mathematical constant π.

- A literal `CHARACTER` constant is written with apostrophes as delimiters. Thus `'C'`, `'CAT'`, `'Monkey'`, and `'pi equals 3.1415927'` are examples of `CHARACTER` constants. A character constant has a length. The lengths of the four constants just given are 1, 3, 6, and 19. Note that the apostrophe is not part of the value of the constant but blanks appearing inside the apostrophes are. If an apostrophe must appear in a `CHARACTER` constant as in O'Malley it is expressed `'O''Malley'`; i.e., the embedded apostrophe is written twice.

- A literal `LOGICAL` constant is written `.TRUE.` (for the value *true*) or `.FALSE.` (for the value *false*). Thus an assignment statement assigning the `LOGICAL` variable L the value *false* would be written:

```
      L = .FALSE.
```

6.3.9   Operators

The arithmetic operators are: + (addition); - (subtraction); * (multiplication); / (division); ** (exponentiation). The relational operators are: `.LT.`
(less than); `.LE.` (less than or equal); `.EQ.` (equal); `.NE.` (not equal); `.GE.`
(greater than or equal); `.GT.` (greater than). Relational operators are used
with operands of type `INTEGER`, `REAL`, `DOUBLE PRECISION`, `CHARACTER`, or
`CHARACTER*n`. The value of a relational expression has type `LOGICAL`. Thus
the expression

```
      X .LT. Y
```

has the value *true* or *false*.

The logical operators are: `.NOT.` (negation); `.AND.` (logical *and*); `.OR.`
(logical *or*). Logical operators are used with operands of type `LOGICAL`; for
example,

```
      (X .LT. Y) .OR. (X .LT. Z)
```

6.3.10   Evaluation of expressions

The evaluation of expressions is done in the way we normally expect in mathematical work. Thus in the arithmetic expression

```
      X + Y*Z
```

the multiplication is performed first then the addition. Parentheses are used
to group subexpressions; for example,

```
      (X + Y)*Z
```

where now the addition is performed first, then the multiplication. The arithmetic operators thus have an order of precedence. The operator with the highest order of precedence is performed first within any parenthesis-free subexpression; the operator with the lowest order of precedence is performed last.
The order of precedence from lowest to highest is (+ -) (* /) **. The grouped
operators have the same precedence level. In the expression

```
A + B*C**3
```

the evaluation proceeds as follows: raise C to the power 3, multiply the result by B, and then add A. If two operations are at the same level, they are performed in left-to-right order, excepting exponentiation that is done right-to-left (C**2**3 is equivalent to C**(2**3) = C**8).

The order of precedence for logical operators from lowest to highest is: .OR. .AND. .NOT. .

Relational operators all have the same level.

Across the various types of operators the order of precedence from lowest to highest is: logical, relational, arithmetic. Thus

```
X + Y .LT. Z .OR. A .LT. B
```

is equivalent to

```
((X + Y) .LT. Z) .OR. (A .LT. B)
```

The latter form is preferable since it makes the order of evaluation explicit.

6.3.11 Statements

Each statement is usually written on a single line but if the statement is too long to fit on a line it may be extended onto one or more (up to nineteen) lines. The statement must begin in column 7 of the line or to the right of column 7 and cannot extend beyond column 72 (another atavism, frequently ignored by compilers). Many Fortran systems accept a tab character at the front of the line as equivalent to 7 or more spaces; otherwise you must explicitly type at least 7 blanks at the front of the line. A long statement can extend to the next line, with the continuation line beginning at column 7 or to the right of column 7 and not extending beyond column 7. The continuation line has a non-blank character, except 0, in column 6 to identify it as a continuation line.

6.3.12 Statement order

The declarations, which are non-executable statements, precede the executable statements within a subprogram. The PROGRAM, FUNCTION, or SUBROUTINE statement must come first in the subprogram. The `IMPLICIT` statement, if it is used, should come next. Then type statements, `PARAMETER` statements, and `EXTERNAL` statements follow. It is important (See section 6.4 for more information on the meanings of these statements.) to note that the type of a symbolic constant must be declared in a type statement before it is given a value in a `PARAMETER` statement. Also, a symbolic constant should be given a value in a `PARAMETER` statement before it is used; for example,

```
INTEGER N
PARAMETER (N = 30)
REAL A(N)
```

is the correct order for these three statements. For the sake of clarity it is a good idea to group declaration statements of the same kind together.

Executable statements (Assignment, `CALL`, `CONTINUE`, `DO`, `GOTO`, `IF`, `OPEN`, `READ`, `RETURN`, `STOP`, `WRITE`) follow the declarations. `FORMAT` statements may appear anywhere within the subprogram. For clarity, it is a good idea to group them in a single place, say just before `END`.

Fortran has implicit typing: names beginining with the letters I, J, K, L, M, and N are implicitly typed as `INTEGER`; names beginning with any other letter are implicitly typed `REAL`. This atavism, which may once have had some convenience, is a source of programming errors. Implicit typing can be turned off with the statement

```
IMPLICIT NONE
```

which must precede any type declarations. With this statement in place, the type of every variable must be declared explicitly, as in C.

6.3.13   Comments

A comment line is identified by the letter `C` or an asterisk `*` in column 1 of the line.

6.3.14   Blanks

Strictly speaking blanks are ignored by the compiler, however it is unwise to put meaningless blanks in a program or to not use them when they could improve legibility. Thus

```
RE AL X, Y  and   REALX,Y
```

are allowed but

```
REAL X, Y
```

is clearer.

6.3.15   Intrinsic functions

Certain common functions like the square root are part of Fortran. They are called *intrinsic functions*. A list of some of these functions appears in table 6.1. The type `DOUBLE PRECISION` can replace `REAL` everywhere in this table.

6.4   Description of statements

In this section we systematically describe the statement types, proceeding in alphabetical order. First a few words about the notation we use in these descriptions.

6.4.1   Notation

To describe the syntax, or form, of statements we use a certain notation and conventions that are described below.

1. Special characters (except as noted below) and capitalized words appear in statements exactly as shown.

2. Lower case letters and words stand for objects defined elsewhere.

3. Square brackets are used to indicate optional items.

Name	Definition	Type of parameter	Type of result
`ICHAR(C)`	Convert to integer	`CHARACTER`	`INTEGER`
`CHAR(K)`	Convert to character	`INTEGER`	`CHARACTER`
`ABS(X)`	Absolute magnitude	`INTEGER` or `REAL`	`INTEGER` or `REAL`
`MOD(J,K)`	Remainder of J/K	`INTEGER`	`INTEGER`
`SQRT(X)`	Square root	`REAL`	`REAL`
`EXP(X)`	Exponential function	`REAL`	`REAL`
`LOG(X)`	Natural logarithm	`REAL`	`REAL`
`SIN(X)`	Trig. sine function (X in radians)	`REAL`	`REAL`
`COS(X)`	Trig. cosine function (X in radians)	`REAL`	`REAL`
`TAN(X)`	Trig. tangent function (X in radians)	`REAL`	`REAL`
`ASIN(X)`	Trig. arcsine	`REAL`	`REAL`
`ACOS(X)`	Trig. arccosine	`REAL`	`REAL`
`ATAN(X)`	Trig. arctangent	`REAL`	`REAL`

Table 6.1: A partial list of Fortran intrinsic functions.

4. An ellipsis ... is used to denote one or more repetitions of an item.

5. Lower case words or phrases that appear in the syntax descriptions are in bold letters (e.g., **name**) when they appear in the running text.

Thus the syntax of a `CALL` statement is described by the expression:

```
CALL name [(parameter [, parameter]...)]
```

The form of **name** is described elsewhere (a letter followed by letters or digits, possibly with a maximum of six characters). The outermost pair of square brackets implies that this statement is valid:

```
CALL MYSUB
```

The innermost pair of square brackets followed by the ellipsis imply that inside the parentheses there are one or more **parameters** separated by commas. Thus the following are all valid:

```
CALL SUB2(X, Y)
CALL SUB3(X, 1.0)
CALL SUB4('MYNAME', W**2, A(J))
```

6.4.2   Assignment statement

Syntax:

```
variable = expression
```

Purpose: Assign the value of **expression** to **variable**.
Examples:

```
(1)     X = Y + 3.2*Z
(2)     C = 'A String'
(3)     L = X .LT. Y
(4)     U(K) = U(K)*EXP(SQRT(2.0/W))
```

Remarks:

1. The type of `expression` and the type of `variable` must be the same excepting between numeric types, where `REAL` and `INTEGER` can be paired, and between character types, where `CHARACTER` types of different length can be paired. Thus in example (1), `X` must be `REAL` or `INTEGER`; in example (2), `C` must be `CHARACTER` (of any length); in example (3), `L` must be `LOGICAL`; and in example (4), `U` must be `REAL` or `INTEGER`.

2. In the case

```
integer_variable = real_expression
```

the integer part of `real_expression` is assigned to `integer_variable`. Thus in the statement `K = -3.95` the value assigned `K` is `-3`.

3. In the case

```
real_variable = integer_expression
```

the integer part of `real_variable` is assigned the value of `integer_expression` and the fractional part of `real_variable` is assigned the value zero.

4. In the case

```
character_variable = character_expression
```

we may have the length of `character_variable` less than the length of `character_expression`. In this case, characters are chopped from the right end of `character_expression`. If the length of `character_variable`, is greater than the length of `character_expression`, the excess space on the right end of `character_variable` is filled with blanks.

In the above, `DOUBLE PRECISION` may replace `REAL`.

6.4.3   CALL statement

Syntax:

    CALL name [(parameter [, parameter]...)]

Purpose: Execute the subroutine `name`.

Examples:

    (1)   CALL SUB1
    (2)   CALL SUB2(X, Y)
    (3)   CALL SUB3(X, 1.0)
    (4)   CALL SUB4('MYNAME', W**2, A(J))

Remarks:

1. A `parameter` may be any of the following: `variable`, `expression`, `subroutine_name`, `function_name`, `array_name`.

2. If `parameter` is `subroutine_name` or `function_name` then an `EXTERNAL` statement must declare the `subroutine_name` or `function_name`. The `EXTERNAL` statement must be located in the same program unit as the `CALL` statement.

3. The `parameters` must agree in number and type with the `parameters` in the corresponding `SUBROUTINE` statement; i.e., the k-th `parameter` in each list must have the same type and each list must have the same number of `parameters`.

4. The `parameters` appearing here are called `actual` parameters to distinguish them from the `formal` parameters appearing in the corresponding `SUBROUTINE` statement.

6.4.4   COMMON statement

Syntax:

    COMMON [/name/] common_item [, common_item]...

Purpose: Share data between program units.
Examples:

(1)   COMMON X, Y
(2)   COMMON /PARAMS/ A, B, C

Remarks:

1. `common_item` may be a simple variable or an array name.

2. If the **name** part is absent as in example (1), the statement is called a *blank* COMMON statement, otherwise it is called a *labelled* COMMON statement.

3. If two program units have labelled COMMON statements with the same **name** then `common_items` in corresponding positions refer to the same data regardless of whether or not they have the same name. The two COMMON statements should have the same number of `common_items` and corresponding `common_items` should have the same type.

4. If two program units have blank common statements then `common_items` in corresponding positions refer to the same data regardless of whether or not they have the same name. The two COMMON statements should have the same number of `common_items` and corresponding `common_items` should have the same type.

5. If one COMMON statement follows another in the same program unit and both have the same **name** or both are blank COMMON statements then the lists of `common_items` are concatenated; e.g.,

        COMMON /CPARMS/ X, Y
        COMMON /CPARMS/ Z

   is equivalent to

        COMMON /PARMS/ X, Y, Z

   The latter form is preferred because it is clearer.

6.4.5 CONTINUE statement

Syntax:

```
CONTINUE
```

Purpose: This is a null statement, it doesn't do any computation.
Examples:

```
(1)    10    CONTINUE
```

Remarks:

1. This statement is often used with a label as the last statement in a DO
   loop or as the target of a GOTO statement. Example (1) shows a CONTINUE
   statement with a label of 10.

6.4.6 DO statement

Syntax:

```
DO  do_variable = expressn_1, expressn_2 [,expressn_3]
```

Purpose: Controls repeated execution of a sequence of statements.
Examples:

```
(1)          DO J = 1, 20
                X(J) = 0
             END DO
```

```
(2)          DO  K = 0, N, 2
                WRITE(*,*) K, K**2, K**3
             END DO
```

```
(3)          DO  J = 1, 100
                P(J) = 1
                DO  K = 1, N
                    P(J) = P(J) + SQRT(J*K)
                END DO
             END DO
```

Remarks:

1. The DO statement causes the sequence of statements following the DO statement, up to the matching END DO, to be executed repetitively. This sequence of statements is called the **range** of the DO. If expressn_3 is absent it is assumed to have the value 1. Initially, do_variable is assigned the value of expressn_1; and the iteration count is given the value

$$\max(\lfloor (\text{expressn\_2} - \text{expressn\_1} + \text{expressn\_3})/ \text{ expressn\_3} \rfloor, 0)$$

If the iteration count is not zero the range is executed. After each execution of the range the do_variable is incremented by expressn_3, and the iteration count is decremented by 1; then the range is executed again and this continues until the iteration count reaches 0, at which point the iteration is terminated.

2. do_variable is a simple variable of type INTEGER, and the types of expressn_1, _2, _3 are INTEGER.

3. The effect of example (1) is to set the values of X(1), X(2), ..., X(20) equal to zero.

4. The effect of example (2) is to write, on successive lines, the values: 0 0 0; 2 4 8; 4 16 64; .... The last line has the values N, N**2, and N**3 if N is even; otherwise it has the values (N-1), (N-1)**2, and (N-1)**3.

5. The effect of example (3) is to evaluate expressions:

```
1 + SQRT(1) + SQRT(1*2) + ... + SQRT(1*N);
1 + SQRT(2*1) + SQRT(2*2) + ... + SQRT(2*N);
        ... ;
1 + SQRT(100*1) + SQRT(100*2) + ... + SQRT(100*N).
```

These values are assigned to P(1), P(2), ..., P(100), respectively. Example (3) illustrates that one DO can be contained in the range of another DO. This so-called nesting of DO statements can be arbitrarily deep.

6. No statement in the range is permitted to change the value of do_variable, expressn_1, expressn_2, expressn_3.

7. Execution of the range must begin with executing the DO; that is, execution of a statement in the range by jumping to it from outside the range, using a GOTO, is forbidden. On the other hand, it is permitted to jump out of the range using a GOTO.

6.4.7   END statement

Syntax:

    END

Purpose: Marks the end of a program unit.
Examples:

    (1)   END

Remarks:

1. Every program unit must have this statement as its last statement.

6.4.8   EXTERNAL statement

Syntax:

    EXTERNAL name [, name]...

Purpose: Declares names of FUNCTION and SUBROUTINE subprograms that are passed as parameters in calls to subprograms.
Examples:

    (1)   EXTERNAL MYFUNC, MYSUB

Remarks:

1. name is the name of a FUNCTION or SUBROUTINE appearing as an actual argument in a call to a subprogram.

2. Every FUNCTION and SUBROUTINE name used as an actual parameter in a call to a subprogram must appear in an EXTERNAL statement in the program unit in which it is so used.

6.4.9   FORMAT statement

Syntax:

```
FORMAT (edit_descriptor [, edit_descriptor]...)
```

Purpose: Defines the input or output format (number of columns used, floating-point form, etc.) of values of `iolist` items. (cf., `READ` and `WRITE` statements.)

Examples:

```
(1)     99  FORMAT(1X, I10)
(2)     98  FORMAT(I10, 5X, E15.8, 5X, F10.2)
(3)     97  FORMAT(A, 3(2X, I5))
(4)     96  FORMAT(A, 2X, I5, 2X, I5, 2X, I5)
```

Remarks:

1. A repeatable `edit_descriptor` is one of: `Iw`, `Fw.d`, `Ew.d`, `A`. These edit descriptors are associated with `iolist` items: `I` with items of type `INTEGER`, `F` and `E` with items of type `REAL`, `A` with items of type character. `F` is used to specify conventional decimal format (e.g., `0.003956`), `E` is used to specify floating-point format (e.g., `0.3956e-02`). The lower-case letters `w` and `d` denote unsigned integers: `w` specifies the width, number of columns, occupied by the item; `d` specifies the number of digits after the decimal point. The width of an item associated with `A` is the length of the `CHARACTER` type. The value of `d` is ignored when reading `REAL` values; it is only meaningful when writing.

2. A repeatable `edit_descriptor` may be preceded by an unsigned integer (viz. `3I10`) denoting multiple descriptors. Thus `3I10` and `I10, I10, I10` are equivalent.

3. A nonrepeatable `edit_descriptor` is one of: `nX` / `nP`. These descriptors are not associated with `iolist` items. `X` denotes a blank, `/` denotes end of line, `P` denotes a scale factor. The edit descriptor `3PE15.7` prints a floating-point value with 3 places before the decimal point. For example, writing the value `-0.01255` with the edit descriptor `E15.7` produces

```
-0.1255000E-01
```

Using the edit descriptor 3PE15.7 results in

```
-125.5000000E-04
```

4. The FORMAT statements in examples (3) and (4) are equivalent, example (3) being a more compact form of example (4).

6.4.10 FUNCTION statement

Syntax:

```
[type] FUNCTION name ([parameter [, parameter]...])
```

Purpose: Declares a subprogram to be a FUNCTION subprogram. It is the first statement in the subprogram.

Examples:

```
(1)     FUNCTION FUN1(X)
(2)     REAL FUNCTION FUN2(X1, X2)
(3)     CHARACTER*8 FUNCTION FUN3(CHR1, CHR2, XYZ)
```

6.4.11 GOTO statement

Syntax:

```
GOTO label
```

Purpose: Jump to the statement labelled label and resume executing statements there.

Examples:

```
GOTO 50
```

Remarks:

1. GOTO statements should be used with care. Indiscriminate use of these statements results in programs with tangled control paths that are hard to understand.

6.4.12   IF statement

Syntax:

```
IF (logical_expression) THEN
    [statement]...
ENDIF
```

or

```
IF (logical_expression) THEN
    [statement]...
ELSE
    [statement]...
ENDIF
```

or

```
IF (logical_expression) THEN
    [statement]...
ELSEIF (logical_expression) THEN
    [statement]...
[ELSEIF (logical_expression) THEN
    [statement]...]
ELSE
    [statement]...
ENDIF
```

Purpose: Allows conditional execution of a sequence of statements.

Examples:

```
(1)   IF (X .LT. Y) THEN
          WRITE(*,*) 'X IS LESS THAN Y'
      ENDIF

(2)   IF (A(J) .GT. A(J+1)) THEN
          T = A(J)
          A(J) = A(J+1)
```

```
        A(J+1) = T
     ENDIF

(3)  IF (X .LT. Y) THEN
        WRITE(*,*) 'X IS LESS THAN Y'
     ELSE
        WRITE(*,*) 'X IS GREATER THAN OR EQUAL TO Y'
     ENDIF

(4)  IF (ABS(X-Y) .LE. ABS(X)*EPS) THEN
        GOTO 20
     ELSE
        Y = X
        J = J + 1
     ENDIF

(5)  IF (C .EQ. 'A') THEN
        CALL SUBA(X)
        A(1) = A(1) + 1
     ELSEIF (C .EQ. 'B') THEN
        CALL SUBB(X)
        A(2) = A(2) + 1
     ELSEIF (C .EQ. 'C') THEN
        CALL SUBC(X)
        A(3) = A(3) + 1
     ELSE
        CALL ERROR(X)
        A(4) = A(4) + 1
     ENDIF
```

Remarks:

1. Execution of this statement proceeds as follows. If the value of
   logical_expression is true then the sequence of statements following
   THEN and before ELSE, ELSEIF, or ENDIF (whichever appears first) is executed. If the value of logical_expression is false then the statements

following `THEN` and before `ELSE`, `ELSEIF`, or `ENDIF` (whichever appears first) are skipped and:

- if `ELSE` comes first then the sequence of statements following `ELSE` is executed;

- if `ELSEIF` comes first then, if the associated `logical_expression` is true, the statements following `THEN` are executed; otherwise they are skipped.

- if `ENDIF` comes first then the statements immediately following `ENDIF` are executed.

2. In example (1), the `WRITE` statement is executed if and only if the value of `X` is less than the value of `Y`.

3. In example (2), the sequence of three statements in the body of the `IF` is executed if and only if the value of `A(J)` is greater than the value of `A(J+1)`.

4. In example (3), the message `X IS LESS THAN Y` is written if and only if the value of `X` is less than the value of `Y`; otherwise the message `X IS GREATER THAN OR EQUAL TO Y` is written.

5. In example (4), the `GOTO` is executed if and only if the value of the absolute magnitude of `(X-Y)` is less than or equal to the product of the absolute magnitude of `X` and the value of `EPS`.

6. In example (5), the subroutine `ERROR` is called if and only if the value of `C` is not equal to 'A', or to 'B', or to 'C'.

7. Every `IF` must be terminated with an `ENDIF`.

6.4.13   IMPLICIT statement

Syntax:

```
IMPLICIT type(range [, range]...) [, type(range [, range]...]
```

Purpose: To associate a type with all names starting with a particular letter, or range of letters, excepting names of intrinsic functions.

Examples:

```
(1)     IMPLICIT CHARACTER*32 (C), REAL (I-L, N)
(2)     IMPLICIT NONE
```

Remarks:

1. `type` is a Fortran type (`REAL`, `LOGICAL`, etc.).

2. `range` is a single letter or a pair of letters separated by a dash.

3. Example (1) declares all variables with names starting with the letter C to have the type `CHARACTER*32` and all variables with names starting with the letters I, J, K, L, N to have the type `REAL`.

4. Example (2) declares no variables to have an implicit type; that is, all variables must be explicitly typed.

5. If a name is explicitly typed as in

```
REAL CENTER
```

then this type declaration overides the effect of an implicit type declaration. Thus in a program unit containing the declaration line shown in example (1) and this `REAL` declaration, `CENTER` would have they type `REAL` but `COURSE` would have the type `CHARACTER*32`.

6. The scope of this statement is the program unit in which it appears.

7. Programming errors associated with wrong types can be more easily detected by using the `IMPLICIT` statement shown in example (2) in each program unit and explicitly declaring the types of all variables.

6.4.14   OPEN statement

Syntax:

```
OPEN (unit_spec [, FILE = 'file_name'] [, STATUS = 'status'])
```

Purpose: Associates a file with a unit number used in a READ or WRITE statement.

Examples:

```
(1)     OPEN (3, FILE = 'MYDATA', STATUS = 'OLD')
(2)     OPEN (UNIT = 7, FILE = 'MYOUT', STATUS = 'NEW')
(3)     OPEN (8)
```

Remarks:

1. unit_spec is an integer, a unit number alone, or a unit number preceded by UNIT = (cf., example (2)).

2. file_name is the name of the file that is to be associated with the unit number given in unit_spec. Thus, referring to example (1), if a subsequent READ statement had the form

   ```
   READ(3, 99) X
   ```

   then the value of X would be read from the file named MYDATA. In this situation, case is important; that is, the file name is MYDATA, not mydata.

3. status is NEW, OLD, SCRATCH, and UNKNOWN. NEW is used for files that are to be created, OLD is for files that already exist. Thus, the statement in example (2) might be used in conjunction with a WRITE statement of the form

   ```
   WRITE(7, 98) RESULT
   ```

   to write the value of RESULT on the new file MYOUT, but it could not be used in conjunction with a READ statement of the form

```
READ(7, 98) VALUE
```

which presupposes the existence of the file MYOUT. SCRATCH is used for files that are only temporary, for example to save some data during a computation; they are removed when program execution is terminated, or when the file is closed. UNKNOWN is processor dependent; in some systems (e.g., DEC Fortran) the system tries OLD and if it cannot find the file it creates a NEW file. The default status is OLD.

4. The form used in example (3) is for creating a scratch file to hold intermediate results during a computation. It is destroyed when the program stops.

6.4.15   PARAMETER statement

Syntax:

```
PARAMETER (name = expression [, name = expression]...)
```

Purpose: Gives a name to a constant. Thus it allows you to use PI instead of 3.141592653 and to be protected against accidentally changing PI.

Examples:

```
(1)    PARAMETER (PI = 3.141592653, ALPHA = 'ABC')
(2)    PARAMETER (MAXVAL = 100, MINVAL = MAXVAL-50)
```

Remarks:

1. The type of name must agree with the type of expression.

2. A name used in a PARAMETER statement cannot have its value changed during execution of the program. An attempt to change it normally causes an error message.

3. expression is either a constant, or an expression containing constants. In the latter case the expression must be of type INTEGER.

4. Previously named constants can be used in expression (cf., example (2)).

5. It is recommended that named constants be used rather than literal constants. Programming errors are likely to be reduced in cases where the constant is used more than once (using a name assures that the same value is used everywhere). Also, it is easy to modify the value of the constant in editing the program because its value appears only in the `PARAMETER` statement (cf., type statement, remark 9).

6. Any variable used in a `PARAMETER` statement must have had its type previously declared.

6.4.16   PROGRAM statement

Syntax:

```
PROGRAM name
```

Purpose: Gives the name **name** to the main subprogram.
Examples:

```
(1)     PROGRAM MYPROG
```

Remarks:

1. This statement is optional. If it is used then it must be the first statement of the main subprogram, and in this case, **name** is the name of the main subprogram. If it is not used, then the main subprogram has the default name `MAIN`.

2. It is recommended that this statement be used to improve program readability.

6.4.17   READ statement

Syntax:

```
READ (unit, format [, END = label]) iolist
```

Purpose: Reads data from a file or the keyboard.
Examples:

```
(1)      READ (5, 99) X
(2)      READ (5, 99) X1, X2, C(I)
(3)      READ (*, 99) X
(4)      READ (*, *) X
(5)      READ (7, *) W1, A(K)
(6)      READ (7, *) K, (A(J), J = 1, 5), Y
(7)      READ (7, '(A)') C
(8)      READ (9, *, END = 100) K, A(I,J)
```

Remarks:

1. `unit` is an integer, greater than zero, that has been defined as the unit number for a file by an `OPEN` statement. Thus, given that the `OPEN` statement

   ```
   OPEN(5, FILE = 'MYDATA', STATUS = 'OLD')
   ```

   has already been executed, the effect of the statement in example (1) is to read one value from the file `MYDATA` and assign it to `X`. Similarly, the effect of the statement in example (2) is to read three values from `MYDATA` and assign them to `X1`, `X2`, and `C(I)`, respectively.

2. `unit` may also be *. In this case, the data is read from the keyboard. Thus the effect of the statement in example (3) is to read one value entered from the keyboard and assign it to `X`. When reading from the keyboard, execution of the `READ` statement is not completed until the **return** key has been depressed: depressing this key signals the end of the line.

3. `format` is a `label` on a format statement or it is *. Examples (1)–(3) illustrate the first alternative; examples (4)–(6) illustrate the second alternative. In the second alternative, a default format specifier is used for each item in the `iolist` consistent with the type of the item; the phrase **list directed input** is used to describe the style of input determined by using this alternative.

4. `format` may also be a character constant as illustrated in example (7). In this case, the string of characters is treated just as if it appeared in a `FORMAT` statement. The effect of executing the statement in example (7) is the same as the effect of executing

```
        READ(7, 88) C
 88     FORMAT (A)
```

5. `label` denotes a labelled statement in the same program unit as the `READ` statement. If the `READ` statement is executed and there is no more data on the file, i.e., the end of the file has been reached, then the effect of `END = label` is to cause a `GOTO label`. Thus the effect of executing the following when all data has been read from unit 9 is to `GOTO` the statement X = 0.

```
        READ (9, *, END = 100) K, A(I,J)
        ...
 100    X = 0
```

6. `iolist` is a list of one or more variables separated by commas; these variables are to be assigned the values read. An `iolist` element can be an implied `DO`, as illustrated in example (6). The effect of the statement in this example is the same as the effect of

```
        READ (7, *) K, A(1), A(2), A(3), A(4), A(5), Y
```

7. There are many other options for the `READ` statement. Consult your computer manual for further information.

6.4.18   RETURN statement

Syntax:

```
        RETURN
```

Purpose: Return control to the calling subprogram from a subprogram.
Examples:

   (1)   `RETURN`

Remarks:

1. This statement is used only in `SUBROUTINE` or `FUNCTION` subprograms. When it is executed, it causes execution of the subprogram to stop and returns control to the program unit that called the subprogram.

2. If the control path in a subprogram reaches an `END` statement, the `END` has the same effect as `RETURN`.

6.4.19   STOP statement

Syntax:

     `STOP`

Purpose: Stops program execution and prints an optional message on the screen.
Examples:

   (1)   `STOP`

Remarks:

1. When this statement is executed, the program stops.

6.4.20   SUBROUTINE statement

Syntax:

     `SUBROUTINE name [(parameter [, parameter]...)]`

Purpose: Declares a subprogram to be a `SUBROUTINE`. This statement must be the first statement in the subprogram.
Examples:

```
(1)        SUBROUTINE MYSUB
(2)        SUBROUTINE MYSUB1(X, Y)
```

Remarks:

1. parameter is a name and name may identify a simple variable, an array, a FUNCTION subprogram, or a SUBROUTINE subprogram.

2. When the subroutine is called, the actual parameters must match the parameters in the SUBROUTINE statement (called the formal parameters) in number and type (cf., CALL statement).

6.4.21   Type statements

Syntax:

> *type* var [, var]...

Purpose: Defines the type of a variable.

Examples:

```
(1)        INTEGER X1, X2, K
(2)        REAL ROOT, SOLN
(3)        CHARACTER C1, C2, LET
(4)        CHARACTER*20 MESSG1, MESSG2
(5)        INTEGER X1, X2(50)
(6)        REAL ROOTS(4), A(10, 20)
(7)        CHARACTER C1(10)
(8)        CHARACTER*80 LINES(50)
(9)        LOGICAL TEST, BOOL(10)
(10)       REAL A(N,*), Z
(11)       DOUBLE PRECISION HPSOLN, Z
```

Remarks:

1. *type* may be REAL, DOUBLE PRECISION, INTEGER, LOGICAL, CHARACTER, or CHARACTER*n where n is an integer in the range (1, 2, ..., 127).

2. `var` may be a `name` or an array declarator.

3. All items in the `var` list have the specified type. Thus, in example (1), `X1`, `X2`, and `K` are defined to have the type `INTEGER`.

4. In example (4), `MESSG1` and `MESSG2` are defined to be character strings of length 20.

5. The type `DOUBLE PRECISION` is used to declare variables whose values are about twice as accurate as those declared `REAL`. Eight bytes are used to store a value of type `DOUBLE PRECISION`; four bytes are used to store a value of type `REAL`.

6. In example (6), `ROOTS` is defined to be a one-dimensional array of 4 elements, each element being of type `REAL`; and `A` is defined to be a two-dimensional array consisting of 10 rows and 20 columns having elements of type `REAL`.

7. In example (8), `LINES` is defined to be a one-dimensional array of character strings each of length 80.

8. The declaration in example (10) would appear in a `SUBROUTINE` or `FUNCTION` subprogram. It declares `A` to be an array of `N` rows and an indefinite number of columns; `N` would have to appear as a formal parameter in the `SUBROUTINE` or `FUNCTION` statement.

9. The sequence

```
PARAMETER (N = 10, M = 20)
REAL A(N, M)
```

is equivalent to

```
REAL A(10, 20)
```

The former is longer but is preferred when the row and column dimensions are frequently used in the program text because modification of `N` and `M` is simpler – they only have to be changed in the `PARAMETER` statement, not every place 10 and 20 appear in the program text.

6.4.22   WRITE statement

Syntax:

```
WRITE (unit, format) iolist
```

Purpose: To write values on a file or the screen.
Examples:

```
(1)      WRITE (5, 99) X
(2)      WRITE (5, 99) X1, X2, C(I)
(3)      WRITE (*, 99) X
(4)      WRITE (*, *) X
(5)      WRITE (7, *) W1, A(K)
(6)      WRITE (7, *) K, (A(J), J = 1, 5), Y
(7)      WRITE (7, '(A)') C
(8)      WRITE (9, *) K, A(I,J)
```

Remarks:

1. unit is as defined for the READ statement, except that * denotes the screen. Thus in example (3), the value of X is written on the screen.

2. format is as defined for the READ statement.

3. iolist is as defined for the READ statement except the items listed have their values written on the designated unit or screen.

6.5   Reading and writing

6.5.1   Reading

List directed input (cf., READ statement, Remark 3) should be used for reading data. The other alternative, FORMAT directed input, is troublesome and likely to result in errors. The remaining discussion in this section is concerned with list directed input.

When the READ statement is executed, the data on one line of the input file is read. The correspondence between the data on the line and items in the `iolist` is left-to-right. For example, assume that the next line to be read from the input file is

```
'My Data is' 3.96 5
```

and that the following READ statement is executed:

```
READ(5,*) MESSG, X1, K1
```

The effect is that the string constant `'My Data is'` (without the quotes of course) is assigned to MESSG (which must be of type CHARACTER); the value 3.96 is assigned to X1 (which must be of type REAL); the integer value 5 is assigned to K1 (which must be of type INTEGER).

As the example above illustrates, a blank is used to separate the data on the input line. One or more blanks may be used, thus

```
'My Data is'        3.96     5
```

would give the same result when read by the above READ statement.

The number of items read from the input line is just the number of items in the `iolist`. Extra items are ignored. Thus if we read the two lines

```
12.1  -13.2E-5  44.0
-3.1  445.2      99.9
```

using the two statements

```
READ(5,*) X1, X2
READ(5,*) Y1, Y2, Y3
```

then the values assigned are as follows:

```
X1 = 12.1,  X2 = -13.2E-5
Y1 = -3.1, Y2 = 445.2, Y3 = 99.9
```

Note that the value 44.0 on the first line is not read.

If there are more items in the `iolist` than on the line of data, the excess items are not be assigned values. Thus if we read

```
12.1  -13.2E-5 44.0
```

with the statement

```
READ(5,*) X1, X2, X3, X4
```

then the values assigned are as follows:

```
X1 = 12.1, X2 = -13.2E-5, X3 = 44.0
```

and the value of X4 is unchanged.

When a character constant is read and the length of the constant differs from the length of the CHARACTER type of the item in the iolist, the rules are like those for the assignment statement (cf., Assignment statement, Remark 4); i.e., characters are chopped from the right end of the constant if it is too long, and blanks are filled in on the right if it is too short.

6.5.2   Writing

While list directed output can be used with the WRITE statement, FORMAT directed output is the preferred mode since it allows more control over the form of the output. With list directed output, the format is the default format provided by the system. This is usually adequate for a quick look at the results, but not adequate for nice presentations of results in tables. The rest of the discussion is concerned with FORMAT directed output.

A WRITE statement normally writes one line of output. The connection between the iolist and the list of format edit descriptors can be described as follows. Assume we have a pair of statements of the form:

```
     WRITE(5, 99) iolist
99   FORMAT(descriptors)
```

For each item in iolist there must be a corresponding edit descriptor in descriptors. The correspondence is left-to-right. The beginning of execution of the WRITE statement initiates format control. Format control proceeds from left to right through the descriptors. Each action of format control depends on the next edit descriptor and the next item in the iolist. If format control encounters a nonrepeatable edit descriptor it performs the action specified by that descriptor. If it encounters a repeatable edit descriptor it writes, in the output file, the value of the corresponding item in iolist.

To illustrate this, consider the pair of statements

```
        WRITE(6,99) X1, X2, C, K1, 'the end'
99      FORMAT(1X, E15.7, 2X, F15.7, 2X, A, 2X, I5, 2X, A)
```

The output line consists of a blank in column 1, the value of X1 (assumed REAL) in columns 2-16, blanks in columns 17 and 18, the value of X2 (assumed REAL) in columns 19-33, blanks in columns 34 and 35, the value of C (assumed CHARACTER*10) in columns 35-45, blanks in columns 46 and 47, the value of K1 in columns 48-52, blanks in columns 53 and 54, and the string constant 'the end' (without the quotes) in columns 55-61. The value of X1 is written in floating-point form with seven digits after the decimal point (e.g., $-.1234567e+02$), the value of X2 is written in ordinary decimal form, i.e., *fixed point* form with seven digits after the decimal point (e.g., -12.3456789); the value of K1 is written in integer form (e.g., 29). If the number of characters required to express the value is less than the field width, as is the case for X1, X2, and K1, the left end is padded with blanks to fill out the field; thus the value is always *right-justified* (i.e., moved as far to the right as possible) in the field.

WARNING: When writing numbers that are small or large compared with 1 you should use the E format descriptor. Inexperienced programmers often use F instead. If it is used to write small numbers, the number written may be 0 even though the actual value is not zero but simply too small to show up with this format descriptor. For example, if you try to write the value 10^{-12} with the format specification F10.7 the number printed will be zero. With E12.4 the correct value will be printed. If the number you try to print is too large then the result is system dependent. Sometimes a block of asterisks is printed.

6.6 Examples

We close this with two short example programs. The first of these, CIRCLE, reads from a file the radii of circles and computes their circumference and area. This program is shown in figure 6.1. It illustrates opening, reading and writing files, and symbolic constants. The second, SORT, illustrates the use of control statements. It is shown in figure 6.2. Note the backward counting in the DO. It also illustrates the use of an array, and also how to declare an array that starts with row index 0, rather than the default value 1. Finally, it shows the value of a parameter statment: if you need to sort more numbers, you need only reassign the value of the parameter NX. For more examples, look in the textbooks already referenced.

```
      PROGRAM CIRCLE
* This program reads a number, the radius of a circle,
* from the file 'circle.dat', computes the area and
* circumference of the circle, and writes the results on
* the file 'circle.out'. It repeats these steps until
* all of the numbers on the file 'circle.in' have been
* read.
      IMPLICIT NONE
      REAL RADIUS, CIRCUM, AREA, PI
      PARAMETER(PI = 3.1415927)
* Open input file.
      OPEN(UNIT=7, FILE='circle.in', STATUS='OLD')
* Open output file.
      OPEN(UNIT=8, FILE='circle.out', STATUS='NEW')
* Write headers on output file.
      WRITE(8,*) 'Program CIRCLE output'
      WRITE(8,99)'RADIUS', 'CIRCUM', 'AREA'
* Begin main loop.
 10   CONTINUE
          READ(7,*,END=100) RADIUS
          CIRCUM = 2*PI*RADIUS
          AREA = PI*RADIUS*RADIUS
          WRITE(8,98) RADIUS, CIRCUM, AREA
          GOTO 10
* End main loop, write completion message to stdout and stop.
 100  WRITE(*,*) 'Program CIRCLE done.'
      STOP
 98   FORMAT(3(1PE15.7,1X))
 99   FORMAT(9X,A,10X,A,12X,A)
      END
```

Figure 6.1: The CIRCLE program. A simple example showing the reading and writing of files.

```
          PROGRAM SORT
* This program reads a list of integers and sorts them
* using a simple insertion sort algorithm. The maximum
* number of integers allowed is 20.
          IMPLICIT NONE
          INTEGER NX
          PARAMETER(NX = 20)
          INTEGER I, J, LAST, N(0:NX)
* Open input file.
          OPEN(UNIT=7, FILE='sort.in', STATUS='OLD')
* Open output file.
          OPEN(UNIT=8, FILE='sort.out', STATUS='NEW')
* Write header on output file.
          WRITE(8,*) 'Program SORT output'
* Read the input file and sort on the fly.
          READ(7,*) N(1)
          DO I = 2,NX
             READ(7,*,END=20) N(0)
 LAST = I
 DO J = I, 1, -1
    IF(N(0) .LT. N(J-1)) THEN
        N(J) = N(J-1)
             ELSE
        N(J) = N(0)
        GOTO 10
             ENDIF
          ENDDO
 10       CONTINUE
       ENDDO
* End reading input file and sorting.
 20    CONTINUE
       WRITE(8,99) (N(I), I=1,LAST)
* End main loop, write completion message to stdout and stop.
       WRITE(*,*) 'Program SORT done.'
       STOP
 99    FORMAT(I10)
       END
```

Figure 6.2: The SORT program. An example of an insertion sort program showing the use of control statements.

II Tools

7 Elements of Matlab

7.1 What is MATLAB?

MATLAB is an interactive system for matrix computations. It has a simple command language that allows you to easily multiply and invert matrices, solve systems of linear equations, and perform many other operations on rectangular arrays of numbers. It is also easy to plot data on the screen or printer with MATLAB.

MATLAB is often used interactively as if it were a very powerful hand calculator. But you can also use MATLAB in a programmable mode; you may write scripts for it just as you do for other command languages. You can also write your own functions, and these can be invoked interactively or from scripts or from other functions.

The examples in this document were run on UNIX workstations; both a Sun 3/60 under the SunView window environment and a DECstation 5000/200 under the X-Window system were used. A basic knowledge of UNIX is assumed for the remainder of this chapter. MATLAB is available on other platforms, including PC's; the following MATLAB material applies to those platforms as well.

7.2 Getting started

These notes are intended to get you started, providing only the bare essentials. The *MATLAB User's Guide* [MathWorks 92b] and the *MATLAB Reference Guide* [MathWorks 92a] are the basic manuals.

## Chapter 7

7.2.1   Bringing up MATLAB

If you have to login to a different machine than the server in order to run MATLAB and you are using an X terminal, make sure the DISPLAY environment is set properly. Without the proper DISPLAY setting, the figure window cannot appear on your screen. This can be set by typing the command

> setenv DISPLAY *yourterminalname*:0

from the UNIX shell.

If this does not work, you may need to type

> xhost + *remotemachinename*

where *remotemachinename* is the name of the remote machine with MATLAB installed on it. In some cases, you may need to do this before logging into the remote machine. This should enable that machine to display on your local screen.

Once the DISPLAY environment is set correctly, get into the directory from which you wish to use MATLAB. Start MATLAB by typing the command

> matlab

from the shell. The basic MATLAB environment is activated, and your window should appear as shown in the top of figure 7.1. Use the quit command to exit MATLAB.

7.2.2   Standard help

Notice the message on the initial MATLAB window:

```
Commands to get started: intro, demo, help help
Commands for more information: help, whatsnew, info, subscribe
```

Each of these facilities may be entered by typing the appropriate name. When help is entered, a list of topics appears. To narrow the choice, just enter

> help *aparticulartopic*

```
% matlab

                    < M A T L A B (R) >
          (c) Copyright 1984-1993 The MathWorks, Inc.
                    All Rights Reserved
                        Version 4.1
                        Jun 10 1993

Commands to get started:  intro, demo, help help
Commands for more information:  help, whatsnew, info, subscribe

>> ...
    .
    .
    .
>> quit

 0 flop(s).

%
```

Figure 7.1: MATLAB window: A sample session.

and helpful information on *aparticulartopic* is brought to the screen.

The `info` command provides the address of The MathWorks, Inc.; it also tells you how to obtain more information on MATLAB.

The `terminal` command lists the graphics terminals capable of running MATLAB.

Typing `demo` brings up MATLAB *Expo*; this is a mouse-driven facility with MATLAB demonstrations. These demos include examples, games, and snazzy graphics. Expo was completely implemented using MATLAB with the MATLAB *UI* (User Interface) tools. Try a few of the demos to see what MATLAB can do.

7.3   Some examples

This section demonstrates some basic matrix and plotting commands. As you read about each, type in the statements, as printed in `this font`, followed by a carriage return. MATLAB prints out each variable as it is assigned.

The constructs and syntax in these statements will be described in detail in section 7.4. This section merely provides some examples to give you the flavor of MATLAB.

7.3.1   Simple matrix operations

The following statements produce matrices A and B:

```
A = [ 1 2; 3 5]
B = [ 4 5; 6 7]
```

The matrices made by these statements are:

$$A = \begin{pmatrix} 1 & 2 \\ 3 & 5 \end{pmatrix}, \quad B = \begin{pmatrix} 4 & 5 \\ 6 & 7 \end{pmatrix}$$

You can create new matrices by using A and B in expressions. The statements

```
C = A + B
D = A * B
```

produce the matrices

$$C = \begin{pmatrix} 5 & 7 \\ 9 & 12 \end{pmatrix}, \quad D = \begin{pmatrix} 16 & 19 \\ 42 & 50 \end{pmatrix}$$

Unlike some other programming languages, the MATLAB multiplication operator * performs correct matrix multiplication when working with two matrix operands; this is *not* an elementwise operation.

The statement

```
E = A'
```

makes E the transpose of A; i.e.,

$$E = \begin{pmatrix} 1 & 3 \\ 2 & 5 \end{pmatrix}$$

And the statement

```
F  =   A * A'
```

makes F the product of A and its transpose; i.e.,

$$F = \begin{pmatrix} 5 & 13 \\ 13 & 34 \end{pmatrix}$$

The statements

```
Y = [1; -1]
X = A \ Y
```

give the solution to the equation

$$A * X = Y$$

that is,

$$X = \begin{pmatrix} -7 \\ 4 \end{pmatrix}$$

In order to get a feel for the notation here, think of the backslash operator, \, as denoting division from the left so that A \ Y in MATLAB is equivalent to the mathematical expression $A^{-1} \times Y$.

We can also use the backslash operator to solve a system with a rectangular coefficient matrix. In the case that the coefficient matrix A has more rows than columns, MATLAB returns the least squares approximation to the solution.

7.3.2 Simple plots

The statements

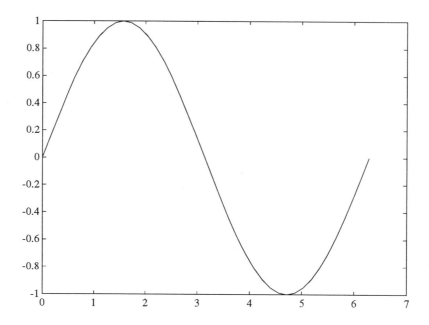

Figure 7.2: Plot of sine function on $[0, 2\pi]$.

```
U = 0:pi/20:2*pi
W = sin(U)
plot(U,W)
```

generate a plot of the sine function[1] on the interval $[0, 2\pi]$ as shown in figure 7.2. The first statement creates a vector of 41 values beginning at zero, in increments of $\pi/20$, the last value being 2π. The second statement produces a vector of 41 values equal to the sines of the 41 values in U. The last statement makes a plot of the curve whose abscissae (the x-axis values) are given by the values of the elements of U and whose ordinates (the y-axis values) are given by the elements of W. Note that the name pi in a MATLAB statement denotes a constant equal to π.

Type in these three statements. Observe that when values are assigned to a vector, those values are printed out across the screen. Extra lines are used

---

[1]The use of functions in MATLAB is described in more detail in sections 7.5 and 7.6.

if needed, and the columns are numbered.

Notice that a new window is generated by the first plotting command; this graphics window is called the *figure window*. If you are using an X terminal, the mesh grid outlining the MATLAB figure window may appear first, allowing you to place it anywhere on the screen. Use the mouse to drag it to your preferred location and then click the lefthand button of the mouse.

The figure window remains until you exit your MATLAB session or until you use the `close` command. Any additional plotting commands reuse this figure window, unless you open a new figure window with the `figure` command. On most windowing systems, the figure windows can be closed, reopened, moved, or resized, in the same manner as any other window.

This is a very simple plot. You may feel the need to label the axes and provide a title. This is not difficult. MATLAB commands for labelling plots and commands for controlling the size of the axes and grid are covered in section 7.8. Image processing is discussed there as well.

7.4   Short outline of the language

This section provides information about the basic syntax and semantics for MATLAB commands. For additional information, use the `help` command or see your MATLAB manual.

7.4.1   Types

Fundamentally there is one type, a rectangular array of numbers. There are no type declarations. The dimensions of an array are determined by the context.

Nevertheless, it is convenient to think of three types in the language: *scalar*, actually an array consisting of one row and one column; *vector*, actually an array consisting of one row and c columns, or an array consisting of r rows and one column; and *matrix*, an array consisting of r rows and c columns.

7.4.2   Names

Names consist of a letter followed by zero or more letters, digits, and underscore characters. Only the first 19 characters are significant. Uppercase and lowercase letters are distinguished; thus, A1 and a1 denote different variables.

7.4.3   Scalar constants

These values are written with an optional decimal point and an optional power of 10. A minus sign is placed at the front of negative values. No blanks are permitted within a value. Examples of legal values are:

```
99   39.24   -0.0075   1.35e-24   0.2E-5   12.0e44
```

Complex numbers are also allowed. If you type

```
z=2-5i
```

at the MATLAB prompt, the response will be

```
z =
   2.0000 - 5.0000i
```

The character j may also represent the value of $\sqrt{-1}$ as in the following:

```
zz = 3 + 2j
```

```
zz =
   3.0000 + 2.0000i
```

7.4.4   Display format

MATLAB has the ability to display the values of variables in several different ways, including short, long, short e, and long e. The default format is called a short format and shows the number to 4 decimal places. For instance, if you type

```
x = 32.75
```

MATLAB responds with

```
x =
   32.7500
```

Should you specify that you wish to use the short display format at this format, MATLAB displays x in the same manner.

```
format short
x


x =
   32.7500
```

The long format has fourteen decimal places.

```
format long
x


x =
  32.75000000000000


  z


z =
   2.00000000000000 - 5.00000000000000i
```

The two e formats give values in scientific form (i.e., floating-point), both long and short:

```
format short e
x


x =
   3.2750e+01


format long e
x


x =
      3.275000000000000e+01
```

It is also possible to display values in hexadecimal or in bank format (with up to two decimal places). Try

```
help format
```

for information on other format options. It is important to know that all values are stored as double precision numbers regardless of the format chosen for output.

7.4.5 Vector constants

A vector constant may be expressed explicitly as in

```
[99  39.24  -0.0075]
```

a vector (1 row, 3 columns) of three elements, or it may be expressed implicitly as in

```
[1:0.5:3]
```

which is equivalent to the expression

```
[1  1.5  2  2.5  3]
```

The expression `1:0.5:3.0` is a *constructor*. The semantics of this constructor are given by:

$$initial\ value : step : final\ value$$

The parameter *step* may be negative, as in

```
[3:-0.5:1]
```

If the *step* parameter is omitted, it is assumed to be 1. The elements of a vector may be separated by one or more blanks, as above, or by commas. Type the statement

```
V = [6:-0.3:3]
```

and observe the resultant vector.

These vectors are called *row* vectors. *Column* vector consist of a single column with one or more rows. A column vector may be defined by the expression

```
[9; -45.4; 0.22]
```

where a semicolon (;) is used to terminate each row. The transpose of a row vector also forms a column vector. Try entering the statements

```
VR = [-5; 0.25; 3.5; 0.0; 6.2]
VT = V'
```

to see the column vectors produced.

7.4.6   Matrix constants

A matrix constant may be expressed by explicitly listing the elements, with rows separated by a semicolon, as in

```
[1  2  3  4; 1  4  9  16; 0.5  1.0  4.5  -8]
```

which is a matrix consisting of 3 rows and 4 columns. The rows of a matrix may be written on separate lines of the input, omitting the semicolon, as in

```
[1 2 3 4
1 4 9 16
0.5 1.0 4.5 -8]
```

A row can be specified with a vector constructor as in

```
[1:4; 1  4  9  16; 0.5  1.0  4.5  -8]
```

Type the statement

```
M = [1:4; 1  4  9  16; 0.5  1.0  4.5  -8]
```

and observe the resultant matrix.

7.4.7   Arithmetic operators

The arithmetic operators are

$$+ \quad - \quad * \quad / \quad \backslash \quad \hat{\ }$$

standing for addition, subtraction, multiplication, right division, left division, and exponentiation. The precedence of these operators is as expected; namely, ^ is done first, *, /, and \ next, and then + and -. Of course, parentheses may be used to alter this operation order.

Addition, subtraction, multiplication, and exponentiation have their usual meanings when applied to matrices, vectors, and scalars. The left division and right division operators act as ordinary division when applied to scalars. Their meaning in matrix operations is defined as follows: A \ B is equivalent to the mathematical expression $A^{-1} \times B$; A / B is equivalent to the mathematical expression $A \times B^{-1}$.

When a period character "." appears in front of an arithmetic operator, it means the operation should be performed element-by-element. For operations with scalars and for the addition and subtraction of vectors or matrices, there is no change in the operation. Recall the 2×2 matrices A and B defined earlier:

```
       A

A =
              1.00           2.00
              3.00           5.00
      B

B =
              4.00           5.00
              6.00           7.00
```

Now consider the following example :

```
C = A .* B
D = A ./ B
```

The matrices computed here are:

$$C = \begin{pmatrix} 4 & 10 \\ 18 & 35 \end{pmatrix}, \quad D = \begin{pmatrix} 0.2500 & 0.4000 \\ 0.5000 & 0.7143 \end{pmatrix}$$

The *MATLAB User's Guide* [MathWorks 92b] refers to these as *array operations*.

Using matrices defined earlier in this chapter, try some of these operations to verify your understanding of them.

7.4.8 Expressions and statements

Expressions are formed in the usual way with parentheses used to denote grouping. MATLAB does a lot of checking; for instance, if you try to do something stupid like multiply a 3×3 matrix by a 4×4 matrix, then MATLAB squawks at you.

Normally, each line you write is an assignment statement as in the examples above. However there are exceptions, as in the use of the `plot` command that appeared in section 7.3.2 and for the control statements described in section 7.6.5.

When you have completed typing in an assignment statement, you get an echo on the screen that shows the value of the expression on the right of the assignment, as we have observed earlier. You can suppress the echo by putting a semicolon at the end of the line. If you type a line containing only an expression, as in

```
A + B
```

then the value is assigned to a default variable `ans`.

A long line of input can be continued on the next line by using an ellipsis as in

```
A = A + B + C ...
      + D
```

Short expressions can be placed on the same line separated by commas

```
x = 4,  y = 3,  z = 4
```

```
x =
              4.00
```

```
y =
              3.00
```

```
z =
        4.00
```

allowing each expression value to be returned in the same order, or they may be separated by semi-colons,

```
x = 4;  y = 3;  z = 4;
```

suppressing the response.

7.4.9   Compatibility

Operations on arrays and vectors must be compatible in the usual sense of matrix algebra. In the expression

```
A * B
```

the number of rows of B must equal the number of columns of A. In the expression

```
A .* B
```

the number of rows of A must equal the number of rows of B and likewise, for the columns.

If x is a scalar, and A is a matrix, then the expressions

```
A * x
A + x
A - x
A / x
```

are all valid; they mean that the indicated operation is to be performed element-by-element with the scalar, yielding an array of the same dimension as A. The expression

    A ^ x

implies that the matrix A is to be multiplied by itself x-1 times.

7.4.10   Matrix references

The usual subscript notation can be used to reference the elements of a matrix. Thus A(2,3) is the element of A in the second row and third column.

   You also can refer to rows of a matrix, columns of a matrix, and blocks of rows and columns. Thus A(:,2) refers to the second column of A. In particular, if A is the matrix defined earlier, then the statement

    X = A(:,2)

gives us the vector

$$X = \left(\begin{array}{c} 2 \\ 5 \end{array} \right)$$

Similarly, A(2,:) refers to the second row of A, i.e., $\left(\begin{array}{cc} 3 & 5 \end{array} \right)$.

   Now, suppose that M is a 12×12 matrix. The expression M(3:5,5:10) refers to a block, or submatrix of M, that consists of the elements in rows 3 through 5 that are also in columns 5 through 10. It is as if you cut out a 3×6 piece of M, as illustrated in figure 7.3.

7.4.11   Relational and logical operators

Relational expressions can be used in MATLAB as in other programming languages, such as Fortran or C.

Relational operators

The relational operators are

    <, <=, >, >=, ==, and ~=

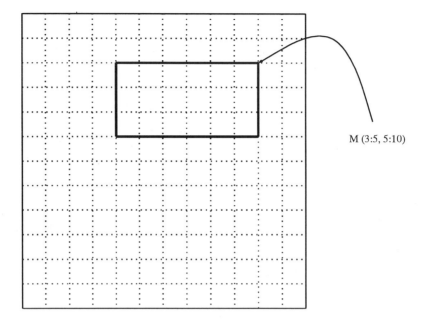

M (3:5, 5:10)

Figure 7.3:   Submatrix of 12 x 12 matrix M.

These can be used with scalar operands or with matrix operands. A one or a
zero is returned as the result, depending on whether or not the relation proves
to be true or false. When matrix operands are used, a matrix of zeros and ones
is returned, formed by componentwise comparison of the matrix elements.

Logical operators

Relational expressions can be combined using the MATLAB logical operators:

```
&, |, and ~
```

meaning AND, OR, and NOT, respectively. These operators are applied element-
by-element.

As in other programming languages, logical operations have lower precedence than relational operations which, in turn, are lower in precedence than
arithmetic operations.

7.5 Built-in functions

There are many built-in functions within MATLAB. You can browse the manuals to see what is available. You can also type

```
help
```

in MATLAB to obtain a list of built-in functions.

The usual math functions are built-in. We have already used the trigonometric function `sin` in the example in section 7.3.2 to produce a simple plot. Those MATLAB commands are repeated here:

```
U = 0:pi/20:2*pi
W = sin(U)
plot(U,W)
```

Both `sin` and `plot` are built-in functions. Notice that the `sin` function has a single argument, the vector U. This function returns a vector of the same length as U where each element is the sine of the value of the corresponding element in U; in other words, $w_1 = \sin(u_1)$, $w_2 = \sin(u_2)$, etc.

In the above example, the `plot` function (or command) has two arguments, U and W; both of these arguments are vectors. MATLAB plots the elements in the first vector U against the elements in the second vector W. The `plot` function can also be used with a single vector argument; in this case, the elements of the vector are plotted against the indices. Type

```
help plot
```

in MATLAB to see what other arguments can be used.

Some built-in functions are rather special. An example is the function `ones` that generates an array of ones; using this, the expression

```
ones(r,c)
```

produces an $r \times c$ array with every element equal to 1. There is a corresponding function `zeros`. Similarly, the expression

```
eye(r)
```

produces the $r \times r$ identity matrix. For more information on any of these functions, use the `help` command.

Some of the functions save a great deal of programming work. For instance, the `polyfit` function is useful for curve fitting and providing polynomial approximations; this is discussed in section 7.5.1. The `eig` function provides the eigenvalues and eigenvectors of a matrix argument; this is described below in section 7.5.2. Section 7.6.4 tells how to write your own functions.

It was noted earlier that MATLAB is case-sensitive. All built-in functions have lower case names.

7.5.1   Polynomial curve fitting

In section 7.3.1, we saw how to find the least squares solution to an overdetermined linear system. Sometimes, the least squares problem is not presented as a matrix problem but rather as a collection of data to be approximated in the least squares sense by a polynomial. One way to determine the coefficients of that polynomial is to set up and solve the appropriate overdetermined linear system. Another way is to call the matlab function `polyfit` to determine those coefficients.

Suppose, for example, that we want to make a polynomial approximation of the function $y = \sin(x)$ in the x-interval $[0, \pi]$ given the values

$$y = [0,\ 0.7071,\ 1.0000,\ 0.7071,\ 0.0000]$$

at the x values

$$x = [0,\ 0.7854,\ 1.5708,\ 2.3562,\ 3.1416].$$

Typing

```
p = polyfit(x,y,2)
```

finds the coefficients

$$p = [-0.3954,\ 1.2420,\ -0.0049]$$

of the quadratic approximating the given data y in the least squares sense. The degree of the approximating polynomial is equal to the third argument of `polyfit`.

In this case, the approximating polynomial is $y = p_1 x^2 + p_2 x + p_3$. Typing

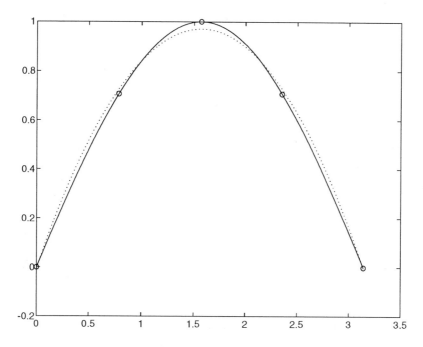

Figure 7.4: A quadratic fit to the sine function on $[0, \pi]$.

```
pvals = polyval(p,x)
```

evaluates this quadratic at the given values of x.

Figure 7.4 shows the the sine function on the interval $[0, \pi]$ with the sampled points marked by circles. The least squares quadratic is shown by the dotted line. This figure was created with MATLAB; how to plot points and how to graph multiple curves on the same plot is covered in section 7.8.

7.5.2 Eigenvalues and eigenvectors

A different MATLAB function makes it easy to compute the eigenvalues and eigenvectors of a matrix. Suppose that we have defined a matrix A. We can compute its eigenvalues by typing eig(A) as in the following example.

```
A = [2 1 0;  1 2 1;  0 1 2];
eig(A)
```

```
ans =

    3.4142
    2.0000
    0.5858
```

In this case, the eigenvalues are the elements of the column vector ans. We can put the eigenvalues into any column vector y by typing y = eig(A).

To also compute its eigenvectors, we must provide a place for MATLAB to store them:

```
[X,D] = eig(A)

X =

    0.5000   -0.7071   -0.5000
    0.7071    0.0000    0.7071
    0.5000    0.7071   -0.5000

D =

    3.4142        0        0
         0   2.0000        0
         0        0   0.5858
```

In this case, the eigenvalues are stored on the diagonal of the matrix D and the corresponding eigenvectors are stored as the columns of the matrix X.

We can check the quality of the computed eigenvalues and eigenvectors by computing the *residual errors*. We will generally find that the computed quantity A*X - X*D is a matrix with very small elements and that A*X(:,j) - D(j,j)*X(:,j) is a vector with very small elements, for $j = 1, 2, 3$. It is generally convenient to express the residual error in terms of the norm of these quantities. In exact arithmetic, these matrices and vectors and their norms would be exactly zero.

7.6 MATLAB scripts and user-defined functions

As mentioned earlier, it is possible to write programs for MATLAB. There are two types of MATLAB programs: *scripts* and *functions*.

A script is a program, containing regular MATLAB commands that could be entered interactively during a MATLAB session. When the name of a script is entered at the command prompt, the script is executed. This means that the commands within the script are executed, affecting the variables in the global workspace.

A function is also a program. As might be expected, a function returns a value, but otherwise it does not affect the variables in the global workspace. Like a script, a function is executed by typing its name at the command prompt. If a function has parameters, they are entered enclosed in parenthesis following the function name.

Both scripts and functions should be stored as files with a `.m` extension, e.g., `myscript.m` or `myfunction.m`. Because of this extension, scripts and functions are typically referred to as *M-files*. Any of the commands discussed above can be used in a MATLAB script or function.

When creating and testing new MATLAB scripts and functions, you may find it useful to have two command windows open: one from which you are running MATLAB and one from which you may be editing the new script or function.

7.6.1 A sample script

The following is a simple script that plots a cosine curve in MATLAB.

```
%  This is a sample MATLAB script
%     that plots a cosine curve

U = 0:pi/20:2*pi
Z = cos(U)
plot(U,Z)      % This statement does the plotting.
```

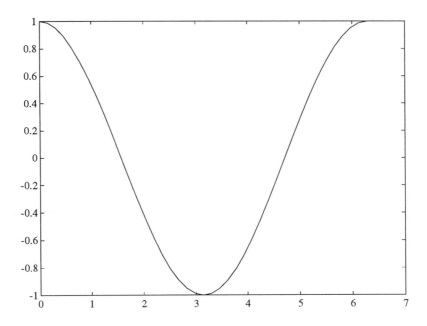

Figure 7.5:   Plot of cosine function on $[0, 2\pi]$.

The first two lines of this script are comments followed by a blank line; see section 7.6.2 for a more detailed discussion on the use of comments and white space in MATLAB scripts. Except for the comment at the end of the last line, the rest of the script is similar to the commands that produced the sine curve in figure 7.2.

To use a script within MATLAB, just enter the script file name without the `.m` extension; typing `myscript` will execute the script `myscript.m`. If the sample script above is stored in a file named `plotcos.m` in the directory from which you are running MATLAB, you need only type

```
plotcos
```

to run the script, causing the appropriate plot to appear in your figure window, as shown in figure 7.5. Note that running this script may alter the values of U and Z.

7.6.2 Comments and white space

For both scripts and functions, the percent symbol % precedes a comment:

```
% This is a comment in MATLAB.
```

The % symbol may be in any position of the line; whatever follows the % is considered to be part of that comment. A comment may even follow a MATLAB command on the same line.

```
plot(U,Z)     % This is a plot command
```

It is useful to place explanatory comments at the beginning of your MATLAB script or function as documentation. If you later type the command

```
help myprog
```

MATLAB responds by printing out the first contiguous block of comments in the script or function stored in `myprog.m`.

Blank lines may be inserted in a MATLAB script or function; this provides white space and promotes the readability of the program. The blank line following the first block of comments in the `myscript` script indicates the end of the lines to be printed by the `help myscript` command.

7.6.3 Continued lines

At times it is desirable to break up a MATLAB command line into two or more separate lines. As discussed above, an ellipsis ... at the end of any MATLAB command line indicates that that line is to be continued onto the next line. This symbol may consist of three or more consecutive periods.

```
S = [ 1   ...
        2;  3   ....
            4 ]
```

The definition of a matrix may require several lines since each line represents a row of the matrix. The line for each row may itself be a continued line.

7.6.4   A sample function

Like a script, a function is a collection of MATLAB commands stored in an M-file. Unlike a script, a function can itself be evaluated. A function may take on a scalar or an array value. A scalar function value can be viewed by typing the file name without the extension, e.g., `myfunction`, or it can be assigned directly to a variable, as in `y = myfunction`. The value of an array-valued function must be assigned to an array variable. User-defined functions may be used not only interactively but also within scripts or other functions.

Note: the name of the function must match the name of the M-file for it. In other words, if you are creating a function named `myfunction`, the file containing the MATLAB commands that define that function must be stored as `myfunction.m`.

A function may require input arguments. Once the arguments have been defined, the function is evaluated by typing its file name (minus the extension) followed by the argument list, e.g., `y = myfunction(arg1, arg2, ..., argn)`. For example, the following function evaluates a cubic polynomial at the larger of the two input arguments.

```
%  This is a sample MATLAB function
%     that evaluates a cubic polynomial
%     at the larger of the two arguments x1 and x2.

function y = mycubic(x1,x2)
x = max(x1,x2);
y = x^3 + 2*x^2 + 1; % This statement determines
                     % the function value.
```

If the function `mycubic` is stored in the M-file `mycubic.m`, we can evaluate the function `mycubic` by typing a series of statements like `z1 = 1; z2 = 2; z = mycubic(z1, z2)`. This series of statements causes the value 17 to be assigned to the variable `z`.

In the following example, the array-valued function `trigfunction` takes an angle `theta` (in radians) as argument and returns both its sine and cosine.

```
%  This is a sample MATLAB function
%     to evaluate the sine and cosine
```

```
%      of the input angle.

function [costheta,sintheta] = trigfunction(theta)
costheta = cos(theta);
sintheta = sin(theta);
```

To evaluate this function, we must assign its value to an array: `[c,s] = trigfunction(0)`. After this call, we see that `c = 1` and `s = 0`.

Function arguments may be manipulated within a function, but input values are the same on exit as on entry.

7.6.5 Control statements

MATLAB contains `for`, `while`, and `if` statements. The syntax of each is illustrated in the examples below. These statements may be used interactively, but are more commonly included with MATLAB scripts or functions.

It is important to recognize that many of the operations that might require one of these statements in a language like C or Fortran do not require them in MATLAB. Matrix multiplication is the most obvious example, since the simple expression

```
A * B
```

produces the multiplication of the matrices `A` and `B` without any looping.

for statement

The sum of all the elements of the vector `V` can be computed using the `for` statement:

```
for  i = 1 : n
   S = S + V(i)
end
```

However, this is more efficiently done by using the built-in `sum` function

```
S = sum(V)
```

The `for` statement may be nested and a constructor with an arbitrary step can be used to define the loop index.

```
total = 0
for  i = firsti : deltai : lasti
   S(i) = 0
   for  j = firstj : deltaj : lastj
      S(i) = S(i) + fun(i,j)
   end
   total = total + S(i)
end
```

where `fun` is some function of `i` and `j`.

`while` statement

A `while` statement can be used to control the number of iterations of a loop:

```
while  err > maxerr
   n = n + 1
   err = funapprox(n,x) - funexact(x)
end
```

A group of `while` statements can be nested and any relational expressions may be used (see section 7.4.11).

`if` statement

```
for  i = 1 : maxrow
   for  i = 1 : maxrow
      if  abs(A(i,j)) < thresh
         A(i,j) = 0
      else
         A(i,j) = sign(A(i,j))
      end
   end
end
```

Further help

The `help` command gives information on these control statements; e.g., type

```
help if
```

Then try

```
help break
```

7.7   Input/output

This section discusses methods for creating input data for MATLAB as well as ways to output the data. Graphical output is covered in section 7.8.

In addition to the commands discussed here, there are a number of MATLAB file input/output functions that resemble those in the C programming language. These include such functions as `fread`, `fwrite`, `fscanf`, `fprintf`, `fopen`, and `fclose`. See the *MATLAB Reference Guide* [MathWorks 92a] or use `help` for more information on using these functions.

7.7.1   UNIX commands within MATLAB

While in MATLAB, it is sometimes useful to run normal UNIX commands. This can be done with the *escape* command (!). For instance, to display the contents of the current directory (when you can't remember the name of your M-file), just type

```
!ls
```

Fortran and C programs can be edited, compiled, and run in the same manner.

```
!vi myprog.f
!f77 -O -o myprog myprog.f
!myprog > myoutput
```

Then the output of these programs can be used as input to MATLAB scripts or can be edited to form M-files to produce plots or other data in MATLAB.

7.7.2   Session log

The `diary` command make it possible to save a log of partial or entire MAT-LAB sessions. If you type

```
diary mylogfile
```

all the lines subsequently appearing in the MATLAB window are saved into a file named `mylogfile`. If the filename is omitted, the name `diary` is used. This feature can be turned off by the command

```
diary off
```

or by exiting MATLAB.

This not only provides a log of your session; it also suggests a method for saving the results to be later edited into another format.

7.7.3   Saving data

An alternate method for storing results from your MATLAB runs is using the `save` command. For example, suppose you have created the following array

```
M = [1:1:3; 10:2:14; 31:3:37; 5:5:15]
```

```
M =
        1      2      3
       10     12     14
       31     34     37
        5     10     15
```

and you wish to store this data for use elsewhere. Just type

```
save mydatafile M /ascii
```

where the `/ascii` option of the command assures the results are in text format. Then the file, `mydatafile`, contains the following:

```
        1.0000000e+00    2.0000000e+00    3.0000000e+00
        1.0000000e+01    1.2000000e+01    1.4000000e+01
        3.1000000e+01    3.4000000e+01    3.7000000e+01
        5.0000000e+00    1.0000000e+01    1.5000000e+01
```

7.7.4   MAT-files

MATLAB data may also be read or stored using *MAT-files* with the `load` and `save` commands. There are some special routines and examples (in both Fortran and C) to assist the user. See the chapter on *Disk Files* in the *Tutorial* section of the *MATLAB User's Guide* [MathWorks 92b].

7.8   Graphics

The sample script given in section 7.6.1 illustrated how to make a plot of the cosine function using the `plot` function as shown in figure 7.5. This plot is a simple two-dimensional X-Y plot drawn in the current MATLAB figure window. Many other types of plots are available in MATLAB. This section introduces you to a number of these and describes how to label, combine, store, and print MATLAB plots.

7.8.1   Types of two-dimensional plots

There are two types of two-dimensional or linear X-Y plots: *line* and *point*. Three-dimensional wire frame and contour plots are also available; these are discussed in section 7.8.5. Polar, logarithmic, semi-log, and bar plots can be employed as well; see the entries on `polar`, `loglog`, `semilogx`, `semilogy`, and `bar` in the *MATLAB Reference Guide* [MathWorks 92a] or type

```
help
```

for more information on these plots.

In the graphical examples discussed earlier in this document, the *line* type of X-Y plot is used. The other type is a *point* plot; points are included as part of figure 7.4. With a line plot, the gaps between the points are filled in smoothly so that you get a continuous curve; in the point type, no fill-in is done so only the points you specify are plotted.

An example of a statement that gives a point plot is

```
plot(U,W,'+')
```

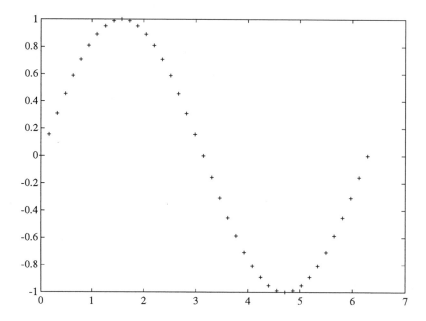

Figure 7.6: Point plot of sine function.

The sine curve appears as 41 distinct points (marked "+"), as shown in figure 7.6. The statement

```
plot(U,W,'+',U,W)
```

gives a point plot for the first plot, and a line plot for the second. Since both curves are the same, the effect is to highlight the points on the curve. This last feature is convenient for showing a least squares fit to experimental data. You can plot the distinct data points and the smooth curve that fits the data in the same picture. More information on graphing two curves in the same plot is given in section 7.8.6.

You can use any symbol in the set {. + * o x} for point plots. You can use any symbol in the set {- -- : -.} for line plots. If nothing is specified, as in most of the examples here, the default line plot (symbol "-") is employed. You can also plot with different colors when using a color monitor; look at

```
help plot
```

for information on specifying color arguments.

7.8.2 Labelling plots

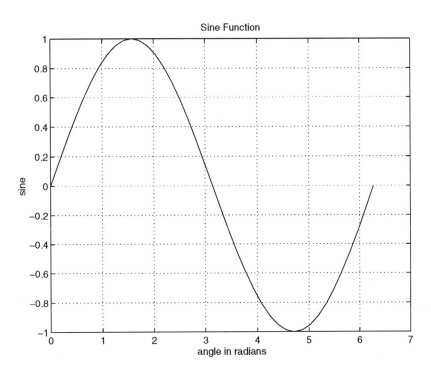

Figure 7.7: Labeled plot of sine function.

The picture can be labelled, the axes can be labelled, and you can display grid lines in the coordinate system. The following script does all of this for the sine plot previously shown in figure 7.2:

```
%  This MATLAB script plots a sine curve

U = 0:pi/20:2*pi
W = sin(U)
plot(U,W)                        % produces basic plot
```

```
title('Sine Function')          % places title at top
xlabel('angle in radians')      % labels x-axis
ylabel('sine')                  % labels y-axis
grid                            % adds grid marking
```

Notice that the labelling commands are given after the plot is created. The result is shown in figure 7.7.

It is also possible to place text on the graph while it is in the figure window by using the mouse. The command

```
gtext('Your text')
```

makes a crosshair appear on the window containing the plot, as in figure 7.8(a). Just move the mouse to the desired location and click; the label containing Your text appears there, with the first letter placed in the inside corner of the northeast quadrant of the crosshair. The final result is in figure 7.8(b).

7.8.3 Handle Graphics

Plots can also be labelled or manipulated in other ways using what MATLAB calls *handle graphics*. When the figure window is formed, it is assigned a unique number called its *handle*. This permits more than one figure window to exist at a time.[2] If you type

```
gcf
```

the value returned contains the handle of the current figure window.

A figure window is provided with a set of default properties. You can use the figure window handle to redefine any of these properties. As an example, we can alter some of the properties of the current figure window as follows:

```
handfig = gcf
set (handfig, 'Position', [0, 0, 300, 280])
```

The first line stores the handle of the current figure window in the variable handfig. The second line moves the window to the bottom left corner of the screen and resizes it to be 300×280 pixels. Type

---

[2]For more information on multiple figure windows, see section 7.8.6.

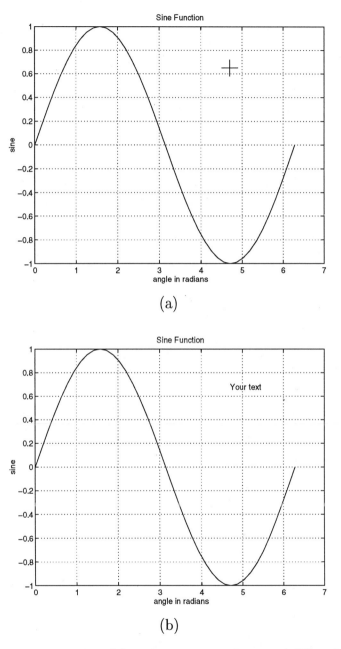

Figure 7.8: Sine function plot: (a) with `gtext` crosshair; and (b) with label entered by `gtext`.

```
help gcf
help set
help get
```

to learn to alter other figure window properties.

In MATLAB, every graphical object has a handle. The list of graphical objects includes the screen itself, the figure windows, and the axes of the plot within the figure windows. For instance, the handle of the root screen is the integer zero. Images, lines, surfaces, and text along with user interface controls and menus are also graphical objects. This hierarchy of graphical objects with handles allows the user to control the MATLAB graphics environment as he wishes.

The `gca` function returns the handle of the axes object in the current figure window. Using this and the `axes` function, the ticknames or ticks on the axes of the current figure may be modified. The older `axis` function can be used for a similar purpose. See the *MATLAB Reference Guide* [MathWorks 92a] or type

```
help gca
help axes
help axis
```

for more information.

7.8.4   Hardcopy plots

When a plotting command is executed, the plot appears in the active figure window. This is the current figure window for all plots. Later plots may erase this plot, unless a new figure window is created or control is given to another window. However, if you type

```
print
```

on some systems, a hardcopy of the plot in the current figure window is printed on your default printer.

You can also save a copy of the plot with the `print` statement. For example,

```
print myplot -dps
```

translates the current plot into *PostScript* and stores it in a file named `myplot.ps`.

It is possible to convert the plot into a color or an encapsulated PostScript file by using other options of the `print` command. For example,

```
print mycplot -dpsc
```

generates the color PostScript file, `mycplot.ps`, and

```
print myeplot -deps
```

produces the encapsulated PostScript file, `myeplot.eps`. The PostScript files can be printed out from the UNIX shell with the `lpr` command

```
lpr myplot.ps
lpr mycplot.ps
```

and both the PostScript and encapsulated PostScript files can be incorporated into figures within the text of a paper.

Many other options are available for this command. Refer to the section on `print` in the *MATLAB Reference Guide* [MathWorks 92a] for more information.

7.8.5 Three-dimensional plotting

At times, three-dimensional grid plots or contour plots are desired. MATLAB provides some facility for these.

Three-dimensional grids

Mesh plots show a three-dimensional surface as a mesh or wire frame surface. Here the x and y values merely provide the size of the grid; a $rank(x) \times rank(y)$ matrix gives the values of the grid points of the surface – one value for each xy point. Thus, the only needed argument to the `mesh` command is that matrix.

The plot in figure 7.9 is created by the following script:

```
%  This MATLAB script plots a 3-D sine curve as mesh

U = 0:pi/20:2*pi;
```

```
X = ones(size(U'))*U;
Y = U'*ones(size(U));
W2 = sin(X) + sin(Y);

mesh(W2)
title('Mesh plot:  sin(X) + sin(Y)')
```

A surface plot with shading can be obtained by adding the following two commands:

```
surf(W2)
title('Surface plot:  sin(X) + sin(Y)')
```

This is shown in figure 7.10.

If we use the fill3 function,

```
fill3(X, Y, W2, U)
title('3-D polygon plot:  sin(X) + sin(Y)')
```

we no longer have a surface. Instead, the grid is broken up into three-dimensional polygons, one per column. This is displayed in figure 7.11. Consult the manual or type help for a more detailed explanation of the fill3 and related functions.

Contour plots

The command to produce a contour plot of a surface works in much the same way as the mesh and surf commands. The plot in figure 7.12 is generated by the same script as in section 7.8.5, substituting the following lines for the original mesh and title commands.

```
contour(W2)
title('Contour plot:  sin(X) + sin(Y)')
```

7.8.6   Multiple plots

Often there are times when you need to observe more than one plot at the same time. You may wish to plot more than one function on the same plot. You may wish to have more than one plot in the same window. You may even wish to have more than one figure window containing plots.

Mesh plot: sin(X) + sin(Y)

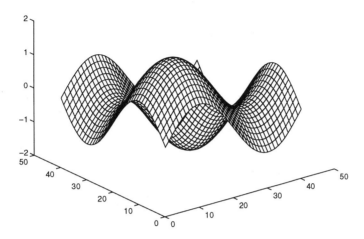

Figure 7.9: Mesh plot of $\sin X + \sin Y$.

Surface plot: sin(X) + sin(Y)

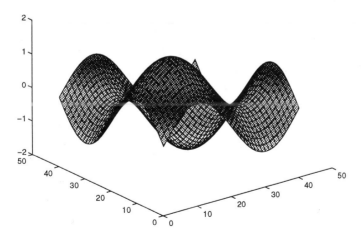

Figure 7.10: Shaded surface plot of $\sin X + \sin Y$.

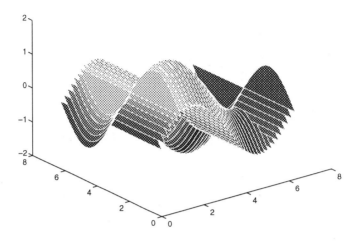

Figure 7.11:   Three-dimensional polygon plot of $\sin X + \sin Y$.

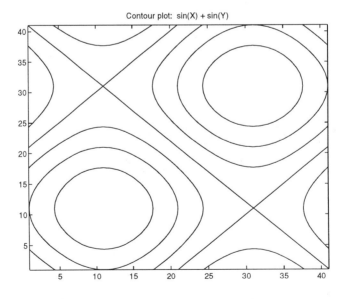

Figure 7.12:   Contour plot of $\sin X + \sin Y$.

Multiple functions in a plot

Suppose you want two curves plotted on the same graph as in figure 7.4. The script below graphs the sine and cosine curves together. This plot is shown in figure 7.13.

```
%  Plots both sine and cosine curves together

U = 0:pi/20:2*pi
W = sin(U)
Z = cos(U)
plot(U,W,U,Z)
```

The pattern illustrated here holds true in general, and the vector of abscissae (U in this example) need not be the same in each case. Thus

```
plot(X1,Y1,X2,Y2,X3,Y3)
```

plots the three curves $(f(X1, Y1), g(X2, Y2), h(X3, Y3))$ in the same plane.

An alternate method for putting one plot on top of another is to use the hold command. This tells MATLAB not to erase the contents of the figure window until the command

```
hold off
```

is entered. In this way, any number of plots can be placed within the same plane with the axes remaining constant.

Some built-in functions provide multiple plots. The surfc function combines a surface plot with a contour plot of the same three-dimensional object. Figure 7.14 shows the effect of this function on the same sine function used in figures 7.9 through 7.12.

Multiple plots in a window

Occasionally, it is useful to have more than one plot within in the figure window. This can be done; an example is shown in figure 7.15 of drawing both the mesh and contour figures shown in figures 7.9 and 7.12 in one window. The two plot statements that produced this example were preceded by the function named subplot, as follows

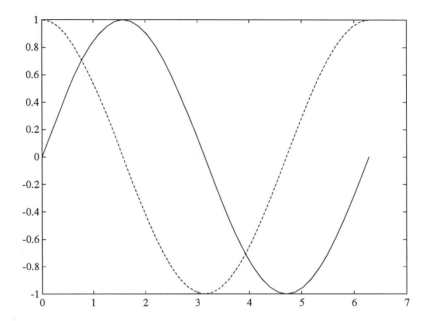

Figure 7.13: Plot of sine and cosine.

```
subplot(2,1,1), mesh(W2)
subplot(2,1,2), contour(W2)
```

There are three numeric arguments to subplot; these are all positive numbers. The first number specifies the vertical number of plots desired in the figure window; the second specifies the horizontal number of plots. The last number tells in which subplot region the plotting command (plot, mesh, contour) is to plot.

Figure 7.16 contains four plots in a single figure window. The MATLAB script that produced these plots is below:

```
%  This script uses the 'sphere' and 'cylinder'
%  functions to produce subplots of a sphere and
%  the cylinder constructed with one of the
%  columns of the matrix defining its y-axis.
```

Combined surface and contour plot: sin(X) + sin(Y)

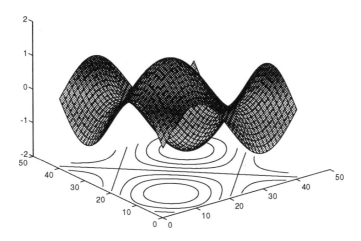

Figure 7.14: Combined shaded surface plot and contour plot of $\sin X + \sin Y$. A color version of this figure may be found on color plate 1.

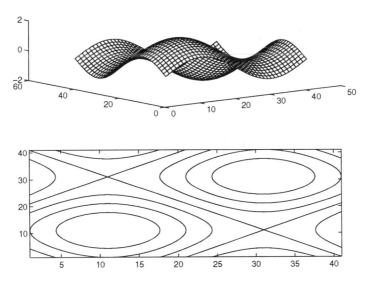

Figure 7.15: Subplots of mesh and contour plots.

```
[Sxx, Syy, Szz] = sphere(16);
subplot(2,2,1), mesh(Sxx, Syy, Szz)
title('Mesh plot:  sphere(16)')

subplot(2,2,2), surfl(Sxx, Syy, Szz)
title('Shaded Surface plot:  sphere(16)')

[Cxx, Cyy, Czz] = cylinder(2+Syy(8,:));
subplot(2,2,3), mesh(Cxx, Cyy, Czz)
title('Mesh plot: cylinder(2+Syy(8,:))')

subplot(2,2,4), surf(Cxx, Cyy, Czz)
title('Surface plot:  cylinder(2+Syy(8,:))')
```

There are two built-in functions used in this script: `sphere` and `cylinder`. The `sphere` function returns three matrices containing the coordinates of the sphere; in this case, each of the matrices will be of size 17×17. The `cylinder` function is similar, but we have specified the radius of the cylinder to be 2 plus the elements of the middle row of the matrix that defines the y-coordinates of the sphere.

Multiple figure windows

The `figure` command is used to create additional figure windows. If you type

```
hnum = figure
```

a new figure window is activated, and the handle number of the new figure window is assigned to hnum. If you wish to make a new figure window with a particular handle (say 123), just type

```
figure(123)
```

The `figure` command can also be used to specify other parameters for the new window. These include the size of the window, the title of the window, the position of the window on the screen, and the background color. See the *MATLAB Reference Guide* [MathWorks 92a] or type

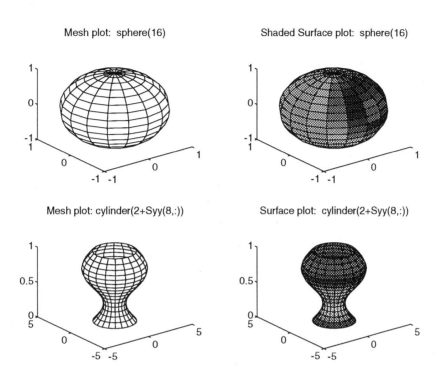

Figure 7.16: Subplots showing the mesh and shaded surface plots of a sphere determined by three 17×17 matrices of coordinates (named Sxx, Syy, and Szz) and the mesh and surface plots of a cylinder whose width is two plus the y coordinates of the sphere. A color version of this figure may be found on color plate 1.

```
help figure
```

for more information.

7.8.7 Creating images

Besides producing various plots, it is possible to create an image in a figure window with MATLAB. These images may be colored by different color maps. Refer to the manuals or type

```
help image
help colormap
help ColorSpec
```

to find out more about these possibilities.

7.9   That's it!

Well, that is a brief overview of MATLAB. Be sure to do all of the examples given here while you are on the computer. This gets you started.

We have tried to make this short so that you would be able to quickly get into using MATLAB. On the other hand, this has forced us to leave out a lot of stuff that is potentially useful. Now it is up to you to learn more from the manuals and on-line documentation.

8 Elements of IDL

IDL (**I**nteractive **D**ata **L**anguage) is a complete, structured language developed by Research Systems, Inc. (RSI). With language constructs similar to those of Fortran, it allows interactive data analysis and visualization. Two-dimensional (X-Y) plots, three-dimensional (X-Y-Z) plots, surface plots, contour plots, shaded plots, mesh plots, even bar charts, are all possible using IDL. Operators include matrix multiplication as well as finding the minimum and maximum of vectors or arrays. Procedures and functions exist to perform a number of other operations, including Fourier transforms and interpolation. This language can be used interactively or to create application programs.

This chapter introduces you to using and programming IDL. It is only a beginning; refer to the IDL manuals for more advanced and complete information.

A system closely related to IDL is PV-WAVE, now produced by Visual Numerics. PV-WAVE is built on the same kernel as IDL.

Although IDL is available on a number of systems, the remainder of this chapter assumes a UNIX interface. Best use of the chapter can be made if you work through the examples shown.

8.1 Getting started

Before starting IDL for the first time, put the following command in your `.login` or `.cshrc` file:

```
source /usr/local/idl/idl_setup
```

This is required if you are working on SGI Indigos, otherwise IDL may not function properly. You need to execute this command (or log out and log back in again) for it to take effect. If this does not work, check with the lab assistant or systems manager to obtain the correct path.

```
% idl
IDL. Version 2.2.2 (ultrix mipsel).
Copyright 1989-1992, Research Systems, Inc.
All rights reserved.  Unauthorized reproduction prohibited.
Site:  100.
Licensed for use by:  University of Colorado  1/15/92


IDL> ...
    ⋮
IDL> exit
% ...
```

Figure 8.1: Sample IDL session run on a DECstation 5000/240.

If you are using the X-window system, be sure that the DISPLAY environment is set correctly to the terminal you are using. This is done by typing the command

$$\text{setenv DISPLAY } \textit{yourterminalname}:0$$

to allow the plot window to appear on your screen. If setting the *DISPLAY* variable is not permitted, you may first need to type

$$\text{xhost } +\textit{remotemachinename}$$

where *remotemachinename* is the name of the remote machine with IDL installed on it. In some cases, you may need to do this before logging into the remote machine.

IDL can be invoked by typing idl at the shell prompt. Typing the command, exit, at the IDL prompt terminates an IDL session. Figure 8.1 shows the beginning and ending of a sample session with IDL, as run on a DECstation. All of the examples in this chapter were run on a DECstation 5000/240.

8.1.1 Demos

To learn more from this chapter you should use a "hands-on" approach. If you haven't already done so, now is the time to get started. Set up the necessary paths and environments for IDL on your terminal.

A number of demonstrations are available. To observe the power of IDL, along with some of its many applications, here are two such demos you can try.

HPSC demos

The High-Performance Scientific Computing (HPSC) Laboratory at the University of Colorado at Boulder has a set of demos based on the data generated by members of previous HPSC classes.[1] To run these demos, you must first get into the correct directory for your machine:

```
cd .../Demos/IDLdemo
```

Then just type

```
rundemo
```

and follow the instructions. You may later wish to look at the main IDL programs that perform the demos; for example, two of these are `chain.pro` and `advect.pro` and are stored in the same directory.

XDEMO

The `XDEMO` is part of the IDL package that works with the X-window system. To start this feature running, enter IDL and type `XDEMO` at the IDL prompt:

```
IDL> XDEMO
```

or

```
IDL> xdemo
```

---

[1] This software is available through the anonymous ftp site at `cs.colorado.edu` in the `/pub/HPSC` directory.

as IDL is not case-sensitive. The main demo window should appear with four items to choose from with the mouse:

```
Quit
Examples
What is IDL
Help
```

Clicking on `Help` brings up a window with a short explanation of the demo package. The `What is IDL` button provides a brief description of IDL and its applications. If you click on `Examples`, a menu that lists a number of examples appears. Try any and all that interest you. Choosing `Quit` terminates the demonstration.

8.1.2   Help

As with most tools, there is a `HELP` facility. When you type the `help` command within IDL,

```
IDL> help
```

it returns the current state of your IDL workspace: the names, types, and values of your variables as well the amount of storage you are using. An example of the response to `HELP` is shown in figure 8.2. The IDL statements used in this example to define the given variables are explained below.

A second type of help is the question mark ?. If you type a question mark at the IDL prompt, you may obtain information on specific routines, whether intrinsic or from the User Library. When using the X-window system, a menu appears in response to the question mark. The names of routines are given in the left side of the window and can be scrolled up and down to find a particular routine. Routines or keywords from different libraries can be displayed by clicking on the appropriate diamond in the top part of the window. Clicking on a displayed routine name or keyword brings up information on that routine within the large subwindow. Again this can be scrolled.

Enter IDL and type " ? ". Browse the lists of available functions and click on any that appear interesting.

```
% idl
IDL. Version 2.2.2 (ultrix mipsel).
Copyright 1989-1992, Research Systems, Inc.
All rights reserved.  Unauthorized reproduction prohibited.
Site:  100.
Licensed for use by:  University of Colorado 1/15/92


IDL> help
% At $MAIN$ .
Code area used:  0.00% (0/16384), Data area used:  0.05% (2/4096)
# local variables:  0, # parameters:  0
Saved Procedures:
Saved Functions:

IDL> a = 10
IDL> b = a + 5.0
IDL> s = [1,2,3,4,5]
IDL> t = [1.0,2.0,3.0,2.0,1.0]
IDL> help
% At $MAIN$ .
Code area used:  0.00% (0/16384), Data area used:  1.03% (42/4096)
# local variables:  4, # parameters:  0
A INT = 10
B FLOAT = 15.0000
S INT = Array(5)
T FLOAT = Array(5)
Saved Procedures:
Saved Functions:

IDL> exit
% ...
```

Figure 8.2:   Sample usage of **help** within an IDL Session.

IDL Type	Size
Byte	1 byte
Integer	2 bytes
Longword Integer	4 bytes
Floating-point	4 bytes
Double precision floating-point	8 bytes
Complex floating-point	8 bytes
String	0 to 32,767 bytes
Structure	any combination

Table 8.1: IDL data types.

8.2   Exploring the basic concepts

8.2.1   Data types

Variables in IDL can be of various types, much like those in Fortran. The possible IDL types with their sizes are shown in table 8.1.

8.2.2   Constants

IDL constants are similar to those in Fortran. Numbers that include a decimal point, such as 3.25 or .00001, are assumed to be real or floating-point values. Without a decimal point, they are assumed to be integer. Values may also be expressed in scientific notation, as in 4.031E-6 or 6.732D15; the first of these values is a single precision, floating-point value, while the second is a double precision value.

As in Fortran, the intrinsic function, COMPLEX, is used to combine the real and imaginary parts of a complex value, e.g., COMPLEX(2.0,3.7) or $2 + 3.7i$. Other conversion functions are available, such as FIX, FLOAT, and DOUBLE; these behave as the Fortran functions of the same names.

Strings should be enclosed in apostrophes, ', or quotes, ". The null string may be represented by '' or "". Octal and hexadecimal constants are also allowed. Refer to the *IDL User's Guide* [RSI 90] for more information.

8.2.3 Variable names

Variable names can be one to fifteen characters in length. The first character must be alphabetic; the remaining characters may be letters, digits, dollar signs ($), or underscores (_). To differentiate system variables from user variables, all system variable names begin with an explanation point, !.

As in Fortran, no distinction is made between upper and lower case letters; the variable, x1, is the same as the variable, X1. Referring back to figure 8.2, notice that the assignment statements for the variables, a, b, s, and t, all used lower case letters; however, the output of the HELP procedure listed these variables in upper case letters.

8.2.4 Scalar variables

Declaring a scalar variable is simply done by assigning the value of an appropriate expression to that variable's name. For example, the following IDL statements

```
IDL> A = 10
IDL> B = A + 5.0
IDL> PRINT, A, B
```

produce the output:

```
10    15.0000
```

As can be seen by this example, assignment statements in IDL look the same as those in Fortran.

The first assignment statement

```
IDL> A = 10
```

sets the variable, A, to the integer value, 10. This value, 10, is assumed to be integer since it has no decimal places. Thus, the variable, A, is dynamically defined by this statement to be an integer variable.

The second statement

```
IDL> B = A + 5.0
```

declares the variable, B, to be a floating-point variable. This is because B takes on the value of the integer variable, A, incremented by the floating-point value, 5.0. In general, the result of an expression is of the same type as the highest precision value in that expression. It is legal to combine integer and real values in the same expression.

The third statement

```
IDL> PRINT, A, B
```

is a PRINT command to print out the value of these two variables. This command is really a procedure name followed by two arguments. These items must be separated by commas.

If you refer back to figure 8.2, you see that the variables, A and B, were defined using the same assignment statements there. And the output of the HELP procedure as to the types and values of those two variables agrees with the discussion here.

Caution: You need to be aware that the type of a variable may change dynamically with each assignment. This is demonstrated in the following set of commands. In IDL, type the following:

```
IDL> C = 2
IDL> D = 0.5
IDL> PRINT, C, D
IDL> HELP
IDL> C = C * C
IDL> PRINT, C, D
IDL> HELP
IDL> C = C * D
IDL> PRINT, C, D
IDL> HELP
```

Examine the HELP descriptions of the variables. You should be able to explain any changes in the value and type of each variable.

Plate 1

Combined surface and contour plot: sin(X) + sin(Y)
2
1
0
-1
-2
50
40
30
20
10
0
0
10
20
30
40
50

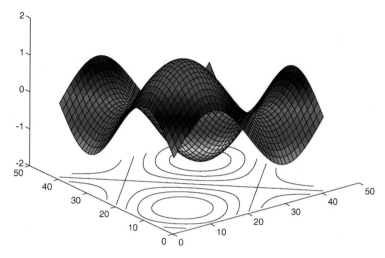

Figure 7.14

Mesh plot: sphere(16)
Shaded Surface plot: sphere(16)
Mesh plot: cylinder(2+Syy(8,:))
Surface plot: cylinder(2+Syy(8,:))
1
0
-1
0.5
5
-5

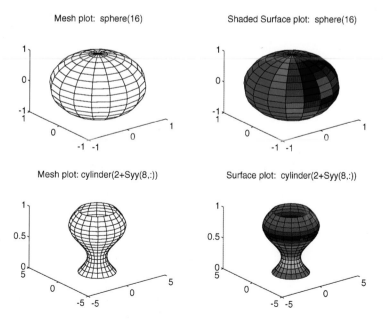

Figure 7.16

Plate 2

Figure 10.3

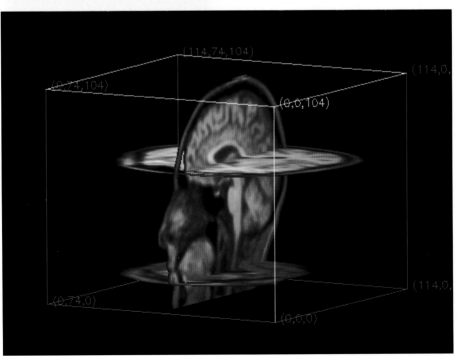

Figure 10.25

Plate 3

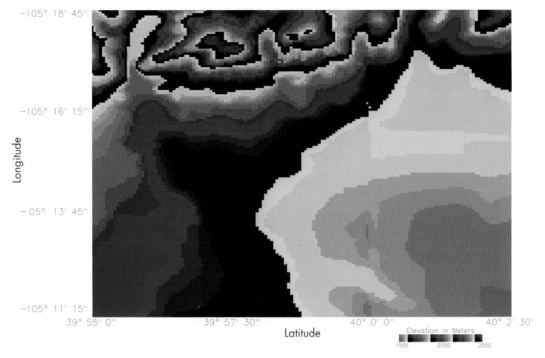

Figure 10.17

Figure 10.20

Plate 4

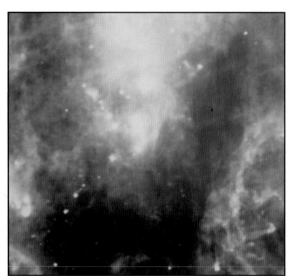

Figure 10.5

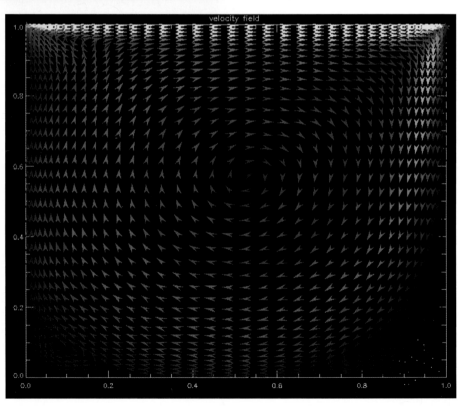

Figure 10.22

8.2.5 Operators, expressions, and statements

For the most part, IDL expressions match those in Fortran. All the usual arithmetic operators, +, -, *, /, and ^ (replacing the Fortran **), can be applied to both scalar and array variables. There are also some additional operators, such as, # for matrix multiplication and MOD. Parentheses may be used as well as relational and logical operators. Table 8.2 gives a list of all the available operators and their precedence. Notice that the logical operators do *not* have dots around them, as they do in Fortran.

Priority	Operator	Operation
First (Highest)	^	Exponentiation
Second	*	Multiplication
	/	Division
	MOD	Modulus
	#	Matrix Multiplication
Third	+	Addition
	−	Subtraction
	<	Minimum
	>	Maximum
	NOT	Logical Negation
Fourth	EQ	Logical Equality
	NE	Not Equal
	LE	Less than or Equal
	LT	Less than
	GE	Greater than or Equal
	GT	Greater than
Fifth (Lowest)	AND	Logical AND
	OR	Logical OR
	XOR	Exclusive OR

Table 8.2: IDL operator precedence.

IDL expressions are formed using a combination of variables and constants separated by these operators in a meaningful way. As in Fortran, assignments are made by putting a variable name on the left-hand side of the statement and an expression on the right-hand side of the statement with an equality sign, =, between the two sides, such as

```
IDL> B = A + 5.0
```

The relational operators and the boolean operators work in interesting ways. Recall that A has the value 10 and B is 15.0. Try typing in the following command:

```
IDL> PRINT,  A < B,  A LE B,  A > B,  A GE B,  NOT A
```

This should yield the result:

```
        10.0000    1   15.0000    0    -11
```

There are a few important points to note: First, expressions can be used as arguments to the PRINT procedure and to most other procedures and functions. Secondly, the minimum and maximum operators display the minimum or maximum as appropriate. Next, the relational operators return 1 or 0; these are the values that represent TRUE or FALSE. Finally, the Boolean operators expect 0 or 1 as the usual arguments and return unexpected results if used otherwise. Normally, a 0 is FALSE, and a non-zero number is TRUE. In the above case, the expression, NOT A, resulted in the incorrect answer.

Long expressions can carry over onto a new line. The continuation is signaled by putting a dollar sign, $, at the end of each line that needs to be continued, e.g.,

```
IDL> W = B + 5.0 + Z^2 - FLOAT(A MOD 2)    $
IDL>         - 6.0 * C / D
```

To get some practice in using IDL expressions, select five numbers: two integers and three floating-point values. Use IDL to calculate their average. Write down the command you used to calculate the average, as well as the answer. Make sure you pick numbers that make this a less-than-trivial problem.

8.2.6 Vector variables (one-dimensional arrays)

There are two primary ways of creating arrays. The first method is to specify the values by initializing the variables, and the second method is to use a function to create an array of a given size and then read or compute its values.

Implicit array declaration

As is done for scalar variables, arrays can be declared by the assignment of values. For example, the following commands

```
IDL> S = [1, 2, 3, 4, 5]
IDL> T = [1.0, 2.0, 3.0, 2.0, 1.0]
IDL> PRINT, S, T
```

give the results:

```
1    2    3    4    5
1.0000    2.0000    3.0000    2.0000    1.0000
```

Thus, as in shown in figure 8.2, S is now declared as an integer array of five elements, while T is a real array, also containing five values. Just as in Fortran 90, the operators described in the previous section may work on an entire array, as well as on scalars. The array operations are done elementwise. Thus, the statement

```
IDL> PRINT,  S + T,  S < T,  S - 1,  T * 0.5,  S EQ T
```

produces the output

```
2.0000    4.0000    6.0000    6.0000    6.0000
1.0000    2.0000    3.0000    2.0000    1.0000
0    1    2    3    4
0.5000    1.0000    1.5000    1.0000    0.5000
1    1    1    0    0
```

To get practice using arrays in IDL, try multiplying and dividing the arrays S and T. Make sure that the results are correct.

Explicit array declaration

The second method of creating arrays is to use explicit array-creation functions provided by IDL. This is the equivalent of typed DIMENSION statements in Fortran. For example, the commands

```
IDL> G = INTARR(7)
IDL> H = FLTARR(5)
```

create an array variable, G, with seven integer elements, and an array variable, H, with five floating-point elements. In other words, the built-in function, INTARR, creates an integer array of the size given by the argument(s) passed to it. Similarly, FLTARR declares a real array. Array-declaring functions for other types, such as BYTARR or DBLARR, are also available.

If you now print out the values of the arrays, G and H,

```
IDL> PRINT, G, H
```

you see that the initial values of the array elements are set to zero.

```
     0     0     0     0     0     0     0
   0.0000    0.0000    0.0000    0.0000    0.0000
```

Notice that there is a difference between integer and floating-point zeros.

It is also important to know that IDL considers the lowest subscript to be 0 and the highest subscript, one less than the number of elements in the given dimension: 4 for H or 6 for G. In the numbering of subscripts, IDL is more like C than Fortran.

Arrays may also be assigned values when they are declared explicitly. The IDL index generation functions provide this capability. Try typing in the following commands:

```
IDL> R = INDGEN(4)
IDL> PRINT, R
```

Here the INDGEN function replaces INTARR and declares R to be an integer array with four elements. At the same time, the elements of this array are assigned values equal to their index within the array; i.e.,

```
R[0] = 0
R[1] = 1
R[2] = 2
R[3] = 3
```

Similarly, the command

```
IDL> P = 0.25 * FINDGEN(5)
```

creates a five-element, floating-point array with the following values:

```
P[0] = 0.00
P[1] = 0.25
P[2] = 0.50
P[3] = 0.75
P[5] = 1.00
```

8.2.7 Multi-dimensional arrays

Multi-dimensional arrays can be created in a similar fashion. For example, the following statements form multi-dimensional arrays using three different methods:

```
IDL> M1 = [ [1,2,3], [3,4,5] ]
IDL> M2 = BYTARR(3,2)
IDL> M3 = INDGEN(4,3,2)
```

Implicit declaration

The first statement

```
IDL> M1 = [ [1,2,3], [3,4,5] ]
```

shows an implicit declaration. M1 is declared to be an integer array, with two rows and three columns, by direct assignment of the two vectors. If you now type

```
IDL> PRINT, M1
```

the results are

```
    1       2       3
    3       4       5
```

When subscripting two-dimensional arrays, the first coordinate refers to the column and the second coordinate refers to the row. This is the reverse of linear algebraic notation. If you type out specific elements of the array (remembering that the subscripts start at zero),

```
IDL> PRINT, M1(0,0)
    1
IDL> PRINT, M1(0,1)
    3
IDL> PRINT, M1(1,0)
    2
IDL> PRINT, M1(1,1)
    4
IDL> PRINT, M1(2,0)
    3
IDL> PRINT, M1(2,1)
    5
```

you should see the form of the array. Unlike Fortran, the elements of an array are stored in *row-major* order,[2] meaning that the elements of M1 are stored as:

```
    1       2       3       3       4       5
```

or as

$$M1_{0,0} \quad M1_{1,0} \quad M1_{2,0} \quad M1_{0,1} \quad M1_{1,1} \quad M1_{2,1}$$

with the leftmost subscript varying first. So, if you type

```
IDL> PRINT, M1(4)
```

the response with be the fifth element of the array (since the first element is numbered zero) or

---

[2]Graphical images are also stored as two-dimensional arrays and, by convention, are displayed in row-major order.

4

It is legal to only use one subscript when referring to an element of a multi-dimensional array.

Explicit declaration

The second array-declaring statement

```
IDL> M2 = BYTARR(3,2)
```

is an explicit declaration and uses the BYTARR function to make M2 a 3×2 byte array. When specifying arrays with the *typ*ARR functions, the first argument refers to the size of the first dimension, the second argument to the second dimension, and so on, as in Fortran DIMENSION statements. Now if you type

```
IDL> PRINT, M2
```

the output is

```
    0       0       0
    0       0       0
```

Thus, arrays of type BYTE are also initialized to zero by the BYTARR function.

Initialization by index generation

The third type of array declaration

```
IDL> M3 = INDGEN(4,3,2)
```

uses the built-in function, INDGEN, to explicitly declare a three-dimensional integer array while initializing the 24 elements with a sequence of integers, starting with the value, zero. If you print out this array,

```
IDL> PRINT, M3
```

you can see how the elements of the array have been initialized:

```
0      1      2      3
4      5      6      7
8      9     10     11

12     13     14     15
16     17     18     19
20     21     22     23
```

Recall that arrays are referred to in *row-major* order. The statement, PRINT, M3, illustrates this order; the value of an element indicates its position in storage. For example, the element with value 0 is the first element stored in the array, etc. For M3, the storage coordinates are:

```
(0,0,0) (1,0,0) (2,0,0) (3,0,0)
(0,1,0) (1,1,0) (2,1,0) (3,1,0)
(0,2,0) (1,2,0) (2,2,0) (3,2,0)

(0,0,1) (1,0,1) (2,0,1) (3,0,1)
(0,1,1) (1,1,1) (2,1,1) (3,1,1)
(0,2,1) (1,2,1) (2,2,1) (3,2,1)
```

Subscripting

As in Fortran 90, arrays can be manipulated in whole or by specifying subsections. For example, individual elements or ranges can be used:

```
IDL> PRINT, M1(2,1)
    5
IDL> PRINT, M2(*,0)
    0      0      0
IDL> PRINT, M3(0:1,2,0)
    8      9
```

The first PRINT statement

```
IDL> PRINT, M1(2,1)
```

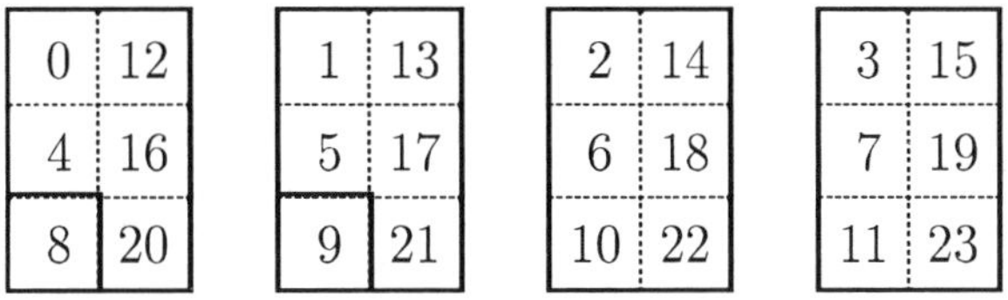

0	12
4	16
8	20

1	13
5	17
9	21

2	14
6	18
10	22

3	15
7	19
11	23

Figure 8.3: Subsection M3(0:1,2,0) of M3.

refers to a single element of M1 and so prints out a single value. An * in an array subscript specification implies the entire range of that dimension of the given array. So, the second PRINT statement

```
IDL> PRINT, M2(*,0)
```

asks for all the values of the first column of the M2 array:

```
        0        0        0
```

A *subsection* of an array refers to a portion of the array. The subsection requested by the third PRINT statement

```
IDL> PRINT, M3(0:1,2,0)
```

is shown in figure 8.3.

On your own, split up M3 at the z-dimension. Assign the 0th dimension to variable M30, and the 1st dimension to the variable M31. Double check your results.

8.2.8 Linear algebra operations

Two-dimensional arrays may represent matrices and one-dimensional arrays may be vectors. Since IDL purports to be an **I**nteractive **D**ata **L**anguage, it must be able to manipulate all kinds of data, including matrices. Thus, there exist a number of operators and functions to deal with matrices.

Operating on vectors and matrices

As mentioned above, all the scalar operators and most functions can work on arrays or matrices. For instance, examine and then try the following commands and check that your responses match the ones given here.

```
IDL> M1 = [ [1,2,3], [3,4,5] ]
IDL> PRINT, TOTAL(M1)
        18.0000
IDL> PRINT, SQRT(M1)
        1.00000         1.41421         1.73205
        1.73205         2.00000         2.23607
IDL> PRINT, EXP(M1)
        2.71828         7.38906         20.0855
        20.0855         54.5981         148.413
IDL> PRINT, COS(M1)
       0.540302        -0.416147       -0.989992
      -0.989992        -0.653644        0.283662
```

We can consider operations on vectors as well:

```
IDL> V2 = [4, 3]
IDL> PRINT, V2 * V2
        16        9
IDL> V3 = [1, 2, 1]
IDL> PRINT, V3, V3*V3, TOTAL(V3 * V3)
        1       2       1
        1       4       1
      6.00000
```

The last expression, TOTAL(V3 * V3), provides the dot product of V3 with itself.

Matrix multiplication

Recall that the # sign is used as an operator for matrix multiplication. For example, assuming that M1 has been defined as above in sections 8.2.7 and 8.2.8, type in the following statements; these should return the given responses:

```
IDL> M1T = TRANSPOSE(M1)
IDL> PRINT, M1, M1T
            1         2         3
            3         4         5
            1         3
            2         4
            3         5
IDL> M1SQ = M1 # M1T
IDL> PRINT, M1SQ
           10        14        18
           14        20        26
           18        26        34
```

The first statement uses the intrinsic function, TRANSPOSE, to create a new 3×2 array, M1T, the transpose of M1. The next statement prints out both matrices. The third statement uses the matrix multiplication operator, #, to multiply M1 by its transpose. And the last statement prints out the final result, a 3×3 symmetric matrix.

Whoops! A 3×3 result for multiplying a 2×3 matrix by a 3×2 matrix?

To quote from the *IDL User's Guide*, when discussing the matrix multiplication operator, #:

> The notion of columns and rows in IDL are reversed from that of linear algebra although their treatment is consistent.

In section 8.2.7, you printed out the individual elements of M1. If you ignore the normal IDL subscripting and look at this with linear algebraic notation in mind, then the array, M1, can be thought of as the matrix

$$M1 = \begin{pmatrix} 1 & 3 \\ 2 & 4 \\ 3 & 5 \end{pmatrix}$$

where $M1_{2,0}$ refers to the element in the third row and first column. If you think of M1 and M1T in this manner, then the result of the matrix multiplication makes sense. You also have to pretend that when a full matrix is printed, as in

```
IDL> PRINT, M1
          1          2          3
          3          4          5
```

what appears on the screen is the transpose of the matrix, not the matrix itself.

To make sure this works, let's try a few more examples:

```
IDL> PRINT, M1T # M1
         14         26
         26         50
IDL> PRINT, M1 # V2
         13         20         27
IDL> PRINT, M1 # V3
% Operands of matrix multiply have incompatible
                dimensions: M1, V3.
IDL> PRINT, V3 # M1
          8
         16
```

Note that the matrix multiplication operator, #, treats a vector as either a column vector or a row vector, depending on which it needs.

Linear algebra routines

Besides the intrinsic TRANSPOSE function, there exist a number of other procedures and functions to solve systems of linear equations and to determine eigenvectors and eigenvalues. Several of these are listed here:

- INVERT – INVERTs a matrix;

- LUBKSB – uses BacKSuBstitution to solve a system of linear equations with the LUDCMP routine;

- LUDCMP – performs an LU DeCoMPosition;

- MPROVE – iteratively iMPROVEs the solution vector of a linear system of equations; uses LU decomposition;

- SVBKSB – performs a BacKSuBstitution to solve a system of linear equations using arrays created by SVD, a Singular Value Decomposition routine;

- SVD – performs a Singular Value Decomposition on a matrix;

- TQLI – finds eigenvalues and eigenvectors of a real, symmetric matrix, using the QL algorithm with Implicit shifts;

- TRED2 – reduces a real symmetric matrix to TRidiagonal form, using Householder's method;

- TRIDAG – solves a TRIDiAGonal system of linear equations.

Most of these commands use the algorithms from *Numerical Recipes in C: The Art of Scientific Computing* [Press et al 88]. Be sure to refer to the *IDL User's Guide* or use the ? help function for more information on these and other routines.

8.2.9   System variables

IDL has a number of built-in variables. As mentioned above, the names of these system variables all begin with the character !. The values of the variables are set by IDL, but may be changed by the user, if desired.

Some of these system variables represent numeric values, such as !PI and !DPI, the single and double precision values of π. Two more numerical system variables are !DTOR, the factor for converting degrees to radians, and !RADEG, the factor for converting radians back to degrees.

Other system variables are used to control portions of the IDL environment. For instance, the variable, !PROMPT, defines the prompt you receive from IDL at the beginning of each new command. If you type the command

```
IDL> PRINT, !PROMPT
```

the response is

```
IDL>
IDL>
```

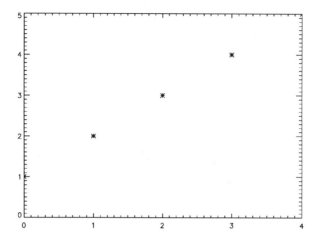

Figure 8.4:   IDL point plot of S.

The first line is merely printing out the current prompt, as requested. The second is the prompt for the next command. Like other system variables, the value of !PROMPT can be changed. If you type

```
IDL> !PROMPT = 'myidl:'
```

the response is

```
myidl:
```

This is merely the altered prompt, requesting the next IDL command.

It is even possible to create your own system variables with the command, DEFSYSV. See the *IDL User's Guide* for more information.

8.3   Plotting

Being a graphics language, IDL provides commands for producing many types of graphs or plots. Some of these are intrinsic to IDL; others have been written by users and exist in the IDL User's Library[3] that has become part of the IDL package. We introduce you to a few in this section.

---

[3]The User Library includes many other useful and varied mathematical and statistical modules. Refer to the IDL manuals for more information.

8.3.1 Two-dimensional plotting

The main command for two-dimensional (X-Y) plotting is the intrinsic PLOT procedure. There are other plotting commands as well, such as OPLOT, and the logarithmic scale plotting procedures: IO_PLOT, OI_PLOT, and OO_PLOT.

For example, try typing in the following set of commands, with S and T defined as in section 8.2.6:

```
IDL> S = [1, 2, 3, 4, 5]
IDL> PLOT, S, PSYM=2
```

In a new window called the plot window should appear a plot similar to the one in figure 8.4. The first argument of the PLOT command is S; so it is a plot of the array, S, that is made. Since just S is plotted, IDL plots the indices of the array versus the values of the elements. Thus, the five points plotted are (0,1), (1,2), (2,3), (3,4), and (4,5). The keyword, PSYM, tells the PLOT command which symbol is to be used to denote the points; in this case, the number 2 means an asterisk (*) is to be used. Some other possible values of PSYM are shown in table 8.3.

PSYM	Symbol
1	+
2	*
3	.
4	Diamond
5	Triangle
6	Square
7	X

Table 8.3: IDL plotting symbols (values of PSYM).

If no value for PSYM is designated, a line is drawn to connect the points. So if you type the following PLOT command:

```
IDL> PLOT, S, TITLE="Example Plot"
```

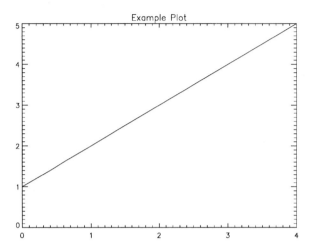

Figure 8.5:   IDL line plot of S.

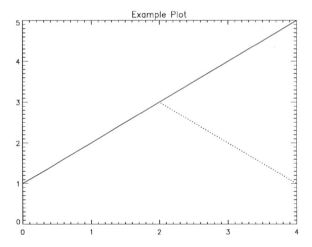

Figure 8.6:   IDL line plot of S with T.

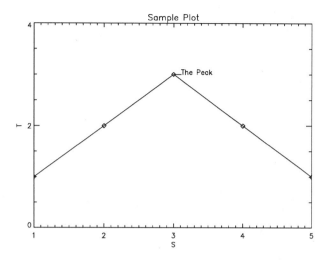

Figure 8.7: IDL plot of S versus T.

the plot now contains a line connecting the points, instead of asterisks at the given points. This plot is shown in figure 8.5. The keyword TITLE provides the text for a title at the top: Example Plot. Similar keywords XTITLE and YTITLE can be used to provide titles for the X-axis and the Y-axis.

LINESTYLE	Type of Line
0	Solid
1	Dotted
2	Dashed
3	Dash Dot
4	Dash Dot Dot Dot
5	Long Dashes

Table 8.4: IDL plotting line styles (values of LINESTYLE).

Now type in the following commands, using the array T as previously defined in section 8.2.6:

```
IDL> T = [1.0, 2.0, 3.0, 2.0, 1.0]
IDL> OPLOT, T, LINESTYLE=1
```

As you can see, OPLOT stands for OverPLOT and does not destroy the previous plot. Here the line connecting the values from the array T is plotted on top of the existing plot for the array S. The resultant plot should appear similar to the one in figure 8.6. Ignoring types, the first three values of both arrays are the same, and so the first part of the two lines are drawn on top of each other.

Notice that the keyword LINESTYLE has been used. The default value of LINESTYLE is 0; this denotes a solid line. In this example, the values of the array T are to be connected with a line style of type 1; this is a dotted line. Table 8.4 shows other possible values for the LINESTYLE keyword.

Of course, it is also possible to plot S against T. This is done with the following commands:

```
IDL> PLOT, S, T, TITLE='Sample Plot',   $
IDL>               XTITLE='S', YTITLE='T'
IDL> XYOUTS, 3.0, 3.0, '__The Peak'
IDL> OPLOT, S, T, PSYM=4
```

This plot should be similar to the one shown in figure 8.7. Notice that the first argument array is used for the X-values, while the second provides the Y-values. The PLOT command has also included titles for the plot itself and for the X-axis and the Y-axis. The built-in procedure used in the second command XYOUTS writes the string (given as the third argument) beginning at the desired X and Y coordinates (the first and second arguments). The third command uses the OPLOT procedure to mark each point on the graph with a diamond shape.

Finally let's try to plot an example of a sine and a cosine wave, using the following commands:

```
IDL> X = (6 * !PI) * FINDGEN (101)
IDL> PLOT, SIN (X / 100.0), TITLE="Sine and Cosine Plots"
IDL> OPLOT, COS (X / 100.0), LINESTYLE=2
```

Using the IDL function FINDGEN, the first command creates a 101-element, floating-point, one-dimensional array where each element is initialized to the value of its index, from 0.0 to 100.0. Then each element of this array is multiplied by 6π to form the array X containing 0, 6π, 12π, ..., 600π. Notice the use of the system variable !PI in the expression.

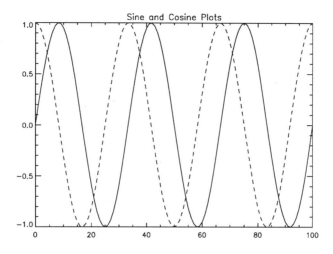

Figure 8.8: IDL plot of sine and cosine waves.

The second command plots the sine wave. The only argument of the `PLOT` procedure is the given array expression, `SIN (X/100.0)`; so the values of this expression are plotted against the indices of the array: 0 to 100. A title is included as part of the `PLOT` statement.

The third command draws the cosine curve over the' sine curve using a dashed line, since the `LINESTYLE` is set to 2. This plot is shown in figure 8.8.

There are many other important keywords associated with the plotting and image processing commands, and it is vital that you read the *IDL User's Guide* to gain familiarity with them. There are too many to describe here. The keywords allow you to specify ranges and orientations, titles and plotting symbols, and many other useful functions

8.3.2 Surface plotting

Two-dimensional arrays can be visualized in a three-dimensional fashion by using a variety of techniques, such as contour plots, wire mesh surfaces, and shaded surfaces. For this type of plot, we consider two-dimensional arrays such that each element of an array represents a z-coordinate of the grid. The element in the first row and first column of the array is the value of one corner grid point, and the element in the last row and last column is the value of the

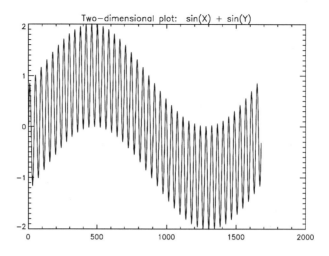

Figure 8.9: IDL normal X-Y plot of W = sin X + sin Y.

grid point in the opposite corner.

As an example, consider the following construction of the array W:

```
IDL> U = !PI*(FINDGEN(41)-1.0)/20.0
IDL> X = FLTARR(41,41)
IDL> Y = FLTARR(41,41)
IDL> FOR I=0,40   DO   X(I,*) = U
IDL> FOR I=0,40   DO   Y(*,I) = U
IDL> W = SIN(X) + SIN(Y)
```

The first command creates a vector U containing 41 elements; these elements are

$$[0, \pi/20, \pi/10, \pi/5, \ldots, 39\pi/20, 2\pi].$$

The array X has each row set to U, and Y has each column set to U. The final matrix W is the sum of the sine of the elements in the two arrays. If you type

```
IDL> PLOT, W
```

a plot like that in figure 8.9 appears. But this two-dimensional plot does not clearly provide the shape denoted by the elements of the array. Now type the command

```
IDL> SURFACE, W
```

A figure like that is figure 8.10 appears. This mesh surface provides the shape of the data contained in the array.

If the mesh plot is not clear enough, you can request that the surface be shaded. The following command does this:

```
IDL> SHADE_SURF, Z
```

This shaded surface is displayed in figure 8.11. This is especially nice on a color monitor but also provides shading on a black and white terminal. You can even alter the colors. The example below loads in a color table and plots the shaded surface.

```
IDL> LOADCT, 3
IDL> SHADE_SURF, Z
```

The procedure LOADCT from the User's Library loads one of sixteen predefined color tables, numbered 0 to 15. Color table 3 is called the *Red temperature* table. If you just type

```
IDL> LOADCT
```

with no parameters, a listing of the other color tables appears. Once this new color table has been loaded, the second command replots the shaded surface using the new colors.

Sometimes surface plots are not sufficient. Portions of the surface may be hidden. Hence contour plots can also be made. If you type the following command

```
IDL> CONTOUR, Z
```

a contour plot of the data, like that in figure 8.12, appears in your plot window.

A combination of a contour and a surface plot can be done with the User Library procedure SHOW3. This often makes the data clearer. Just type the command

```
IDL> SHOW3, Z
```

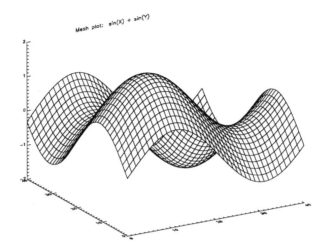

Figure 8.10:   IDL surface plot of $W = \sin X + \sin Y$.

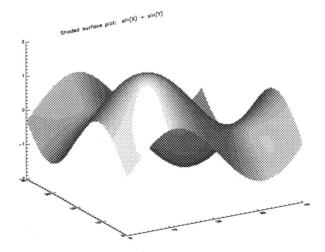

Figure 8.11:   IDL shaded surface plot of $W = \sin X + \sin Y$.

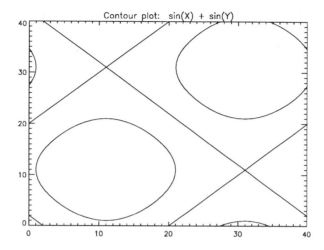

Figure 8.12:   IDL contour plot of $W = \sin X + \sin Y$.

and a plot similar to that in figure 8.13 appears. Notice that the mesh surface seems to be enclosed in a box. The top of this box shows the contour of the surface, while the bottom provides an image[4] of the data.

The orientation of these plots can be changed. Enter the command

```
IDL> SURFACE, Z, AX=50, AZ=25
```

The two keywords AX and AZ provide the rotation. AX is the number of degrees that the plot is to be rotated along the X-axis toward the front of the screen. AZ similarly provides the rotation of the z-axis in a counterclockwise direction. The result of this command should match the plot given in figure 8.14. Compare this plot with the one in figure 8.11. The AX and AZ keywords can be used with any of these plotting procedures.

8.3.3   Image processing

An *image* (also called a *raster* image) is a two-dimensional array of *pixels*. Each pixel represents the color and intensity of that particular point in the overall image. An eight-bit (one-byte) pixel allows 256 $(= 2^8)$ colors to be

---

[4]See section 8.3.3 for more information on images and image processing.

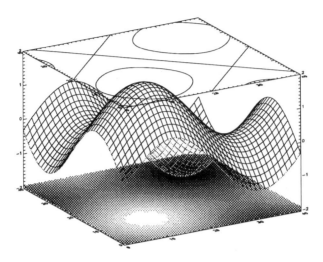

Figure 8.13:   IDL plot of W = sin X + sin Y, created with the SHOW3 procedure.

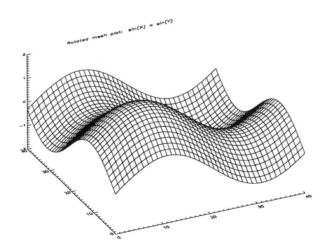

Figure 8.14:   IDL rotated surface plot of W = sin X + sin Y.

shown in the image. Color translation tables are used to assign colors for each possible pixel value.

For instance, type in the following commands:

```
IDL> IMG = BYTARR(512,512)
IDL> FOR I=0,511  DO  IMG(I,0:511)=I
IDL> TV, IMG
```

The first command declares the 512×512 array of bytes that holds image data. The next command is a `FOR` loop that sets all the values of each column of the array to the index of that column, 0 to 511. The `FOR` statement is discussed more fully in section 8.4.4. Notice that subsection subscripting is used for the array `IMG` in this statement. The last command uses the procedure `TV` to display the image array onto the given display device. You should see a vertically-striped image in the display window with each color used twice.

If you are using a color monitor and would like to run through the different available color tables, enter

```
IDL> LOADCT, 0
IDL> LOADCT, 1
IDL>  ...
IDL> LOADCT, 15
```

This series of commands redisplays the image for each of the sixteen color tables. The color table for 0 is the same as black and white. Some monitors require that the mouse cursor be within the image window for the color to appear.

Other image processing commands to perform scaling (`TVSCL`) or zooming in on an image (`TVZOOM`) are available. The *IDL User's Guide* discusses these and additional commands in more detail.

Often image processing requires access to image data. Specific examples of reading in and using data are shown in a later section.

8.3.4   Hardcopy output

The IDL default output graphics device is the plot window on the current `DISPLAY` environment. This window could be generated using the X-window system, Sun workstations, TEKtronix terminals, or Hewlett-Packard graphics. To check what device is currently in use, enter the command

```
IDL> PRINT, !D
```

The system variable !D is a structure that represents the current device. The first entry in the structure gives the code for the given device. Unlike other systems variables, !D cannot be changed by direct user assignment statements; however the SET_PLOT procedure can alter its value.

There are times when you would prefer to have hardcopy output of your plots instead of just seeing them in the graphics window. Plots and images can be written to *PostScript* files. This is how the figures used in this chapter were obtained. The SET_PLOT command can be used to tell IDL where to send the output. For example if you type the following commands:

```
IDL> SET_PLOT, 'PS'
IDL> PLOT, S
```

then the plot of S and any plots made following the execution of these commands are converted to a PostScript form, instead of appearing in the graphics window. This data is stored in a file named idl.ps upon exiting IDL. Note that the argument PS is a string and so should be enclosed within apostrophes or quotes. If you would just like the plot of S to be stored in a particular PostScript file, say plot.ps, then use the following commands:

```
IDL> SET_PLOT, "PS"
IDL> DEVICE, FILENAME = "plot.ps"
IDL> PLOT, S
IDL> DEVICE, /CLOSE
```

The DEVICE procedure handles the current graphics device that is set by the SET_PLOT procedure. So the second command above opens a PostScript file named plot.ps that is be used instead of the default idl.ps. Any plots made after the execution of this command are placed in this file. The fourth command closes the file plot.ps that now contains only the plot of S. Any more plots made after the fourth command are placed in the default file idl.ps.

You can print PostScript files on the printer using the following commands at the UNIX prompt.

```
% setenv PRINTER hpsc
% lpr -Dpostscript plot.ps
```

or just

```
% setenv PRINTER hpsc
% lpr plot.ps
```

The first command sets the `PRINTER` environment to the printer called `hpsc`. Be sure to replace the `hpsc` by the name of your PostScript printer. The second command sends the file `plot.ps` to the printer. The `-D` option tells what type of data is in the file to be printed. On most PostScript printers, this is the default and so would not need to be included.

Create a two-dimensional array. The first dimension should be net income; the second dimension should be the year. The array should be initialized so that it represents the net income of some imaginary millionaire from 1984 to 1995. Make up your own numbers; negative numbers are acceptable. Using this data, plot a graph with years labeled on the X-axis. Include titles for the top of the graph and both axes. Save the output to a file, and print out the graph on the printer.

8.4 Programming in IDL

So far all the examples have been interacting directly with IDL. As mentioned above, it is possible to write IDL programs; such programs all have the file extension `.pro`. In many ways, IDL programming is similar to Fortran programming. In this section, we first discuss some types of IDL statements. Then we present some sample programs.

8.4.1 Constants, variables, and expressions

Assignment statements, expressions, constants, variables, and arrays are used in IDL programs in the same way they are used interactively. So you already know a lot about IDL programming.

In addition, IDL allows `COMMON` blocks as in Fortran. For more information, see the *IDL User's Guide*.

8.4.2   Comments

One important part of any programming language is the ability to comment the code. A comment in IDL is any string that follows a semi-colon (;), and the semi-colon can be placed in any position on a line of code. If a comment takes more than one line, each line must be preceded by a semi-colon.

8.4.3   The end of the program

As in Fortran, all IDL programs must have an END statement as the last line of the code. Leaving out this END statement is a frequent error in writing IDL programs.

8.4.4   Control statements

Of course one of the purposes of writing a program is to do something a little more complex and repetitive than you would care to do interactively. And for this, you need control statements. IDL contains the control statements you expect: IF, CASE, FOR (similar to the Fortran DO), REPEAT, and WHILE statements.

IF statements

The IF statements behave the same as those in many other languages. The form of a simple IF-THEN statement

```
IF  (X GT 0.0)  THEN  Y = SQRT(X)
```

or a simple IF-THEN-ELSE statement

```
IF  (X GT 0.0)  THEN  Y = SQRT(X)  ELSE  Y = 0.0
```

should appear quite familiar. However when a branch of an IF is to include more than one statement, these statements must be enclosed in BEGIN-END blocks. The following lines provide an example of this:

```
IF  (X GT 0.0)  THEN  BEGIN
    Y = SQRT(X)
    PRINT, X, 'IS GREATER THAN ZERO'
ENDIF  ELSE  BEGIN
    Y = 0.0
    PRINT, X, 'IS NOT GREATER THAN ZERO'
ENDELSE
```

Each BEGIN-END block is considered as a single statement even though it includes many statements. The IF part of the ENDIF is optional as is the ELSE part of the ENDELSE, but including them assists in matching the limits of the blocks.

FOR statements

The FOR statement is used like a Fortran DO statement. The general form is

```
FOR  index = start,last[,incr]  DO  statement
```

Notice that the loop index is defined in the same manner as a Fortran DO loop index. The DO must follow the index definition; forgetting this is a common IDL programming error. An example of the simple FOR statement follows:

```
FOR  I = 0,N  DO  C(I) = FLOAT(I)/2.0
```

Another example was used in section 8.3.3. If the body of the loop contains more than one statement, it needs to be enclosed in a BEGIN-END block as shown in the following example:

```
FOR I=0,N,3   DO  BEGIN
    C(I) = FLOAT(I)/2.0
    PRINT, I, C(I)
ENDFOR
```

Again the FOR part of the ENDFOR is optional.

REPEAT statements

In addition to the FOR statement, IDL includes two other types of iterative
loop statements. The first of these is the REPEAT statement. A piece of code
including a simple REPEAT statement is shown here:

```
K=1
REPEAT  K=2*K  UNTIL  K GT 1000
```

This loop terminates when K is 1024 or 2^{10}. Again when the loop body contains
more than one statement, a BEGIN-END block must be used

```
K=8192
REPEAT  BEGIN
        Z(0:K-1) = ....
        K=K/2
ENDREPEAT  UNTIL  K EQ 1
```

and the REPEAT part of the ENDREPEAT statement is optional.

WHILE statements

The last iterative IDL statement is the WHILE statement. This example of the
WHILE statement

```
K=1
WHILE  K LT 1000  DO  K=2*K
```

has the same effect as the simple one illustrating the REPEAT statement in
section 8.4.4. WHILE statements can also be used to control reading input files.
(This is covered more fully in section 8.5.3.)

```
WHILE  NOT EOF(12)  DO  READ, X, Y, Z
```

Once more, BEGIN-END blocks are required to close loop bodies of more than
one statement

```
K=8192
WHILE  K GT 1  DO  BEGIN
    Z(0:K-1) = ....
    K=K/2
ENDWHILE
```

and the `WHILE` part of the `ENDWHILE` statement is optional.

It is important to notice that the syntax of the `WHILE` statement includes a `DO`, as does the `FOR` statement. Watch out for this common source of programming errors.

8.4.5 Procedures

IDL *procedures* are similar to Fortran subroutines. The first statement of a procedure must be a `PRO` statement:

PRO *yourproc, Param1, Param2, ..., ParamN*

and the last statement must be an `END` statement. The `PRO` statement provides the name of the procedure *yourproc* and the formal parameters *Param1* through *ParamN*. Keywords can also be used in defining parameters:

```
PRO  MYPRO, I, J, KEY=K
```

Calling `MYPRO` is done within your IDL program

```
MYPRO, III, KEY=L2, B1
```

in the same manner as calling other IDL procedures. In this call, the value of `III` is passed to `I`, `L2` is passed to `K`, and `B1` is passed to `J`. The end of the procedure or a `RETURN` statement returns control to the calling program.

Procedures should be placed near the beginning of a `*.pro` file, before the main program.

8.4.6 Functions

IDL *functions* are similar to Fortran functions. The first statement of a function must be a `FUNCTION` statement:

FUNCTION *yourftn, Param1, Param2, ..., ParamN*

and the last statement must be an END statement. The FUNCTION statement is used in the same manner as the PRO statement is for procedures. The RETURN statement returns control and a value to the calling program. Functions defined in your program may be used in any expression with the arguments contained within parentheses.

Functions should be placed near the beginning of a *.pro file, before the main program.

8.4.7   Compiling an IDL program

Once you have written an IDL program, you should compile and test it. This is done with the .RUN command:

IDL> .RUN *yourprog*

where *yourprog*.pro is the file containing your IDL program. Be sure to include the period before the word, RUN. Do *not* include the .pro extension portion of the file name containing the program.

This command first compiles the program, printing out messages if it finds any errors. When the program compiles successfully, it is automatically executed. Of course the execution may generate execution error messages.

Once a program has been compiled, the executable code is saved for the duration of the IDL session. To rerun the program without recompiling it, use the .GO command:

IDL> .GO *yourprog*

This executes the program from the existing executable file.

A number of other options are available for running and compiling programs. For instance, it is possible to obtain program listings. A program that has stopped because of an error can be restarted at that point. Programs can even be single-stepped for debugging purposes. Refer to the *IDL User's Guide* for more information.

8.4.8 Sample program: creating a test image

The following program displays a test image on the screen, using the TV procedure.

```
;  test image
WINDOW, XSIZE=1000, YSIZE=1000
A = BYTARR(1000,1000)
FOR  I=0,999,50  DO  A(I,0:999) = 255
FOR  I=0,999,50  DO  A(0:999,I) = 255
TV, A
END
```

The WINDOW procedure brings up a new display window of width, XSIZE, and height, YSIZE. The other commands in this program should be familiar to you.

A window can also be given an index from 0 to 9. Several display windows can be present at any time. The WDELETE procedure can be used to remove a display window.

Using your favorite editor, type in the above program and store it into a file called test.pro. (To get into UNIX from IDL, you either have to exit IDL or use the $ escape character, as described in section 8.5.1.) When you are back at the IDLprompt, type the command

```
IDL> .RUN TEST
```

to see the test image. To save a PostScript version of the image, use the SET_PLOT, 'PS' command and rerun TEST, as follows:

```
IDL> SET_PLOT, "PS"
IDL> DEVICE, FILENAME = "myimage.ps"
IDL> .RUN TEST
IDL> DEVICE, /CLOSE
```

This saves a black and white copy in the file named myimage.ps. If you want to save a color PostScript version, you need to use the /COLOR parameter on the first DEVICE statement.

8.5   Input/output

Since IDL can be used both as an interactive language and as a programming language, there exists a number of ways to read and write data files within IDL. There are also methods to save and restore information from interactive sessions.

8.5.1   UNIX commands within IDL

While in an IDL session, it is sometimes useful to perform a normal UNIX command without exiting IDL. This can be done using the dollar sign, $; this symbol acts as an *escape* command to IDL. For instance, typing

```
IDL> $ ls
```

prints out the current contents of your working directory.

8.5.2   Session log

Occasionally, it is useful to keep a record of what you are doing during your IDL session. This could be for later perusal or could be edited to form an IDL program. Suppose you type the following commands:

```
IDL> JOURNAL, 'mydiary.pro'
IDL> C = [3,7,-1,6]
IDL> PRINT, ALOG(C)
IDL> JOURNAL
```

The first JOURNAL command opens a file named mydiary.pro and begins to record the interactive session there. Notice that the filename, mydiary.pro, is a string and must be enclosed with apostrophes or quotes. If you had typed the JOURNAL command with no argument, the default output file would have been named idlsave.pro. The last command (and second JOURNAL command) closes the session log file.

Now print out the contents of the file named mydiary.pro. It should be similar to the following:

```
; IDL Version 2.2.2 (ultrix mipsel)
; Journal File for schauble@hartree
; Working directory: /tmp_mnt/nordsieck/nordsieck/....
; Date: Thu May  7 16:38:39 1992

C = [3,7,-1,6]
PRINT, ALOG(C)
;      1.09861      1.94591         NaN      1.79176
; % Program caused arithmetic error: Floating illeg....
```

The lines beginning with semi-colons are like comments. These include a heading for the session file as well as all of the responses IDL made while the session was being recorded.

In this example, the last line was an IDL error message, caused by trying to compute the logarithm of -1. The value printed for this element is NaN, meaning Not a Number.

The commands, SAVE and RESTORE, can be used to save the values of variables being used in an IDL session and then later restore them for use in a new session.

8.5.3 Input/Output statements

Data to be read and analyzed is often written to the data file by a different program in a different language. Yet, we need to be able to read these files. Similarly, we often want to produce output files that might be read by other programs.

IDL allows input from such files and produces output files. As in Fortran, each file is given a *Unit* number. Standard input is assumed by the READ procedure and standard output by the PRINT procedure, as you have seen in a number of examples in this document. These and other input/output commands are discussed in this section.

Standard I/O

Standard input, output, and error files are opened when IDL starts executing. Standard input usually refers to the terminal keyboard while standard output and error use the terminal screen.

Each of these is also assigned a logical unit number, one that does not conflict with user unit file numbers. Standard input may be referred to as unit number 0. Standard output is unit number -1 and the unit number for the standard error file is -2.

As mentioned above, the `PRINT` statement uses standard output. We have seen a number of examples of this command. The standard syntax is

>   PRINT, *List*

where *List* represents the list of variables, constants, or expressions whose values are to be written out to standard output.

In a similar fashion, the `READ` statement is used to read input data from standard input. The general form of this statement is

>   READ, *List*

where *List* is the list of variables to be read. Should any variable in the list be an array, the `READ` command expects all the elements of the array to receive input values. For example, when executing the following two statements,

```
R = FLTARR(4)
READ, R
```

IDL expects four input values, one for each element of R.

Opening and closing files

Before any file, other than standard input/output, can be used, it must be opened. In IDL, there are three ways to open a file. An explanation and general format for each of the possible three commands is given below:

- `OPENR`, *Unit, File[, Recordlength]* – This command `OPEN`s a file named *File* on unit number, *Unit*, for Reading. `OPENR` is used to open read-only files. This is usually used for input files.

- `OPENW`, *Unit, File[, Recordlength]* – This command `OPEN`s a new file named *File* on unit number, *Unit*, for Writing. This file can be used for both input and output.

- OPENU, *Unit, File[, Recordlength]* – This command OPENs an existing file named *File* on unit number, *Unit*, for Updating. This file can be used for both input and output.

For each of these commands, *Unit* is the logical unit number assigned to the given file; it can be any number between 1 and 128. *File* is the name of the file being opened; it is considered a string constant and should be enclosed in apostrophes or quotes.

Of course, once a file has been opened, it may be closed. For instance, you may write a file and then wish to close it and read it back in. There is only one command to close a file; this command identifies the file by its unit number.

        CLOSE, Unit

Reading files

There are four input commands. Three input procedures are similar to Fortran READ commands:

- READ, *List* – This command reads in values for the given *List* of variables from standard input.

- READF, *Unit, List* – This command reads in values for the given *List* of variables from the file on the given *Unit* number.

- READU, *Unit, List* – This command reads in values for the given *List* of variables from the file on the given *Unit* number. However, the input file is assumed to contain *unformatted* data; that is, data in a binary, not ASCII, format.

There is a way to include the equivalent of Fortran FORMAT statements in all three of these input procedures. See the *IDL User's Guide* for more information.

There is also an input function, ASSOC. Basically, this function ASSOCiates an array with a binary or unformatted file. This is an efficient way of retrieving stored image data. The ASSOC function completes the association between the array and the file:

> ASSOC (*Unit*, *typ*ARR(*Columns*,*Rows*))

The first parameter provides the *Unit* number; this number should have been assigned to the input file when it was opened, using an OPENR procedure. The second parameter describes the array to which the file is to be associated.

For example, consider the following code:

```
IDL> ONEIMAGE = BYTARR (400, 400)
IDL> OPENR, 10, 'IMAGES'
IDL> Q = ASSOC (10, BYTARR(400,400))
IDL> ONEIMAGE = Q(0)
```

The first command defines ONEIMAGE to be a 400×400 byte array. The second command opens a file named IMAGES on unit 10 for reading. The third instruction associates this file, IMAGES, with a byte array of the same size as ONEIMAGE. The last command assigns the first image (or set of data to fill the array) to ONEIMAGE. The following command

```
IDL> ONEIMAGE = Q(1)
```

assigns the second set of data or image to ONEIMAGE.

Writing files

It is possible to write data files using three different commands.

- PRINT, *List* – This command is the one you have been using throughout this document. It prints the values of the expressions contained in *List* to standard output; this is normally the terminal screen.

- PRINTF, *Unit*, *List* – This command writes the values of the expressions contained in *List* to the file on the given *Unit*.

- WRITEU, *Unit*, *List* – This command writes the values of the expressions contained in *List* to the file on the given *Unit*, in an unformatted or binary manner.

There is a way to include the equivalent of Fortran FORMAT statements in all three of these output commands. See the *IDL User's Guide* for more information.

8.5.4 Sample program: reading in an image

Type in the following section of code into a file called READ_IMAGE.PRO in the current directory. To do this, use a UNIX editor.

```
;    read_image
PRO  READ_IMAGE, FILENAME, NX, NY, A
OPENR, 1, FILENAME
A = BYTARR (NX, NY)
P = ASSOC (1, BYTARR(NX,NY))
A = P(0)
CLOSE, 1
RETURN
END
```

Notice that this is just a procedure definition; it is not a complete program. To include this code within a larger program, use the @ character:

```
@ READ_IMAGE
```

When IDL compiles your program and finds this line, it looks for a file named READ_IMAGE.PRO and include it in your program at this point.

Most of the code should be self explanatory. One special item to note is the use of the ASSOC function that associates a variable array with the data in a file; thus, when the variable is used in the program, the data is loaded dynamically. This lowers the overhead of reading data, especially for large-sized data files.

8.6 Using IDL efficiently

This section contains practical suggestions to use for your own data manipulation and processing.

- **Use array operators instead of loops.** The two statements below have the same result:

```
FOR j=0,511 DO FOR i=0,511 DO A(j,i) = 5
```

```
A = 5
```

- **Avoid IF statements.** The two statements below have the same result:

  ```
  FOR i=0,N DO IF B(i) GT 0 THEN A(i)=A(i)+B(i)
  ```

  ```
  A = A + (B GT 0) * B
  ```

- **Use IDL functions and procedures.** IDL provides a lot of useful functions; you should make sure that you know them. For example, the following segment of code could be replaced by a shorter segment, as shown.

  ```
  SUM = 0
  FOR I = 0,99 DO SUM = SUM + ARRAY(I)
  ```

  is the same as:

  ```
  SUM = TOTAL( ARRAY(0:99) )
  ```

- **Access large arrays correctly.** Since IDL stores arrays in column-major format, it is important to reference them in the same order. For example, in the following two lines of code, the latter is more efficient:

  ```
  FOR X=0,511 DO FOR Y=0,511 DO A(X,Y) = ...
  ```

  ```
  FOR X=0,511 DO FOR Y=0,511 DO A(Y,X) = ...
  ```

8.7 Summary

Here we have only begun to consider the many features and power of IDL.

Animation is possible by creating a set of frames or plots that can then be run quickly in sequence. Resetting the system variable, !P.MULTI allows more than one plot in a display window at a time.

A large library of routines exists for IDL and is constantly being enlarged. For instance, a map of the world resides within such a library. If you type the following command,

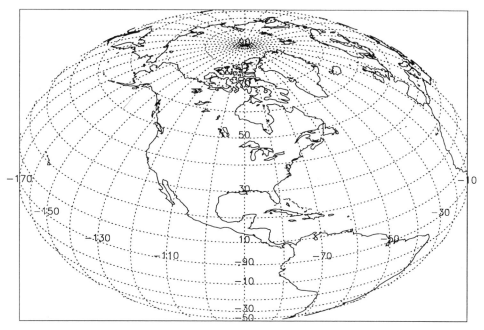

Figure 8.15:   IDL Orthographic projection of North America, as created with the IDL Map Projection package.

```
IDL> MAP_SET,40,-90,0,/ORTHOGRAPHIC,    $
IDL>            /GRID,/CONTINENT,/LABEL
```

the map shown in figure 8.15 appears on your screen. Other versions of this map are available as well, such as Mercator.

For more information on IDL mapping commands and other features of IDL, refer to the *IDL User's Guide* [RSI 90], *Introduction to IDL* [RSI 88], or *IDL Basics* [RSI 92].

9 Elements of AVS

AVS (**A**pplication **V**isualization **S**ystem) is a powerful graphical programming language used for scientific visualization. Several visualization solutions for large data sets are built into the AVS program. Additional solutions can be easily developed using a graphical interface to combine various functional modules. AVS provides a number of standard modules, and it is also possible to develop specialized application-specific modules in a conventional language such as Fortran or C.

This chapter serves as an introduction to using and programming AVS. The first part of the chapter discusses basic concepts, including how to set up and start the program. The second section goes through the steps of developing a graphical program using standard as well as specialized modules with the *Network Editor*. The third section describes different methods for displaying a geometric object using the *Geometry Viewer*; this is one of the types of Data Viewers used by AVS and is one of the most important subsystems of AVS. The last part of the chapter describes using prepackaged AVS applications. This chapter does not discuss how to develop your own custom modules.

9.1 Basic concepts

In this part of the chapter, we explore the basic concepts and requirements of AVS, including how to start the tool. This is meant to be a "hands-on" process; therefore, you should follow the instructions and work through the examples below. Explanations, instructions, and trouble-shooting information are given as required.

AVS is a *click-and-point* tool. Rather than entering specific commands, you must use your mouse to perform and control the operations you desire. Occasionally you may need to type in the pathname for a file or directory.

Version 3 of DEC AVS running on a DECstation 5000/240 was used to develop and test the examples for this chapter; it is distributed by Digital Equipment Corporation. There may be slight differences between this version and the one you are using. It is a good idea to keep a log of any such differences for later reference.

9.1.1   Prerequisites

Background

In order to successfully complete this chapter, you need to have some prior knowledge of the UNIX environment and the X-window system graphical interface. Specifically, knowledge of shell variables and experience with using a windowing system is very important.

Hardware requirements

AVS can only be run on a system that supports a color monitor (at least 8 bit-planes); therefore, you need to have access to such a machine. It is not necessary for AVS to be physically present on this machine; `rlogin` can be used to remotely execute the program. In this case, the *DISPLAY* environment variable needs to be set correctly, by using the command

`setenv DISPLAY` *hostname*`:0`

where *hostname* is the name of the machine that is used for display and user-interface I/O. If setting the *DISPLAY* variable is not permitted, you may first need to type

`xhost +`*remotemachinename*

where *remotemachinename* is the name of the remote machine with AVS installed on it. In some cases, you may need to do this before logging into the remote machine.

For the purpose of this chapter, we assume that AVS is present on a mounted filesystem on the workstation.

Modules and data files

This chapter uses some custom modules and data files that have been developed for the High-Performance Scientific Computing course.[1] Hence, you should also have access to a machine that has these files loaded in the correct directories. Generally, this machine is the same machine from which AVS is run.

9.1.2 Running AVS

Start AVS with the following command:

```
avs -swrender &
```

This initializes and starts up AVS, running it in the background.[2] After a few moments, the main menu window should appear on the screen as shown in figure 9.1.

If AVS does not seem to start up properly or the main menu window does not appear, check for the following problems:

- AVS is not in your execution path. Check with the lab assistant to find out what the path should be.

- The environment variable DISPLAY is not set correctly.

- The X-server does not allow the remote machine to open a window on your screen. Use xhost to grant correct authorization.

9.1.3 AVS Subsystems

AVS is functionally divided into a number of parts, each of which is a Data Viewer: the *Graph Viewer*, the *Geometry Viewer*, the *Image Viewer*, the *Volume Viewer*, and the *Network Editor*. Previously-defined AVS applications are

---

[1]These modules are available through the anonymous ftp site at cs.colorado.edu in the /pub/HPSC directory.

[2]The -swrender option is needed on most machines without special rendering hardware.

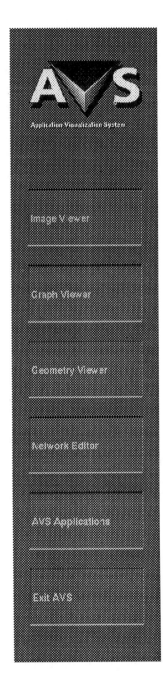

Figure 9.1: DEC AVS main menu window.

available through the *AVS Applications* subsystem as well. These subsystems are all listed in the main menu. Usually, only one is active at any given time.

The data viewers are subsystems of AVS that allow users to visualize their data quickly and easily as long as the data are in one of the various formats that AVS accepts. No programming need be done if one of these subsystems is used; just select a data file and, after the data are loaded, select a mechanism for its visualization. We discuss an important viewer, the *Geometry Viewer*, in section 9.3.[3]

While the data viewers offer a wealth of options for manipulating and displaying the data set, they are limited by the fact that the methods they use are fixed. The Network Editor allows much greater flexibility by providing a way to customize all aspects of the visualization process, from the format of the input data to the method of preprocessing the data to the method of displaying the data.

9.2 AVS graphical programming: the Network Editor

In this section, we describe the process of using and developing graphical programs in AVS. An AVS *network* is a graphical program; it is used for visually displaying data. Such programs are called networks because each is a set of interconnected functions or *modules*. You can create a network by using the mouse to connect the desired modules to input, rearrange, and then display your data. Use the Network Editor to select which modules to use and to interconnect them to specify the data flow.

The AVS main menu window allows you to select which subsystem to enter. To enter the Network Editor, move the mouse pointer to the block labeled `Network Editor`. When this block becomes highlighted, click the **LMB**.[4] Two new windows representing the Network Editor programming environment should appear.

---

[3]The first implementation of AVS consisted only of the Geometry Viewer and was only intended to be a demo.

[4]**LMB** is an acronym for Left Mouse Button. We also use the acronyms **MMB** and **RMB** to indicate the middle and right mouse buttons.

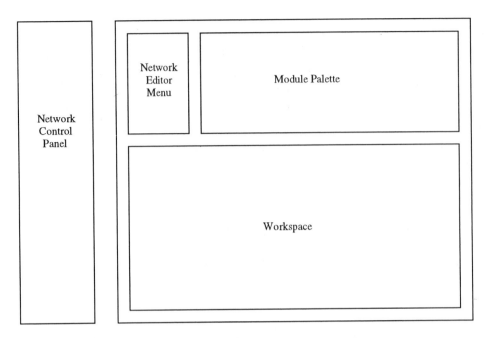

Figure 9.2: The format of the DEC AVS Network Editor windows.

The Network Editor display is divided into four subsections as shown in figure 9.2. A brief description is given below. Don't worry if everything is not immediately clear; more explanations are given as the chapter proceeds.

The AVS Network Editor screen is divided into two parts. On the lefthand side is a window providing the *Network Control Panel*; on the righthand side is the development window with tools for building and manipulating networks.

The *Network Control Panel* contains controls attached to activated modules; these controls can be whatever that particular module provides as part of its interaction with the user. The controls for each module are arranged to form a *Control Stack*. This is simply a method to prevent the user from being overwhelmed by seeing the controls for all the modules at once; only the controls for any one module are visible at any given time.

A *stack* in AVS is an implementation of a menu abstraction. Only one item (or submenu) in a stack can be selected at once, and the submenu items (in this case, controls) are visible below the stack. Initially, the module control stack is empty.

On the top lefthand side of the development window is another stack of menu options: *Network Tools*, *Module Tools*, etc. Initially, the *Network Tools* submenu is selected, and consequently the *Network Editor Menu* appears below the stack. This menu provides the user with a way to save and load network programs. The *Module Tools* submenu includes options for saving and loading individual modules.

Immediately to the right of this stack is the *Module Palette*, containing a list of available modules. These are divided into four types as explained in section 9.2.1. Because of limited space on the screen, not all of the available modules can be displayed. To view modules currently not displayed, click on the arrows at the top of each column with the **LMB** or the **RMB**. Clicking on the *up* arrow scrolls the column up, and clicking on the *down* arrow scrolls the column down. Note that the modules are ordered alphabetically in each column.

The *Workspace* is the bottom half of the development window. This is the actual area where networks (or AVS programs) are put together.

9.2.1 Modules

A module can be viewed as a *black box* that takes in some input and generates output. AVS comes installed with a diverse number of modules; these can be divided conceptually into the following types:

- *Data Input.* These modules read in or transform raw data into an internal format that AVS can understand.

- *Filters.* These modules convert the data to a slightly different form. For example, a filter module might sample and downsize the data, or it might filter out unwanted values.

- *Mappers.* These modules convert data into a given visual representation that can be directly displayed using the display modules available.

- *Data Output.* These modules can be used in a variety of ways to convert data representations into actual pictures or to write as output to some other medium (such as disk).

These subdivisions are not strictly defined, but serve more or less as guide-lines. Some modules might fall into more than one category, and others might not belong (strictly) to any given category.

Each module has a number of inputs, a number of outputs, and any number of controls. The inputs can either be from data files or from the outputs of other modules. Similarly, the outputs may be displayed on the screen or sent to data files or piped as input to other modules. The controls for each module allow specification by the user of exactly how the module should be used. For example, a data input module might have a file browser as a control, allowing the user to select an input file. This is the way a module interacts with the user; different modules usually have different numbers and types of controls.

A small square is embedded in the righthand side of the rectangular icon for each module. When you click on this square with the RMB, a window containing a description of the module appears. Try looking at some of these descriptions before you go on; most of the unfamiliar terms are explained further in this chapter.

9.2.2   Module Data Types

AVS has a number of specific data types that are used for the internal representation of the user's data. The input and output ports of modules can only read and display data of these types. The module ports are color-coded to represent specific types; a few of these are described below:

- **Red** = *Geometry*. A *geometry* is a mathematical description of an object. The object is defined in terms of lines, polygons, etc. Information about colors, lights, rendering modes, etc. can also be included.

- **Yellow** = *Colormap*. A *colormap* is a conversion table for converting integer index values to pixel (i.e., color) values.

- **Light Blue** = *Pixmap*. A *pixmap* is an image stored in terms of pixel values.

- **Orange** = *Unstructured Cell Data*. This is commonly used in finite element analysis. Our examples rarely use this data type.

- **Multicolor** = *Field*. This is a very flexible data type, capable of storing many kinds of scientific data. It is a generalization of an array structure. Each color specifies information such as dimensionality, type, vector length, etc.[5]

Observe the many modules already present in the Module Palette. The colored input ports are located on the top of each module, while the output ports are on the bottom.

9.2.3   Creating a Network Program

Writing a program with the Network Editor consists of picking a set of modules and connecting them. The following example demonstrates the method by which to do this.

In this example, we wish to animate a number of particles, given data files that describe a vector field (for example, a body of turbulent fluid) and initial positions of the particles in the field. Data files have been created from a simulation of water motion in a cubicle enclosure for this example. This visualization application is also discussed in chapter 10 on Scientific Visualization.

The custom module *advection* has been written specifically for this application. It is responsible for reading in the data files, calculating the positions at various time steps, and creating the frames to be displayed.[6] The actual display is done by a standard AVS module called *display image*.

Our network (or graphical program) looks something the one shown in figure 9.3.

Selecting Modules   First of all, we need to bring the modules into the Workspace. In the column labeled *Data Input* in the Module Palette, you

---

[5]The *image* data format, where individual red-green-blue values can be specified for each pixel, is technically a field.

[6]Both the module and the data files are available via the HPSC anonymous ftp site.

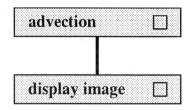

Figure 9.3: Simple Advection network.

should see a module labeled *advection*.[7] To copy this module to the Workspace, click on the module, hold the **LMB** down, and drag the module towards the Workspace. Once the module outline is in the Workspace, release the **LMB**. The controls associated with the module are now visible in the Network Control Panel. Notice that there are three controls: a dial labeled `Time step`, and two buttons labeled `Reset` and `Sleep`. The `Sleep` button is highlighted, indicating that the advection module is paused.

Now drag the display image module from the *Data Output* column in the Module Palette, until you have an image similar to that shown in figure 9.4. Note that a display window associated with the display image module appears on the screen. The window is currently blank.

Connecting Modules In order for data to pass from one module to the next, the modules must be connected. To connect them, move the mouse pointer to the multicolored bar on the advection module; click and hold the **MMB**. A line connecting the two modules appears. While holding down the **MMB**, move the mouse pointer to this line until it turns a different color. Now release the **MMB**. These two modules are now connected; this means that data output from advection flows into the display image module. This is confirmed by the fact that the display window now has a regular grid displayed on it; the yellow dots represent the particles in the vector field.

As mentioned in section 9.2.2, the colored bars on the modules indicate ports attached to the modules for the purpose of data input and output. The

---

[7]You might need to click on the up/down arrows on top of the column to bring the module name into view. Since there are usually more modules that can be displayed in the limited space of the Module Palette, the arrows are used to scroll up and down the module list; this list is always in alphabetical order.

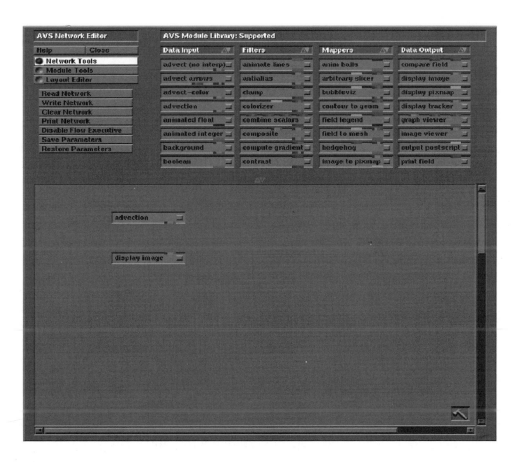

Figure 9.4: The DEC AVS Network Editor window with an unconnected network in the Workspace area.

multicolored bar at the bottom of the advection module means that it is an output port of type *image*, and the multicolored bar at the top of the display image module means that it is an input port of type image. In this manner, color-coding is used to indicate the type of data that can flow from or to a module. This is helpful not only in identifying the modules but also as an error-checking device; the Network Editor does not allow incompatible module ports to be connected.

Running Your Program   The program (or network) is running as soon as a module is placed on the Workspace, yet nothing appears to be happening. This is because the advection module is paused. The animation may be started by clicking the `Sleep` button in the Network Control Panel. If the controls for the advection module no longer appear in the Network Control Panel, click on `advection` in the *Top Level Stack*, and they should reappear.

Notice how the modules in the network are highlighted as they execute. You may move the display image window if it covers part of your network in the Workspace. One frame of this animated display is shown in figure 10.6 in chapter 10.

You may stop the animation by clicking the `Sleep` button again. You may also click on the `Reset` button to reset the grid.[8] The `Time step` dial may be used to speed up the animation; however, the larger the time step, the less accurate the model.

The controls for the display image module may also be used. For instance, the `monochrome` button causes the image to be only black and white.

Modifying a Program   One important feature of AVS is that the user can modify a network by disconnecting and removing modules. Here we modify the advection network by connecting the advection module to a different display module.

We would like to use a display module called *display pixmap* to provide the output. However, since the advection module outputs image data, we need to use a module that can convert the data to type *pixmap*. Therefore, we want a network that looks like the one shown in figure 9.5.

---

[8]It is advisable to first pause the animation, and then reset the grid.

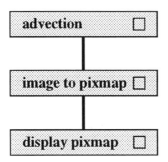

Figure 9.5: Alternate Advection network.

To build this network, first disconnect the display image module. The procedure for disconnecting modules is the same as for connecting modules, except that the **RMB** is used instead of the **MMB**. You may now delete this module by dragging it with the **LMB** to the *hammer* icon in the bottom-right corner of the Workspace. Deleting a module causes all the controls and windows associated with it to be deleted as well.[9]

Only the advection control should remain in the Network Control Panel. If the advection module is still flashing in the Workspace, it is still executing. You should click on `Sleep` and `Reset` before continuing.

Now drag the *image to pixmap* and display pixmap modules to the Workspace. The former can be found in the *Mapper* column of the Module Palette, while the latter is in the Data Output column. Notice that as each module is brought to the Workspace, the controls associated with it appear in the Network Control Panel. Recall that the controls for a particular module can be accessed by clicking on the appropriate menu item in the Top Level Stack in the Network Control Panel. Bring up the controls for the advection module.

Connect the modules as needed and rerun the animation by clicking on the `Sleep` button, as before. Notice that the image only takes up part of the display pixmap window.

Further experimentation   Experiment with the network by deleting the display pixmap module and connecting the output from the image to pixmap

---

[9]You should normally delete unused modules since they consume system resources.

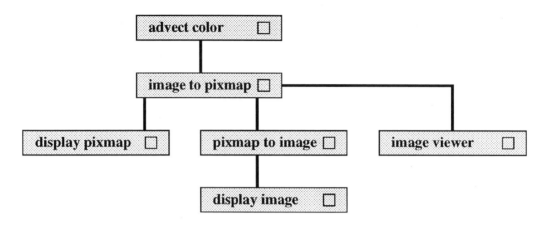

Figure 9.6: Advection network with three display windows.

module to a new module *pixmap to image*. Then you can use the display image module to view the network.

In fact, the advection module can be connected to more than one display window at the same time. Try the network shown in figure 9.6; this uses a different advection module, *advect_color*. The output port for this module is also of type image. This network opens three different display windows. You may need to rearrange them in order to watch all three at once. Click on **Reset** and **Sleep** to start them executing. Since three display windows are active, the execution may be noticeably slower.

This module colors the particles according to their velocity. Fast moving particles (across the top of the displays) are red; slower ones (in the middle and corners) are blue.

9.3   The Geometry Viewer

The *Geometry Viewer* is an important subsystem of AVS that can be used to manipulate objects of *geometry* type. The Geometry Viewer can be entered either from the Main Menu or from any other subsystem; for example, the Network Editor may request the Geometry Viewer by using the **Data Viewers** menu on the top of the Network Control Panel.

9.3.1 Viewing standard geometry objects

As an example, we use the Geometry Viewer to look at a geometric model provided with your version of AVS. If you are continuing from the previous example, exit the Network Editor by clicking on the Exit button on top of the Network Control Panel. This automatically deletes any display windows.

From the AVS main menu, select the Geometry Viewer. The *Control Panel* for the Geometry Viewer is shown in figure 9.7. This submenu contains a number of controls. From the Control Panel, click on the button labeled Read Object; this action brings up a file browser that you can use to select an appropriate file. Pick any data file in this directory with the file extension .geom. After the file is selected, a new display window appears with the object shown inside it.

Transforming Objects. Objects can be rotated, scaled and translated using the mouse.[10] To select the object to be transformed, just click on it using the **LMB**.

Move the mouse pointer to the display window and while holding down the **MMB**, drag the pointer slowly to the left. You should be able to observe the object rotate along one of the coordinate axes. The exact rotation depends on the position of the mouse as well as the direction of the rotation; this is analogous to using a **SpaceBall** device. The object behaves as if it were attached to a *virtual* sphere whose center is at the center of the display window; the mouse control moves that part of the sphere *protruding* out of the display window. Experiment with rotating the object until you get a viewpoint that you are satisfied with. Don't worry if the object rotates out of the window; you can press the Normalize button on the Geometry Viewer Control Panel to bring it back into view.

Similarly, holding down the **RMB** and dragging the mouse pointer translates (or moves) the object in the direction specified.

Finally, the object can be scaled to be larger or smaller. To do this, hold down the SHIFT key and the **MMB** together while dragging the mouse pointer

---

[10]It is not necessary to enter the Geometry Viewer if one only needs to transform the objects; the transformation can be done using the mouse in any window that is being displayed by the Geometry Viewer or the *render geometry* module.

Figure 9.7: DEC AVS Geometry Viewer Control Panel.

in the display window.

You may choose to display more than one object. Just click on a second data file in the *File Browser*. The second object is placed on top of the first one.

More techniques The Geometry Viewer allows detailed control over aspects of visualization such as lighting, cameras, textures, etc. One interesting effect is that of perspective; this makes objects farther from the observer appear smaller.

To switch on the perspective mode, first click on the `Cameras` menu selection in the middle of the Geometry Viewer Control Panel. This brings up a submenu of choices. Click on the `Perspective` option. You should see a difference in how the object is being displayed. The color bars at the bottom of the *Cameras Control Panel* may be used to alter the background color of the display window.

You may change the color of the object by editing its properties with the Geometry Viewer controls. Use the `Edit Property` control, available under the `Objects` stack. The translucency of the object is controlled by the `Trans` slider in the same window.

Also, change the color of the light in the lighting model for the scene using the `Lights` stack. Notice how the light and intrinsic color interact to produce a new shade.

9.3.2 Using the Geometry Viewer within the Network Editor

The Geometry Viewer can also be applied to displays generated by programs running in the Network Editor.

Exit the Geometry Viewer and return to the AVS main menu. Now re-enter the Network Editor and create the *read molecule* network shown in figure 9.8. The *read molecule* module is another custom module developed for this course. It displays a configuration of a number of atoms forming a molecule.

After you have connected the network, click on the `Read Molecule` button in the Top Level Stack to bring up a file browser window. From this window, choose a molecule data file from the

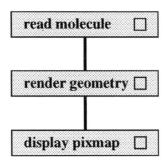

Figure 9.8: Read Molecule network.

/usr/avs/local/data/molecules/set1

directory. The number of atoms contained in each data file is indicated by the name.

At the top of the Network Control Panel, there is a button labelled Data Viewers. This allows access to the data viewers (the Geometry Viewer, the Image Viewer, or the Graph Viewer) from within the Network Editor. Click on this button and a submenu of the three data viewers is displayed. Pick the Geometry Viewer.

Using the same techniques you learned for the Geometry Viewer with your chosen object in section 9.3.1, alter the appearance of the display. Hitting the Normalize button causes the molecule to fill the display window. You may rotate the molecule to find a better viewpoint. You may alter the background color and lighting. You may change the colors of the particles by using the Edit Property. The Subdivisions slider bar at the bottom of the Geometry Viewer panel modifies the shape of the particles. It is well worth your while to experiment with these controls to see how they affect the display.

To leave the Geometry Viewer and return to the Network Editor, just click on the Close button at the top of the Geometry Viewer Control Panel.

9.4   AVS applications

It is possible to make an existing AVS network into an AVS Application. This has been done with the advection network. You may wish to experiment with this network as an AVS application, as well as the other applications currently installed with your version of AVS.

Click on the `Exit` button at the top of the Network Control Panel to leave the Network Editor. This returns you to the AVS main menu. Now click on the `AVS Applications` button. The *AVS Applications Control Panel* appears, listing several options.

Clicking on `Advection` brings up an advection display window and the Network Control Panel. The development window containing the Network Editor, Workspace, and Module Palette is not shown as this is an application that has already been developed. You may start, stop, and reset the advection animation as before.

Clicking on `Exit` returns you to the AVS Applications window. Try some of the other applications.

9.5   Further reading

For more information, read chapters 1 and 2 of the *DEC AVS User's Guide* [DEC 90b]. These chapters explain the basic concepts of AVS in further detail. They also describe the various data types available in AVS.

The Network Editor is fully described in chapter 6 of the *DEC AVS User's Guide*, and a detailed description of the Geometry Viewer is available in chapter 5.

You might also want to glance through the *DEC AVS Developer's Manual* [DEC 90a] to get an idea of the various modules that are available, along with their inputs, outputs, and associated controls.

III Scientific Visualization

10 Scientific Visualization

10.1 Definitions and goals of scientific visualization

Visualization of Scientific Data describes the application of graphical methods to enhance interpretation and meaning of scientific data. *Visualization of Scientific Data* is abbreviated to *Scientific Visualization* or *Visualization* throughout this chapter. Scientific data can be derived from various sources, including measuring instruments, or may be obtained as a result of scientific computations performed on supercomputers. However, data do not become useful until some (or all) of the information they carry is extracted. The goal of scientific visualization is to provide concepts, methods and tools to create expressive and effective visual representations from scientific data. Such visual representations convey new insights and an improved understanding of physical processes, mathematical concepts and other quantifiable phenomena expressed in the data [Pang 93]. Together with quantitative analysis of data, such as offered by statistical analysis, image and signal processing, visualization attempts to explore ALL information inherent in scientific data in the most effective way. Therefore, scientific visualization is expected to enhance and increase scientific productivity.

Concepts and tools of scientific visualization are based on other disciplines: psychology/perception and human factors offer a scientific basis to understand human visual performance, its abilities and limitations. Experts in computer graphics provide algorithms and tools to transfer numerical data values into pictures. Artists and graphic designers offer their knowledge of aesthetics and other design issues to increase interpretability of visual representations. Scientists define their needs to explore scientific data and thus drive the quest for visual exploration. Scientific visualization provides concepts, methods and tools from existing disciplines to best use human abilities and computer algorithms for the display of scientific data.

On the other hand, there is a fine difference between the goals of visualization and goals of other disciplines or subdisciplines. While psychology and perception are important for understanding abilities and limitations of a scientist viewing a picture, basic principles of perception theories and the awareness of visual illusions, such as the Hermann grid or the Müller-Lyer illusion [Sekuler & Blake 85], do not fully explain the complex visual information present in a three-dimensional vector field visualization .

While image processing exclusively deals with images, it uses a limited number of visual representations (gray values or color pixel displays; shaded surfaces) to visually express the result of numerical algorithms. The chosen visual representations are not of essence to image processing, but rather the development of the underlying techniques, such as filters, geometric corrections, or image compression.

Computer vision is concerned with the computerized extraction of information from images [Boyle & Thomas 88]. It is therefore not of concern to Computer Vision how the human viewer extracts information from a picture, but rather how the computer may accomplish a similar task and initiate a certain action dependent on the result.

The field of computer graphics provides tools to design pictures from symbolic or numeric descriptions and to interact with these pictures [Hill 90]. Computer graphics is concerned with the development of algorithms (and their efficiency) to create pictures on a computer display. While computer graphics works hand in hand with visualization, it is not concerned with pictures on displays once their appearance is satisfactory. The extraction of meaning from the picture in the human mind is not of concern to this field.

User interface issues have developed in parallel, but separately, from computer graphics [Olsen 92]. The wide availability of bitmapped graphics gave access to new visual appearances of user interfaces that have cumulated in the widespread use of windows and widgets. However, user interface design is not applying its methods to the understanding of processes and data, but rather to the ease-of-use of programs. In a similar way, human factors take into account the problems humans encounter when working with machines, not the output from these machines.

In having reviewed a series of areas that add to the understanding of visualization, we can conclude that scientific visualization cannot be replaced by existing disciplines, but it offers more than the sum of knowledge derived from these separate disciplines. It has therefore become a discipline of its own.

The aim of this chapter is to discuss meaningful visual representations of scientific data, by providing a framework of underlying concepts in visualization. Even though the examples are focused on scientific data, concepts and techniques discussed here apply to other types of data as well, such as engineering data, financial data, computer software and hardware performance data.

10.2 History of scientific visualization

In the late 1980's, data rates increased sharply both from measuring devices (such as data collecting missions in space or medical instruments) and as a result of computations on fast computers (such as supercomputers). In addition to the increase in the number of measuring devices, their resolution was multipled. National supercomputers allowed access to virtually any scientist in the United States for large calculations. Computers and measuring devices add gigabytes (10^9) of data to the already existing amounts of data on a daily basis. Knowing that the trend to increase available data volumes will only continue, there is clearly a need to search for more efficient ways of dealing with large amounts of numbers.

Computer graphics received a boost in the mid 80's through the development of improved and faster graphics hardware. New raster graphics techniques replaced the previous technology of limited, slower vector graphics. Combined with powerful and affordable processors, personal graphics workstations emerged. The saying *A picture is worth a thousand words*, surpassed its promise: one picture might express several megabytes of data values.

In 1986, the National Science Foundation sponsored an advisory panel on *Graphics, Image Processing and Workstations* to make recommendations in response to the needs developed by high data rates and the opportunity of using the new generation of graphics workstations. The widely published report produced by the panel [McCormick et al 87] called for new tools in a new field termed *Visualization in Scientific Computing*, or short *Scientific*

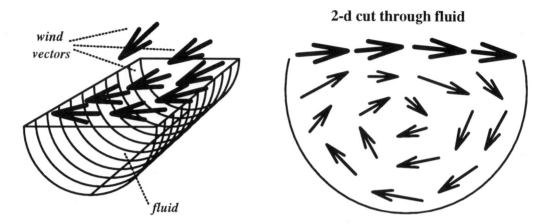

Figure 10.1: Wind velocity vectors over the surface of a fluid body cause particles in the fluid to move.

Visualization. Since 1987, a multitude of new applications have confirmed the necessity and power of this new methodology.

10.3   Example of scientific visualization

The following example shows various graphical ways to effectively visualize vector fields and the movement of particles.

Wind velocity vectors over the surface of a fluid body cause movement inside the fluid. We can derive the velocity of the fluid at control points. In order to observe the movement of the fluid better, we cut one two-dimensional (2D) slice through the three-dimensional (3D) body of fluid (figure 10.1).

Examples for effective graphical displays of the movement in the fluid along the 2D cut include:

- plotting vectors (using length, color, shape of arrows to distinguish magnitudes) to show the velocity of the fluid.

- animate the actual motion of particles suspended in the fluid.

An observer should be able to visually derive answers from questions such as:

- Is the movement symmetric?

- Where are the strongest/weakest movements for time step $t = t_k$?

- Where are the strongest/weakest movements over time?

- How do the particles distribute over time?

Different graphical representations answer different questions that an observer might have.

The following figures were created to display the movement of the fluid using features of three different visualization tools: IDL (**I**nteractive **D**ata **L**anguage; AVS (**A**pplication **V**isualization **S**ystem); and MATLAB. Figure 10.2 shows lines depicting information about strong and weak movements for one time step. Long rods indicate strong movements in the area surrounding the center, but not in the center or at the border. Figure 10.3 provides arrows signifying the direction of the velocities; the lengths of the arrows denote the magnitude of the velocities. Figure 10.4 is a combination of the two previous figures: the arrows denote the direction of the velocity; and both the size of the arrowheads and the length of the arrows indicate the magnitude of the velocity. Figure 10.5 shows only arrowheads: while the direction of velocity vectors is indicated by the arrowhead, color depicts their magnitude (short vectors received blue colors, medium vectors increased from green to red, and long vectors received orange and yellow colors). This visual representation clearly enhances the symmetry of the data set. Figure 10.6 is one frame of an animation of the data; it shows the positions of the particles after a certain time.

10.4 Concepts of scientific visualization

10.4.1 Mapping *numbers to pictures*

Scientific Visualization is essentially a mapping process from one domain (a real phenomena) into another domain (numbers) into yet another domain (pictures) and further into a fourth domain (the subjective interpretation of the viewer) as shown in figure 10.7 [Domik & Gutkauf 94].

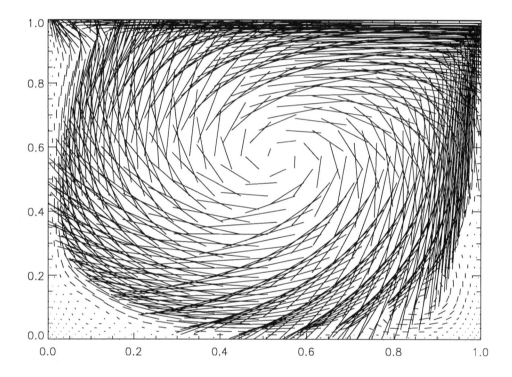

Figure 10.2: Lines depict information about strong and weak movements for one time step. Long rods indicate strong movements in the area surrounding the center, but not in the center. (This figure was created by Salim Alam using IDL.)

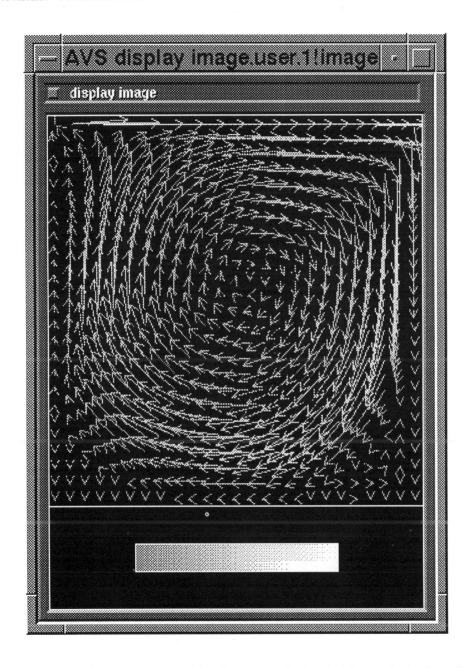

Figure 10.3: Arrows indicate the direction of the velocity and the length of the arrows denote the magnitude of the velocity. A color version of this figure may be found on color plate 2. (This figure was created by Paul Pinkney using AVS.)

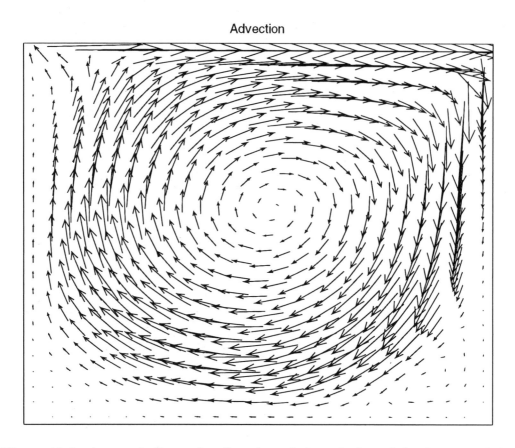

Figure 10.4: Arrows indicate the direction of the velocity while the size of the arrowheads and the length of the arrows indicate the magnitude of the velocity. (This figure was created by Anna Szczyrba using the `quiver` function in MATLAB.)

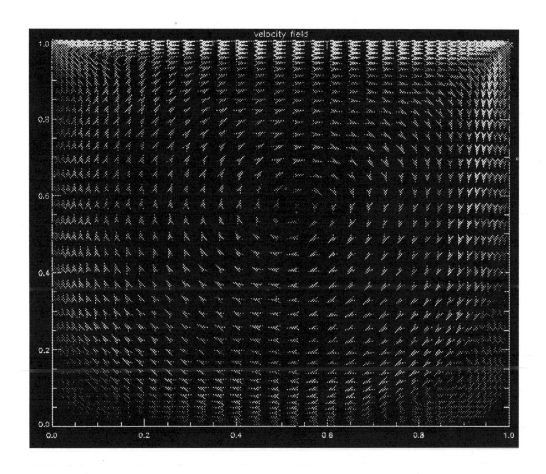

Figure 10.5: Velocity is shown by arrowheads: direction of velocity vectors is indicated by arrowhead; color depicts magnitude of vector (blue, green, red, yellow indicate growing vectors). This visual representation clearly enhances the symmetry of the data set. A color version of this figure may be found on color plate 4. (This figure was created by Wolfgang Schildbach using IDL.)

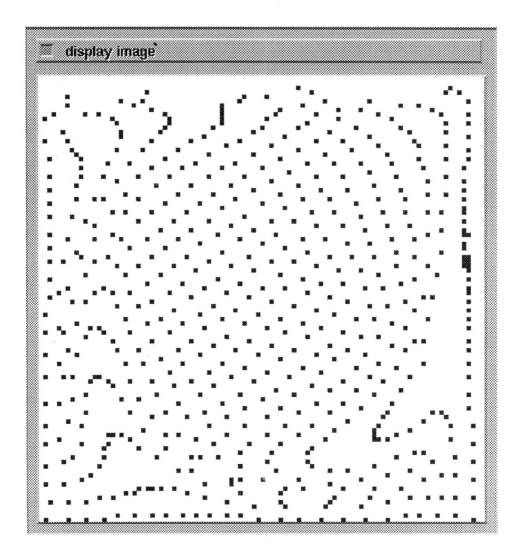

Figure 10.6: Distribution of particles after a certain time. To observe the changes in their positions, an animation is most effective. (This figure was created by Salim Alam using AVS.)

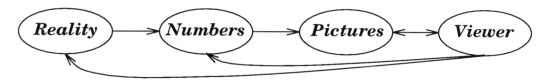

Figure 10.7: Visualization mapping process.

A scientist looking at a picture uses the picture as a *vehicle of thinking* [McKim 80], but intents to interpret the meaning of the *numbers* (or the real phenomena itself) expressed in the picture. The picture activates mental processes such as the perception of spatial relationships, the discovery of patterns or anomalies in large data sets, or the intuitive comprehension of complex processes. These mental processes are obviously different from the ones activated when interpreting numbers without the help of pictures.

In the following sections, we suggest a strategy to create expressive and effective pictures to ensure correct and meaningful interpretation by the viewer.

10.4.2 Expressiveness

The process of data interpretation becomes one step removed from the actual data themselves. If the mapping of *numbers to pictures* is not performed carefully, pictures might not express the true meaning of the underlying numbers, and therefore lead to misinterpretation of scientific facts. Examples of a non-intentional artifact might be a color coding of data values that produce abruptly changing hues from continuously increasing numbers. A picture is called *expressive* [Mackinlay 86] if it expresses the characteristics of the underlying data values, nothing more or less.

Expressiveness is strongly influenced by the structure of data (section 10.6) and their type of data values.

10.4.3 Effectiveness

If the mapping process is not performed purposefully, the resulting images might not be effective for the interpretation aims the scientist has in mind and therefore they are not useful. The visual representation of a two-dimensional data set in form of isolines (contour lines) is very effective when identifying local maxima, but very ineffective when trying to locate south-facing slopes. Effectiveness is strongly influenced by the interpretation goals of the scientist (section 10.10).

10.4.4   Subjectivity of interpreter

Visual cues are graphical elements of a picture that we visually separate out as a single entity, e.g., shape/form, a line and its orientation, or color [Keller & Keller 92]. Interpretation of visual cues can be subjective. The meaning of visual cues depends on culture, education, experience, and individual abilities and disabilities of the viewer. For instance, various interpretations of an underlying order of hue (hue does NOT have an inherent order in human perception) are strongly influenced by education. The sequence of hues

green - yellow - red - white

is interpreted as a sequence of increasing numerical values for most geographers and geologists. In order to reach the same conclusion, an astronomer would expect a color coding of

red - yellow - white.

The first sequence of hues relates to the color coding in a map (green meadows in flat areas, less growth at higher elevations as yellow, red rocks, white snow) to which geographers and geologists have adapted; the second sequence relates to the brightness of stars in a telescope (red for faint stars, yellow for brighter stars, white for strongest brightness), something astronomers can better relate to when interpreting visual cues.

10.4.5   A picture is the summary of visual cues

If the numerical data to be visualized are very simple, e.g., a list of numbers, mapping the numbers to one type of visual cue might suffice. For instance, the numbers 13, 2, 15, 17, 29, 10 can be expressed as a histogram (using bars – positions, length – as visual cues) as demonstrated in figure 10.8.

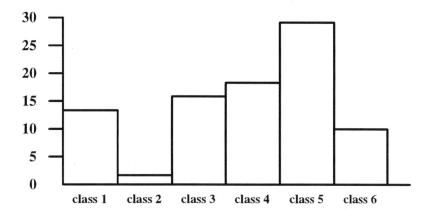

Figure 10.8: Histogram showing distribution of 6 classes.

In most cases, numerical data to be visualized might consist of complex data structures and many parameters. For example, data for aerodynamic research might consist of pressure (scalar values at 3D location), deformation (vectors distributed in 3D space), and shape (e.g., airplane). To visualize all available data, various visual cues need to be used. Each individual parameter needs to be mapped onto one or more such cues (e.g., shapes of arrows in 3D space; color; shaded rendering of airplane): the resulting picture is thus *a summary of visual cues*. In order to further the understanding of the relationship between parameters, the mapping should result in a *coherent* picture, where the picture could also be described as one entity.

10.5 Visual cues

Visual cues are elements of a picture. We can produce visual cues with the aid of our computer graphics tools. Perceptual counterparts of visual cues are called *perceptual elements*, and relate to our perception of visual cues. Examples of visual cues/perceptual elements are below:

1. spatial: position, motion

2. shape: length, depth, area, volume, thickness

3. orientation: angle, slope

4. density

5. color: color (hue, brightness, saturation), contrast

Complex visual cues are combinations of simple cues, e.g., a realistic picture of a natural scene is composed of objects of various shapes, sizes and colors at various positions.

10.5.1   Innate reactions to visual cues

Some visual cues are natural for us to interpret: if we increase the brightness on a series of objects, we interpret a natural ordering of the information from *low to high*, or *less to more*, or similar. The hue *blue* makes objects/information appear farther away, cooler, and lower than the hue *red*, that seems to relate to objects/information that are nearer, warmer, and higher. In part, the difference of our perception of blue vs. red is influenced by biological facts, as you could read in chapter 2 of the Lab Manual (Computer Graphics and Visualization).[1]

10.5.2   Acquired reactions to visual cues

Interpretation of other visual representations are acquired through education, e.g., the interpretation of street signs, international travel signs, isolines or iso-surfaces. Once we learn the meaning of these representations, this knowledge usually stays with us. The use of natural visual cues is advantageous, because it reduces the danger of misinterpretation. However, natural visual cues are usually too simple to use for the representation of complex information contents. Acquired visual cues are often powerful, but yet simple to interpret, once the knowledge to do so has been acquired.

10.5.3   Illusory visual cues

Some visual representations are known to fool the viewer: e.g., the illusory triangle, the Hermann grid, equal brightness steps, or simultaneous contrast.

---

[1]The Lab Manual chapters for the HPSC course are available via anonymous ftp at the `cs.colorado.edu` site in the `/pub/HPSC` directory.

Look into chapter 2 of the Lab Manual to review some of these illusions. In order to avoid the danger of illusions, one must beware of their occurrence.

10.6 Characterization of scientific data

Scientific data can be classified according to their *semantics*, e.g., data representing

1. temperature dependent on location and time;

2. the DNA structure of a bacterium;

3. a digital elevation map;

4. surface of a space shuttle

usually carry different visual appearances. In order to effectively represent scientific data, knowledge about their semantics is important.

Data values can also be classified by data type as *nominal*, *ordinal* and *quantitative* [Mackinlay 86]. Nominal values describe members of a certain class, e.g., Iron, Magnesium, Calcium, Copper, Zinc]; no ordering can be imposed on this class, e.g., Iron is not larger or higher or earlier than Calcium. Visual cues that naturally relate to nominal data values are hue and position.

Ordinal values are related to each other by a sense of order: low density in growth; medium density in growth; high density in growth. Visual cues used to express ordinal values should depict this order, such as density, brightness, position, or size. If color is used, a color bar must be present.

Quantitative values carry a precise numerical value; scalar fields are often expressions of quantitative values, such as a three-dimensional MRI (Magnetic Resonance Imaging) data set. Even though color is often used to visualize quantitative data sets, this is in general a very imprecise visualization of the underlying values. The value in displaying quantitative information via color is that this transforms it to ordinal information (color bar or equivalent explanation must be present) and lets us quickly pick out low, medium or high values.

Visual representation of data is also strongly influenced by the *syntax* of the underlying data. The scientific data above may be syntactically represented by different dimensions and data structures, such as:

1. three-dimensional grid containing a floating-point number at each grid point for a series of time intervals: $y_t = f_t(x_1, x_2, x_3)$;

2. positions in 3D space, each position carrying information of its coordinate values and a category descriptor (type of molecule): $y_t = (x_1^i, x_2^i, x_3^i, x_4^i)$;

3. a two-dimensional, even-gridded, array of integer values: $y = f(x_1, x_2)$.

We need to classify the underlying structure of a data set in an easily comparable format. The following denotations of data sets and basic visualization techniques of section 10.7 are based on [Brodlie et al 92].

Data sets D are described as $D_n^{mC}(d)$, where

n describes the dimensions of the data set(s)

m describes the number of data sets defined over the same dimensions or the length of a tuple in a point data set

C describes the category of data (P for single points, S for scalar data, V for vectors, T for tensors)

d provides, if necessary, a more detailed description of C. In the case of V or T, d defines the length of vector or size of matrix, respectively; in the case of P or S, q (quantitative), o (ordinal), or n (nominal), or a combination of these data types combined by '+', can be specified.

Valid data sets are, for example, D_2^S (two-dimensional scalar data set), or $D_2^{3S}(2q + n)$ (three data sets defined over the same two dimensions; two data sets contain quantitative values, one contains nominal values); $D_3^{V_3}$ (one vector field in three-dimensional space with each vector of length three); $D^{7P}(5q + o + n)$ (point data set with each point described by seven values, five of which are of quantitative type, one ordinal and one nominal).

10.6.1 Points

A set of points $[P_1, P_2, P_3, \ldots]$ is denoted as D^{mP}. An example is $D^P = [1, 5, 8, 33.4, 15]$. The sets D^{2P} (e.g., $[(1,2), (3,4), (5,5), 4.4, 8)]$) and D^{3P} (e.g., $[(1,2,3), (4,5,6), (4,5,6.8,0), (6,0,4.5)]$) are pairs and triplets of numbers, respectively. Expressive visualizations for point data sets are scatter plots ($m \leq 3$) and glyphs (see section 10.7.2). In the case of position information, as in molecular dynamics, an obvious visual presentation of the information is a map of suitable objects (e.g., spheres) to the indicated positions.

10.6.2 Scalars

Scalars are usually samples of a continuous function. In this chapter, we are assuming that we sample a function in equidistant steps in each dimension in order to receive a discrete data set. In reality, sampling is often done in random order. In such a case, we assume that data can be resampled in equidistant steps. Data sets D based on continuous functions can be characterized by functions f of independent (x) and dependent (y) variables:

$$y_i = f_i(X), \quad where \quad X = (x_1, x_2, x_3, \ldots, x_n); i = [1, \ldots, m]$$

The so represented functions can take on various shapes: $y = f(x_1, x_2)$ is a two-dimensional, scalar function, and the data set it produces is denoted by D_2^S; in this chapter we usually work with D_1^S, D_2^S and D_3^S.

In the one-dimensional (1D) case, D_1^S, data are sampled from a one-dimensional function, $y = f(x)$. Positions of the data values are therefore determined from the variable x. Typical representations are line drawings, scatter plots, or histograms.

D_2^S is data sampled from a two-dimensional function, $y = f(x_1, x_2)$. A typical example is a digital topographic map (= digital elevation model), where x_1, x_2 denote the sampling locations (usually an even two-dimensional grid) and y represents the elevation at this location. Examples of visual representations in this category are contour lines, wire frame models, shaded surfaces, and images. The latter category, images, denotes the visual representation of each y value as one pixel on the screen, either in gray shades or color.

Several scalars can be available at the same location, e.g., $y_1 = f_1(x_1, x_2)$ and $y_2 = f_2(x_1, x_2)$, where y_1 denotes height and y_2 denotes density of growth.

The resulting data set may be defined as D_2^{2S} $(D_2^{2S}(q+o))$ in our notation. A valid representation of this entity is a two-dimensional wire frame plot to indicate y_1 by surface height and the use of color (presence of color bar necessary to display ordinal data!) to indicate y_2 within the same picture. This not only allows the representation of both variables at the same time, it also enhances the understanding of the relationship between y_1 and y_2: if lowest elevations have the densest growth and highest elevations have the least dense growth, this becomes obvious in the picture.

Often a series of data measurements are conducted over one area. This leads to data sets defined as D_2^{mS}. Visual cues that can be superimposed in a meaningful way (such as hue and brightness; color and shaded view; image and contour lines) are limited. With larger m, glyphs become very useful to indicate each individual scalar but also relate all scalars belonging to the same location in the two-dimensional space to each other.

Data sets of type $D^{(m+2)P}$ can be converted to data sets of type D_2^{mS} (and vice versa). D_2^{mS} can be treated as a special case of $D^{(m+2)P}$ in that two values of the $(m+2)$ tuple are treated as spatial positions.

D_2^{mS} can also be seen as a three-dimensional scalar field (D_3^S), with m overlapping 2D slices. Possible visualizations for this entity therefore also include overlapping (with or without regard to hidden lines/surfaces) of visualizations fitting for D_2^S.

Similar to D_2^S, data that are sampled from a three-dimensional function, $y = f(x_1, x_2, x_3)$, is denoted as D_3^S. Tools to visualize a regular three-dimensional lattice of points have recently been developed in computer graphics. There are basically two approaches to finding a visual representation of three-dimensional points:

1. if the importance is on clusters of data points, the view of the surface of these clusters might suffice (e.g., isosurfaces); or

2. if the importance is on the visual representation of every single data point, the volume directly needs to be displayed (e.g., ray tracing of volumes). Furthermore, subsets in the form of two-dimensional slices of the volume can be displayed as suggested for D_2^S, or as suggested for D_2^{mS}, if more than one slice is to be displayed simultaneously.

Note the similarity of D_3^{mS} and D^{mP} as noted for the two-dimensional case above.

10.6.3 Vectors

The expression $y = f(x_1, x_2, x_3)$ might describe a function of vectors, with a resulting data set of $D_3^{V_2}$, if each independent variable is of length two. Vectors need to be defined by the dimension of each vector as well as by the dimensionality of the data set containing vectors: $D_2^{V_3}$ defines a set of three-dimensional vectors in a two-dimensional plane.

Possible displays for $D_2^{V_2}$ are arrows, indicating length and direction of each two-dimensional vector.

Similar to the scalar cases, $D_2^{mV_2}$ indicates the availability of m vectors, each of dimension 2, at any 2D position of interest. Animations are expressive visualizations for this category, usually expressing a time series of vector data information.

Three-dimensional vector visualization, $D_3^{V_3}$ or $D_3^{mV_3}$, uses 3D shapes of arrows, particle traces and streamlines to display available data. Again, animation can be used to express time or other information.

10.6.4 Tensor fields

An n-dimensional tensor field is denoted as $D_n^{T_k}$, with k describing the order of the tensor. One example of a symmetric, second order tensor display is given in [Haber & McNabb 90]: first, the tensor's principal directions and magnitudes are calculated. A cylindrical shaft is oriented along the major principal direction; the color and length of the shaft indicate the sign and magnitude of the stress in this direction. An ellipse wraps around the central portion of the shaft and its axes correspond to the middle and minor principal directions of the stress tensor. The color distribution of the disk indicates the stress magnitude in each direction.

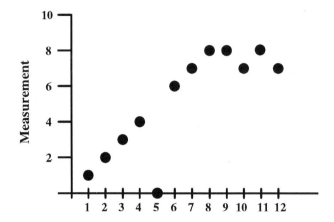

Figure 10.9: One-dimensional scatter plot using position along the y-axis to visualize data values.

10.7   Visualization techniques

10.7.1   Scatter plots

Scatter plots use position as their primary visual cue. A one-dimensional scatter plot uses positions along one axis to indicate values of D^P, e.g.,

$$D^P \;=\; [P_1 = 1;\; P_2 = 2;\; P_3 = 3;\; P_4 = 4;\; P_5 = 0;\; P_6 = 6;$$
$$P_7 = 7;\; P_8 = 8;\; P_9 = 8;\; P_{10} = 7;\; P_{11} = 8;\; P_{12} = 7]$$

might be measurements taken independently.

The scatter plot in figure 10.9 uses a linear mapping between data values and y-position; often a logarithmic scale is being used. For any non-linear mapping (and, if appropriate, also for linear mappings), the mapping function [data value → position on plot] should be annotated.

A two-dimensional scatter plot relates pairs of points in a two-dimensional coordinate grid. A data set ($D^{2P}(2q)$) might be given as

$$[(3,3),(3,4),(4,4),(5,4),(5,5),(6,4),(6,5),(7,5),(7,8),(8,8),(9,9),(12,8)]$$

and represent in a linear manner the relationship of measurements of 12 randomly picked leaves (figure 10.10).

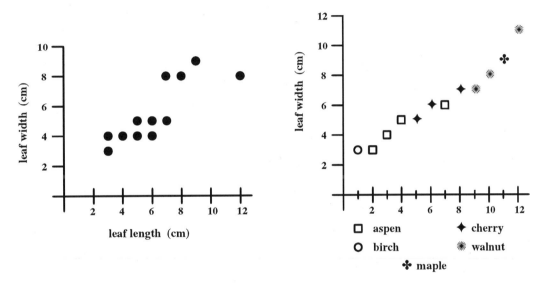

Figure 10.10: (left). Two-dimensional scatter plot to express $D^{2P}(2q)$.

Figure 10.11: (right). One-dimensional scatter plot with symbols to express $D^{2P}(q + n)$.

A similar data set, $D^{2P}(q + n)$, is expressed in figure 10.11. It depicts leaf width (q) and leaf type (n). For nominal information, symbols or color can be very effective to indicate an additional dimension. For ordinal or quantitative information, size or orientation might be the more appropriate indicator. The influence of data type on effective visual cue is discussed in more detail in section 10.6.

If measurements of leaf length, width, and thickness of leaves are available, we receive triples of numbers instead of pairs: $D^{3P}(3q)$. Three-dimensional scatter plots are ambiguous on two-dimensional screens. However, if the view point is mobile (e.g., the view is constantly rotating around the scattered points), a good three-dimensional impression of the *point cloud* can be obtained. If points on the screen are represented by larger objects, *depth-cueing* (loss of reflection with distance from viewer), perspective viewing (decrease of object size with distance from viewer) is recommended. Other possibilities are the use of stereo equipment or even virtual reality environments to observe the location of each point in 3D space.

10.7.2   Glyphs

Sometimes several variables or constants are used to describe each dimension in a data set. Such data sets are of form D^{mP} (e.g., auto parts; computer hardware or software performance measurements; census data), or D_2^{mS} (e.g., velocities of particles in 2D space at m time steps; absorption of light on 2D surface for m different wavelengths) or D_3^{mS}.

Glyphs (called *textons* in perception research) have been invented to specifically express such complex data sets. One glyph is composed of individual parts or segments. Each part or segment is seen as one visual cue and relates to one data value. Data values that should appear in the same spatial position on the screen are mapped into individual parts of the glyph. One glyph is usually identified by the viewer as **one** figure (relating to the sum of all parts of the glyph). In our example in figure 10.12 we look at 9 data items, each consisting of a quadruple of data values. The data set is therefore of form D^{4P}, and appropriate visualizations might include color and symbols combined with a two-dimensional scatter plot. The glyph we are using for a visual representation of the data sets is composed of four individual parts: a

Visual entities

Numerical data

$P_1 = [2, 1, 0, 1]$

$P_2 = [1, 1, 0, 1]$

$P_3 = [0, 0, 1, 2]$

$P_4 = [2, 0, 2, 2]$

$D^{4P}(3\,o + n) =$ $P_5 = [1, 1, 0, 0]$

$P_6 = [2, 1, 0, 2]$

$P_7 = [0, 0, 1, 2]$

$P_8 = [2, 0, 0, 2]$

$P_9 = [2, 1, 1, 1]$

	2	1	0
m = 1	◯	○	○
m = 2	│	│	│
m = 3	◯	◯	◯
m = 4	◡	—	◠

Visualization of data

Figure 10.12: The use of glyphs to depict a data set of type D_m^P.

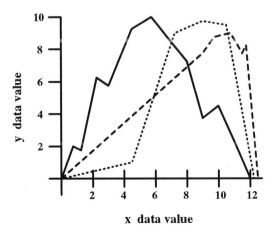

Figure 10.13: Three line graphs distinguished by line style.

pair of circles (large, medium or small, depending on the first data value in the quadruple); one vertical line (long, medium or short, depending on the second data value); brightness (white, medium gray or black, depending on the third data value) and curve (concave, straight or convex, depending on the fourth data value). These four visual entities form a face that can be recognized as a new single entity as a smiling, sad, mean or unintelligent face. The search for similar quadruples is now reduced to matching similar faces.

Faces have shown to be powerful glyphs for small data sets [Chernoff 73]. Glyphs for larger data sets are suggested by [Picket & Grinstein 88] and by [Beddow 90].

Looking beyond visualization, one can also use senses other than our visual sense to express data values. Smith et al [Smith et al 90] and Scaletti and Craig [Scaletti & Craig 91] use sound as an additional visual cue.

10.7.3   Line graphs

Line graphs are used to display continuous information and are therefore an effective visual representation of scalar data sets of form D_1^S and of form D_1^{mS}. For D_1^{mS}, m lines–differing by color, style or thickness – can be displayed on the same output media (see figure 10.13).

10.7.4   Histograms, pie charts

A histogram looks like a discrete line graph (see figure 10.8). The area of each rectangle (length of each histogram bar) is of special meaning to the observer, usually relating to the number of occurrences of an event. The location of each histogram bar usually relates to a certain category or class of information (data). The total area of all rectangles also carries meaning in a histogram, namely to describe the number of all occurrences in all categories.

If information is compared to some total number, then a constant comparison between each data value and the whole data set should be possible, such as in figure 10.14.

For this type of visual representation, pie charts as depicted in figure 10.15 (circles to depict the *whole*) are typically in use.

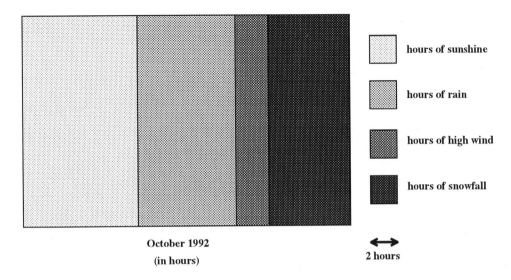

Figure 10.14: Relating information to the total sum.

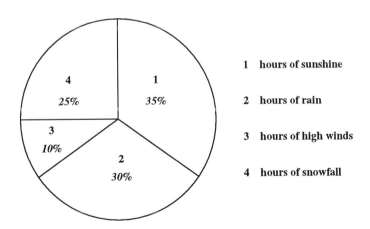

Figure 10.15: Relating information to the total sum: pie chart.

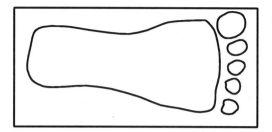

**Appropriate threshold value to
show outline of foot and toes in
infrared image.**

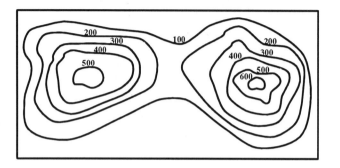

**Lines connecting equal
heights in topographic
map. The closer the
lines, the steeper the
mountain in that area.**

Figure 10.16: Contour lines (isolines).

10.7.5   Contour plot (isolines)

For functions $y = f(x_1, x_2)$ that produce data sets of type D_2^S, contour plots
can be used to show lines of constant values (= threshold values) inside the
two-dimensional data set: $y = c$. The full data set can be visually expressed by
contour lines $y = [c_1, c_2, c_3, ..., c_n]$, where distances between threshold values
are usually equidistant (figure 10.16). Contour lines are also called isolines
because they connect points of equal values.

10.7.6   Image display

Image displays are frequently chosen as visual representations of quantitative
data of type D_2^S. For raster devices, image displays are a straightforward
mapping of each data point along a two-dimensional grid into a pixel (= picture
element) on the screen. To indicate the data value at the pixel location, gray
levels or color can be chosen. If color is used, simultaneous use of a color bar
(see section 10.8 on *Annotations*) is mandatory. Figure 10.17 displays data of

a digital elevation model of the area surrounding Boulder, Colorado as a color image.

A combination of contour lines and images are often used to enhance interpretability of data. Redundancy of visual techniques produce, in general, desirable results. Figure 10.18 shows a digital elevation model of the same area around Boulder overlaid with isolines.

10.7.7 Surface view

Wire frame drawing is a fast drawing method to depict surfaces of data sets of type D_2^S. The numeric values of the data set are treated as elevations, and visually depicted as a *terrain*, showing peaks for local maxima and valleys for local minima in the data set (figure 10.19). The surface of the terrain is drawn as if it were made of wire; hidden surfaces are often not eliminated to add efficiency to the drawing algorithm. Projection of the three-dimensional surface onto the two-dimensional screen allows a natural interpretation of the data values, similar to standing on a mountain and observing the surrounding terrain.

In a similar but more time-consuming algorithm, the surface can be shaded by an artificial light source to lend more realism to the display (figure 10.20). In this case hidden surfaces are always removed, and only front views are visible. Shading does not necessarily involve color: the amount of light reflected is a function of light source, viewpoint and surface characteristics, and may be expressed by the brightness of a pixel on the screen.

In all surface views, but specifically if data are noisy or characterized by high-frequency data values, animation of the scene supports the interpretation. Moving the view point around the three-dimensional scene gives visibility to otherwise hidden parts of the surface view. Surface views are recommended for smooth data sets. In noisy data sets, visibility can be strongly diminished.

10.7.8 Color transformations

Data sets of type D_2^{3S} can be displayed as one single color image, if each single data set D_2^S is mapped into one *color dimension*. Color dimensions are independent (orthogonal) characteristics of color models, e.g., red, green,

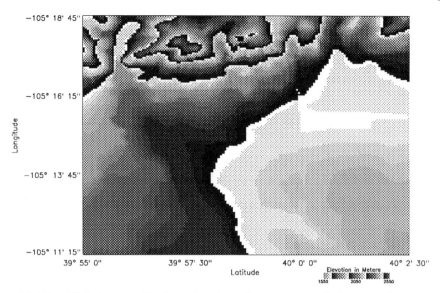

Figure 10.17: This image display showing the contour of the terrain around Boulder, Colorado was created with IDL, using data derived from the `denver-w` and `greeley-w` files available at the USGS/EROS Data Center anonymous FTP server for the United States Geological Survey. A color version of this figure may be found on color plate 3.

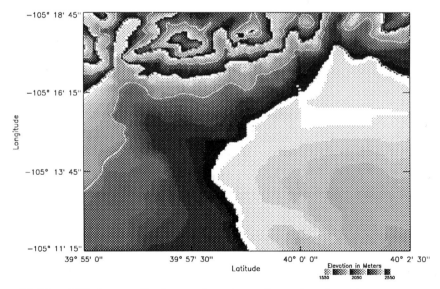

Figure 10.18: IDL image display and contour lines showing the area around Boulder, using the same data as in figure 10.17.

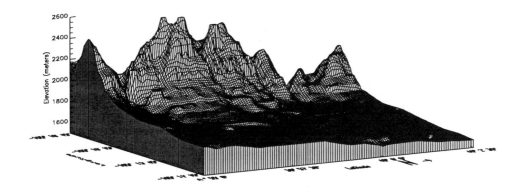

Figure 10.19: Wire frame surface of the area surrounding Boulder, Colorado. This was created with IDL with the same data set used in figures 10.17 and 10.18.

Figure 10.20: Shaded surface view of area around Boulder, Colorado created with IDL from the same data used in figures 10.17, 10.18, and 10.19. A color version of this figure may be found on color plate 3.

blue in the RGB color cube, or hue, saturation, value in the HSV model (e.g., [Foley et al 90] or chapter 2 of the Lab Manual). For an example, look at the RGB color transformation in figures 10.21 and 10.22 below. Figure 10.21 shows four separate images (astrophysical data collected at different wavelengths), from which three were chosen to be superimposed on the screen: the lower right image was displayed on the red phosphors; the lower left image on the green phosphors; and the upper left image on the blue phosphors of a workstation monitor. The resulting image in figure 10.22 displays values from the color gamut on the available output device.

An interpretation of color transformations is fairly simple, if the underlying color model is well understood. In our case, yellow objects indicate high data values in the *red image* and in the *green image* at the corresponding spatial location; dark blue objects indicate low values in the *red image* and *green image* but high data values in the *blue image*.

10.7.9   Isosurfaces

In data sets of type D_3^S or similar, we assume that data are available in a regular lattice in 3D. When using isosurfaces to visually represent volumetric data, the assumption is given that traceable *objects* are hidden inside the 3D volume. Similar to isolines, surfaces of constant values are identified and illuminated, shaded and projected onto the two-dimensional screen. Figure 10.23 shows one isosurface of an MRI (Magnetic Resonance Imaging) data set.

10.7.10   Ray tracing of volumes

If data values inside a volumetric data set, D_3^S, are not expected to form solid objects, a view of the whole volume is usually preferred. This is done by assigning opacity and color to each volume element (= voxel) and displaying all of the elements inside a volume. The projection from three dimensions to two dimensions is done via ray casting (ray tracing) from a point-of-view through the volume. Each voxel touched by the same ray adds to the resulting color of its representative image point. Each ray results in one image pixel to be displayed on a two-dimensional screen. Ray-traced images of volumes (see figure 10.24) can assume a transparent, cloud-like appearance (translucency).

Figure 10.21: Astrophysical data from NASA's Infrared Astronomy Satellite (IRAS) mission of 1983 varying only in their spectral response.

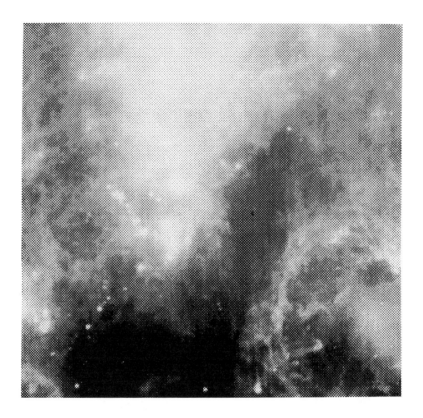

Figure 10.22: RGB color transformation of the three IRAS data sets of figure 10.21. A color version of this figure may be found on color plate 4.

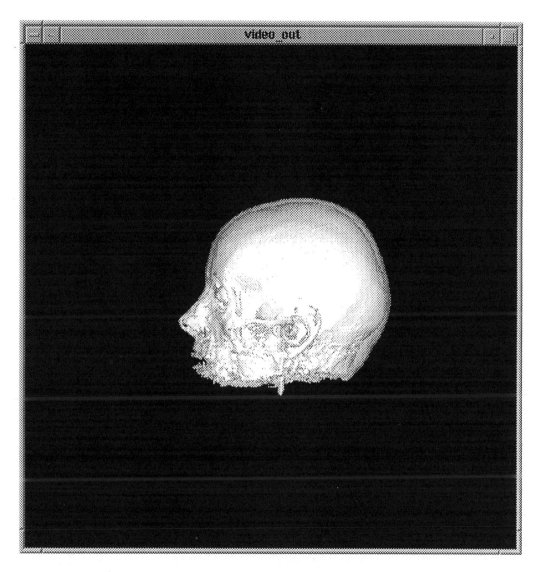

Figure 10.23: Rendered isosurface of MRI data created by IDL. Data for this figure may be obtained via anonymous ftp at `omicron.cs.unc.edu`, provided through the courtesy of Siemens Medical Systems, Inc., Iselin, NJ.

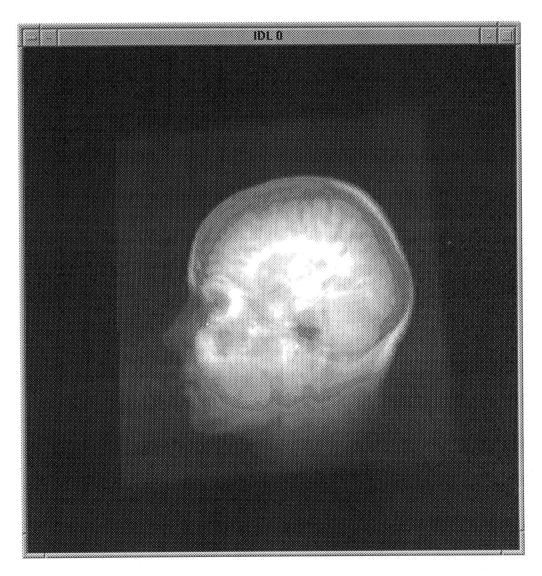

Figure 10.24:   Volume visualization (*translucent display*) from the same data sets as depicted in figure 10.23. (This figure was created by Michael Kreutner using IDL.)

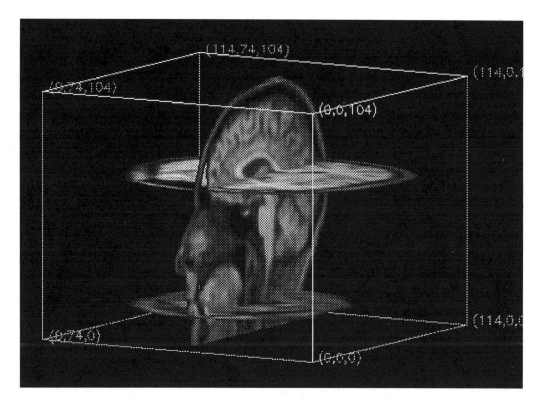

Figure 10.25: Use of IDL's data slicer. Three slices show subsets of a D_3^S data set. A color version of this figure may be found on color plate 2.

10.7.11 Data slicers

Any data sets of higher order can be explored by observing subsets of the data. This is specifically meaningful with volumetric data, where a view of a two-dimensional slice through the three-dimensional volume provides a view of the inside of the data set. Figure 10.25 shows three slices through a data set of type D_3^S.

10.7.12 Arrows

Data sets of type $D_2^{V_2}$ can be displayed as arrows in a two-dimensional plane. An arrow is a symbol that can, at the least, display dual information: one type of information (e.g., magnitude) is expressed by the length of the arrow; and one type of information (e.g., direction) is expressed by the slope of the arrow. Other visual cues can express additional information, e.g.,

- starting location of arrow

- thickness of shaft

- color of arrow.

The danger in this type of graphical display is that some of the vectors might get enormously long and interfere with vectors in the surroundings. (For example, see figure 10.2.) If scaling of the vectors is used, so that there is sufficient distance between neighboring arrows, some of the arrows might become too small for visual interpretation. Alternative displays of vectors, such as shown in figure 10.5, may be very useful.

It might be of advantage to combine a vector display with a scalar visualization, by extracting one scalar information from the vector field, e.g., magnitude, and superimposing the arrows with an appropriate visual representation of the scalar field, e.g., image display.

Visual representation of data sets of type $D_2^{V_3}$ are an extension of the above explained technique. Arrows are rendered as three-dimensional shapes that point into or out of the display surface. Similarly, $D_3^{V_3}$ is visualized by displaying shapes of three-dimensional arrows inside a three-dimensional volume. For a more realistic display, depth-cueing (reflection loss with distance) should be used.

10.7.13   Streamlines and particle tracks

Streamlines or particle tracks trace a direction, e.g., the direction of a flow computed for an application in computational fluid dynamics (CFD). This method can be seen as an extension of arrows, in that arrows are added onto each other and create polylines (= a set of contiguous straight line segments approximating a curve). This method is useful to depict vector data over a certain time. This method can be used in two as well as three dimensions.

10.7.14   Animation

Animation can become a visual cue as well. Because of its analogy to nature (things change over time), time is often expressed in form of an animation. An example might be the distribution of carbon monoxide in Denver: the data set of type D_3^S, showing a snapshot of the distribution at 5 am, is depicted as a translucent volume. The distribution of each following hour is now displayed in consecutive frames, using animation to move from one picture to the next. Animation shows a continuous sequence of visual representations, allowing the viewer to observe changes between pictures. Changes between pictures might contain as much information for the scientist as the pictures themselves.

10.8   Annotations

A visualization might be unreadable because certain information is omitted. If color is used, a *color scale* (usually a bar relating color and corresponding data values) must be present. Furthermore, the scaling of world coordinates (the coordinate system used to describe the problem) to screen coordinates must be documented by a *scale bar* if there is any relevance in this information. If spatial dimensions are expressed on the screen, *orientation* must be presented to the viewer: e.g., a North arrow, or an indication of the vertical and horizontal directions. The use of animation to express a time series (or spectral series) of data must be annotated by a *time indicator* (or *spectral indicator*) or any other appropriate indicator specifying the relationship of picture and data variable. Too much clutter on the screen distracts from essential information, but omitting explanatory notes and cues might make visual representation useless.

10.9   Interactivity

Interactivity describes actions the user is able to perform on graphical representations. Interactivity enhances the value of visualizations by letting the user explore data and steer visualization processes. Examples of interactions are

- using the mouse to click on objects on the screen to request (often numerical) information about these objects;

- changing parameters of data sets to control physical phenomena;

- moving/rotating objects in three-dimensional space (e.g., three-dimensional scatter plot) to lend a 3D effect to two-dimensional screens.

10.10   Interpretation goals to pursue with visualization

Besides syntax and semantics of data sets to be visualized, the interpretation goal of the scientist should influence the choice of visual representation. The interpretation goal defines the task a scientist has set out to do with the help of visualization: e.g., identify local maxima in the data set; observe the behavior of one variable (y_1) in relationship to another one (y_2); observe symmetry in the data set; etc. [Wehrend & Lewis 90].

While a contour plot is an effective visual tool to identify areas of a certain threshold value, an image display of the same data set is ineffective for the same purpose. In general, a surface plot is more effective in classifying slopes in a data set than an image display. However, an image display has no hidden surfaces and can thus depict even noisy data. Clever use of color tables can enhance image displays for various tasks, e.g., quickly locating values of specific characteristics.

It is impossible to list all interpretation aims and give hints about effective visual representations, because options of interpretation aims are too numerous. However, the *visualizer* must carefully consider the use of the visual representation BEFORE time is spent in encoding the data.

10.11   Quantitative versus qualitative data interpretation

Clearly, the picture does not always tell us the whole story about underlying information. Quantitative data interpretation gives us precision that is often necessary to prove/disprove a theory. A mean or median value, or maximum/minimum numbers in a large array may characterize the data set sufficiently and more accurately than any pictorial information. On the other hand, pictures are supplementary to numerical information in that they convey information contained in the data, but not obvious when interpreting numbers. Approximate symmetry in a large data set might be hard to derive from numbers alone, but can be obvious when looking at a visual representation. Likewise, similarities, patterns, or sudden deviations might be effectively represented in the picture, but hardly noticeable in the numbers.

For most scientific purposes, a combination of pictures and numbers is needed to take advantage of all information that can be extracted from a data set. This fact bears a strong message on writing software for visualization purposes: do not lose the numbers while creating pictures! In other words, if a scientist clicks on various molecules on his/her screen in order to receive quantitative information about the molecules, the information sought for is not the shading value or reflection constant of the spheres representing the molecules, but rather information about the chemical character of the molecule.

IV Architectures

11 Computer Performance

11.1 Introduction and background

Today's computers include personal computers, workstations, mainframes, and supercomputers. Some perform well for word-processing; some have excellent graphical capabilities; and some are designed for efficient number-crunching. Given an application, how does one decide which computer to use or to purchase? How can one determine the efficiency of a program running on a certain architecture? How can one judge the performance of a computer?

The major reason for measuring the performance of a computing system is for predicting its performance on some computations you wish to do. But why do you want to predict performance? Perhaps you are interested in purchasing a system. Before investing the money, you would be wise to study the performance of the system. Of course vendors usually supply you with plenty of performance data, but it is best if you supplement this information with your own observations. Not only are you less likely to be biased, but if you understand how to do these kinds of measurements, you are likely to come up with a more accurate measure *for the way in which you intend to use the system.* The latter point is important because the speed of high-performance computing systems tends to be very dependent on the nature of the jobs run on them. The system may perform very well on one class of problems and be a complete bust on another.

Another reason why you would want to predict performance is in striving for efficiency in a computation. From measurements of arithmetic speeds on simple problems, you can get some idea of the expected performance on the problem at hand. If the actual performance on your problem falls short of what you expect, then you want to understand why, and you probably take steps to improve the performance. Not only does this save you money, it is likely to expose mistakes that you made, lead you to a better algorithm, and in the end, allow you to do more work better.

Predicting computer performance is difficult. It is more of an art than a science, and here we want to begin learning that art. The reason why it is difficult is that many factors beyond the basic operation speeds influence the performance. One can measure the time it takes to add, multiply, and divide numbers — though even this is rather tricky — but using these measurements to predict the performance of a major computation like simulating the flow of fluid in a pipe is almost impossible. At best you can predict a bound on the performance, that is, the best performance you can hope to achieve. What you actually achieve is another matter. To understand why this is so and to learn how to deal with it, you need to understand the many factors that can influence performance. This chapter provides a starting point for this.

In section 11.2 of this chapter, we define computer performance and present methods for measuring CPU speed and computer performance. In section 11.3, we cover benchmarking, including a discussion of several of the commonly used benchmarks. In section 11.4, we explain the effect of compiler optimizations on performance results, while in section 11.5 we talk about the effect of some architectural factors on performance. Finally in section 11.6, we introduce additional performance issues related to vector and parallel computing. In particular this chapter is directed toward the study of computer performance as related to high-performance computers in a UNIX environment.

11.2   Computer performance

By *computer performance* we mean the effectiveness of the operation of a computer, usually on a given application or benchmark. This notion of performance includes speed, cost, and efficiency.

A number of factors affect the performance of a computer:

- CPU execution time for

    — register operations

    — integer operations

    — floating-point operations

    — string operations

- Memory access time to read and write data

  — in cache

  — in main memory

  — in auxiliary memory

- Memory size

- Systems and file service

- Compiler

- Input/Output

- Communication, especially with regard to parallel computers

In this chapter, we are concerned mostly with the first of these: CPU performance or speed, in particular, as it relates to the speed of floating-point operations. High-performance scientific computing implies creating numerically intensive programs and running them on those computers best able to execute them efficiently.

11.2.1 Determining the speed of a CPU

The CPU speed of a computer is related to the performance of that machine, but it is not the same thing. The CPU speed tells us how fast the machinery of the CPU works; the performance metrics tell us how fast the CPU can execute given tasks.

Recall the following definitions:

- **millisecond (msec):** one-thousandth (0.001 or 1×10^{-3}) of a second.

- **microsecond (μsec):** one-millionth (0.000001 or 1×10^{-6}) of a second.

- **nanosecond (nsec):** one-billionth (0.000000001 or 1×10^{-9}) of a second.

- **Hertz (Hz):** one cycle per second; an international unit for frequency.

- **MegaHertz (MHz):** one million cycles per second.

A CPU clock cycle is the finest unit of operation we consider within the CPU. The speed of the CPU clock may be defined as follows:

- **Clock rate:** the number of clock cycles executed per second (usually in units of MHz)

- **Clock speed:** the length of time it takes to execute a single CPU clock cycle (usually in units of msec or nsec).

These two definitions are the reciprocals of each other. If the rate of a CPU clock is 50 MHz, then the length of a single clock cycle for that CPU is 20 nsec. And if one clock cycle occurs every 12.5 nsec, then the clock rate is 80 MHz.

Table 11.2 on page 365 includes the clock rate of some current workstations. The processor cycle times of a few Cray models are shown in table 11.3 on page 391 while the cycle times of several parallel computers are given in table 11.6 on page 399. Notice that the range of processor speeds for the workstations is not very different from the large supercomputers.

In order to determine the performance of a computer, we need to be able to time its execution on various tasks. The following terms are used to describe different types of times related to computer execution.

- **user time:** The amount of time the CPU is active for a given process excluding operating system overhead.

- **system time:** The amount of time the operating system is active for the given process.

- **elapsed time:** The difference between the time when the timed process begins on the given processor and the time that it ends. This is the total time the process required to run. Since most computers allow multiprocessing, the *elapsed time* for a process is usually more than the sum of the *user time* and the *system time*.

- **wall clock time:** Another term for elapsed time.

- **response time:** Still another term for elapsed time, since this is the length of time the user waits for the execution of his program to complete.

One additional term related to computer execution time is **throughput**: the amount of work (jobs or operations) that the computer can complete in a given amount of time.

Since each machine is different and timing utilities may behave differently on different machines, timing the execution of programs can become a difficult procedure. This section describes some tools available for this process.

In UNIX environments, there are many tools for measuring time. The UNIX command `man -k time` produces a long list. The following utility programs can be used to assist in timing program executions. There are advantages and disadvantages to each.

UNIX `time` command: In both the Bourne shell (*sh*) and the C shell (*csh*), the UNIX `time` command times the execution of another command, producing the elapsed, user, and system times upon the completion of the command. For instance in the Bourne shell,

```
time latex perf
```

compiles the LaTeX file named `perf.tex` and then reports on the time this process took:

```
11.7 real        5.3 user        0.3 sys
```

This information shows that the total user time for this operation is 5.3 seconds while the system time is only 0.3 seconds. The elapsed time is 11.7 seconds. Other executions of the same command may produce slightly different results.

In the C shell, the result of the `time` command is in a different format. It also provides additional information. For example, the command

```
time latex perf
```

executes the `latex perf` command and then prints a line showing the time of execution:

```
5.3u 0.5s 0:31 18% 0+1124k 97+18io 129pf+0w
```

Reading this line from left to right, the figures should be interpreted in the following manner. The first notation `5.3u` means that the total user time is 5.3 seconds while the second figure tells us the system time is only 0.5 seconds. The elapsed time for this operation is 31 seconds. In this example, the ratio of user time to elapsed time is 18% (5.3/31). The remaining figures provide information on input, output, and storage requirements for the program execution.

The user time is the same in both cases since the same command is being executed under both shells. But notice the difference in the two elapsed times: 11.7 seconds versus 31 seconds. In additional tests, three more executions of the `time latex perf` command in the Bourne shell resulted in elapsed times varying from 6.4 seconds to 10.7 seconds, while three more executions of the `time latex perf` command under C shell produced elapsed times from 8 seconds to 12 seconds. This difference in the elapsed times may be caused by changes in the load of the machine executing the command. Also notice that the sum of the user time and the system time is usually less than the elapsed time. These results suggest the dubious value of using elapsed time for careful performance measurements, particularly in a multiprogramming environment.

The main problem with the `time` command in either shell is that the resultant times are only correct to one-tenth of a second. Since most current CPU cycle times are in nanoseconds, millions of cycles must be executed to produce a non-zero time.

Another problem is that the timing is done outside the application being timed. In other words, the `time` command first records the current time, and then it starts the execution of the application. When the application has finished executing, the final time is recorded, and the differences are displayed in the appropriate format. This does not permit the timing of a small set of instructions within a program or the timing of an inner loop of a program; only the execution of entire programs can be timed with this command.

UNIX `date` command: The UNIX `date` command returns the current date and time (correct to the current second) when entered at the UNIX prompt. It can also be called within a C program using the `system` function. By employing string functions to separate out the values for the hours, minutes, and seconds from the returned string expression, it is possible for this command to record times during the execution of a program and thus to time inner portions of the program.

While this utility has the advantage of providing time values during the execution of a program, it has the disadvantage of only being correct to seconds. Again this means that tens to hundreds of millions of cycles must be executed to produce a non-zero time.

Most UNIX Fortran compilers provide a similar function.

The Fortran programming language has traditionally been used for high-performance scientific computing As a result, most Fortran compilers provide additional timing routines to assist the scientific and performance programmer:

ETIME: There is an intrinsic Fortran procedure named `ETIME` in many Fortran compilers; a double precision version of this is named `DTIME`. These procedures return the time that a process has used since it started. Thus it is as if a private clock, private to a given process, is started when that process starts; it runs only while that process is running, stopping temporarily when the process is idle and resuming when the process becomes active. Thus it is not the same as wall-clock time because it excludes the effects of multiprogramming. In fact, `ETIME` gives the two components of the time used by the process, user time and system time.

For a DECstation 5000/240, the single precision time given by `ETIME` is in microseconds. In section 11.2.2, we show how to determine how often an `ETIME` clock is advanced. On the DECstation 5000/240, it is advanced once every 4 msec. This is a finer clock than either the UNIX `time` or `date` commands and allows the timing of smaller sets of instructions.

Warning: Different Fortran compilers define the parameters for `ETIME` differently. Also different units of time may be measured on different machines.

SECOND: Some Fortran compilers, including the Cray compiler, contain a function named `SECOND` that provides the elapsed time used by the calling process. Like `ETIME`, this function returns a finer unit of time than do time and date. On the Cray Y-MP, the SECOND timer ticks once for every processor clock cycle; in section 11.2.2 we determine that it ticks once every 6 nsec. Timing done with this function is even more accurate than the others discussed above.

SYSTEM_CLOCK: The SYSTEM_CLOCK function may be used to return the elapsed time used by a process. It is included in both Fortran 90 and High Performance Fortran (HPF), and it returns integer data from the real time clock of the processor on which the program is executing.

The three output parameters for this function are COUNT, COUNT_RATE, and COUNT_MAX; each of these is optional. The first parameter COUNT provides the current value of the processor clock; it ranges in value from 0 to COUNT_MAX. The parameter COUNT_MAX is the maximum integer value that can be held in the real time clock. The parameter COUNT_RATE returns the number of processor clock counts per second to provide the *granularity* of the processor clock; this may be in thousandths or hundredths meaning that the clock is accurate to milliseconds or tens of milliseconds. Both COUNT_MAX and COUNT_RATE are dependent on the real time clock used by the given processor.

The *granularity* of a clock relates to the size of the smallest interval measured by the clock; it is the time between instants when the clock time is advanced one unit. A fine granularity implies a very small interval (perhaps in nanoseconds) while a coarse granularity implies a larger interval of time (such as seconds or tenths of seconds).

11.2.2 Determining clock granularity

It is important to know the granularity of the clock for a timing routine. The finer the clock granularity, the more accurate the information obtained by the routine. A coarse clock granularity requires longer sequences of instructions to be timed in order for them to be measured by the clock correctly.

Given a timing routine such as ETIME on a certain machine, how does one determine the granularity of the clock used by that routine? This can be an intricate procedure, dependent on the architecture and the routine. We provide two examples in this section.

```
* mytime.f
      SUBROUTINE MYTIME(TIME)
* Time for DEC 5000/240, values are in seconds.
      EXTERNAL ETIME
      REAL T(2),TIME
      TIME = ETIME(T)
      RETURN
      END
```

Figure 11.1: A Fortran timing routine for the DEC 5000/240 that uses ETIME.

ETIME on the DEC 5000/240

First we create a timing routine called MYTIME that uses ETIME as is shown in figure 11.1. The value assigned to TIME is the user time plus system time elapsed since the start of execution of the program. We do not use the values T(1) and T(2) that provide user time and system time, respectively.

To determine the precision of the value returned by MYTIME on the DEC 5000/240, we run an experiment in which we time a program segment that executes in an interval somewhat shorter than the granularity of the clock. The program used is given in figure 11.2. The subroutine MY_DATE_TIME also called by this program merely provides a date and time stamp for the run of the program.

This program makes 20 independent measurements of the time to execute the loop in which the sum

$$1 + 1 + 1 + \ldots + 1$$

is computed. The output from this program is shown in figure 11.3.

Notice that the times in this output are either 0.003906 seconds or zero seconds. The second column of the output is simply a sanity check: it confirms that the inner loop was executed 10,000 times. When the recorded time is zero, the clock reading before the 10,000 additions was the same as the clock reading following the 10,000 additions. This means the ETIME clock did not change during that time; it did not tick. The 10,000 additions were performed between ticks of the ETIME clock. But when the recorded time is not zero, we know that the clock must have ticked at least once. Since there are zero times between

```
* timetst.f:  Test clock granularity
      IMPLICIT NONE
      INTEGER J,K,SUM
*********************************************************************
* J,K - loop indices.
* SUM - variable to accumulate sum.  (OUT)
*********************************************************************
      REAL T1, T2
*********************************************************************
* T1,T2 - time readings.  (difference OUT)
*********************************************************************
      CHARACTER*10 DATEBUF, TIMEBUF
      CHARACTER*40 PROGRAM_LABEL
      PARAMETER(PROGRAM_LABEL='timetst, 28 aug 1992, ldf/cjcs')
*********************************************************************
* DATEBUF - Current date.  (OUT)
* TIMEBUF - Current time.  (OUT)
* PROGRAM_LABEL - Name, creation date, author.  (OUT)
*********************************************************************
      EXTERNAL MY_DATE_TIME, MYTIME
*********************************************************************
* MY_DATE_TIME - a function returning current date/time.
* MYTIME - a function returning time.
*********************************************************************
* Write header.
      WRITE(*,*) PROGRAM_LABEL
      DATEBUF = ' '
      TIMEBUF = ' '
      CALL MY_DATE_TIME(DATEBUF, TIMEBUF)
      WRITE(*,*) 'RUN DATE_TIME ', TIMEBUF, ' ', DATEBUF
      WRITE(*,*)
      WRITE(*,*) 'TIME(sec), SUM'
      WRITE(*,*)
* Main loop to time 20 samples of 10000 additions.
      DO K=1,20
          CALL MYTIME(T1)
          SUM = 0
          DO J=1,10000
              SUM = SUM + 1
          ENDDO
          CALL MYTIME(T2)
          WRITE(*,999) T2-T1, SUM
      ENDDO
* Done.
      STOP
 999  FORMAT(1X,1PE12.5,1X,I7)
      END
```

Figure 11.2: A Fortran program to test the granularity of the **ETIME** clock routine for the DEC 5000/240.

```
timetst, 28 aug 1992, ldf/cjcs
RUN DATE_TIME  14:10:31    01-Sep-92

TIME(sec), SUM

0.00000E+00    10000
0.00000E+00    10000
0.00000E+00    10000
3.90600E-03    10000
0.00000E+00    10000
0.00000E+00    10000
0.00000E+00    10000
0.00000E+00    10000
0.00000E+00    10000
0.00000E+00    10000
3.90600E-03    10000
0.00000E+00    10000
0.00000E+00    10000
0.00000E+00    10000
0.00000E+00    10000
0.00000E+00    10000
0.00000E+00    10000
3.90600E-03    10000
0.00000E+00    10000
0.00000E+00    10000
```

Figure 11.3: Output from `timetst.f` as run on the DEC 5000/240.

the nonzero times, we can guess that the clock ticked only once. And given that the nonzero times are all the same, we can presume that the granularity of the ETIME clock is that value: $0.003906 = \frac{1}{256}$ seconds. You can confirm this by increasing the number of iterations of the loop from 10,000 to a higher value. As you gradually increase, you start to see times of 0.007812, 0.023436, and so on. Note that $0.003906 \approx 2^{-8}$, $0.007812 \approx 2 \times 2^{-8}$, etc.

When you run this experiment several times, you do not get identical results; the locations of the nonzero values change because the *ticks* of the clock are initiated by an independent process. Thus if your process starts immediately before a clock tick you find a nonzero value early in the list, but if your process starts shortly after a clock tick your first nonzero value appears later.

These results show that the granularity of the clock on the DECstation 5000/240 is about 4 msec. To measure the time of execution of a block of code, the clock is read immediately before entering the block and again when leaving the block. The difference between the times is the time spent in the block, and it should be evident that error in this measurement is less than 8 msec. Thus if we time a block of code in this way and find that the execution time is 1 sec, the relative error in this measurement is less than 0.8%.

SECOND on the Cray Y-MP

The function SECOND is a timing tool that is used with Fortran on the Cray. A revised version of the MYTIME function for the Cray is shown in figure 11.4.

As with the ETIME function used by the DECstation timing routine, the next step is to determine the granularity of the clock used by the Cray SECOND function. If you try to use the DECstation version of the program timetst.f in figure 11.2, you may not get the same kind of results. The granularity of the SECOND timer on the Cray is much smaller than the granularity of the ETIME routine on the DECstation; in fact, more than one tick of the Cray SECOND timer occurs during a single integer add on the Cray.

Consider the program timetst_cray.f shown in figure 11.5. Since the Cray Y-MP is a vector processor (see section 11.6 for more about vector processors), the compiler tries to optimize the code by including vectorizations (see section 11.4 for more on compiler optimizations). We need to make sure

```
* mytime_cray.f
* Time for Cray Y-MP, values are in seconds.
*****************************************************************
      SUBROUTINE MYTIME(T)
      REAL T
      EXTERNAL SECOND
*****************************************************************
* T - the total elapsed time.
* SECOND - the Cray Fortran routine for returning elapsed time.
*****************************************************************
      T = SECOND()

      RETURN
      END
```

Figure 11.4: A Fortran timing routine for the Cray Y-MP that uses SECOND.

that the code being timed is not vectorized. For this reason, the compiler directive CDIR$ NOVECTOR is placed just before the loops containing the scalar operations to be timed; this tells the compiler not to perform any vectorizations on this code. To avoid having the compiler remove the loop as part of the optimization, we have placed a WRITE statement inside the loop; this forces the loop to be executed.

The output of this program is shown in figure 11.6. Observe that none of the times shown in the output are zero. Instead they are all in the neighborhood of 2.7 to 2.9 msec. However, consider the differences between the times; these are all multiples of 6 nsec. Since the cycle time of the Cray Y-MP as shown in table 11.3 on page 391 is 6 nsec, we can presume that the clock granularity is 6 nsec.

Comment: the times given in this output are much larger than an integer add time should be. This is because only a single add is timed, and that time includes the overhead of calling and returning from the SECOND function. If we wanted a more accurate timing of the integer add operation – instead of trying to determine the clock granularity, we should include enough repetitions of the operations to be timed to make the calling overhead insignificant. We discuss this further in section 11.2.4.

```
* timetst_cray.f:  Test clock granularity
      IMPLICIT NONE
      INTEGER K,SUM
**********************************************************************
* K - Loop index.
* SUM - Accumulating sum of timed additions.  (OUT)
**********************************************************************
      REAL T1, T2, SECOND
      EXTERNAL SECOND
**********************************************************************
* T1, T2 - Beginning and ending times for each add operation.
*        (difference OUT)
* SECOND - Cray timing function.
**********************************************************************
      CHARACTER*10 DATEBUF, TIMEBUF
      CHARACTER*40 PROGRAM_LABEL
      PARAMETER (PROGRAM_LABEL = 'timetst_cray, 7 sep 92, ldf/cjcs')
**********************************************************************
* DATEBUF - Current date.  (OUT)
* TIMEBUF - Current time.  (OUT)
* PROGRAM_LABEL - Name, creation date, author.  (OUT)
**********************************************************************
* Write header.
      WRITE (*,*) PROGRAM_LABEL
      DATEBUF = ' '
      TIMEBUF = ' '
      CALL MY_DATE_TIME(DATEBUF, TIMEBUF)
      WRITE (*,*) 'RUN DATE_TIME ', TIMEBUF, ' ', DATEBUF
      WRITE (*,*)
      WRITE (*,*) 'TIME(sec), SUM'
      WRITE (*,*)

* Timed loop of 20 adds.
CDIR$ NOVECTOR
      SUM = 0
      DO K=1,20
         CALL MYTIME(T1)
         SUM = SUM + 1
         CALL MYTIME(T2)
         WRITE(*,999) T2-T1, SUM
      ENDDO

* Done
      STOP
 999  FORMAT(1X,1PE12.5,1X,I7)
      END
```

Figure 11.5: A Fortran program to test the granularity of the SECOND clock routine for the Cray Y-MP.

```
timetst_cray, 7 sep 92, ldf/cjcs
RUN DATE_TIME  15:30:40    09/07/92

TIME(sec), SUM

   2.90400E-06        1
   2.82600E-06        2
   2.79600E-06        3
   2.79600E-06        4
   2.79600E-06        5
   2.79600E-06        6
   2.79600E-06        7
   2.79600E-06        8
   2.83800E-06        9
   2.80800E-06       10
   2.82600E-06       11
   2.79600E-06       12
   2.79600E-06       13
   2.79600E-06       14
   2.79600E-06       15
   2.82600E-06       16
   2.79600E-06       17
   2.79600E-06       18
   2.79600E-06       19
   2.81400E-06       20
```

Figure 11.6: Output from `timetst_cray.f`.

11.2.3   Measures of computer performance

The performance of one computer is better than the performance of another computer when the same application program executes faster on the first computer than on the second. In other words, computer performance is measured by the reciprocal of execution speed. However, many factors besides the CPU speed of a machine affect the execution speed of any given application program. These factors may produce conflicting performance results when a machine is tested using multiple applications.

The following units have been developed over the years to describe measurements of performance:

- **mips:** Originally this term referred to the speed of a computer in terms of the number of millions of instructions per second it could perform. For example if a set of 15,000 instructions runs in 3 msec on a certain machine, then the mips for that application running on that machines is 5; that is,

$$\frac{15,000 \text{ instrs}}{3 \text{ msec}} = \frac{15,000,000 \text{ instrs}}{3 \text{ secs}}$$
$$= \frac{5,000,000 \text{ instrs}}{1 \text{ sec}}$$
$$= 5 \text{ mips}$$

The reader should note that the mips measurement is highly dependent on the instruction mix used to produce the execution times. A program executing 100,000 instructions with a high percentage of simple integer operations runs in a shorter amount of time than a program executing the same number of instructions with a high percentage of complex instructions such as floating-point operations. Hence two different programs with the same number of instructions can produce different mips measurements on the same machine.

Another problem with the accuracy of the mips performance measurement is that some architectures have a simple instruction set and others are designed with a more complex instruction set. While programs running on a simple instruction set architecture may compile into more instructions than the same program on the more complex instruction set architecture, simpler instructions usually execute more quickly than complex instructions. Thus it can be difficult to compare the mips of two different machines even when using the same program.

The term mips is still used in the manner described above. However it is also used to describe the results of the MIPS[1] benchmark (see section 11.3). In this context, the millions of instructions per second are correlated to the performance of a VAX 11/780.

- **mops:** This term defines the speed of a computer in terms of the number of millions of operands per second it can handle. These operands may refer to register values or to data located in the cache or in the memory of the machine. The machine instructions for some processors contain only one operand per instruction; in this case, the mops for the processor should be the same as the mips. Other processors may have two, three, or more operands in an instruction; the mops for these machines are greater than the mips.

  Consider the above example where a set of 15,000 instructions runs in 3 msec on a certain machine. Assume that each instruction for that machine contains three operands. Then the mops is 15 for this application; that is,

$$\frac{15,000 \text{ instrs} \times 3 \text{ operands/instr}}{3 \text{ msec}} = \frac{45,000,000 \text{ operands}}{3 \text{ secs}}$$
$$= 15 \text{ mops}$$

- **CPI:** This term is used to define the speed of a computer in terms of the number of clock cycles per instruction.

  Continuing with the above example, assume that the clock rate of the given computer is 20 MHz. This means that each clock cycle takes 50 nsec. The earlier examples have shown that this machine executes at 5 mips; that is, it executes five million instructions per second or one

---

[1]MIPS also refers to a family of RISC architectures developed by MIPS Computer Systems, Inc., a subsidiary of Silicon Graphics, Inc. This family includes the MIPS R2000, MIPS R3000, MIPS R4000, and MIPS R10000 microprocessor chips.

instruction per 200 nsec. Hence each instruction requires an average of 4 clock cycles, and the machine operates at 4 CPI on average.

- **flop:** This common performance unit defines the speed of a computer in terms of the number of `floating-point` operations it can perform in one second.

If we return to the same example and assume each instruction performs one floating-point operation, then the number of floating-point operations is the same as the number of instructions. So the machine operates at 5,000,000 flops.

However it is highly unlikely that all the instructions are floating-point operations; for example, some may be integer operations that control loops or perform subscript calculations. If we instead assume that only four-fifths of the instructions are floating-point operations, then the machine operates at 4,000,000 flops; that is,

$$\frac{4}{5} \text{ floating-point operations/instr} \quad \times \quad 5,000,000 \text{ instrs/sec}$$
$$= \quad 4,000,000 \text{ flops}.$$

The main problem with this performance measurement unit as we have defined it is that not all floating-point operations require the same number of clock cycles. For instance, a floating-point division may take 4 to 20 times as many cycles as a floating-point addition. Thus the flop must be more precisely defined any time it is used to measure machine performance. In some contexts, all floating-point operations are included in the count of flops. In others, only the multiplications and additions are included. (The LINPACK benchmark uses this definition of the flop.) Yet another definition of flop proposed by C.B. Moler [Golub & Van Loan 83] includes some time for access of array elements in addition to multiplications and additions. This flop is defined by

$$s = s + a(i,k) * b(k,j).$$

All of these definitions are valid ones, but you must be careful to state which one you are using.

The flop is an especially popular measure of machine performance on scientific and other mathematical programs, but it is not a reasonable measure for programs or benchmarks using few or no floating-point operations.

- **megaflop (Mflop):** A speed of a computer as the number of Millions of `floating-point` operations it can perform in one second.

  For instance, the machine in the example above operates at 4 Mflops.

- **gigaflop (Gflop):** Another multiple of the *flop*, this term defines the speed of a computer as the number of billions of `floating-point` operations it can perform in one second. As displayed in table 11.1, most of today's supercomputers are capable of executing in Gflops.

- **teraflop (Tflop):** Still another multiple of the *flop*, this term defines the speed of a computer as the number of trillions of `floating-point` operations it can perform in one second. This is the rate that most new supercomputer designs are attempting to achieve.

- **theoretical peak performance:** While not a *unit*, this term is also used to describe the computer performance of a machine. This is simply the maximum number of flops that a machine can obtain theoretically. It is almost never achieved. According to Dongarra [Dongarra 94],

  > The theoretical peak performance is determined by counting the number of floating-point additions and multiplications (in full precision) that can be completed during a period of time, usually the cycle time of the machine.

Full precision in this context denotes double precision arithmetic that uses 64 bits on most architectures. However some machines do not follow the IEEE arithmetic standard. For instance, the Cray Y-MP uses 64 bits to represent a single precision value; in this case, the single precision results would be used for the theoretical peak performance.

Table 11.1 provides a short list of supercomputers with the name of the machine series, the name of the manufacturer, and the theoretical peak performance given in Mflops.

Machine	Manufacturer	Number of Processors	Theor. Peak Mflops
CM-2	Thinking Machines	2,048	20,000
CM-5	Thinking Machines	16,384	2,000,000
Cray-2	Cray Research	8	4,000
Cray Y-MP	Cray Research	16	15,000
Cray T3D	Cray Research	2,048	307,000
Cray-3	Cray Computer Corp	16	16,000
IBM ES/9000	IBM	6	2,700
IBM 9076 SP1	IBM	64	8,000
Intel iPSC/2	Intel	128	250
Intel iPSC/860	Intel	128	5,120
Intel Delta	Intel	512	20,480
Intel Paragon	Intel	4,000	300,000
KSR1	Kendall Square Research	32	1,300
KSR2	Kendall Square Research	5,000	400,000
MP-1	MasPar	16,384	550
MP-2	MasPar	16,384	2,400
NEC SX-A	Nippon Electric Co	4	22,000

Table 11.1: A list of selected supercomputers with their manufacturer's name and their theoretical peak performances in Mflops and their maximum number of processors. This information was taken from [Dongarra 94] and [van der Steen 94].

```
         CALL MYTIME(T1)
         DO I=1, NREP
*           put R1 repetitions of "op" here
         ENDDO
         CALL MYTIME(T2)
         DO I=1, NREP
*           put R2 repetitions of "op" here
         ENDDO
         CALL MYTIME(T3)
         OPTIME = ((T2-T1) - (T3-T2))/(NREP*(R1-R2))
*           OPTIME is the estimated cost of "op" in seconds
```

Figure 11.7: Outline of timing code segment.

Still other units of computer performance measurement refer to specific benchmarks programs. Such units include *SPECmarks* and the other definition of *mips*. These are discussed in the next section.

11.2.4   Timing elementary operations

In this section we discuss how to time the elementary arithmetic operations on the DECstation 5000/240. This system uses a MIPS R3000 CPU and R3010 FPU chipset [Kane 88, DEC 91], operating at 40 MHz (25 nsec cycle time). Thus we can expect that arithmetic operations require some small multiple of 25 nsec. Although we describe this timing procedure for the DECstation 5000/240, the basic concepts apply to the timing of any very fast operation on any machine.

As we observed previously, the granularity of the clock we use to measure time on the DECstation is only about 4 msec. So how do we measure an arithmetic operation time that we expect to be much smaller than the granularity of the clock? The basic idea is simple: we measure the time T it takes to do a large number N of the operations; then the operation time is given by T/N. The details are best explained with the outline of the code segment for the timing shown in figure 11.7.

The statement CALL MYTIME(T1) is used to get an initial reading of the time T1. After completing the computation we are measuring, we obtain the new time T2 with the statement CALL MYTIME(T2). Then the difference between the readings (T2 - T1) is the elapsed time. Now this elapsed time includes not only the time to do the NREP × R1 operations, but also the overhead for the loop control arithmetic. However the latter should not be included in the operation time as it is just an artifact of the technique we use to measure the time.

Removing the loop overhead time is the reason for the second time measurement: the measurement starting with the statement CALL MYTIME(T2) and ending with the statement CALL MYTIME(T3). Since the amount of loop control arithmetic is the same for both DO loops, the time for the loop control arithmetic is canceled out by the expression ((T2-T1) - (T3-T2)); this expression contains only the time for performing NREP*(R1 - R2) repetitions of "op". Thus the statement

OPTIME = ((T2-T1) - (T3-T2))/(NREP*(R1 - R2))

gives the time for one execution of "op".

To get correct timings, we need to be sure that the compiler does not alter our program when it tries to improve or *optimize* the code.[2] Specifically, we need to use values of R1 and R2 that are not altered in any way by the optimization stages of the compiler. If these values are too small, the optimizing compiler may modify the loop structure and thereby change the amount of loop control arithmetic from what you think it is; if the values are too big, the code is clumsy with an unnecessarily large loop body. After some experimentation, we have found that the values of 16 for R1 and 8 for R2 work well for the DECstation.

As an example, we measure the cost of double precision, floating-point addition X = X + Y. It is useful to report the average operation time and the average rate of executing the statement in Mflops.

Figure 11.8 shows only the most important parts of the program. Here R1 and R2, defined in the code outline above, are 16 and 8 respectively; the variable DUP represents the difference between R2 and R1. The variable DT represents the elapsed time for DUP×NREP executions of the statement X = X+Y. The granularity of the clock causes an error in DT that could be as large

---

[2]Compiler optimization is the subject of section 11.4.

```
* add.f
* Double precision add test
            .
            .
            .
      DOUBLE PRECISION X, Y
            .
            .
            .
      SUMTIME = 0
            .
            .
            .
      DO J=1,NSAMP
         CALL MYTIME(T1)
         DO I=1,NREP
* 16 copies of timed op X = X+Y follow.
            X = X+Y
                .
                .
                .
            X = X+Y
         ENDDO
         CALL MYTIME(T2)
         DO I=1,NREP
* 8 copies of timed op X = X+Y follow.
            X = X+Y
                .
                .
                .
            X = X+Y
         ENDDO
         CALL MYTIME(T3)

* Compute elapsed time and update variables.
         DT = (T2-T1)-(T3-T2)
         SUMTIME = SUMTIME + DT
      ENDDO

* Compute and write average time and Mflops.
      AVGTIME = SUMTIME/(NSAMP*NREP*(R1-R2))
      MFLOPS = 1/(AVGTIME*1.0E6)
            .
            .
            .
```

Figure 11.8: Outline of `add.f`, a program to determine the time required for one double precision addition on a DECstation 5000/240.

```
add1, 31 aug 1992, ldf/cjcs
RUN DATE_TIME  10:32:12      03-Sep-92

X, Y, NREP, NSAMP, DUP
MINTIME(sec), MAXTIME(sec), AVG.OP.TIME(sec), MFLOPS, X

1.00000E-01  1.00000E-01  1200000        20    8
 7.2261E-01  7.3042E-01  7.5475E-08  1.3249E+01  5.7600E+07


1.00000E-01  1.00000E-01  1800000        20    8
 1.0859E+00  1.0937E+00  7.5557E-08  1.3235E+01  8.6400E+07
```

Figure 11.9: Sample output from add.f.

as 16 msec; notice how DT is computed and remember that each time (T1, T2, T3) in the expression on the right-hand side could be in error by almost 4 msec. By observing the minimum and maximum values of DT shown in the output, we can verify that the difference between these values is no more than 16 msec. If we find that the difference is larger than this, then it is most likely due to an unusually heavy load placed on the system by other users; in this case, it would be best to wait for a quieter time to do the measurements.

Figure 11.9 shows the output from a sample run of this program. The loop parameters have been chosen (NSAMP = 20 for both runs with NREP = 1,200,000 and 1,800,000) so as to yield a value of DT in the neighborhood of 1 sec. Given that the measurement error should be less than 16 msec, this choice promises a maximum relative error of about 1.5% in the value reported for the average operation time represented by the variable AVGTIME.

If you are not careful in constructing the loop body, the compiler can change the loop behind your back when it optimizes the code! For example, if you use a sequence of statements in a loop such as

```
        DO I=1,N
            X = Y + Z
            X = Y + Z
            X = Y + Z
        ENDDO
```

then the compiler can recognize that the first two statements are ineffective. Indeed the whole loop is ineffective and the compiler may remove it and replace it with the single statement

```
        X = Y + Z
```

Whether or not this happens depends on the compiler and the level of optimization. To prevent this from happening, we use a loop of the following form

```
        X = 0
        DO I=1,N
            X = X + Y
            X = X + Y
            X = X + Y
        ENDDO
```

All of these statements are effective since they contain a *recurrence*;[3] that is, each statement recomputes a new value for X using the old value. Of course a *really* smart compiler could figure out that 3*N*Y is the final value of X and eliminate the loops entirely, but fortunately for us the current compiler is not so smart. Note, by the way, that the final value of X is printed in the rightmost column of the output as a sanity check.

---

[3]See chapter 12 for more about recurrences.

11.3   Benchmarks

In this section, we discuss the usage of benchmarks, various categories of benchmarks, the description of some common benchmarks, and methods for obtaining copies of benchmark codes and results.

11.3.1   Purpose of benchmarks

In discussing performance, the word *benchmark* refers to a set of programs or program segments that are used to measure performance. The time required for a computer to execute the benchmark provides the performance measure.

Benchmarks can be used to compare the performance of two or more machines. After all, when determining which computer to use or purchase for a given application, it is the performance of that computer on the given application that is important. If users are unable to test their applications on different machines, they should use the results of the benchmark programs that most resemble their applications run on those machines.

11.3.2   Types of benchmarks

Benchmarks are written for many purposes: to check the CPU performance of a computer, the file server performance, the input/output interface, the network, the communication speed, etc. Some benchmarks are user-written; that is, they are written by the user needing the information and are internal to the user's organization. However the code for many commonly used benchmarks is available over the Internet; most of these codes have been modified and fine-tuned over the years and can be used *with care* to produce performance data that can be compared with existing libraries of numbers corresponding to the performance of other machines run with the same benchmark programs. Still other benchmarks belong to special organizations that test new computers and publish the results; such results are considered less biased than those produced by the computer manufacturers.

If a benchmark is taken from existing code, it is termed a *kernel* benchmark. In other words, the benchmark is created from the kernel of an application program, an essential part of the program, such as the main loop or main subloop in the code. A kernel benchmark may be a loop from the original application or a simplified version of the original code; nevertheless it is intended to be representative of the application program. Other benchmark programs are *real* application programs.

When a benchmark is written specifically to represent a statistical sample of operations or program statements, it is called a *synthetic* benchmark. The code for a synthetic benchmark exists for no other purpose.

Some benchmarks may be classified as *novelty* or *toy* benchmarks. These small programs are based on algorithms that do not represent normal application programs. Game or puzzle-solving programs are included in this category of benchmarks, as well as sorting routines.

11.3.3   Internet sites for benchmarks

There is an electronic storage facility, called `netlib`, that includes many common benchmarks, including both the source code and the performance results for a number of machines. It is maintained by the Oak Ridge National Laboratory in Tennessee. Other mathematical software and information is kept at this site as well. This material can be accessed on the Internet[4] in several ways:

electronic mail: A listing of the contents of this library can be obtained by electronic mail to `netlib@netlib.att.com` or `netlib@ornl.gov` with a single line as the message body:

> `send index`

anonymous ftp: To reach `netlib` by anonymous ftp, type

> `ftp netlib.att.com`

and change to the directory called `/netlib`.

xnetlib: In an X-windows environment where `xnetlib` is installed, the command

---

[4]See [Krol 92] for more on the Internet.

       `xnetlib`

brings up an interactive window directly connected to `netlib`. Send mail to `netlib` with the message body `send index from xnetlib` for simple installation instructions.

mosaic: Mosaic is an interface to the World Wide Web (WWW); the mosaic address

       `http://www.netlib.org`

is an alternate way to reach `netlib`. Clicking on `The NetLib Repository` provides a listing of the available files. And clicking on `Performance Database Server` provides access to compiled data on performance.

Copies of this library are located at alternate sites all over the world.

The National Institute for Standards and Technology (NIST) also stores most of the common benchmarks. To obtain a listing of these benchmarks, send electronic mail to `nistlib@cmr.ncsl.nist.gov` with a single line as the message body:

       `send index`

This repository does not include a library of current benchmark results.

There also is a USENET newsgroup named `comp.benchmarks` that discusses the purpose and value of existing benchmarks, publishes results of many benchmarks, introduces new benchmarks, and often supplies source code for benchmarks. Further information about the benchmarks mentioned here as well as other benchmarks can be found in that newsgroup.

In addition, a number of the publically available benchmarks (and some results) can be found at the following `ftp` sites:

       `sony.com:/pub/benchmarks`
       `ftp.nosc.mil:/pub/aburto`
       `netlib.att.com:/netlib/benchmarks`

Some `mosaic` sites exist as well.

Workstation	Clock	SPECmarks			LINPACK
	MHz	INT92	FP92	89	Mflops
DEC 5000/200	25.0	19.5	26.7	18.5	3.7
DEC 5000/240	40.0	27.9	35.8	32.4	5.3
DEC 3000-500 AXP	150.0	84.4	127.7	126.1	30.0
DEC 10000-660 AXP	200.0	106.5	200.4	184.1	43.0
HP 9000/750	66.0	48.1	75.0	77.5	24.0
HP 9000/735	99.0	109.1	167.9	146.8	41.0
IBM RS/6000-350	41.6	35.4	74.2	73.7	19.0
IBM RS/6000-980	62.5	73.3	134.6	126.3	38.0
IBM Power2-990	71.5	126.0	260.4	—	140.0
SGI Indigo R3000	33.0	22.4	24.2	26.3	4.0
SGI Indigo R4000	50.0	57.6	60.3	70.2	12.0
SGI Crimson	50.0	61.7	63.4	70.4	16.0
Sun SPARC 2	40.0	21.8	22.8	25.0	4.0
Sun SPARC 10/30	36.0	45.2	54.0	57.3	9.3
Sun SPARC 10/40	40.0	50.2	60.2	—	10.0

Table 11.2: Performance data based on the LINPACK and SPEC benchmarks for selected workstations. The SPECmarks quoted came from various articles posted on `comp.benchmarks` or from manufacturers' brochures; the LINPACK values came from [Dongarra 94].

11.3.4 Common benchmarks

Some common benchmarks are described in the following subsections. Table 11.2 on page 365 provides a short list of high-performance workstations with the name and clock rate of the machine, followed by the performance rated in SPECmarks for the SPEC benchmark and in Mflops for the LINPACK benchmark. Both of these benchmarks are described in further detail below. The SPECmarks quoted are from various articles posted on `comp.benchmarks` or from manufacturers' brochures; the LINPACK values are from [Dongarra 94].

LINPACK

In the 1970s, a set of Fortran subroutines was developed to solve systems of linear equations. This set of routines is called LINPACK [Dongarra et al 79]. Since LINPACK was popular in many areas of scientific computing, it is logical that one of its routines has become the source of a kernel benchmark for comparing the floating-point performance of different machines. The LINPACK benchmark measures the speed of solving a system of 100 simultaneous linear equations using software from LINPACK in double precision arithmetic. Results are sometimes given for single precision arithmetic as well.

LINPACK was the first linear algebra software package to use the Basic Linear Algebra Subprograms 1 (BLAS1). The BLAS1 is a set of subroutines and functions that perform the common scalar-vector operations such as multiplying a vector by a scalar and vector-vector operations such as the dot product. The BLAS1 are written in Fortran 77, but many computer manufacturers also provide a version of the BLAS1 routines written in the assembly language specific to their machine. These assembly language routines are designed to run as fast as possible. Thus it is easy to optimize a Fortran 77 program that uses the BLAS1 routines by replacing the compiled Fortran 77 BLAS1 with these fast routines, and many of the reported LINPACK results have been produced using them.

While LINPACK is still used for benchmarking purposes, it is no longer recommended for solving linear algebra problems. Released in 1992, LAPACK [Anderson et al 92] replaces both LINPACK and EISPACK [Smith et al 76], the collection of routines for eigenvalue problems introduced in 1970. LAPACK makes use of the newest, most accurate algorithms in numerical linear algebra. It also makes efficient use of memory by dividing large matrix operations into smaller block matrix operations. LAPACK is built on the BLAS1, 2, and 3. The BLAS2 and 3 perform matrix-vector and matrix-matrix operations, respectively, and may also be available in optimized versions. LAPACK and all of the BLAS are available via `netlib`.

More information on the LINPACK benchmark, including both code and results, is also available from `netlib`.

Livermore Loops

The Livermore Loops, also known as the Livermore Fortran Kernels (LFK), are twenty-four Fortran kernels derived from a group of application programs by Lawrence Livermore National Laboratory. An earlier version included only fourteen kernels. Some of the kernels are as simple as an inner (dot) product or a matrix multiply; others include a search loop from a Monte Carlo routine, a two-dimensional explicit hydrodynamics fragment, and an incomplete Cholesky routine with conjugate gradient. Like the LINPACK benchmark, results are provided in Mflops, but the count of floating-point operations has been normalized to account for the disparity in the execution time of different types of operations. A similar version of the kernels exists in C.

More information on the benchmark, including both code and results, is available from `netlib`. Also see [McMahon 86].

Whetstones

This benchmark is designed to measure floating-point performance. It is a single program originally written in Algol-60, and there is a Fortran version as well.

The type and frequency of statements in this program were chosen to represent the type and frequency of statements from numerically intensive programs at the National Physical Laboratory in England based on statistics gathered in 1970. These statements are translated to instructions for an imaginary computer called *Whetstone*. Hence this is a synthetic benchmark.

It is publically available by ftp at

`netlib.att.com:/netlib/benchmark/whetstone`

More information concerning the Whetstone benchmark is available from `netlib`.

Dhrystones

This benchmark is designed to determine the performance of system or integer programs, such as compilers; it does not model numerically intensive applications. It was designed by Reinhold Weicker and is written in Ada, C, and

Pascal. Two versions exist: V1.1 and V2.1; results from the latter version are more reliable.

Like the Whetstone benchmark, the use of statements and data types in this benchmark program are based on statistics. However the figures concerning the use of various programming language features for the Dhrystone benchmark were derived from many published sources in the early 1980s. Hence it is another synthetic benchmark.

The Dhrystone code consists of many small loops with integer or logic statements. One problem with this benchmark is that the small loops often fit into the cache of the machine being tested. This means that the results may show a better performance than the user should expect with larger programs. [Weicker 84]

Again these benchmarks are publically available by ftp from

```
netlib@ornl.gov
ftp.nosc.mil:pub/aburto
```

More information is available from `netlib`.

MIPS

The MIPS rating for a computer is determined from a set of small programs known as the MIPS benchmark. The ratio of the geometric mean of the performance of these programs to that for the same programs run on the VAX 11/780 provides the MIPS rating. The VAX 11/780 is defined as a MIPS-1 machine, meaning that it is capable of executing one million instructions per second. If a machine is rated at 10 MIPS, it supposedly performs the given benchmarks ten times faster does than a VAX 11/780.

These benchmark results can be deceiving because many of the VAX runs were made a number of years ago. Therefore these ratings compare the execution of programs compiled and run on machines with recent compilers that include all the latest optimizing techniques (see Section 11.4 for more information of optimizing compilers) against the execution of the same programs compiled with an older compiler on the VAX. If the VAX compiler were updated to include these techniques and the newly compiled programs rerun on the VAX 11/780, it is likely that the VAX execution times would decrease.

This in turn, would reduce the MIPS ratings of the newer machines with good optimizing compilers.

SPECmarks

The System Performance Evaluation Cooperative was created in the 1980s as a subgroup of NCGA (National Computer Graphics Association). The founders of this group included representatives from Hewlett-Packard, DEC, MIPS, and Sun Microsystems. This group maintains sets of benchmark programs to rate the performance of new machines.

The SPECmark value is the ratio of the geometric mean of the execution time for a selected set of benchmark programs on the given machine to that for the same set of programs on a VAX 11/780 using elapsed time. The first set of benchmark programs is called SPEC89 and contains ten programs. A later set SPEC92 is made up of twenty programs: six integer programs and fourteen floating-point programs. The SPEC89 result for a machine includes the average speed for both the integer and the floating-point programs; the SPEC92 result is broken into two parts: the SPECint92 result and the SPECfp92 result.

These three benchmarks were designed to test CPU performance. Two later benchmarks were developed by the SPEC group: SDM to test a UNIX Software DevelopMent workload and SFS for benchmarking a Systems level File Server workload.

Unlike other benchmarks, the SPEC benchmark programs are carefully controlled by the personnel within the SPEC group; they run the tests and publish the results in their own periodical. This provides a more uniform and less biased set of results.

The SPECmark values for many machines are also published by their manufacturers. Occasionally output from various SPEC tests shows up on the `comp.benchmarks` newsgroup. Table 11.2 shows the SPECmark values for some workstations extracted from the file `/pub/spectable` available from the anonymous ftp site: `ftp.cdf.toronto.edu`. Notice how the difference between the SPECint92 and SPECfp92 values provides more information than is shown by the SPEC89 value.

To see the latest list of SPEC benchmarks, retrieve `/pub/spectable` from the anonymous ftp site: `ftp.cdf.toronto.edu`. To obtain more information on the SPEC benchmark programs, contact

> SPEC
> c/o NCGA
> 2722 Merrilee Drive
> Suite 200
> Fairfax, VA 22031

or send an electronic mail message to `spec-ncga@cup.portal.com`.

Shallow Water code

The Shallow Water benchmark is a kernel from a Fortran weather prediction application program used by the National Center for Atmospheric Research (NCAR). The purpose of the benchmark is to compare the performance of supercomputers on this code. Results are given both in elapsed time and in Mflops [Sadourney 75, Hoffmann et al 88].

SPICE

SPICE (Simulation Program with Integrated Circuit Emphasis) is a circuit-level simulator developed in the early 1970s at Berkeley. Many versions of it are still being used to assist in the analysis of processor design today. Parts of the code for this simulator are now used for benchmarking purposes to measure a floating-point workload. This is considered to be a real benchmark program as compared to a kernel benchmark; complete routines are used for the benchmark programs. The results are given as elapsed time or as Mflops.

The most recent version of this benchmark is `spice 3f2` and can be purchased from Berkeley. It is available in both C and Fortran. To order a software catalog that includes this and other items, send an electronic mail message to `ilpsoftware@berkeley.edu`.

NAS parallel benchmarks

A set of benchmarks developed by NASA Ames Research Center for the purpose of comparing the performance of parallel machines is known as the NAS Kernels or the NAS Parallel Benchmarks. This is made up of eight kernels, ranging from an integer sort program to what they call an *embarrassingly* parallel program (since it maps easily to a parallel implementation with almost no serial code and almost no communication). The user is allowed to alter the basic benchmark code to make the best use of the architecture of the given parallel machine.

The results are provided as ratios of the time taken to execute each kernel on the computer being tested to the time taken for the same program executed on a Cray Y-MP with a single processor. Current NAS performance data can be found via the `mosaic` home page:

> `http://www.nas.nasa.gov/RNR/Parallel/NPB`

For more information, see [Bailey et al 94] or contact

NAS Parallel Benchmarks
NAS Systems Division
Mail Stop 2588
NASA Ames Research Center
Moffett Field, CA 94035

11.4   The effect of optimizing compilers

One of the goals of the original Fortran compiler was to optimize the generated assembly code as much as possible. In the late 1950s when Fortran was designed, most programmers wrote code in assembly language. To do so, they were required to understand not only the assembly language, but also the organization of the machine on which it ran. Many of these programmers included all types of tricks in their code (modifying instructions in memory, using the same data location for more than one value, storing data within instructions to save memory space, etc.) in order to squeeze out the best performance for both execution speed and memory space. It was difficult for these programmers to believe that a compiler could generate as efficient a program as they could. The early Fortran compilers were forced to meet this challenge.

The fact that Fortran came close to its original goal and has improved in performance ever since is as much of the reason for its use in high-performance scientific computing as its portability. It quickly became the language for mathematical software such as LINPACK and many commercial and research software libraries. As the language matured, more features were added to make both performance and scientific programming easier and more efficient. This in turn made it even more popular among the scientific computing community.

Code that has been generated by an optimizing compiler should run faster than code that has not been optimized. The higher the level of optimization, the faster the code should execute. Therefore as part of recording benchmark results, it is essential to identify the compiler and the level of optimization used for the compilation. It is quite possible for separate benchmarking runs on the same architecture to produce disparate results, simply because different compilers or optimizations are used to create the benchmarking object code.

Recall that the BLAS routines may be available in a fast assembly code version on some machines. Other assembly code libraries for common operations exist as well. These might be accessed by the compiler with a library option pointing to the given library of routines; they might also be automatically included when requesting a high level of optimization from a compiler. Compiled programs that include such routines run more efficiently than similar programs with the same routines compiled from high-level language source code.

11.4.1   Possible compiler optimizations

The goal of compiler optimization is to generate efficient assembly code. This is done at different levels and stages during the compilation process.

One optimization technique is the recognition of *common subexpressions* Consider the Fortran statements taken from an advection code shown in figure 11.10(a). The variable names and expressions used in this loop closely resemble those in the formulas describing this operation, making the code more understandable for the human reader. The subexpression

```
0.5D0*SIN(2.0D0*PI*(I+J)*H)
```

is used in two of the statements in the loop. A local optimizing stage of the compiler may generate code that first creates a temporary variable containing the common value and then substitutes it into the statements as given in figure 11.10(b). Simply calculating this common subexpression beforehand and then substituting its computed value into the two statements removes four multiplications, one addition, and one SIN function evaluation from the code without altering any of the results of the computation. Since these statements fall within a loop that executes n^2 times, $4n^2$ multiplications, n^2 additions, and n^2 evaluations of the SIN function have been eliminated.

Further optimization is possible by examining the loop for more common subexpressions. For instance, the value of N/2.0D0 is used twice within the loop. So we may add a statement to compute that value and use the new variable name in the original expressions, saving one division per loop iteration. Since the value of N does not change during the execution of the loop, it is possible to remove the computation of N/2.0D0 from the body of the loop and place it ahead of the loop. In this manner, it only executes once, not n^2 times. In fact, it is possible to create statements to compute all of the following subexpressions

```
CSBE1 = 2.0D0*PI*H
CSBE2 = -16*H*H
CSBE3 = N/2.0D0
CSBE4 = E*H
```

and put them before the loop as the values of these variables are not modified during the loop. This is shown in figure 11.10(c). By this simple *reordering of statements*, the common subexpressions are only computed once, before the loop begins. This eliminates $5(n^2 - 1)$ multiplications and $2(n^2 - 1)$ divisions in addition to the operations already saved by the use of the first common subexpression. Reordering of statements can be used in other ways to reduce execution costs as well.

Loop unrolling is another optimization technique. By repeating the body of a loop several times within that loop, the number of iterations needed for the loop is divided by the number of body repetitions. In turn, this reduces the number of clock cycles spent updating the loop variable and checking for the end of the loop; in other words, it reduces the *loop overhead*. For example,

```
      DO J=0,N-1
        DO I=0,N-1
          U(I,J) = 0.5D0*SIN(2.0D0*PI*(I+J)*H)
          V(I,J) = -0.5D0*SIN(2.0D0*PI*(I+J)*H)
          PHI(I,J,B1) = EXP(-16*H*H*((I-N/2.0D0)**2 +
     $              (J-N/2.0D0)**2))*SIN(E*I*H)
        ENDDO
      ENDDO
```

(a)

```
      DO J=0,N-1
        DO I=0,N-1
          CSBEXP = 0.5D0*SIN(2.0D0*PI*(I+J)*H)
          U(I,J) = CSBEXP
          V(I,J) = -CSBEXP
          PHI(I,J,B1) = EXP(-16*H*H*((I-N/2.0D0)**2 +
     $              (J-N/2.0D0)**2))*SIN(E*I*H)
        ENDDO
      ENDDO
```

(b)

```
      CSBE1 = 2.0D0*PI*H
      CSBE2 = -16*H*H
      CSBE3 = N/2.0D0
      CSBE4 = E*H
      DO J=0,N-1
        DO I=0,N-1
          CSBEXP = 0.5D0*SIN(CSBE1*(I+J))
          U(I,J) = CSBEXP
          V(I,J) = -CSBEXP
          PHI(I,J,B1)= EXP(CSBE2*((I-CSBE3)**2 +
     $              (J-CSBE3)**2))*SIN(CSBE4*I)
        ENDDO
      ENDDO
```

(c)

Figure 11.10: An example of common subexpressions. (a) The original code in the
initw2d1.f subroutine used to initialize array values within a larger advection code.
(b) Computing one subexpression separately and substituting its value in the other
statements. (c) Computing additional subexpressions outside the loop.

consider the simple loop shown in figure 11.11(a). The result of executing this loop should be the same as that result from executing the code displayed in figure 11.11(b). In the second block of code, the loop from the first example has been unrolled six times. This means that the loop overhead should be one-sixth of that for the first example. The second loop takes care of the last few elements when N is not divisible by six. Loop unrolling also produces more instructions in the execution code; thus more instruction is space required.

Optimizing compilers keep track of where variables are assigned values and where they are used in the program. For instance, in the following code segment

```
W = 3.5
WSQ = W * W
WHALF = W / 2.0
```

the variable W is *assigned* the value 3.5 in the first statement, and it is *used* in the second and third statements to assign values to WHALF and WSQ. The use of an *unassigned* variable (a variable that has not been assigned a value) may cause a compiler warning message. Likewise some variables are assigned values, but never used. This may be due to an error (as in the case of unassigned variables) or may be the result of a program modification. When an *unused variable* is found, an optimizing compiler may remove the entire statement that defines the variable by simply not generating code for it. In other words, unused variables might not be computed.

A compiler may *remove a loop* if it determines that the result of the computation in that loop is not needed later in the program. For instance if no elements of the array A are used in the program following the loop in the examples in figure 11.11, there is no need to execute that loop; so the compiler removes the loop. However if just one element of the array is used later, perhaps by printing out the first element as a double check at the end of the program, the loop is not removed.

A loop may also be replaced by a simpler computation. Consider the loop in the example shown in figure 11.12(a). Since the number of iterations for this loop is constant, the compiler can tell that the value of X after the completion

```
     DO I=1,N
        A(I) = B(I)
     ENDDO
```

(a)

```
     NN = (N/6) * 6
     DO I=1,NN,6
        A(I) = B(I)
        A(I+1) = B(I+1)
        A(I+2) = B(I+2)
        A(I+3) = B(I+3)
        A(I+4) = B(I+4)
        A(I+5) = B(I+5)
     ENDDO
     DO I=NN+1,N
        A(I) = B(I)
     ENDDO
```

(b)

Figure 11.11: An example of loop unrolling. (a) The original simple loop. (b) The loop unrolled to a depth of six.

```
     DO I=1,100
        X = X + 2
     ENDDO
```

(a)

```
     X = X + 200
```

(b)

Figure 11.12: An example of loop replacement. (a) A replaceable loop. (b) A statement that produces output equivalent to that of the loop.

of the loop should be 200 more than it was at the beginning of the loop. Thus it may simply replace the loop by the statement given in figure 11.12(b). *Replacing the loop* with a single statement eliminates 99 additions plus the loop overhead.

Another trick of optimizing compilers is to replace a procedure call by the procedure code itself; this is called *in-line expansion*. For example, consider the pieces of Fortran code given in figure 11.13(a). Since this subroutine is so simple, it makes sense to put it directly into the program, replacing the CALL statement and removing the subroutine as shown in figure 11.13(b). This eliminates the cost of a procedure call at the expense of increasing the size of the program object code.

11.4.2 Assembly language

The programmer must be aware of all these tricks of optimizing compilers when writing programs to measure performance. The statements to be timed or measured may be combined, moved, modified, replaced, or removed by the compiler, and the generated code might not resemble the code to be measured at all. The best way to be sure that unwanted changes have not been made by the compiler is to examine the assembly code itself. This is not as difficult as it may seem.

Consider the Fortran program displayed in figure 11.14(a). If we compile this with no optimization and look at the assembly code, we should be able to see the difference between the two methods of computing the value for X. Figure 11.14(b) shows the compilation of this program with the -S option to produce an assembly code listing file (with a .s extension).[5] It also shows the results of an execution of the program to check that the program runs correctly. Both the compilation and run were done on a Sun SPARCstation 10. The assembly code listing contains 118 lines. The code generated for the loop is shown in figure 11.15(a). The opcode ld is a load operation; the first of these loads the value of 200 into floating-point register 1 (f1) and the second

---

[5]The -S option works for many Fortran compilers in the UNIX environment such as the UNIX f77 compiler for the Sun SPARCstations and the SGI Indigos; but it does not work for others, such as the DEC Fortran compiler. Check your Fortran manual to find the correct option for your compiler.

```
    ...
    CALL MYSUB (A, B, N)
    ...
    SUBROUTINE MYSUB (X, Y, M)
    INTEGER I, M
    REAL X(M), Y(M)

    DO I=1,M,2
       X(I) = Y((I+1)/2)
    ENDDO

    RETURN
    END
```

(a)

```
    ...
    DO I=1,N,2
       A(I) = B((I+1)/2)
    ENDDO
    ...
```

(b)

Figure 11.13: An example of inline expansion. (a) The subroutine call with the subroutine. (b) Inline expansion replaces the subroutine call.

```
        PROGRAM EXAMPL

        REAL X
        INTEGER I

        X = 0
        DO I=1,100
           X = X + 2
        ENDDO
        WRITE(*,*) X

        X = 0
        X = X + 200
        WRITE(*,*) X

        END
```

(a)

```
        % f77 -S exampl.f -o exampl
        exampl.f:
         MAIN exampl:
        % exampl
           200.000
           200.000
```

(b)

Figure 11.14: An example Fortran program with compilation and execution. (a) The sample program. (b) Compiling and executing the program.

```
        L17:
                sethi       %hi(L2000001),%o0
                ld          [%o0+%lo(L2000001)],%f1
                ld          [%17+-0xffc],%f2
                fadds       %f2,%f1,%f2
                st          %f2,[%17+-0xffc]
                add         %16,0x1,%16
                cmp         %16,0x64
                bg          L20
                nop
                b           L17
                nop
        L20:
```

(a)

```
                sethi       %hi(L2000002),%o0
                ld          [%o0+%lo(L2000002)],%f4
                ld          [%17+-0xffc],%f5
                fadds       %f5,%f4,%f5
                st          %f5,[%17+-0xffc]
```

(b)

```
                sethi       %hi(L2000000),%o0
                ld          [%o0+%lo(L2000000)],%f3
                st          %f3,[%17+-0xffc]
```

(c)

Figure 11.15: Portions of the Sun SPARCstation 10 assembly code listing for the
Fortran program given in figure 11.14. (a) Assembly code generated for the DO loop.
(b) Assembly code generated to add 200 to X. (c) Assembly code to initialize X to
zero.

loads X into f2. The **fadds** is a floating-point add operation that adds the values of the two floating-point registers and places the result back into f2. The contents of f2 are then stored in the location for X in main memory by the st operation. The remainder of the code in the loop adds one to the loop index, compares it to the final value for I, and branches back to statement L17 (if the loop is to be repeated) or to statement L20 (if the loop is finished). The **nop**'s (no operation) are fillers for dead cycles in the operation of the processor.

Figure 11.15(b) shows the corresponding code for the addition statement. Here the value of 200 is loaded in to f4, and X is put into f5. The result of the addition is also in f5 and is stored back into X. It should be clear that this code takes fewer cycles.

Furthermore this example shows that if you have an idea of how the generated code might appear, you can locate the loops and other statements within an assembly code listing. For instance, in both cases, the assembly code statements above are preceded by the statements in figure 11.15(c). These are the statements that initialize X to the value zero. In both cases, the sample code statements are followed by **call** statements to set up the **WRITE**'s.

This same program was also compiled with an assembly code listing on an SGI Indigo containing a MIPS processor. This compiler includes the original Fortran statements as comments in the assembly code, making it easier to read. The assembly code for the first initialization of X and for the DO loop is given in figure 11.16. The operation codes are different, but it is still possible to understand what is going on. The floating-point register operands begin with $f, while general purpose register operands only have the $. The line numbers from the original Fortran program have been included to aid the programmer.

It does take a little practice to read assembly code, but it is not too difficult. You do not need to be able to write assembly code to be able to understand most of it. Manuals explaining the assembly codes for different processors are usually available from the manufacturer if needed.

11.4.3   Compiler optimization levels

Most compilers have different levels of optimization from no optimization to as much optimization as possible for the compiler. In some cases, the highest level

```
#   6           X = 0
        li.s    $f4, 0.0000000000000000e+00
        s.s     $f4, 28($sp)
        .loc    2 7
#   7           DO I=1,100
        .loc    2 7
        li      $14, 1
        sw      $14, 24($sp)
        .loopno 0 0
$32:
        .loc    2 8
#   8             X = X + 2
        l.s     $f6, 28($sp)
        li.s    $f8, 2.0000000000000000e+00
        add.s $f10, $f6, $f8
        s.s     $f10, 28($sp)
        .loc    2 9
#   9           ENDDO
        lw      $15, 24($sp)
        addu    $24, $15, 1
        sw      $24, 24($sp)
        bne     $24, 101, $32
```

Figure 11.16: Part of the SGI Indigo assembly code listing for the Fortran program given in figure 11.14. This includes the first initialization of X to zero and the DO loop. This assembly code listing should be compared to the SPARC assembly code shown in figure 11.15. At first glance they appear quite different. However they both use labels (L17 versus $32); it is possible to guess at the meaning of many of the operation codes from the neumonics; and both general purpose registers and floating-point registers are distinguishable as operands.

of optimization may not be perfected, and the compiler writers may warn of possibly incorrect results using that level under certain circumstances. The -O option ('O' for Optimization) for a UNIX compiler requests the default level of optimization; this level may have been included during the installation of the compiler and may be different on different machines. If the -O is immediately followed by an integer, a particular level of optimization is requested, with 0 or 1 being the lowest optimization level. For instance, the -O2 option requests the second level of optimization.

When measuring performance of a machine on a certain program or set of programs, the same compiler and the same level of optimization should be used for all compilations. Easy duplication of reported results is important. Therefore all the compiler and system parameters associated with any performance measurements must be recorded.

11.5 Other architectural factors

As we already mentioned, the performance of a computer is also affected by architectural features other than the processor in the CPU. In this section, we consider some of those features.

11.5.1 Memory hierarchy

The need to access data values and instructions from memory is one bottleneck in computing speed. Large memories tend to be slow to access. The average time to read a data item from a large memory is greater than the time required for arithmetic and logical operations; in fact, it can be a few orders of magnitude greater. Fast memories tend to be small; they cannot hold all the values needed for a reasonably-sized program. Fast memories are also more expensive than large, slow memories.

To demonstrate the memory bottleneck, consider the following example. Suppose a machine is capable of adding two numbers together within three clock cycles. Also assume that the machine requires five clock cycles to fetch each of the two operands and another five to store the result. This means that an addition on a machine with only a main memory takes more than the three

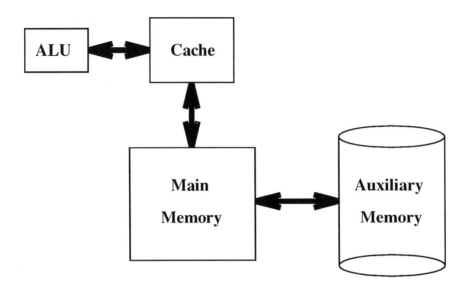

Figure 11.17: The structure of a basic memory hierarchy.

clock cycles; it requires at least 18 cycles.

Computer design seeks to overcome this problem with a hierarchy of memory devices. The idea is to place the data that is accessed more often into small, fast memories while storing the rest in slower memories. The basic memory hierarchy of a computer is thus broken up into main memory, auxiliary memory, and cache, as shown in figure 11.17. The arithmetic-logical unit (ALU) of the CPU contains the register file; this may be considered a form of small, fast, expensive memory.

ALU register accesses are usually completed in one clock cycle. The cost of a cache access is usually 2 to 4 times the cost of accessing a register value. To access data from main memory may require from 5 to 10 CPU clock cycles. And the cost of accessing auxiliary memory may be as much as 20,000,000 nsec or a few million clock cycles.

When referring to memory hierarchy components, the term *latency* is used to mean response time, and *bandwidth* is used instead of throughput. The basic definition of bandwidth is the maximum number of bits (or bytes) per second that can move along a channel from one computer component to another. This can also be called the communication rate or the transfer rate.

The memory hierarchy is effective because of the *principle of locality of reference*. *Spatial* locality infers that data items stored near each other in the address space of the memory tend to be accessed near each other in time. *Temporal* locality implies that the set of data items referenced over a fixed time window changes slowly with time. This means that the data items being accessed by a program at a given time will probably be accessed again in the near future as might their neighboring data items. Putting the most recently accessed data items into a fast memory so that they can be accessed again quickly should make efficient use of that small memory. Further, the contents of this small, fast memory should be dynamic, slowly changing over the course of the execution of the program.

A *cache* is a small, fast memory that is sometimes placed on the same chip with the CPU. It holds copies of recently accessed words that, according to the principle of temporal locality, are likely to be accessed again. Data items from a cache can be accessed quickly. The reuse of words in the cache saves multiple main memory accesses that require more clock cycles. Values can also be written directly to the cache and copied to main memory at a later time, avoiding the delay often caused by writing to memory.

The size of a block of data that is copied from (or to) the main memory to (or from) the cache in one memory access is called a *cacheline*; a cacheline may contain 8 to 128 bytes of data, depending on the machine. In other words, a cacheline usually contains more than the one data item being accessed. This means that some of the neighboring data items are also placed into the cache at the same time as the desired data item. According the the principle of spatial locality, these neighboring items may be needed as well.

When a data value is requested from memory, the cache is first checked to see if the value is there. If it is, a *cache hit* has occurred, and the value is obtained from the cache. If the data is not in the cache, this is called a *cache miss*, and the cacheline containing the needed value is fetched from the main memory and put into the cache; this requires a table of current cache entries called the *cache table* to be updated. The requested data value is then retrieved from the cache. Any later requests for that data value will be cache hits and can be fetched from the cache. The previous cacheline destroyed by the new cacheline needs to be written back to memory first only if it has been modified and not yet written to main memory.

The *hit ratio* of a process running on a given cache is the average of the number of cache hits divided by the number of accesses:

$$h = \frac{\#\text{hits}}{\#\text{accesses}}.$$

The *miss ratio* is $1 - h$. These provide a measure for the performance of the cache.

Most computers today have two caches: one for data values and one for instructions. Loops are the backbone of most high-performance scientific computing, and the instruction caches are usually able to store all the instructions for a loop iteration, thus reducing instruction fetch time. Some machines also have two or more levels of caches between the CPU and the main memory.

The *main memory* contains data and instructions for each executing program in units of *pages*. The size of a page may be from 256 to 8192 bytes, depending on the machine. The page size is a multiple of the cacheline size for a given machine. The *memory access time* of a machine is the time that is required for a data value or an instruction to be fetched from the main memory; this is the cost of reading the cacheline containing that data item from main memory into the cache. Writing a cacheline from the cache to main memory usually requires a similar amount of time.

Most computers today employ *virtual memory*. Executing programs are assigned a limited number of pages in main memory; that memory is shared by other processes and the operating system. *Auxiliary memory* (usually a disk) is the location of the rest of the data and program instructions for each user, not just those pages currently being used by each program. If a program needs to access some data or instructions not currently in the main memory, we say that a *page fault* has occurred; the program halts while one of its pages is written out to the auxiliary memory and the new page containing the desired data items is read into the main memory. This involves writing out the old page to disk (if it has been modified), reading in the new page from the disk, and updating a table of current page entries known as the *page table*; all of this takes considerably more time than a simple memory access. Sometimes the system even swaps out the executing program, replacing it with another program that is ready to go, until the new page has been installed.

When the set of data items currently being used by the executing program cannot all fit into the cache at the same time, the process is slowed by the constant fetching and replacing of cachelines containing the needed data. This situation is called *thrashing*. Thrashing may also happen between main memory and auxiliary memory when the number of pages assigned to a process is not great enough to hold the set of pages being used by that process. Hence, the size of a cacheline, the size of the cache, the size of a page, and the number of pages in main memory may all have an effect on the overall performance of the memory hierarchy of the machine. Thus, in addition to the various access times of the different elements of the memory hierarchy, the sizes of these components are factors in the performance of the machine itself as are the page size or cacheline size used by each component.

11.5.2 Pipelined processors

One of the early methods for speeding up the execution of a program allows the next instruction to be fetched from memory while the current instruction is being executed. Pipelined processors take this idea a bit further. One instruction may be fetched while a second one is being decoded, a third one is fetching its needed operands, a fourth one is performing the desired operation, and a fifth instruction is writing its result. This is diagrammed in figures 11.18 and 11.19. In other words, the different parts involved in the execution of one instruction are broken up into stages; each stage may operate concurrently with the other stages on an independent instruction within the processor instruction pipeline. In many respects, this is similar to an assembly line in a factory. If each stage in the pipeline only takes one clock cycle, the processor is capable of executing one instruction per cycle; that is, it has CPI = 1. However, it is not possible to keep the pipeline full all of the time because of branch instructions and other factors; so CPI = 1 represents an ideal (peak) value.

Early forms of pipelined processing appeared in the IBM/360 Model 91 and in the CDC 6600. Among the more recent examples of pipelined processors are several RISC (Reduced Instruction Set Computer) processors, including the IBM RS/6000 processor, the line of MIPS processors, the HP-PA processors, and the Sun SPARC processors.

Some CPU architectures contain more than one instruction pipeline and so allow more than one type of instruction to execute at the same time. If a ma-

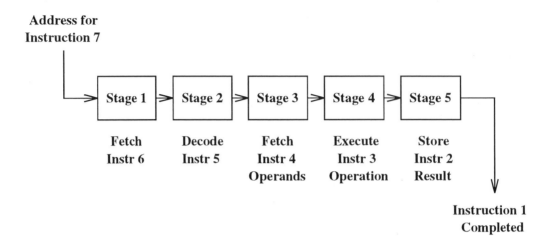

Figure 11.18: The structure of a basic processor pipeline.

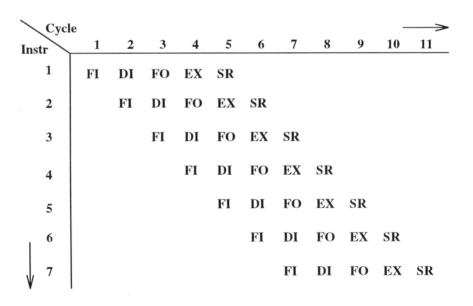

Figure 11.19: Overlapping the stages of instruction execution in a basic processor pipeline.

chine is capable of executing more than one instruction per CPU clock cycle, it is called a *superscalar* computer. For example, the IBM Power2 CPU contains an instruction-pipelined, RISC processor with multiple, separate pipelines for integer operations and for floating-point operations; it is capable of completing two integer operations, two floating-point add operations, and two floating-point multiplications during the same CPU clock cycle, thus yielding an ideal processing speed of CPI = 6.

11.6 Vector and parallel computers

Supercomputers employ various forms of vector and parallel architectures to achieve their very high speed. The measurement of the performance of these machines needs to include factors related to these architectural features. In this section, we briefly describe the basic design of some supercomputers and the effect of these designs on their performance. This information is included here for completeness. More thorough discussions of the architectures and the performance of these machines can be found in chapters 12, 13, and 14.

11.6.1 Vector processors

A *vector computer* or *vector processor* contains a set of special arithmetic units known as *vector* or *arithmetic pipelines*. These pipelines are used to process the elements of vectors or arrays efficiently by overlapping the execution of different parts (*stages*) of an arithmetic operation on different elements of the vector. Again, this is similar to an assembly line where the arithmetic operation is performed in stages: each element in the pipeline is handled, concurrently and independently, at a different stage of the arithmetic operation. The Convex C3880 and the Cray Y-MP are vector processors.

A *vector* is an one-dimensional array of values. A vector processor operates on entire vectors instead of on the more conventional scalar operands. Some of these machines contain *vector register* to hold many elements of a vector at a time; on the Cray Y-MP, a vector register holds 64 elements. The vector registers feed the values of the vector elements into the arithmetic pipelines, allowing the pipeline to perform one arithmetic operation per time unit (usually one clock cycle).

Pipelining of operations permits parts of those operations to be done in parallel. However, the parallelism is not complete until the first few vector elements have filled the pipeline. This *priming* of the pipeline accounts for part of the *startup time* or overhead in getting a vector operation flowing through the pipeline.

The following terms are used to characterize vector processor performance:

- $\mathbf{R_n}$ For a vector processor, the number of Mflops obtainable for vector operands of length n.

- $\mathbf{R_\infty}$ The asymptotic value of $\mathbf{R_n}$ as $n \to \infty$. Note that the startup time becomes completely negligible as $n \to \infty$; hence $\mathbf{R_\infty}$ represents the peak speed for a vector operation.

- $\mathbf{n_{1/2}}$ The length m of a vector such that $\mathbf{R_m}$ is equal to $\mathbf{R_\infty}/2$. This is called the *half-performance length* of the machine and represents the length of a vector needed to produce half the peak performance.

Table 11.3 summarizes the characteristics of four Cray models while table 11.4 compares the performance of those four Crays. These figures are based on data from [Dongarra 94] and [Hockney & Jesshope 88].

For more information on vector processors and their performance, see chapter 12 on vector computing.

11.6.2   Parallel computers

A parallel computer is a machine with two or more connected processors that may operate in parallel. Such a machine is also called a multiprocessor. These machines may be broken up into two main types: MIMD (**M**ultiple-**I**nstruction, **M**ultiple-**D**ata stream) and SIMD (**S**ingle-**I**nstruction, **M**ultiple-**D**ata stream).[6] In other words, the processors of a MIMD multiprocessor may

---

[6]These two terms are from Flynn's taxonomy. The other two classifications are SISD (**S**ingle-**I**nstruction, **S**ingle-**D**ata stream) and MISD (**M**ultiple-**I**nstruction, **S**ingle-**D**ata stream). The first of these includes the common von Neumann computer; this machine performs one instruction at a time on one set of data. The second set is considered to be empty. [Hockney & Jesshope 88]

Characteristic	Cray Model			
	1	**X-MP**	**Y-MP/832**	**C90**
Year Introduced	1976	1983	1989	1990
Max # Procs	1	14	8	16
# Vector Regs	8	8	8	8
Clock Cycle	12.5 nsec	8.5 nsec	6.0 nsec	4.2 nsec
Max Memory	1Mbytes	16 Mbytes	128 Mbytes	16 Gbytes
Word Size	64 bits	64 bits	64 bits	64 bits
Vector Reg Size	64 words	64 words	64 words	64 words

Table 11.3: A comparison of the characteristics of some Cray vector supercomputers, based on data from [Dongarra 94] and [Hockney & Jesshope 88].

Performance	Cray Model						
	1	**X-MP**		**Y-MP/832**		**C90**	
# Processors	*1*	*1*	*4*	*1*	*8*	*1*	*16*
Clock Cycle (nsec)	12.5	8.5	8.5	6.0	6.0	4.2	4.2
Peak Performance (Mflops)	160	235	940	333	2667	952	15238
LINPACK, $n = 1000$ (Mflops)	110	218	822	324	2144	902	10780
Shallow Water (Mflops)	—	—	560	—	1532	—	—
$\mathbf{R}_\infty$ (Mflops)	22	70	70	—	—	15000	15000
$\mathbf{n}_{1/2}$	18	53	53	—	—	650	650

Table 11.4: A comparison of the performance of some Cray vector supercomputers, partially based on data from [Dongarra 94].

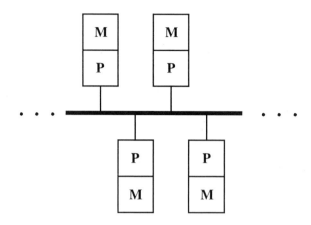

Figure 11.20: A distributed-memory MIMD multiprocessor.

act independently of one other on different data, while the processors of a SIMD machine perform the same instruction at the same time but on different data. The category of MIMD multiprocessors is also divided into two subtypes: those with distributed memory and those with shared memory. Look at chapters 13 and 14 for more information on these multiprocessors.

Benchmarks for the purpose of comparing the performance of various parallel computers are still being developed. The most commonly-used benchmarks include the NAS Kernels and a version of the LINPACK benchmark for solving a linear system of equations with a coefficient matrix of size 1000×1000. Both of these benchmarks may be altered by the user to take advantage of architectural features of the given parallel machine. Table 11.6 on page 399 provides results from both of these benchmarks for some of the parallel machines discussed below. Another set of benchmarks for parallel machines is being developed by the Perfect Club, a group of representatives from manufacturers and universities.

The USENET newsgroup named `comp.parallel` includes discussions of current and new parallel architectures, as well as performance measurements on these machines. Announcements of related conferences and workshops are included here as well.

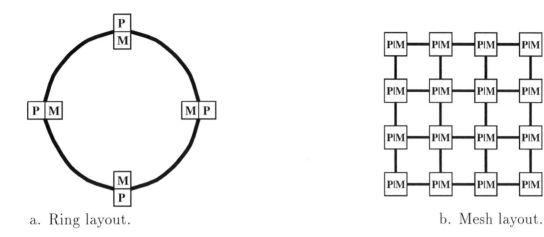

a. Ring layout.   b. Mesh layout.

Figure 11.21: Two possible layouts of a distributed-memory MIMD multiprocessor.

Distributed-memory MIMD multiprocessors

Each of the individual processors of a distributed-memory multiprocessor (DM-MIMD) has a memory unit associated directly with it; that is, each processor has its own local memory. A processor together with its associated memory is called a *node* and is connected to the other nodes by some type of network as displayed in figure 11.20. The nodes of a DM-MIMD multiprocessor send messages to the other nodes and receive messages from them in order to distribute and obtain data. For this reason, these machines are also called message-passing MIMD computers

The network joining the nodes of a DM-MIMD multiprocessor is constructed in various ways; two of these are displayed in figure 11.21. The nodes may each be connected to two neighboring nodes to form a ring. Or they may be connected in a rectangular mesh structure, like the Intel Paragon.

The network of nodes that make up a DM-MIMD multiprocessor is often connected to a *front end* or *host* machine. This is usually a workstation that handles all the input and output for the multiprocessor; it is also used to edit, compile, and store programs.

The cost of communicating data between different nodes of a DM-MIMD multiprocessor is an important element affecting the performance of this type

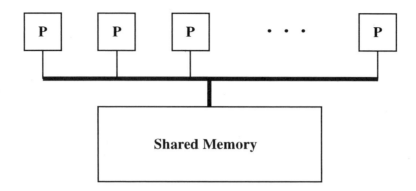

Figure 11.22: A shared-memory MIMD multiprocessor.

of machine. In the early Intel iPSC/1 hypercube, it took about 1,000 times as many clock cycles to pass a data value between two nodes as it did to access that value from local memory in one node. While the communication costs have improved since then, so have the costs for computation; thus communication costs are still an important performance factor.

Another example of a DM-MIMD multiprocessor is the CM-5 by Thinking Machines Corp. Each of its nodes contains four vector processors; hence, this machine can also be described as a distributed-memory/vector-processor MIMD multiprocessor (DM/V-MIMD).

For more information on DM-MIMD multiprocessors, see chapter 13. The theoretical peak performance for some of the machines mentioned above is given in table 11.5 on page 395. Table 11.6 on page 399 compares benchmark results for these machines.

Shared-memory MIMD multiprocessors

A shared-memory MIMD multiprocessor (SM-MIMD) has a single memory that may be accessed by any and all of the processors. Early SM-MIMD machines provided access to the memory by a bus with a wide bandwidth; this connects all the processors to the memory in a linear manner, as shown in figure 11.22. Among such machines are the Encore Multimax and the Sequent Balance. The Alliant FX/8 is a similar machine, with the addition

Machine	Manufacturer	Type of Multiproc	Max Procs	Theor. Peak Mflops
iPSC/860	Intel	DM-MIMD	128	5,120
SP1 (9076)	IBM	DM-MIMD	64	8,000
Paragon	Intel	DM-MIMD	4,000	300,000
Cray T3D	Cray Research	DM-MIMD	2,048	307,000
CM-5	Thinking Machines	DM/V-MIMD	16,384	2,000,000
C3800	Convex	SM/V-MIMD	8	960
Symmetry	Sequent	SM-MIMD	30	1,300
Cray Y-MP	Cray Research	SM/V-MIMD	8	2,600
Cray-3	Cray Comp Corp	SM/V-MIMD	16	16,000
KSR1	Kendall Square	S/DM-MIMD	1088	43,500
KSR2	Research	S/DM-MIMD	5000	400,000
MP-1	MasPar	SIMD	16,384	550
MP-2	MasPar	SIMD	16,384	2,400
CM-2	Thinking Machines	SIMD	65,536	31,000

Table 11.5: A list of distributed-memory MIMD, shared-memory MIMD, and SIMD multiprocessors with their maximum number of processors and their theoretical peak performances with 64-bit floating-point numbers. See [van der Steen 94].

of vector processors attached to each processor; it may be called a shared-memory/vector-processor MIMD multiprocessor (SM/V-MIMD). The Convex C3 computers are also SM/V-MIMD processors, as are the Cray X-MP and the Cray Y-MP. Again, the theoretical peak performance for some of these machines is shown in table 11.5 on page 395, while table 11.6 on page 399 compares benchmark results for these machines.

An important problem for SM-MIMD multiprocessors is *scalability*, the ability to increase the number of processors significantly with a corresponding increase in performance. As the number of processors increases, so does the amount of traffic on the bus connecting the shared-memory and the processors. Increasing the bandwidth of the bus alleviates this to some extent; however, many programs require the use of common variables by most or all the processors. If the value of one variable is updated by each processor, then the processors must line up and wait for serial access to that variable. This is called *contention*. If each of the processors has a private cache, common variables can be kept there, eliminating part of this problem. However, if any of these variables is modified, then each of the caches must be updated to reflect the change; this is called the *cache coherency* problem. Because of bus traffic limitations, most multiprocessors of this design have a maximum of 20 to 30 processors.

Other MIMD multiprocessors

Some MIMD architectures are built as DM-MIMD multiprocessors, but have software that allows them to behave as SM-MIMD multiprocessors (S/DM-MIMD). Data can be passed from the memory of one processor to the memory of another processor without explicit message-passing, thereby allowing simpler programs. The BBN Butterfly was one of the first of this type of machine; the connection of the processors was done with a butterfly switch network (see [Almasi & Gottlieb 94] or [Hwang 93]). The KSR1 by Kendall Square Research is also a member of this group of multiprocessors; this architecture links 32 processors in a ring and connects the rings in a tree structure.

A *cluster* of workstations connected on a network can also be considered DM-MIMD multiprocessors with longer communication times between the processors. The design of the IBM SP1 originated from a cluster of IBM RS/6000 workstations connected in a network. If all the workstations are of the same type, this kind of multiprocessor is called a *homogeneous* cluster. *Heterogeneous* clusters are composed of different types of workstations connected together in some fashion. The PVM (Parallel Virtual Machine) system was developed to handle the message-passing details for this type of multiprocessor. The MPI (Message-Passing Interface) System is currently being developed as a standard for all message-passing machines, including workstation clusters. See [Geist et al 95] for more details concerning these languages.

SIMD multiprocessors

A SIMD multiprocessor directs all its processors to execute the same instruction at the same cycle, but with different data. Like DM-MIMD multipro-

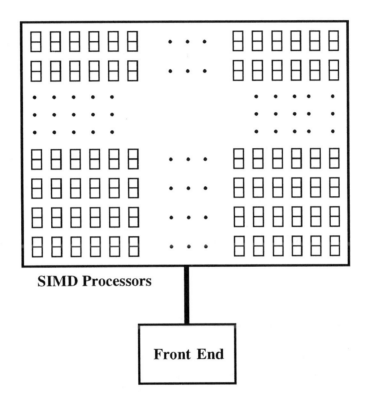

SIMD Processors

Front End

Figure 11.23: A SIMD multiprocessor has two main components: the front end and the bank of SIMD processors. Each SIMD processor and its memory is connected by a network (not shown) to other processors in the SIMD processor bank.

cessors, SIMD machines are often dependent on a front end or host machine that dispatches a sequence of instructions to the SIMD machine. The basic structure of this type of machine is similar to that shown in figure 11.23. Each processor has access to its own private memory containing the data for that processor. Examples of SIMD multiprocessors include the CM-2 from Thinking Machines and the MasPar MP-1 and MP-2. Array processors, such as the ICL DAP, also fall into this category. Table 11.5, page 395, provides the theoretical peak performance for some SIMD multiprocessors, and table 11.6, page 399, compares benchmark results for these machines. For more information on SIMD multiprocessors, see chapter 14 on SIMD computing.

Performance of parallel programs

When considering the performance of a parallel program on a multiprocessor, it is usually compared to the performance of the same program on a single processor of that machine.

Let the symbol T_1 represent the elapsed time of the program run on a single processor. And let T_p represent the time for the program running on p processors of the parallel machine. Then the *speedup* of the program is defined as

$$\textbf{Speedup} \; = \; S(p) \; = \; \frac{T_1}{T_p}.$$

In an ideal situation, every part of the program could run in parallel, and the speedup would be equal to p. Since this is not often possible, a parallel program is considered to be quite good if its speedup is close to p.

A related parameter is the *efficiency* of a parallel program. This is defined as

$$\textbf{Efficiency} \; = \; E(p) \; = \; \frac{S(p)}{p}.$$

The values of $E(p)$ should range between $\frac{1}{p}$ and 1; an ideal value for $E(p)$ would be 1, meaning that $S(p) \; = \; p$ and that the processors are 100% utilized throughout the execution of the program.

Amdahl's Law asserts that speeding up the execution time of one part of a program can at most decrease the entire execution time by the time spent on that part of the program. This means that those operations of a parallel

Machine	Clock Cycle nsec	Number of Procs	LINPACK $n = 1000$ Mflops	NAS Parallel	
				Int Sort	Embar Par
Intel iPSC/860	—	128	219	0.84	4.91
Intel Paragon	20	128	—	0.86	26.02
IBM SP1	16	64	—	3.7	20.79
IBM SP2	—	16	—	3.68	9.71
CM-5	31.25	32	—	0.27	5.88
		128	—	0.96	23.49
		512	—	—	90.47
Convex C3800	16.7	4	425	—	—
		8	795	—	—
Convex C4/XA2	7.41	2	1320	—	—
Cray Y-MP/832	6	1	275	1.0	1.0
		8	2144	6.19	7.95
KSR1	—	32	513	1.1	1.81
		64	—	1.7	4.9
KSR2	25	32	—	1.6	5.1
		64	—	2.94	9.71
MasPar MP-1	83	16,384	—	1.00	1.82
MasPar MP-2	80	16,384	—	1.49	5.63
CM-2	—	65,536	—	0.77	6.71

Table 11.6: Selected benchmark results for a group of distributed-memory MIMD, shared-memory MIMD, and SIMD multiprocessors. Performance figures taken from [Bailey et al 94], [van der Steen 94], and [Dongarra 94].

Amdahl's Law				
	$T_s = 5 \% T_1$		$T_s = 50 \% T_1$	
p	T_p	$S(p)$	T_p	$S(p)$
1	1.0000	1.0000	1.0000	1.0000
2	0.5250	1.9048	0.7500	1.3333
4	0.2875	3.4783	0.6250	1.6000
8	0.1688	5.9259	0.5625	1.7778
16	0.1094	9.1429	0.5312	1.8824
32	0.0797	12.5490	0.5156	1.9394
64	0.0648	15.4217	0.5078	1.9692
128	0.0574	17.4150	0.5039	1.9845

Table 11.7: Values of T_p and $S(p)$ for $T_s = 5\%$ of T_1 ($T_s = T_1/20$) and $T_s = 50\%$ of T_1 ($T_s = T_1/2$), according to Amdahl's Law.

program that must be done sequentially restrict the potential speedup for that program. For example, suppose that 50% of the time to execute a program sequentially is spent executing instructions that could be executed in parallel while the remaining 50% is spent executing instructions that must be executed sequentially. Then even in the best case, where the parallel execution is done in virtually zero time, the total execution time is only reduced by one half. Amdahl's law is a formal statement of this observation.

Let T_s represent the time required for the sequential operations of a parallel program, and let $T_{\parallel}$ represent the amount of time required to complete all the operations that can be done in parallel. The total time for the program to execute on a sequential computer is then

$$T_1 = T_s + T_{\parallel},$$

while the time for it to execute on a parallel computer with p processors is

$$T_p = T_s + \frac{T_{\parallel}}{p}.$$

Substituting the values for T_1 and T_p into the previous equation for speedup yields

$$S(p) \; = \; \frac{T_s \; + \; T_{\parallel}}{T_s \; + \; \frac{T_{\parallel}}{p}}.$$

For a more detailed example, suppose that T_1 is one unit of time. Let T_s be 5% of T_1 and let p vary between 1 and 128. Table 11.7 displays the values for T_p and $S(p)$ for the different values of p. The two last columns in this table show the same figures for the case when T_s is 50% of T_1 (the example discussed in the last paragraph). Figure 11.24 provides plots of these values. Notice that as the number of processors increase, the additional speedup decreases. This is because the speedup has an upper bound of $\frac{1}{T_s}$.

If the speedup of a parallel program exceeds p, the program is called *superlinear*. While Amdahl's Law claims that this should not be possible, some algorithms have achieved this distinction. In particular, several parallel search techniques allow each of the processes to go off in different directions. Some of these processes may be sent off to dead ends, while others are pursuing more profitable paths. As soon as a result is found by one process, it can signal the rest of the processes to terminate their search, thus killing those dead end processes before they would normally have completed on a sequential machine.

Tools for parallel programming

There are a number of tools to help programmers understand and improve the performance of their parallel programs. Most manufacturers of vector and parallel computers provide special operating systems and compilers with extensions that exploit the parallel (and vector) mechanisms of the machine. Both Fortran 90 [Adams et al 92] and HPF (High Performance Fortran) [Koelbel et al 94] are designed to consolidate multiple extensions into a uniform version of the Fortran programming language, allowing additional operations that could be compiled by different machines into parallel or vector operations.

Debugging tools for parallel programs are currently under research. Tracing and visualization tools assist the programmer in studying the execution of a parallel program. *Schedule* is one such tool. During the execution of a program, it collects information in order to provide data-dependency graphs and visual representations of the flow of given tasks through the processors

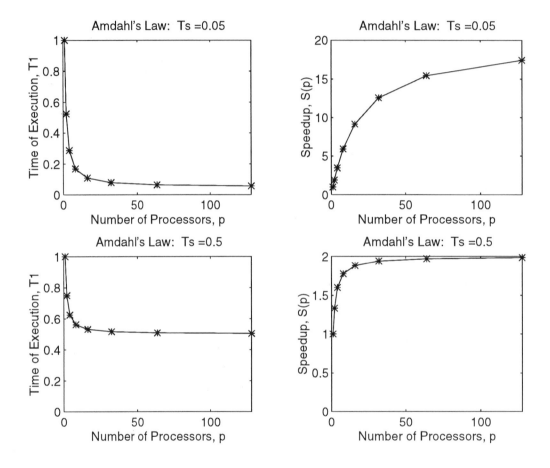

Figure 11.24: Plots of Amdahl's Law for $T_s = 5\%$ of T_1 ($T_s = T_1/20$) and $T_s = 50\%$ of T_1 ($T_s = T_1/2$), where the number of processors varies from 1 to 128.

[Dongarra & Sorensen 87]. These graphs display the allocation of different parallel tasks to different processors and so show where performance might be improved by a reallocation of those tasks.

Another tool called PICL (Portable Instrumented Communication Library) facilitates the collection of traces of send/receive operations in a DM-MIMD machine. These traces can later plotted by a subtool called ParaGraph to show the parallel execution of the traced program. This display can even be animated [Heath et al 90].

11.7 Summary

As we have seen, many factors may influence the performance of a computer. Because of this, it is difficult to compare the performance of computers with different architectures. Just determining the speed of the CPU of a computer is an intricate task. Common benchmarks and performance units have been developed to assist in the process of evaluating computer performance, but they must be used with care. Besides the effect of architectural factors such as pipelining and caching, one must take into account the effect of the compiler used on the benchmarking program. Furthermore, communication and synchronization issues in parallel computers introduce additional factors influencing computer performance. More on this can be found in chapters 12, 13, and 14.

12 Vector Computing

A *vector computer* or *vector processor* is a machine designed to efficiently handle arithmetic operations on elements of arrays, called *vector*. Such machines are especially useful in high-performance scientific computing, where matrix and vector arithmetic are quite common. The Cray Y-MP and the Convex C3880 are two examples of vector processors used today.

The first section of this chapter provides a general overview of the architecture of a vector computer. This includes an introduction to vectors and vector arithmetic, a discussion of performance measurements used to evaluate this type of machine, and a comparison of the characteristics of particular vector computers. A brief history of vector processors is provided as well, with a focus on the Cray vector architectures.

The second section considers special techniques for programming these computers. In particular, it introduces some programming language extensions used in the Cray vector supercomputers; these provide the concepts and general format of language extensions common to vector computer programming.

12.1 General architecture

To understand the concepts behind a vector processor, we first present a short review of vectors and vector arithmetic in this section. We continue by showing the application of these ideas to the hardware in vector processors. We then discuss the performance and history of some vector processors. In particular, we focus on Cray vector processors.

12.1.1   Vectors and vector arithmetic

A *vector*, v, is a list of elements

$$v = \begin{pmatrix} v_1, & v_2, & v_3, & \ldots, & v_n \end{pmatrix}^T.$$

The *length* of a vector is defined as the number of elements in that vector; so the length of v is n. When mapping a vector to a computer program, we declare the vector as an array of one dimension. In Fortran, we declare v by the statement

```
DIMENSION  V(N)
```

where N is an integer variable holding the value of the length of the vector. Throughout this chapter, we use the following terms almost interchangeably: vector, array, list.

Arithmetic operations may be performed on vectors. Two vectors are added by adding corresponding elements:

$$s = x + y = \begin{pmatrix} x_1 + y_1, & x_2 + y_2, & \ldots, & x_n + y_n \end{pmatrix}.$$

In Fortran, vector addition could be performed by the following code

```
DO  I=1,N
   S(I) = X(I) + Y(I)
ENDDO
```

where s is the vector representing the final sum and S, X, and Y have been declared as arrays of dimension N. This operation is sometimes called *elementwise* addition. Similarly, the subtraction of two vectors, $x - y$, is an elementwise operation.

12.1.2   Vector computing architectural concepts

A vector computer contains a set of special arithmetic units called *pipelines*. These pipelines overlap the execution of the different parts of an arithmetic operation on the elements of the vector, producing a more efficient execution of the arithmetic operation. In many respects, a pipeline is similar to an assembly

Step	A	B	C	D	E	F
x	0.1234E4	0.12340E4				
y	-0.5678E3	-0.05678E4				
s			0.066620E4	0.66620E3	0.66620E3	0.6662E3

Figure 12.1: An example showing the stages of a floating-point addition: $s = x + y$.

line in a factory where different steps of the assembly of an automobile, for example, are performed at different stages of the line.

In this section, we discuss how a vector pipeline operates, the advantages of this type of architecture, and other architectural features found in vector processors.

The stages of a floating-point operation

Consider the steps or stages involved in a floating-point addition on a sequential machine with IEEE arithmetic hardware: $s = x + y$.

A: The exponents of the two floating-point numbers to be added are compared to find the number with the smallest magnitude.

B: The significand of the number with the smaller magnitude is shifted so that the exponents of the two numbers agree.

C: The significands are added.

D: The result of the addition is normalized.

E: Checks are made to see if any floating-point exceptions occurred during the addition, such as overflow.

F: Rounding occurs.

Step	τ	2τ	3τ	4τ	5τ	6τ	7τ	8τ
A	$x_1 + y_1$						$x_2 + y_2$	
B		$x_1 + y_1$						$x_2 + y_2$
C			$x_1 + y_1$					
D				$x_1 + y_1$				
E					$x_1 + y_1$			
F						$x_1 + y_1$		

Figure 12.2: Scalar floating-point addition of vector elements.

Figure 12.1 shows the step-by-step example of such an addition. The numbers to be added are $x = 1234.00$ and $y = -567.8$. In deference to the human reader, these are represented in decimal notation with a mantissa of four digits.

Now consider this scalar addition performed on all the elements of a pair of vectors (arrays) of length n. Each of the six stages needs to be executed for every pair of elements. If each stage of the execution takes τ units of time, then each addition takes 6τ units of time (not counting the time required to fetch and decode the instruction itself or to fetch the two operands). So the number of time units required to add all the elements of the two vectors in a serial fashion would be $T_s = 6n\tau$. These execution stages are shown in figure 12.2 with respect to time.

An arithmetic pipeline

Suppose the addition operation described in the last subsection is pipelined; that is, one of the six stages of the addition for a pair of elements is performed at each stage in the pipeline. Each stage of the pipeline has a separate arithmetic unit designed for the operation to be performed at that stage. Once stage A has been completed for the first pair of elements, these elements can be moved to the next stage (B) while the second pair of elements moves into the first stage (A). Again each stage takes τ units of time. Thus, the flow through the pipeline can be viewed as shown in figure 12.3, where the stages of the pipeline addition execute with respect to time as in figure 12.4. (Compare figure 12.2 to figure 12.4.)

Observe that it still takes 6τ units of time to complete the sum of the first pair of elements, but that the sum of the next pair is ready in only τ more

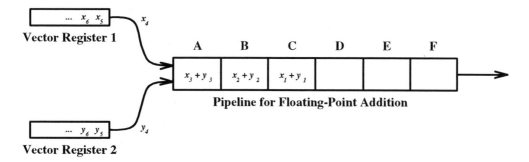

Figure 12.3: Pipeline for floating-point addition of vector elements.

Step	τ	2τ	3τ	4τ	5τ	6τ	7τ	8τ
					Time $\longrightarrow$			
A	$x_1 + y_1$	$x_2 + y_2$	$x_3 + y_3$	$x_4 + y_4$	$x_5 + y_5$	$x_6 + y_6$	$x_7 + y_7$	$x_8 + y_8$
B		$x_1 + y_1$	$x_2 + y_2$	$x_3 + y_3$	$x_4 + y_4$	$x_5 + y_5$	$x_6 + y_6$	$x_7 + y_7$
C			$x_1 + y_1$	$x_2 + y_2$	$x_3 + y_3$	$x_4 + y_4$	$x_5 + y_5$	$x_6 + y_6$
D				$x_1 + y_1$	$x_2 + y_2$	$x_3 + y_3$	$x_4 + y_4$	$x_5 + y_5$
E					$x_1 + y_1$	$x_2 + y_2$	$x_3 + y_3$	$x_4 + y_4$
F						$x_1 + y_1$	$x_2 + y_2$	$x_3 + y_3$

Figure 12.4: Pipelined floating-point addition of vector elements.

units of time. And this pattern continues for each succeeding pair. This means that the time, T_p, to do the pipelined addition of two vectors of length n is

$$T_p = 6\tau + (n-1)\tau = (n+5)\tau.$$

The first 6τ units of time are required to *fill the pipeline* and to obtain the first result. After the last result, $x_n + y_n$, is completed, the pipeline is emptied out or *flushed*.

Comparing the equations for T_s and T_p, it is clear that $(n+5)\tau < 6n\tau$, for $n > 1$. Thus, this pipelined version of addition is faster than the serial version by almost a factor of the number of stages in the pipeline. This is an example of what makes vector processing more efficient than scalar processing. For large n, the pipelined addition for this sample pipeline is about six times faster than scalar addition.

In this discussion, we have assumed that the floating-point addition requires six stages and takes 6τ units of time. There is nothing magic about this number 6; in fact, for some architectures, the number of stages in a floating-point addition may be more or less than six. Further, the individual stages may be quite different from the ones listed in section 12.1.2. The operations at each stage of a pipeline for floating-point multiplication are slightly different than those for addition; a multiplication pipeline may even have a different number of stages than an addition pipeline. There may also be pipelines for integer operations. As shown in figure 12.8, pipelines to perform vector operations on the Cray-1 have from one to fourteen stages, depending on the type of operation performed by the pipeline.

Vector registers

Some vector computers, such as the Cray Y-MP, contain *vector registers*. A general purpose or a floating-point register holds a single value; vector registers contain several elements of a vector at one time. For example, the Cray Y-MP vector registers contain 64 elements while the Cray C90 vector registers hold 128 elements. The contents of these registers may be sent to (or received from) a vector pipeline one element at a time.

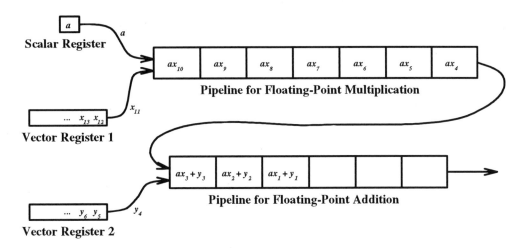

Figure 12.5: Chaining used to compute $ax + y$.

Scalar registers

Scalar registers behave like general purpose or floating-point registers; they hold a single value. However, these registers are configured so that they may be used by a vector pipeline; the value in the register is read once every τ units of time and put into the pipeline, just as a vector element is released from the vector pipeline. This allows the elements of a vector to be operated on by a scalar. To compute $y = 2.5 \times x$, the 2.5 is stored in a scalar register and fed into the vector multiplication pipeline every τ units of time in order to be multiplied by each element of x to produce y.

Chaining

Figure 12.3 is a diagram of a single pipeline. As mentioned in section 12.1.2, most vector architectures have more than one pipeline; they may also contain different types of pipelines.

Some vector architectures provide greater efficiency by allowing the output of one pipeline to be *chained* directly into another pipeline. This feature is called *chaining* and eliminates the need to store the result of the first pipeline before sending it into the second pipeline. Figure 12.5 demonstrates the use of chaining in the computation of a *saxpy* vector operation: $ax + y$, where x

and y are vectors and a is a scalar constant.

Chaining can double the number of floating-point operations that are done in τ units of time. Once both the multiplication and addition pipelines have been filled, one floating-point multiplication and one floating-point addition (a total of two floating-point operations) are completed every τ time units.[1]

Scatter and gather operations

Sometimes, only certain elements of a vector are needed in a computation. Most vector processors are equipped to pick out the appropriate elements (a *gather* operation) and put them together into a vector or a vector register. If the elements to be used are in a regularly-spaced pattern, the spacing between the elements to be gathered is called the *stride*. For example, if the elements

$$x_1, \ x_5, \ x_9, \ x_{13}, \ \ldots, \ x_{4\lfloor \frac{n-1}{4} \rfloor + 1}$$

are to be extracted from the vector

$$(\, x_1, \quad x_2, \quad x_3, \quad x_4, \quad x_5, \quad x_6, \quad \ldots, \quad x_n \,)$$

for some vector operation, we say the stride is equal to 4. A *scatter* operation reformats the output vector so that the elements are spaced correctly. Scatter and gather operations may also be used with irregularly-spaced data.

Vector-register vector processors

If a vector processor contains vector registers, the elements of the vector are read from memory directly into the vector register by a *load vector* operation. The vector result of a vector operation is put into a vector register before it is stored back in memory by a *store vector* operation; this permits it to be used in another computation without needing to be reread, and it allows the store to be overlapped by other operations. On these machines, all arithmetic or logical vector operations are register-register operations; that is, they are only performed on vectors that are already in the vector registers. For this reason, these machines are called *vector-register* vector processors.

---

[1] Conceptually, it is possible to chain more than two functional units together, providing an even greater speedup. However this is rarely (if ever) done due to difficult timing problems.

Memory-memory vector processors

Another type of vector processor allows the vector operands to be fetched directly from memory to the different vector pipelines and the results to be written directly to memory; these are called *memory-memory* vector processors. Because the elements of the vector need to come from memory instead of a register, it takes a little longer to get a vector operation started; this is due partly to the cost of a memory access. One example of a *memory-memory* vector processor is the CDC Cyber 205.

Because of the ability to overlap memory accesses and the possible reuse of vector processors, vector-register vector processors are usually more efficient than memory-memory vector processors. However as the length of the vectors in a computation increase, this difference in efficiency between the two types of architectures is diminished. In fact, the memory-memory vector processors may prove more efficient if the vectors are long enough. Nevertheless, experience has shown that shorter vectors are more commonly used.

Interleaved memory banks

To allow faster access to vector elements stored in memory, the memory of a vector processor is often divided into *memory banks*. *Interleaved* memory banks associate successive memory addresses with successive banks cyclically; thus word 0 is stored in bank 0, word 1 is in bank 1, ..., word $n - 1$ is in bank $n - 1$, word n is in bank 0, word $n + 1$ is in bank 1, ..., etc., where n is the number of memory banks. As with many other computer architectural features, n is usually a power of 2: $n = 2^k$, where $k = 1, 2, 3,$ or 4.

One memory access (load or store) of a data value in a memory bank takes several clock cycles to complete. Each memory bank allows only one data value to be read or stored in a single memory access, but more than one memory bank may be accessed at the same time. When the elements of a vector stored in an interleaved memory are read into a vector register, the reads are staggered across the memory banks so that one vector element is read from a bank per clock cycle. If one memory access takes n clock cycles, then n elements of a vector may be fetched at a cost of one memory access; this is n times faster than the same number of memory accesses to a single bank.

12.1.3 Vector computing performance

For typical vector architectures, the value of τ (the time to complete one pipeline stage) is equivalent to one clock cycle of the machine[2]. Once a pipeline like the one shown in figure 12.3 has been filled, it generates one result for each τ units of time, that is, for each clock cycle. This means the hardware performs one floating-point operation per clock cycle.

Let k represent the number of τ time units the same sequential operation would take (or the number of stages in the pipeline). Then the time to execute that sequential operation on a vector of length n is

$$T_s = kn\tau,$$

and the time to perform the pipelined version is

$$T_p = k\tau + (n - 1)\tau = (n + k - 1)\tau.$$

Again for $n > 1$, $T_s > T_p$.

A *startup time* is also required; this is the time needed to get the operation going. In a sequential machine, there may some overhead required to set up a loop to repeat the same floating-point operation for an entire vector; the elements of the vector also need to be fetched from memory. If we let S_s be the number of τ time units for the sequential startup time, then T_s must include this time:

$$T_s = (S_s + kn)\tau.$$

In a pipelined machine, the flow from the vector registers or from memory to the pipeline needs to be started; call this time quantity S_p. Another overhead cost, $k\tau$ time units, is the time needed to initially fill the pipeline. Hence, T_p must include the startup time for the pipelined operation; thus,

$$T_p = (S_p + k)\tau + (n - 1)\tau$$

or

$$T_p = (S_p + k + n - 1)\tau.$$

---

[2]On some machines, it may be equal to two or more clock cycles.

As the length of the vector gets larger (as n goes to ∞), the startup time becomes negligible in both cases. This means that

$$T_s \rightarrow kn\tau$$

while

$$T_p \rightarrow n\tau.$$

Thus, for large n, T_s is k times larger than T_p.

There are a number of other terms to describe the performance of vector processors or vector computers. The following list introduces some of these:

- R_n: For a vector processor, the number of Mflops obtainable for a vector of length n.

- R_∞: The asymptotic number of Mflops for a given vector computer as the length of the vectors gets large. This means that the startup time would be completely negligible. When the vectors are very long, there should be a result from the pipeline at every τ units of time or every clock cycle. So the number of floating-point operations that can be completed in one second is $1.0/\tau$; dividing this result by one million produces the result in Mflops.

- $n_{1/2}$: The length, n, of a vector such that R_n is equal to $R_\infty/2$. Again for very large vectors, there should be a result from the pipeline at every τ units of time. So, $n_{1/2}$ represents the vector length needed to get a result at every 2τ units of time or every two clock cycles.

- n_v: The length, n, of a vector such that performing a vector operation on the n elements of that vector is more efficient than executing the n scalar operations instead.

Figures 12.6 and 12.7 show the relationship between these terms for a vector machine with $\tau = 6$ nanoseconds and $S_p = 16\tau$. This is an idealized picture; there are slight drops in the rate curve when n is equal to a multiple of the vector-register length.

Table 12.1 provides some performance characteristics for some of the vector computers discussed later in this section. The values of R_∞ and $n_{1/2}$ are for the elementwise multiplication of two vectors. These values and those of other vector computers are also discussed in chapter 11 on Computer Performance.

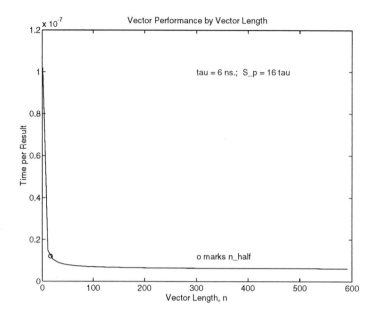

Figure 12.6: Vector performance shown as time per result by vector length.

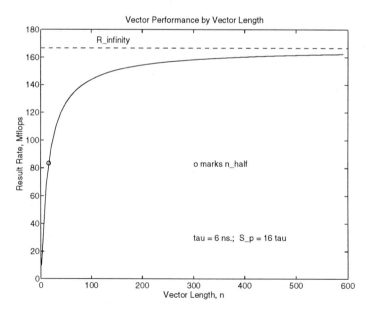

Figure 12.7: Vector performance, R_n, in Mflops by vector length, n.

Performance Characteristics	Year	Clock Cycle (nsec)	Peak Perf (Mflops)	R_∞ $(x \times y)$ (Mflops)	$n_{1/2}$ $(x \times y)$
Cray-1	1976	12.5	160	22	18
CDC Cyber 205	1980	20.0	100	50	86
Cray X-MP	1983	9.5	210	70	53
4 Procs	—	—	840	—	—
Cray-2	1985	4.1	488	56	83
4 Procs	—	—	1951	—	—
IBM 3090	1985	18.5	108	54	high 20's
8 Procs	—	—	432	—	—
ETA 10	1986	10.5	1250	—	—
8 Procs	—	—	10,000	—	—
Alliant FS/8	1986	170.0	6	1	151
8 Procs	—	—	47	1	23
Cray C90	1990	4.2	952	—	—
16 Procs	—	—	15,238	—	650
Convex C3880	—	—	960	—	—
Cray 3-128	1993	2.1	948	—	—
4 Procs	—	—	3972	—	—

Table 12.1: Performance characteristics of vector processing computers using 64-bit floating-point numbers. The expression $x \times y$ refers to the elementwise multiplication of two vectors, x and y. Information in this table was put together with data collected from [Alliant 86], [Almasi & Gottlieb 94], [Dongarra 94], [Hord 90], [Hockney & Jesshope 88], and [van der Steen 94].

12.1.4   The evolution of vector computers

One of the first supercomputers with built-in vector processors was the CDC Star 100. The ideas for this machine were first conceived in 1964 and were based on Iverson's APL programming language [Iverson 62]. Lawrence Livermore National Laboratories contracted to have this machine built for them in 1967, but the first machine was not delivered until 1974. By that time, the magnetic core memory contained in the machine was considered obsolete. The TI-ASC (the Advanced Scientific Computer by Texas Instruments) was a similar machine on the market at about the same time.

The CDC Cyber 205 is based on the concepts originated for the CDC Star 100; the first commercial model was delivered in 1981. This supercomputer is a memory-memory vector machine; it fetches vectors directly from memory to fill the pipelines and stores the pipeline results directly to memory; it has no vector registers. As a result, the startup times for the vector pipelines are large, as is reflected in the large value of $n_{1/2}$ for the CDC Cyber 205 in table 12.1. This machine contains up to four general-purpose pipelines, instead of pipelines designed for specific operations. It also provides both gather and scatter operations. A later shared-memory multiprocessor version of the CDC Cyber 205 is the ETA-10.

More recent vector computers include the IBM 3090, the Alliant FX/8 (a shared-memory multiprocessor with 8 CPU's, each with an attached vector processor), and the computers in the Convex series (each of which may also have multiple CPU's). Perhaps the most successful group of vector processors has been the family of Cray vector computers.

The Cray-1

The first Cray computer, brought out in 1976, was called the Cray-1. It was designed for supercomputing with pipelined vector arithmetic units. Besides being a vector computer, it was the fastest scalar machine at the time that it was introduced. Like the CDC Cyber 205, the Cray-1 provided scatter and gather operations; furthermore, it allowed non-unit strides through vectors using these operations.

The Cray-1, like other Cray vector processors that followed, was a vector-register machine. This type of machine fills the vector pipelines from the vector element values currently in the vector registers. This reduces the time to fill the pipelines (the startup time) for vector arithmetic operations; the vector registers can even be filled while the pipelines are performing some other operation. The vector results may be put back into a vector register after the completion of the operation, or they may be piped directly into another pipeline for an additional vector operation (chaining).

The Cray-1 was the first machine to use chaining. For vector processors that employ chaining techniques, not only is the startup time for each operation smaller than that of a comparable memory-memory machine, but also two floating-point operations may be performed at the same time, thereby doubling the number of Mflops.

Each vector register on the Cray-1 and on most later Cray vector processors contains 64 single precision elements.[3] Each single precision element or word contains 64 bits; this is the equivalent of double precision values on most other machines. The Cray vector processors do not use standard IEEE floating-point representation.

The Cray-1 and later Cray vector processors have twelve different pipelines or *functional units*. These are of four main types:

- Vector pipelines for integer or logical operations on vectors

- Vector/scalar pipelines for floating-point operations using scalars or vectors

- Scalar pipelines for integer or logical operations on scalars

- Address pipelines for address calculations

Figure 12.8 shows how these pipeline groups are distributed. The number of stages in a pipeline is given in parentheses in the box representing that pipeline in the diagram. Notice that a floating-point reciprocal approximation pipeline is used to implement floating-point division; x/y is computed as $x \times y^{-1}$.

One of the problems with the Cray-1 was that it only allowed one memory read and write per clock cycle. Vector processors with this limitation may only read one vector and write one vector result at the same time. When more than

---

[3] An exception is the Cray C90 with vector registers containing 128 elements.

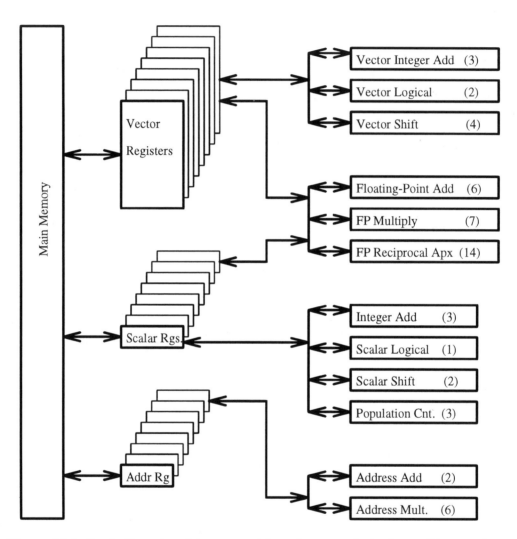

Figure 12.8: Basic Cray-1 architecture with registers and pipelines. The number in parentheses in each pipeline represents the number of stages in that pipeline.

one read (or write) is needed for the same operation, one of the reads (or the writes) must stall waiting for the other to complete. Consider the elementwise multiplication of two vectors of length greater than 64: $s = x \times y$, where $s_1 = x_1 \times y_1$, $s_2 = x_2 \times y_2$, etc. To begin, two vector registers are filled from memory. One contains the first 64 elements of x; the other, the first 64 elements of y. As these vectors are pushed through the pipeline, the results for s go into a third vector register and from there are stored in memory. After the first 64 elements of each vector have been processed, both input vector registers must be refilled from memory and the result vector register must be written to memory. Since only one read and one write can be done per cycle, elements of both input vectors cannot be read in at the same time; so the pipeline has to delay waiting for one of the read operations to finish. This accounts for the drops in the rate curve mentioned earlier.

The Cray X-MP

The Cray X-MP, first delivered in 1983, is faster than the Cray-1. The architecture is similar to that of the Cray-1, but there is more support for overlapped operations along with multiple memory pipelines. The early models of the Cray X-MP have a faster clock than the Cray-1: 9.5 nsec versus 12.5 nsec. Later models of the Cray X-MP have a clock cycle of 8.5 nsec.

The *MP* portion of the name refers to multiprocessing. The Cray X-MP is a shared-memory multiprocessor with each CPU controlling its own set of vector processors. Originally, this machine could be purchased in a 2-processor version and later in a 4-processor version.

The Cray X-MP is capable of two reads and two writes per clock cycle. This allows a faster load and store between the memory and the vector registers and prevents some pipeline delays. Chaining on this machine allows all three floating-point pipelines to operate simultaneously.

The Cray-2

In 1985, the Cray-2 was introduced as a completely redesigned architecture. Available as a multiprocessor with up to four processors, the Cray-2 does not support chaining. With only one memory pipeline per processor, there is a greater memory latency cost. Thus the Cray-2 is an improvement on the Cray X-MP only for those problems that require large amounts of memory.

In 1989, a new company called Cray Computer Corporation and headed by Seymour Cray split from Cray Research, Inc. Each company continued to develop supercomputers using the Cray name. The Cray Computer Corporation followed the ideas derived for the Cray-2.

The Cray Y-MP

Following along the proven lines of the Cray X-MP, Cray Research, Inc. announced the Cray Y-MP in 1987 as a faster rendition of the Cray X-MP with up to eight processors. A later version, called the Cray C90, was announced in 1990, allowing up to sixteen processors.

The Cray Y-MP/M90 was introduced in 1991. This model is more like the Cray Y-MP than the Cray C90 but provides the ability to handle larger memory addresses and has a much larger memory. Partly because of the additional memory, the Cray Y-MP/M90 operates at about half the speed of the Cray C90. Originally limited to 8 processors, the Cray Y-MP/M90 has since been extended to handle up to sixteen processors.

The Cray Y-MP models are the fastest of the Cray models discussed so far. The clock cycle time of both the Cray Y-MP and the Cray Y-MP/M90 is 6 nsec. The Cray C90 has a clock cycle of 4.2 nsec.

Table 12.2 summarizes the characteristics of three Cray machines discussed here: the Cray-1, the Cray X-MP, and the Cray Y-MP/832. Likewise, table 12.3 compares the performance of these three Cray models. Similar tables are provided in the chapter on Computer Performance.

The Cray-3

In the meantime, the group at Cray Computer Corporation continued work begun before the split to develop the Cray-3 (based on the Cray-2). This process took about ten years total, and the machine was never sold commercially. In 1993, one Cray-3 was placed in the supercomputing center at NCAR for research purposes. It contains four processors and has a clock cycle time of 2.1 nsec.

Characteristics	Cray-1	Cray X-MP	Cray Y-MP/832
Year Introduced	1976	1983	1989
Max # of Processors	1	4	8
# Vector Registers	8	8	8
Clock Cycle Time	12.5 nsec	9.5 nsec	6.0 nsec
Max Memory	1 Mbytes	16 Mbytes	128 Mbytes
Word Size	64 bits	64 bits	64 bits
Size of Vector Regs	64 words	64 words	64 words

Table 12.2: Comparison of Cray characteristics, taken from [Dongarra 94].

Performance	Cray-1	Cray X-MP *1 Proc/4 Procs*	Cray Y-MP/832 *1 Proc/8 Procs*
Clock Cycle Time	12.5 nsec	9.5 nsec	6.0 nsec
Peak Mflops	160	235 / 940	333 / 2667
LINPACK Mflops	110	218 / 822	324 / 2144
($n = 1000$)			
Shallow Water Mflops	—	— / 560	— / 1532
R_∞ Mflops	22	70	—
$n_{1/2}$	18	53	—

Table 12.3: Comparison of Cray performance, taken from [Dongarra 94].

The processors of this machine are completely surrounded by a liquid for cooling purposes. When encased in plastic, the appearance of the Cray-3 is similar to a large aquarium with the many wires attached to the processors gently swaying with the flow of the liquid. For this reason, this machine is sometimes called a *goldfish bowl* computer.

12.2   Programming vector computers

Vector computers can perform floating-point operations on vectors of n elements more efficiently than on n scalar variables. To create programs that maximize performance on these computers, you should learn to organize your data as vectors and to use vector operations as much as possible.

Vector computer manufacturers usually provide compilers with additional programming language constructs designed to assist in the handling of vectors or arrays. These constructs also help the programmer to think of a vector as a single entity instead of as a list of separate elements.

The vector computer compilers try to convert scalar code to vector code whenever possible. However these compilers do not know your program and data as well as you do and may be unable to vectorize the code as efficiently as you could. Therefore, it is recommended that you explicitly state the vector operations that you want to take place instead of letting a compiler guess at it. With practice, this will become almost automatic in your programming.

Cray Fortran is an extension of Fortran 77 for specifying vector operations on the various Cray supercomputers. Many of the additional Cray Fortran constructs are now included in Fortran 90 (see [ANSI 91], [Brainerd et al 90], and [Adams et al 92]) and in HPF (High Performance Fortran) (see [Koelbel et al 94]). Cray Fortran is used for the examples in the following discussions, since the constructs are similar to those used by other current vector computers.

This section starts by introducing some of the basic concepts needed for a better understanding of vector computer programming. We then suggest ways to improve the efficiency of vector programs. Finally, we consider issues related to programming a particular type of vector processor, a Cray Y-MP.

12.2.1   Vector programming concepts

There are several terms that frequently come up in relation to vector or parallel programming. These include vector notation, strip mining, recurrences, reduction operations, and dependence analysis. Each of these concepts are discussed in the following subsections.

Vector notation

Vector notation allows the programmer to consider arrays or vectors as whole entities instead of as sets of elements. If the vector x of length n is declared as an array in Cray Fortran by the command

```
DIMENSION  X(N)
```

then it is possible to refer to the entire array at once in any statement as X(1:N) or simply as X. This eliminates the need to consider each element in the array separately, as in a DO loop. For example, the loop to double each element of the vector x

```
DO  I=1,N
   X(I)  =  2.0 * X(I)
ENDDO
```

can be written instead as

```
X(1:N)  =  2.0 * X(1:N)
```

Not only is this simpler, shorter to write, and perhaps more obvious to the programmer, but the Cray Fortran compiler recognizes that this statement is a vector operation and produces the correct object code.

Suppose instead we wanted only to double the odd-subscripted elements of the vector x. This can be stated as

```
DO  I=1,N,2
   X(I)  =  2.0 * X(I)
ENDDO
```

or, in vector notation, as

```
X(1:N:2)  =  2.0 * X(1:N:2)
```

Here the DO loop increment is 2; this is also the third parameter in the vector subscript. This parameter, 2, is also referred to as the *stride* of the vector operation, as we are operating on every second element of the array. Similar notation is used in MATLAB, IDL (Interactive Data Language), Fortran 90, HPF, and Cray Fortran. In all cases, the stride is assumed to be 1 when it is omitted.[4]

Sometimes it is useful to be able to refer to the individual rows or columns of a matrix as vectors. Suppose the matrix A is declared as

```
DIMENSION  A(N,M)
```

Then the expression, A(2,1:M), represents the second row of A. This can also be written as A(2,:). In this case the colon, :, implies that all the elements of that dimension are to be used. Again, similar notation is used in MATLAB, IDL, Fortran 90, and Cray Fortran; however, IDL uses an asterisk, *, instead of a colon.

Strip mining

Strip mining is a method of breaking up a large vector operation into efficient pieces. For example, consider executing the following loop on a Cray:

```
DO  I=1,500
   A(I) = ....
ENDDO
```

where the elements of A are computed sequentially within the loop. Since the vector registers on the Cray Y-MP each contain 64 elements, it is better to break up this loop into smaller segments so that each piece works with a vector of length 64.[5] This gives the vector registers time to be filled with the next vector segment between iterations of the loop. Otherwise, there may be

---

[4]MATLAB assumes the second of the three parameters to be the stride while the other languages use the third parameter as the stride.

[5]For the Cray C90 with 128-element vector registers, you want to use vectors of length 128.

a break in the performance as the vector register empties into the pipeline before the next 64 elements of the vector can be fetched from memory. For example, the loop above may be written as

```
DO  K=1,448,64
   DO  I=K,K+63
      A(I) = ....
   ENDDO
ENDDO
DO  I=449,500
   A(I) = ....
ENDDO
```

Written in vector notation, this code becomes

```
DO  K=1,448,64
   A(K:K+63) = ....
ENDDO
A(449:500) = ....
```

Notice that the vector notation has eliminated the need for the inner loop of the original nested loop. It has also replaced the second loop.

Recurrences

A *recurrence* is a series of computational steps such that the result produced at one step is used in a later step. For instance, the statements

```
DO  I=1,N
   X(I+1) =  2.0 * X(I)
ENDDO
```

contain a recurrence, since the computation of X(3) uses the value of X(2), the computation of X(4) uses the value of X(3), etc. A recurrence can also happen when the variable assigned in a statement appears in the arithmetic expression used to compute that variable, as in

```
Q = Q + X(I)*Y(I)
```

Recurrences are important to recognize in programs for vector processors because they cannot be easily vectorized. In the first example above, the value of X(I) must be computed and stored before X(I+1) can be computed. This code should produce a series of doubled terms. That is, if the original value of X(1) is 1, then X(2) is 2, X(3) is 4, and X(N+1) is 2^N. Thus the first $N+1$ elements of X are $(1, 2, 4, 8, \ldots, 2^{N-1}, 2^N)$.

Suppose we try to vectorize this loop by the following statement:

```
X(2:N+1)  =   2.0 * X(1:N)
```

To understand how this vectorized statement executes, you should think of the right side of the equal sign (2.0 * X(1:N)) as being computed in one operation and the results stored into X(2:N+1) as a second operation. This would mean all the elements of X would first be doubled, and then the doubled values would be stored into the succeeding elements of X. In other words, if the original $N+1$ elements of X are $(1, 2, 3, 4, \ldots, N, N+1)$, the vectorized statement would first double each value, producing $(2, 4, 6, 8, \ldots, 2N, 2N+2)$. Then the each value would be stored into the next element of the array, yielding the final array: $(1, 2, 4, 6, \ldots, 2N-2, 2N)$. This is not the same result as produced by the sequential loop above. In fact, this loop cannot be vectorized.

Reduction operations

A *reduction operation* returns a scalar result from an operation on a vector. One common example of such an operation is the dot product of two vectors: $x \cdot y$. In Fortran, this can be coded as

```
DP = 0.0
DO  I=1,N
  DP = DP + X(I)*Y(I)
ENDDO
```

Because the value of DP is altered at each iteration of the loop, this statement cannot be pushed through a pipeline. Notice that the statement in the loop contains a recurrence. In fact, all reduction operations can be written as recurrences.

One method of handling this problem is to go ahead and compute the $x_i \times y_i$ products using the pipeline. Then each product can be accumulated in a vector, q, where k is the length of q and

$$q_1 = x_1 \times y_1 + x_{k+1} \times y_{k+1} + x_{2k+1} \times y_{2k+1} + \ldots,$$

$$q_2 = x_2 \times y_2 + x_{k+2} \times y_{k+2} + x_{2k+2} \times y_{2k+2} + \ldots,$$

etc. This creates a vector of partial sums instead of the scalar, DP. The sum of the k elements in that vector (the last k partial sums produced from the pipeline) produces the correct value for the dot product.

This could be done by unrolling the loop[6] above. The Cray Fortran code to accomplish this would appear as follows:

```
DIMENSION  DPPARTIAL(K)
...
DPPARTIAL = 0.0
DO  I=1,N,K
  DPPARTIAL(1) = DPPARTIAL(1) + X(I)    *Y(I)
  DPPARTIAL(2) = DPPARTIAL(2) + X(I+1)  *Y(I+1)
  DPPARTIAL(3) = DPPARTIAL(3) + X(I+2)  *Y(I+2)

    ...
  DPPARTIAL(K) = DPPARTIAL(K) + X(I+K-1)*Y(I+K-1)
ENDDO
DO  I=1,K
  DPFINAL = DPFINAL + DPPARTIAL(I)
ENDDO
```

For this reason, special functions are usually provided to handle reduction operations; one such Cray Fortran intrinsic function is named SDOT and is meant to be used to compute single precision dot products. See section 12.2.3 for more on the Cray Fortran intrinsic functions.

Dependence analysis

Two statements are *dependent* if a variable in one statement is used in the other statement. For example, there exists a dependence between the statements

---

[6]See chapter 11 for more information on loop unrolling.

```
A = B + C
     :
D = A * E
```

since the variable A is defined in the first statement and used in the later statement. There is also a dependence within any statement containing a recurrence. And there is a dependence in any statement containing a reduction operation.

When vectorizing a loop, the programmer should be careful to check to see if there exists a dependence within the iterations of the loop; that is to say, if the body of the loop is unrolled, what dependencies would exist between any two statements of the resulting code.

Most compilers for vector or parallel compilers include some dependence analysis of the code. This alerts the compilers to code segments that can or cannot be vectorized or parallelized automatically. This analysis also includes checks for reduction operations and recurrence relations.

For example, the following loop includes a recurrence and should not be vectorized

```
DO  I=1,N-1
  A(I+1) = A(I) + Y(I)
ENDDO
```

because a vectorized version of the loop does not produce the same results as the sequential version. Again, the reasoning is similar to that for the loop discussed in section 12.2.1; you should consider the entire vector operation as being completed before the array values are stored. To see this more clearly, let A be the original or initial vector and let A' be the new vector after the computation. Then the execution of this loop gives

$$A'_2 = A_1 + Y_1$$

$$A'_3 = A'_2 + Y_2$$
$$A'_4 = A'_3 + Y_3$$

$$\vdots$$

$$A'_N = A'_{N-1} + Y_{N-1} \ .$$

Hence,

$$A'_{i+1} = A'_1 + Y_1 + Y_2 + Y_3 + \ldots + Y_i \ ,$$

for $i = 1, 2, \ldots, N - 1$ and

$$A'_N = A_1 + Y_1 + Y_2 + \ldots + Y_{N-1} \ .$$

However, if you try to vectorize this with

```
A(2:N) = A(1:N-1) + Y(1:N-1) ,
```

you obtain the following value:

$$A'_{i+1} = A_i + Y_i \ ,$$

for $i = 1, 2, \ldots, N - 1$. The two values of A'_{i+1} are NOT the same. This difference is due to the dependence between statements in successive iterations of the loop.

For a more detailed discussion of types of data dependence and dependence analysis, see Wolfe's *Optimizing Supercompilers for Supercomputers* [Wolfe 89].

12.2.2 Hints for efficient vector Fortran programming

If you have only done sequential programming before, you may have to develop a slightly different mindset to program efficiently for vector processing. Compilers for vector computers try to vectorize loops whenever possible. But they may not produce the most efficient order for the vectorization. And a compiler may fail to recognize when vectorization can be done.

Here is a short list of suggestions to get you started in vector programming. Again, Cray Fortran is used in the example codes.

1. Most vector-register architectures are designed to handle vectors of a given length;[7] for instance the vector registers for the Crays hold 64 elements of a vector. Using vectors of this length or greater is more efficient than using small vectors. Recall the definition of $n_{1/2}$.

---

[7]Memory-memory vector architectures and some vector-register machines allow variable-length vectors.

2. In a nested loop, the outer loop is not usually vectorized by the compiler; it is the inner loop that it attempts to vectorize. Thus, you should try to arrange your code so that the access along a vector occurs on the inner loop index or so that the longest vector operation occurs in the inner loop. This may require interchanging loops. For example, for $N \gg 5$, the following loop

```
DO  I=1,N
  DO J=1,5
     A(I) = A(I)*F(J)
  ENDDO
ENDDO
```

runs more efficiently if rearranged as

```
DO  J=1,5
  DO I=1,N
     A(I) = A(I)*F(J)
  ENDDO
ENDDO
```

This is because the instructions executed for first version are vectorized into the following code

```
DO  I=1,N
     A(I) = A(I)*F(1:5)
ENDDO
```

This executes the N short scalar-vector multiplications shown below:

$$
\begin{aligned}
A(1) &= A(1) * (F(1), \ F(2), \ F(3), \ F(4), \ F(5)), \\
A(2) &= A(2) * (F(1), \ F(2), \ F(3), \ F(4), \ F(5)), \\
A(3) &= A(3) * (F(1), \ F(2), \ F(3), \ F(4), \ F(5)), \\
&\ \ \vdots \qquad\qquad \vdots \\
A(N) &= A(N) * (F(1), \ F(2), \ F(3), \ F(4), \ F(5)).
\end{aligned}
$$

For the second version, the vectorization is as follows:

```
DO  J=1,5
      A(1:N)  =  A(1:N)*F(J)
   ENDDO
```

Hence, there are only 5 scalar-vector multiplications, and they are on long vectors (meaning they are more efficient). These operations follow:

$$
\begin{aligned}
(\mathtt{A}(1),\ \mathtt{A}(2),\ \ldots,\ \mathtt{A}(\mathtt{N})) &= \mathtt{F}(1) * (\mathtt{A}(1),\ \mathtt{A}(2),\ \ldots,\ \mathtt{A}(\mathtt{N})), \\
(\mathtt{A}(1),\ \mathtt{A}(2),\ \ldots,\ \mathtt{A}(\mathtt{N})) &= \mathtt{F}(2) * (\mathtt{A}(1),\ \mathtt{A}(2),\ \ldots,\ \mathtt{A}(\mathtt{N})), \\
(\mathtt{A}(1),\ \mathtt{A}(2),\ \ldots,\ \mathtt{A}(\mathtt{N})) &= \mathtt{F}(3) * (\mathtt{A}(1),\ \mathtt{A}(2),\ \ldots,\ \mathtt{A}(\mathtt{N})), \\
(\mathtt{A}(1),\ \mathtt{A}(2),\ \ldots,\ \mathtt{A}(\mathtt{N})) &= \mathtt{F}(4) * (\mathtt{A}(1),\ \mathtt{A}(2),\ \ldots,\ \mathtt{A}(\mathtt{N})), \\
(\mathtt{A}(1),\ \mathtt{A}(2),\ \ldots,\ \mathtt{A}(\mathtt{N})) &= \mathtt{F}(5) * (\mathtt{A}(1),\ \mathtt{A}(2),\ \ldots,\ \mathtt{A}(\mathtt{N})).
\end{aligned}
$$

3. It helps to keep the order of Fortran array element storage in mind while writing nested loops. The following loop

```
DO  I=1,N
  DO J=1,N
     D(I,J)  =  C(I,J)*B(I,J)
  ENDDO
ENDDO
```

runs more efficiently as

```
DO  J=1,N
  DO I=1,N
     D(I,J)  =  C(I,J)*B(I,J)
  ENDDO
ENDDO
```

just because Fortran arrays are stored in column-major order as in the following sequence of elements:

```
D(1,1), D(2,1), ..., D(N,1), D(1,2), ....
```

Thus, the vectorized code for the second version

```
DO  J=1,N
  D(1:N,J) = C(1:N,J)*B(1:N,J)
ENDDO
```

is able to handle each column as a vector with a stride of one. This also reduces the memory and caching overhead, since the elements are being read and stored in a sequential manner.

Similar comments can be applied to array storage in other languages, such as C. However, Fortran arrays are stored in column-major order, while arrays in most other languages are stored in row-major order. The above examples should be revised according to the appropriate array storage order.

4. Another problem related to the storage of arrays in memory is that of interleaved memory banks. Recall from section 12.1.2 that the memory of a vector processor may be divided into 2^k memory banks, where words $0, 2^k, 2 \times 2^k, \ldots$ are stored in memory bank 0, words $1, 2^k+1, 2 \times 2^k+1, \ldots$ are stored in memory bank 1, etc. This construction allows 2^k elements of a vector to be read into a vector register within the number of clock cycles required by one memory access, as long as the stride of that vector is 1 or some other odd number.

When the stride of a vector is a multiple of 2, the desired elements of the vector are in evenly-spaced banks. For example, if the stride is 4 and the first element is in bank 3 of eight banks, then the second element is in bank 7, the third element is in bank 3, the fourth element is in bank 7, etc. So instead of eight elements being fetched within one memory access time to fill the vector register, only two of the elements may be read from memory in that time: one from bank 3 and one from bank 7. This reduces the speed of loading the vector register by at least four and may stall the operation of the pipeline being fed by that vector register.

Thus, the programmer should avoid using strides of 2^k, 2^{k-1}, and 2^{k-2}, if at all possible.

Interleaved memory causes a further problem. It is called a *bank conflict* when a machine is trying to access more than one data value in the same memory bank. Suppose two vectors, x and y, both contain an even number, n, of elements that is a multiple of 2^k and suppose that they are stored consecutively in memory; in other words, the last element of x, x_n, is followed immediately by the first element of y, y_1. So x_1 and y_1 are both stored in the same memory bank, x_2 and y_2 are stored in the next memory bank, etc. A vector computation, such as $x + y$, tries to read data from the same banks to fill the respective vector registers. Thus, this type of computation causes bank conflicts when loading the vector registers for both vectors, degrading overall vector performance. One solution to this problem is to always require the dimension of a vector array be an odd number, even if the last element of that array is never used.

5. Compiler dependence analysis is done in a conservative manner and may seem overly cautious. For instance, the following loop can be vectorized and still generate the correct value, despite the apparent dependence.

```
DO  I=2,N
   A(I-1) = A(I) + Y(I)
ENDDO
```

or, in vector notation,

```
A(1:N-1) = A(2:N) + Y(2:N) .
```

However, some compilers recognize only that a dependence exists and do not vectorize the loop. In order to force the vectorization of such a loop by the compiler, you should insert compiler directives before and after the loop to tell the compiler to ignore the dependencies for the duration of the loop or statement. For example, the Cray Fortran directives all start with the characters, `CDIR$`. These characters begin in the first position of the line and may be interpreted as a comment by another Fortran compiler. In the following example,

```
CDIR$ NODEP
      A(1:N-1) = A(2:N) + Y(2:N)
CDIR$ DEP
```

the keyword NODEP in the first directive tells the Cray Fortran compiler to ignore any possible dependencies in the succeeding statements, vectorizing the code if possible. The second directive contains the keyword DEP; this instructs the compiler to worry about dependencies again.

6. Other dependencies can be ignored as well. Consider loops of the form

```
DO  I=1,N
  X(I+1,J) = X(I,K) + Y(I)
ENDDO
```

or in vectorized notation,

```
X(2:N+1,J) = X(1:N,K) + Y(1:N)
```

These can be vectorized if it is known that $J \neq K$. But in most cases, the compiler is unable to make such an assumption and does not vectorize the loop. Hence, the code produced for this loop is serial and much slower than needed. So if you, the programmer, know in advance that J will never equal K, you should enclose this loop within a compiler directive (CDIR$ NODEP) in order to have the vectorization done, ignoring the possible dependence.

12.2.3   Programming a Cray

In this section, we discuss information particular to the Cray vector processors. Many Crays located at supercomputing centers across the United States provide computing time for class accounts via the Internet; the instructor merely needs to request it. Hence, we expect that most users of this text have access to a Cray and that the following information is of use to them.

The operating system for the Cray is called UNICOS. In many ways, UNICOS commands are similar to UNIX commands. However, some are specific to the Cray and the Cray software tools.

Since the Cray Y-MP has a number of processors, it is possible to use it as a parallel computer with tasks of coarse granularity. The Cray Fortran CF77 compiler has options to compile a program for this method of execution. See the Cray *CF77 Compiling Systems, Volume 4* manual for more information on Cray multiprocessing.

More information on the Cray representation of floating-point numbers is included below. This is followed by a discussion of some of the special intrinsic functions available via the Cray Fortran compiler for vector computing.

Cray floating-point representation

As mentioned earlier, the Crays do not use the IEEE representation for floating-point numbers. A single precision number of the Cray is contains 64 bits (not 32 bits); this is the same number of bits as a double precision number on most other machines; however, the IEEE double precision standard demands 11 bits for the exponent and 52 bits for the mantissa while the Cray uses 15 bits for the exponent and 48 bits for the mantissa of a 64-bit single precision number. This means that the Cray single precision values does not exactly match those double precision numbers produced by a machine with the IEEE arithmetic standard.

Cray double precision operations are executed on 128-bit numbers, and they are done in software, not hardware. Such operations may take ten to twenty times as long as the corresponding single precision operations. So only use single precision on the Cray, unless you need the precision provided by 128-bit numbers.

Cray Fortran intrinsic functions

There are a few special Cray Fortran intrinsic routines developed for some common vector operations. We describe three of them here: SDOT, SSUM, and MXM.

- SDOT (N, X, SX, Y, SY) – returns the dot product of N elements of the two vectors, X and Y. If either X or Y is indexed, that subscript tells which element to consider as first in the vector. The third and fifth parameters, SX and SY, represent the strides of the two vectors; they do not need to be the same and probably have a value of one most of the time.

  For example, if

  $$N = 3,$$
  $$X = (1, \quad 2, \quad 3, \quad 4), \text{ and}$$
  $$Y = (0, \quad -2, \quad -1, \quad 0, \quad 2, \quad 1),$$

  then

  $$SDOT (N, X(1), 1, Y(2), 2)$$

  computes the dot product of the two subvectors:

  $$(1, \quad 2, \quad 3) \text{ and } (2, \quad 0, \quad 1).$$

  The resultant value of SDOT should be $1 \times (-2) + 2 \times 0 + 3 \times 1$ or 1.

- SSUM (N, X, SX) – returns the sum of N elements of a vector, X. The third parameter, SX, is the stride.

  Using the values of N and Y defined above,

  $$SSUM (N, Y, 4)$$

  returns the value 2.

- MXM (B, N, C, K, A, M) – returns the $N \times M$ matrix, A, resulting from multiplying two matrices: B * C, where B is a $N \times K$ matrix and C is a $K \times M$ matrix.

In some cases, these functions are the same as the BLAS routines from LINPACK. See the man pages on these routines for further discussion.

For more information on other Cray Fortran functions, see the *Cray Fortran Reference Manual* [Cray 89].

13 Distributed-memory MIMD Computing

13.1 Introduction

Computational problems arising in scientific and engineering applications are often very large. Solving them on conventional computers can be a time consuming process demanding extensive storage. In many cases, *distributed-memory MIMD (DM-MIMD) multiprocessors* can provide both the computational power and the large memory necessary for solving such problems. Recall that MIMD stands for *multiple instruction, multiple data*. A DM-MIMD multiprocessor is a parallel computer in which each processor has direct access to its own local memory only. The processors are interconnected by communication links, and the processors exchange data by passing messages along those links. Because the multiprocessor is a MIMD machine, completely different programs may run on the individual processors at any one time.

Different models of DM-MIMD multiprocessors are distinguished by such factors as the power of the processors, the size of memory, the speed of interprocessor communication, the availability of input/output, and the interconnection pattern of the processors. DM-MIMD multiprocessors have been sold with as few as two and as many as 65,536 processors. The processors have been interconnected as rings, two-dimensional meshes, and tori as well as the more complicated fat-tree, hypercube, and Omega network configurations described in this chapter.

The very simplest DM-MIMD multiprocessor consists of two interconnected *nodes*. Each node is a processor with its associated memory. As an example of how we might use this 2-node multiprocessor, suppose that we need to sum a list of twenty numbers and that those numbers are originally split between the two nodes. To begin, both nodes simultaneously sum their ten local numbers. The nodes then send their sums to each other. Each node then adds its received sum to its own local sum to form the sum of all twenty numbers. This parallel implementation requires roughly half the computation time of a sequential implementation of the sum, and its overall speedup is determined by the cost of the communication step relative to the cost of the arithmetic performed.

In general, the design of a DM-MIMD program takes more thought than did this sample program. The purpose of this chapter is to introduce you to the basics of actual DM-MIMD architectures and to the general techniques needed to write efficient programs for them. In section 13.2, we first review the graphs defining the interconnection pattern of processors in present-day DM-MIMD multiprocessors. (A *graph* is defined to be a set of points or *nodes* interconnected by lines called *edges*.) We then review the evolution of those machines since their introduction in the 1970's and provide information on performance of the various architectures. In sections 13.3 and 13.4, we focus on the hypercube and mesh-connected multiprocessors, respectively. In both cases, we describe efficient programming techniques based on the interconnection graphs. In section 13.5, we briefly discuss programming considerations for some other architectures. Finally, in section 13.6, we work through a sample program for a DM-MIMD multiprocessor with an arbitrary number of processors.

Tutorials on the use of some specific DM-MIMD machines are available by anonymous ftp from `cs.colorado.edu` in the directory `/pub/HPSC`.

13.2   General architecture

13.2.1   Interconnection graphs

The writing of efficient programs for a DM-MIMD multiprocessor generally requires attention to the interconnection pattern of the nodes in that computer. The nodes of the computer can be thought of as lying on the nodes of a particular graph, and the communication links between processors lie on the edges of that graph. In this section, we review some of the graphs on which commercial DM-MIMD multiprocessors are based and give some examples of architectures based on those graphs. More information is provided on these computers in the subsequent sections of this chapter.

Arrays, rings, and tori

Figure 13.1 shows the simplest graphs underlying DM-MIMD multiprocessors. The first diagram is of a linear array of eight processors. In this arrangement, a node has one or two nearest neighbors depending on whether or not it is at the end of the array. Joining the two endpoints converts a linear array into a ring as shown in the second diagram. In this case, all nodes have two neighbors.

Nodes can also be connected to form a $p_1 \times p_2$ two-dimensional array or mesh of processors as shown in third diagram of figure 13.1. Here, $p_1 = 2$ and $p_2 = 4$, and a node has two or three neighbors depending on its location in the mesh. In a mesh with more rows, a node may have up to four nearest neighbors. These neighbors are identified by their relative positions and are called the north, south, east, and west neighbors.

Connecting corresponding nodes on the left and right sides of the two-dimensional mesh and on the top and bottom of the mesh converts it into a three-dimensional torus as shown in the fourth diagram of the figure. In this case, every node has four nearest neighbors.

Increasing the connectivity of the graph of eight processors reduces the maximum distance between any two processors in the graph. In a multiprocessor, this can translate into a decrease in interprocessor communication time. In a linear array, there are $p - 1$ edges between nodes 0 and $p - 1$, and this is the longest path between any two nodes. This is shown by the arrow in the first diagram of figure 13.2. Connecting the ends of the array to form a ring cuts this path in half. If communication is permitted in both directions (i.e., the graph is bidirectional), the longest path is between nodes 0 and $p/2$, and it follows $p/2$ edges as in the second diagram of the figure. In the mesh, the longest path is between nodes at two diagonally opposing corners. It has length $p_1 + p_2 - 2$. When $p = 8$, the longest path between nodes is four for both the ring and the mesh. The advantage of the mesh is more evident for larger numbers of nodes. For example, the longest path in a 16×16 mesh traverses thirty edges while the longest path in a ring of $16^2 = 256$ nodes covers 128 edges.

The torus leads to an even better situation. Note that the torus may be viewed as p_1 horizontal rings of p_2 nodes each interconnected via p_2 vertical rings of p_1 nodes each. Thus, the longest path traveled between node A and node B is, for example, the distance from node A's vertical ring to node B's vertical ring plus the distance around the latter ring to node B. If every ring can be traversed in either direction, the longest path is $p_1/2 + p_2/2$. In the 2×4 torus, the longest path is three. In the 16×16 torus, it is sixteen.

The Intel Touchstone Delta and Paragon are examples of mesh-connected DM-MIMD multiprocessors. The Japanese PAX-9 uses a torus interconnection of processors. The ring and linear array are not popular multiprocessor architectures on their own, but these graphs are often the basis for efficient programming of mesh and hypercube multiprocessors.

Hypercubes

The size of a hypercube is defined by its dimension d. The hypercube graph of dimension d and the multiprocessor based on it are both called *d-cubes*. A cube of dimension $d = 0$ consists of one node. A cube of any dimension can be built recursively beginning with 0-cubes. A 1-cube consists of two nodes and is constructed by joining together a pair of 0-cubes. Similarly, a 2-cube (four nodes) is formed by joining corresponding nodes of a pair of 1-cubes. In general, a d-cube is formed by connecting corresponding nodes in a pair of $(d-1)$-cubes. Figure 13.3 shows one way to form a 4-cube from a set of sixteen 0-cubes. Note that a d-cube always has $p = 2^d$ nodes.

Every node in a d-cube is connected to exactly d others. In particular, if the nodes in a d-cube are assigned d-bit binary identifiers (from 0 through $p - 1$), the d nodes connected to node j can be assigned identifiers differing from j in exactly one bit. Figure 13.4 shows the binary numbering of the nodes in a 2-cube. A d-cube can be constructed by connecting corresponding nodes of two $(d-1)$-cubes in any of d ways. For example, the familiar 3-cube can be made by linking corresponding processors of the two squares (2-cubes) forming its top and bottom, left and right, or front and back faces. In a d-cube, the d neighbors of node j define the d nodes corresponding to node j in the d different $(d-1)$-cubes.

The Cosmic Cube, the Intel iPSC/2 and iPSC/860, and the nCUBE/1, nCUBE/2, and nCUBE/3 are all examples of hypercube multiprocessors. Section 13.3 of this chapter concerns programming techniques for hypercube mul-

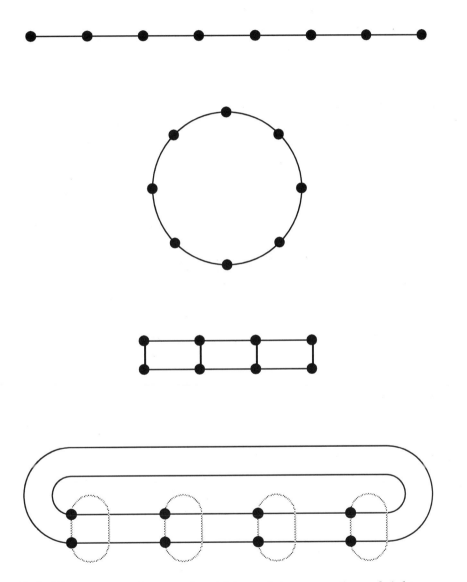

Figure 13.1: Linear array, ring, mesh, and torus interconnections of eight processors.

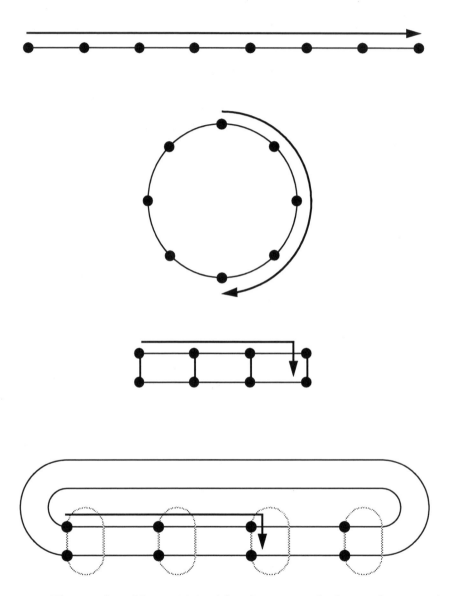

Figure 13.2: The graphs of figure 13.1 with a longest path shown. In general, there is more than one path of greatest length in a graph.

dimension d   d-cube

0

1

2

3

3

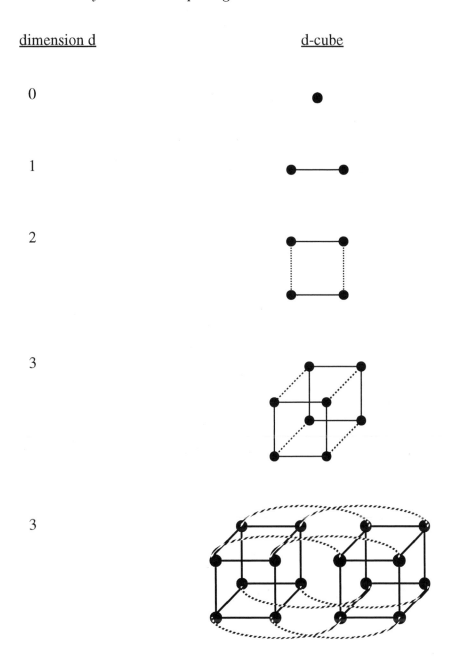

Figure 13.3: The recursive construction of a 4-cube beginning from 0-cubes.

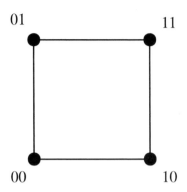

Figure 13.4: A 2-cube with numbered nodes.

tiprocessors.

Trees

Another type of graph used in DM-MIMD multiprocessors is the tree. Some multiprocessors are based on the binary tree in which each interior node (nodes other than leaves of the tree) has up to two children. Others are based on a *4-ary* or *quaternary tree* in which each interior node has up to four children. The nearest neighbors of a tree node are its parents and its children.

A *complete* binary tree (i.e., one in which each node has exactly two children) has height $\log_2 p$ when p is a power of two. A complete quaternary tree has height $\log_4 p$ when p is a power of four. Thus, the longest path between any two nodes of a complete p-node quaternary tree is shorter than it is in a complete binary tree with the same number of nodes. This maximum path length may also be reduced without changing the height of the tree by increasing the number of interconnections in a tree so that some nodes have more than two children. In this case, some nodes have more than one parent. Figure 13.5 shows one way to construct a quaternary tree from a binary tree in this way. (Note that this multi-parent construct is not strictly a tree by most standard definitions even though it is called a tree in the context of machine architectures.)

The Thinking Machines CM-5 is a DM-MIMD multiprocessor based on a quaternary tree in which the nodes have more than one parent. The Teradata

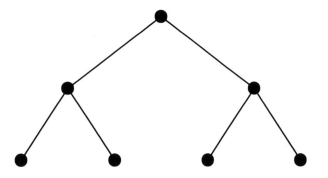

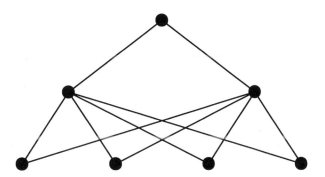

Figure 13.5: A binary tree and a quaternary tree with more than one parent per child node.

DBC/1012 Data Base Computer uses a binary tree interconnection pattern.

Omega networks

An Omega network of dimension $\log K$ has $K(\log K + 1)$ nodes arranged in a $K \times \log K + 1$ two-dimensional array. The nodes, however, are not interconnected as a mesh. Instead, the node at position (i, j) in the array is linked to the node at position (i, j') in the array if and only if the binary representation of j' is formed from a left cyclic shift of the binary representation of j with or without changing the last bit. Thus, the node at position $(3, 2) = (3, (010)_2)$ is linked to the nodes at positions $(3, 4) = (3, (100)_2)$ and $(3, 5) = (3, (101)_2)$, and every node in the Omega network has two nearest neighbors. The Omega network is also sometimes called the butterfly network, the flip network, the baseline, or the reverse baseline network [Leighton 92]. An Omega network is not constructed by using single wire connections but rather by means of 2×2 switches. For more implementation details, see [Hwang 93]. The IBM SP1 and SP2 use a version of the Omega network.

Although the graphs discussed in this section account for those underlying the majority of existing DM-MIMD multiprocessors, there are other possibilities. One example is the pyramidal structure of the EGPA mentioned in section 13.2.2. Processors may also be interconnected by means of a common communication channel or *data bus*. For more information on graphs and their properties in the context of multiprocessor architectures, see [Leighton 92].

13.2.2   The evolution of DM-MIMD computers

Development of DM-MIMD computers began in the early 1970's. The first example, the Tandem NonStop, was delivered in 1976. In the NonStop, nodes were not connected directly to other nodes. Instead, all nodes were connected to a pair of common data buses. Using the data buses, any one node in the machine could communicate directly with any other node. The NonStop was a computer designed especially for database applications [Almasi & Gottlieb 94].

Other designs followed soon after, beginning with the introduction of the Erlangen General Purpose Architecture (EGPA) in 1977 by W. Haendler, F. Hofman, and H. Schneider at the University of Erlangen in Germany. This machine had nodes in a pyramidal structure in which all nodes at the same level of the pyramid formed a torus [Henning & Volkert 85, Wilson 94].

In 1979, the Processor Array eXperiment (PAX-9) was completed by T. Hoshino at the the Institute of Atomic Energy at the University of Kyoto and T. Kawai of Keio University. The PAX-9 was a toroidal array of nine processors. Each node in the array was directly connected to four others. The PAX-9 did not allow the processors to operate fully asynchronously and so permitted a "quasi-MIMD" programming model [Hoshino 86].

The closest ancestor of many of today's DM-MIMD multiprocessors was the Cosmic Cube hypercube multiprocessor built at the California Institute of Technology. Development of this machine began in 1981 and was carried out by a research team headed by computer scientist Chuck Seitz and physicist Geoffrey Fox. The first prototype model went into operation the following year. The Cosmic Cube had 64 nodes, each directly connected to four others. Each node held Intel 8086/8087 processors and 128 Kbytes of memory. The prototype was released in 1982. The prototype Cosmic Cube was followed by improved versions culminating in the Mark III Cosmic Cube completed in 1987. The Mark III uses Motorola 68020 processors [Almasi & Gottlieb 94, Wilson 94].

The Cosmic Cube inspired the development of at least two successful commercial hypercube multiprocessors. The first of these was the Intel Personal Supercomputer (the iPSC, later called the iPSC/1). The first iPSC/1's were delivered in 1985 and have up to 128 nodes. The iPSC/1 nodes have Intel 80286/80287 processors and up to 4.5 Mbytes of memory. Models with vector processors were also available. The hypercube is operated via an Intel System 310AP microcomputer connected to the nodes [Intel 86]. This type of controlling computer is termed the *host*.

In 1985, the first nCUBE hypercube multiprocessor was also released. The nCUBE/1 (also commonly referred to as the nCUBE/ten or nCUBE 3200) had up to 1024 nodes running custom hardware that handled all computation, communication, I/O, and memory management on a single chip. Each node had 64 Kbytes of memory and a 32-bit processor. Its host computer had an Intel 80286 processor and ran a special version of Unix that allowed the nCUBE to emulate a machine with a single distributed file system [Almasi & Gottlieb 94].

The iPSC/1 and the nCUBE/1 have both spawned new generations of hypercube multiprocessors with more powerful processors and more advanced communication. The iPSC/2 hypercube appeared in 1987 using 80386/80387 processors. The iPSC/2 also has separate computation and communication processors on each node. Thus, a message passing through a node that is not its destination passes through the communication processor without impacting the computation. The iPSC/860, which replaced the iPSC/2 in 1990, employs *wormhole routing* by which messages can travel from one processor to any other with almost no delay at the intervening processors. The iPSC/860 uses i860 computation processors. The iPSC/860 and later Intel DM-MIMD multiprocessors all retain the separate communication processors [Dunigan 91].

The nCUBE/2 (also called the nCUBE 6400) was released in 1989 and has up to 8192 nodes and up to 64 Mbytes of memory per node. The nCUBE/2 retains the single chip technology that is the hallmark of nCUBE hypercubes but uses a 64-bit processor and wormhole routing. The nCUBE/2S was released in 1991. Patterned after the nCUBE/2, it has substantially improved computational and communication performance and so represents the second generation of nCUBE hypercubes. The third generation hypercube, the nCUBE/3, was released in 1994. It offers even faster communication and computation and can have up to 65,536 nodes and up to 1 Gbyte of memory per node. The second and third generation nCUBEs have been geared toward the database market [Almasi & Gottlieb 94, Wyckoff 94, nCUBE 94].

The mesh multiprocessor has a more scalable design than does the hypercube. In a hypercube multiprocessor with $p = 2^d$ nodes, a node has d nearest neighbors and so d communication wires. In a mesh-connected multiprocessor of the same size, a node has two, three, or four neighbors and wires, depending on its position in the machine. Regardless of the size of the machine, a mesh node never has more than four nearest neighbors or communication links. Thus, a large mesh multiprocessor can be constructed with less communication hardware than can a hypercube multiprocessor with the same number of nodes. While nCUBE has kept the hypercube architecture for its newest machines, Intel has moved to a mesh-connected architecture.

In 1991, Intel delivered the Touchstone Delta. This single machine, located at the California Institute of Technology, has a 16×32 array of i860 processors and employs wormhole routing of messages. It served as the prototype for the commercial product, the Intel Paragon. The Paragon also uses a two-dimensional mesh interconnection, but its processor is the faster i860XP. Delivery of the Paragon also began in 1991 [Dunigan 92, Dunigan 94, Wilson 94].

While Intel and nCUBE have been particularly successful suppliers of DM-MIMD multiprocessors, they are by no means the only ones. In 1986, FPS delivered its T-series hypercube which used a combination of Weitek floating-point chips and Inmos transputers. In 1987, Ametek completed the mesh-connected Ametek-2010, a descendent of its earlier Ametek S/14 hypercube. Commercial DM-MIMD multiprocessors have not been confined to hypercube and mesh architectures. For example, the Thinking Machines CM-5, a MIMD follower of the SIMD CM-2, is based on a fat tree network and uses SPARC processors. The Cray T3D has DEC Alpha chips interconnected as a torus. The IBM SP1 and SP2 use IBM RS/6000 processors interconnected by means of a high-speed Omega switch [Almasi & Gottlieb 94, Fox et al 88, Hwang 93, Wilson 94].

A wholly different type of DM-MIMD multiprocessor is represented by a cluster of workstations. These sets of interconnected workstations are growing in popularity because of their affordability and the substantial computing power of individual workstations. Thanks to such networking software as PVM [Geist et al 95] workstation clusters can be programmed in the same style as a more traditional multiprocessor. Because the nodes do not have to be identical, a workstation cluster is an example of a *heterogeneous* computing environment. We consider only homogeneous multiprocessors in this chapter.

13.2.3 The communication performance of DM-MIMD computers

The evolution of computers has brought with it a great improvement in performance. Like that of any other computer, the performance of a DM-MIMD multiprocessor may be classified according to such measures as its Mflop rating and its LINPACK benchmark (discussed in chapter 11 on Performance). In addition, the performance of a DM-MIMD computer is determined by the speed of data communication between nodes. In this section, we discuss the communication performance of several machines. In section 13.2.4, we show how the communication and computation speeds interact to determine the overall performance of a DM-MIMD multiprocessor.

Message passing between nearest neighbors

Messages are passed between two nodes when a command to send a message is issued by the originating node and a command to receive a message is issued by the recipient node. The exact message-passing process that follows these commands depends on the specific computer, but it may involve such preliminary steps as the initialization of message buffers on both nodes and the opening of a communication route between them in addition to the actual transfer of the message along the communication links.

If β is the time required to set up the hardware and software for a message transmittal and τ is the time needed to send one byte of data across the wire linking the two nodes, the time required to send k bytes of data from a node to one of its nearest neighbors is given by

$$T_{commun} = \beta + k\tau. \tag{13.1}$$

Typically, the time to send a message is much greater than the time to do a floating-point operation. In particular, $\beta >> \omega \geq \tau$, where ω is the time for a flop. Table 13.1 shows values of β, τ, ω, and the ratio β/ω for some popular DM-MIMD multiprocessors. All times are given in microseconds. In this table, ω is defined to be the time for a double precision (8 byte) floating-point multiply. The first column shows the year of release of the listed machine, although some data were gathered on upgraded models. The last column of the table shows the single processor clock rate.

When the message length is less than 100 bytes, a special message passing protocol is used on the iPSC/2 and the iPSC/860. In these cases, the values of β are roughly halved. The small message β values are listed in parentheses in the table. The communication times for the Paragon are strongly dependent on the operating system, so the operating system name is also listed in parentheses.

This table shows that both computation and nearest neighbor communication performance have increased markedly with the release of each new machine. However, it is still the case that $\beta >> \tau$. This means that it is better to send many bytes of data in one message than to send the same data in many small messages. Programs for DM-MIMD multiprocessors should be designed with this guideline in mind.

Year	Computer	β	τ	ω	β/ω	Clock Rate (MHz)
1985	iPSC/1	862	1.8	43.0	20	8
1987	iPSC/2	697	0.4	6.6	106	16
		(390)			(59)	
1985	nCUBE/1	384	2.6	13.5	28	20
1987	nCUBE/2	200	0.6	1.5	133	20
1988	iPSC/860	136	0.4	0.08	1700	40
		(75)			(938)	
1991	Delta	72	0.08	0.08	1028	40
1991	Paragon XPS35 (OSF)	62	0.01	0.07	886	50
1991	Paragon XPS35 (SUNMOS)	93	0.02	0.07	1328	50

Table 13.1: A comparison of communication and computation times (microseconds) for several hypercube and mesh multiprocessors. Sources: [Dunigan 91, Dunigan 92, Dunigan 94].

Message passing between arbitrary nodes

When the sending and receiving nodes are not nearest neighbors, the cost of data transmission depends completely on how the intermediate nodes along the path of the message handle the message transfer. The early DM-MIMD multiprocessors, such as the iPSC/1, have a single processor per node. This processor handles both computation and communication, so messages coming to a node interrupt any computation in progress even if that node is not the ultimate recipient of the message. Later models, beginning with the iPSC/2 and nCUBE/2, have separate computation and communication hardware that can operate independently. Messages passing through a node are handled by a communication processor without impacting the computation. However, sending a message from one node to a distant node is still more expensive than sending the same message between neighboring nodes as the message is delayed slightly at each intermediate processor. (Throughout this chapter, we refer to a node somewhat imprecisely as "a processor and its associated memory" as the modern nodes actually do have more than one processor.)

The communication performance has improved further with the advent of wormhole routing. With wormhole routing (as implemented on the Intel computers), a preliminary packet is sent from sender to receiver to set up and reserve a communication channel through the intermediate nodes. The message then passes through that channel without delays. The overhead of this circuit switching mechanism is difficult to measure, but it appears to add only a few percent to the total communication time on the newest machines (the Intel Delta and Paragon). That is, on computers with wormhole routing, the time to send a message between distant nodes is roughly the same as the time to send it between neighboring nodes [Dunigan 94].

The speed of internode communication is reflected in the communication bandwidth of a machine: more bytes of data are communicated per second when message transfer time is fast. Table 13.2 shows the communication bandwidth of several multiprocessors measured as a function of message size (8, 1024, and 8192 Kbytes) and the distance it must travel in the computer. A "1 hop" message is a message between nearest neighbors, and a "6 hop" message passes through five intermediate nodes between the sender and receiver. The machines that do not have separate communication processors or do not permit

Computer	8 Kbytes		1024 Kbytes		8192 Kbytes	
	1 hop	6 hops	1 hop	6 hops	1 hop	6 hops
iPSC/1	7	2	482	115	504	401
iPSC/2	21	18	962	850	2248	2164
nCUBE/1	20	6	566	110	643	125
nCUBE/2	50	47	1289	1269	1554	1554
iPSC/860	200	123	1781	1412	2605	2486
Delta	$--$	$--$	~ 5900	~ 5900	~ 11900	~ 11900
Paragon XPS35 (SUNMOS)	$--$	$--$	$--$	$--$	~ 65700	$--$

Table 13.2: The communication bandwidth of several hypercube and mesh multiprocessors. Bandwidths are shown for messages between nearest neighbors (1 hop) and between nodes 6 hops distant (5 intermediate nodes) for three different message lengths. A symbol $--$ means that data were not available. A symbol $\sim$ means that the data are reported to fewer significant figures than are other entries in the table. Sources: [Dunigan 91, Dunigan 92, Dunigan 94].

wormhole routing (the iPSC/1 and the nCUBE/1) suffer a large degradation of bandwidth when the number of hops increases from one to six. The other machines all demonstrate little or no decrease in communication bandwidth for communication between distant nodes.

The passing of large messages

In our expression for T_{commun} in equation (13.1) the size of the message appears only in the term $k\tau$. On most computers, however, the startup time β is also a function of the message size.

As shown in table 13.1, the startup time is reduced on the iPSC/2 and the iPSC/860 when k is very small. On some machines β is also increased when the message size is very large. In this case, the message may actually be sent as a set of smaller packets rather than as one large message, and the sending of each individual packet adds a small amount to the total communication cost. While most of the startup time is incurred for the first packet, the effect of breaking up the message into packets (*packetizing*) is often evident in a plot of communication time between two specific nodes versus message size. Figure 13.6 shows this plot for two hypothetical computers, one that breaks large messages into packets and one that doesn't. The solid line shows $T_{commun} = \beta + k\tau$ plotted as a function of τ when $\beta = 500$ and $\tau = 1$ for a machine that doesn't packetize. As expected, it is a linear function of k. The slope of the line is τ, and its y-intercept is β.

In contrast, the dotted line in figure 13.6 shows the characteristic shape of the curve when the message is divided into 250-byte packets. The solid and dotted lines are collinear until $k = 251$. At this point, the packetizing machine splits the message into two packets–one with 250 bytes and one with the remaining one byte. The sudden step up in time reflects the overhead of sending the second packet. The curve then rises linearly as the second packet grows to 250 bytes. At $k = 501$, the curve takes another step up as the third packet is released. While experimentally determined plots are generally not as smooth as those in the figure, the stairstep shape of the dotted curve are often evident in timings of machines with a measurable packetizing overhead. In particular, you may see it in your timings of programs involving large messages. The packet size is generally moderately large: on the iPSC/1 it is 1024 bytes, on the Delta it is 476 bytes, and on the Paragon it is 1792 bytes.

The overhead of packetizing messages is evident in the bandwidths presented in table 13.2. Specifically, packetizing of messages means that we do not see the linear increase in bandwidth that we might expect from the linear dependence of T_{commun} on message size. This is apparent in the 1-hop data for the nCUBE/2. When message size is increased from 8 to 1024 bytes (a factor of 128), the bandwidth of nearest neighbor communication jumps from 50 Kb/s to 1289 (a factor of only 26). Increasing message size from 1024 to 8192 (a factor of 8) similarly increases the bandwidth by a only factor of 1.2. However, even though efficiency is lost in packetizing, the overall increase in bandwidth with message size reiterates that it is generally better to pass a few large messages than it is to pass many small ones.

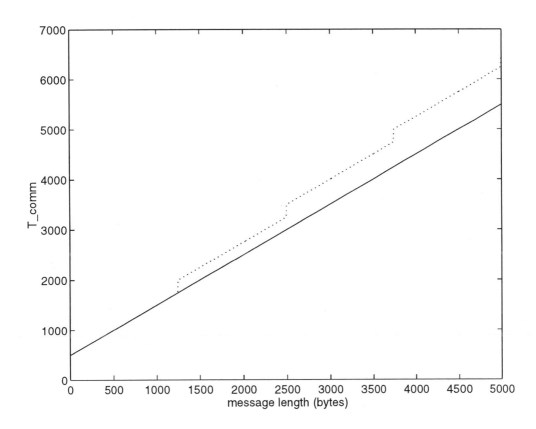

Figure 13.6: The time to send a message between two specific nodes plotted against the length of the message. The solid line shows the characteristic shape of the curve for a machine that does not packetize messages. The dotted line shows the effects of sending packets.

Contention for communication links

A final and very important concern in communication performance is the number of messages traversing a given communication link at any one time. When more than one message passes on a given wire those messages *contend* for that wire. The effects of this contention depend on the computer on which it occurs. For example, if neighboring nodes of an iPSC/1 send messages to each other simultaneously, the time for the exchange of data is roughly $2T_{commun}$: neither the two startups nor the actual transfer of data across the link can be overlapped. Later multiprocessor models permit the overlap of startups for the sent and received messages, and the time for such an exchange (called a *head-to-head send*) approaches $\beta + 2k\tau$.

The problem of contention becomes more complicated when messages sent between various nodes take unknown paths through the multiprocessor. Suppose that a k-byte message is delayed while messages totalling q bytes pass through a link along its path. Then the time it takes for that message to arrive at its destination is increased by time $q\tau$. If q is large, this delay can be substantial. Contention occurs for a communication link whenever the total amount of data attempting to pass through it exceeds the bandwidth of that link [Dunigan 94].

The overall effect of contention on the performance of a parallel program is very difficult to predict in advance, but, in general, it is best to program in a style that limits contention. In sections 13.3.2–13.4.2, we discuss methods for programming mesh and hypercube multiprocessors without contention.

Notice that we have included only node-to-node communication in discussion of communication performance even though some DM-MIMD multiprocessors also have hosts. While the host and node may actually be based on the same type of processor, their function is quite different. A node is wholly dedicated to a single user process while the host is a multiuser machine. Thus, the time required for any operations on the host, including communication, is influenced by the number of other machine users as well as by the background processes running on the host. Furthermore, the link between the host and nodes is a much slower communication path than a link between nodes. Therefore, host-to-node and node-to-host communication is generally substantially less efficient than node-to-node communication. It is best to minimize communication between the host and nodes in any DM-MIMD multiprocessor program, and so we do not study host-to-node or node-to-host communication here.

13.2.4 The overall performance of DM-MIMD computers

The overall performance of a DM-MIMD multiprocessor is determined not only by its communication performance but also by its computational performance. Table 13.1 shows that, like the time required to send a message, the time required for a double precision floating-point multiply has decreased substantially with each new architecture. This is true for the time needed for all other floating-point operations as well. Thus, as both computation speed and communication speed have increased, we can expect that a parallel program will run faster on a new model of a DM-MIMD multiprocessor than on an old one.

However, the faster run time is not the only ingredient of performance. Amdahl's Law tells us that the speedup of a parallel program is limited by the fraction of time spent in operations that can't be implemented in parallel. By extension, the speedup is also limited by the time required for data communication. As no communication occurs when a program is implemented on one node, communication is part of the overhead of a parallel implementation.

To see the effect of data communication on speedup, suppose that we have a serial program with perfectly parallel computation. If no internode communication is needed, the time to run the program on p nodes is just the time to run it on one node divided by p, i.e., $T_p = T_1/p$. If data communication is needed and computation and communication cannot be overlapped, the p-node time increases by the time T_c required for that communication so that

$$T_p = \frac{T_1}{p} + T_c.$$

Recall that the speedup of a parallel program is defined by

$$S = \frac{T_1}{T_p}.$$

To more easily examine the effects of communication on speedup, we first consider its reciprocal

$$R = \frac{1}{S} = \frac{T_p}{T_1}.$$

For our perfect program with data communication, this reciprocal is

$$R = \frac{T_1/p + T_c}{T_1} = \frac{1}{p} + \frac{T_c}{T_1}.$$

If communication is free, $R = 1/p$ and the program has perfect speedup $S = p$. Otherwise, the reciprocal of speedup is determined by the ratio T_c/T_1. The greater the cost of communication compared to the cost of computation, the greater the value of R. That is, the larger the ratio of communication to computation costs, the smaller the speedup S. The last column of table 13.1 shows the ratio of the message startup time β to the time for a double precision floating-point multiply ω. These data clearly show that with an increase in computation speed has come a marked growth in the communication to computation cost ratio. Therefore, while we can expect our parallel program to run much faster on the newer machines, we might also expect it to exhibit a smaller speedup.

The increase in the communication to computation ratio, however, is offset by the substantial improvement in communication capabilities of the newest DM-MIMD multiprocessors. Between the iPSC/1 and the Paragon (SUN-MOS), the ratio β/ω has grown by a factor of 66. At the same time, the communication bandwidth for 8192-byte messages has grown by a factor of 164. To see if any gains have actually been made, it is necessary to examine some computational examples. Standard speedup data is not available to us for a wide range of machines, so we instead examine the speedup of a parallel numerical program as compared to the theoretical peak performance of the machine. This is a somewhat unsatisfying efficiency measure as it mixes issues of efficient programming of the single processor (the tightness of the peak performance) with the issues of parallel programming, but it nonetheless gives us a rough indication of speedup trends.

The parallel program is the solution of the largest linear system that will fit on a given DM-MIMD multiprocessor. This is the Highly Parallel Computing benchmark of [Dongarra 94]. Table 13.3 shows the system size n (number of

p	Computer	n	Measured (Mflops)	Theor Peak (Mflops)	Ratio
1	nCUBE/2	1280	2	2.4	0.83
	iPSC/860	750	24	40	0.60
	Delta	750	24	40	0.60
2	nCUBE/2	1280	4	4.7	0.85
	iPSC/860	1500	58	80	0.73
	Delta	1500	60	80	0.75
8	nCUBE/2	3960	16	19	0.84
	iPSC/860	3000	190	320	0.59
	Delta	3000	230	320	0.72

Table 13.3: The Highly Parallel Computing benchmark for machines with $p = 1, 2$, and 8. Source: [Dongarra 94].

equations), the megaflops measured for that system solution, and the theoretical peak performance of that machine for 64-bit arithmetic. The final column shows the ratio of the measured performance to the theoretical peak. Data are given for machines with $p = 1$, 2, and 8 nodes.

The iPSC/860 was one of the first DM-MIMD multiprocessors with good enough performance to solve realistic scientific problems on a moderate number of processors. Both the iPSC/860 and the Delta use the i860 processor. Between the iPSC/860 and the Delta, the ratio β/ω (for large messages) decreased and the message bandwidth increased. As the table shows, the overall performance of the Delta is even better than that of the iPSC/860, especially as the number of processors is increased.

A comparison of these two machines with the nCUBE/2, however, shows that improvement in communication does not tell the whole story of parallel performance. The data for similarly sized problems when $p = 8$ shows that, while the theoretical peak performance of the Delta is 17 times that of the nCUBE/2, the measured performance increases only by a factor of 14. This is true despite the fact that the communication bandwidth for an 8192-byte message is nearly eight times greater on the Delta than on an nCUBE/2. This is due in part to the almost eight-fold increase in β/ω between the two machines, but it is more greatly influenced by the difficulty of programming

Computer	Mbytes /Node	Problem Size	Gflops
nCUBE/2	4	7776	0.24
iPSC/860	8	12000	2.6
Delta	16	12500	3.5
Paragon (OSF)	32	12000	4.0

Table 13.4: The Highly Parallel Computing benchmark for some 128-node machines. Sources: [Dunigan 94, Dongarra 94].

the i860 processor. (See [Dewar & Smosna 90] for details.) The data for $p = 1$ show that the theoretical peak performance figures are substantially more realistic for the nCUBE/2 than for the i860-based computers.

The main advantage of DM-MIMD multiprocessors is in their large distributed memories and cumulative computing power. The Highly Parallel Computing benchmark demonstrates the full potential of a parallel computer for solving linear systems. This benchmark gives the Gflops attained in solving the largest linear system that will fit on the computer by any stable numerical method. Table 13.4 shows these Gflop ratings and the attainable problem sizes for some of the machines we've examined using 128 nodes. For comparison, a 296-node Paragon (OSF) solves a problem of size 29400 at a rate of 12.5 Gflops [Dongarra 94].

In summary, the performance of a DM-MIMD multiprocessor depends on a variety of interacting factors. The theoretical peak performance depends on such standard concerns as the amount of memory and the speed of the processor. When data must be transferred, it is also determined by the communication bandwidth. The attainable speedup depends foremost on the degree to which the serial program can be divided into independent (and parallel) tasks. It depends further on the number and size of the messages passed and on the ratio of communication and computation times. It is difficult to predict the performance of a real program by looking at any of these interacting factors in isolation, but the performance data we've examined in this section show that a DM-MIMD multiprocessor can be a very powerful tool for many computations.

13.3 The hypercube multiprocessor

13.3.1 The hypercube defined

A hypercube is a distributed-memory MIMD message-passing parallel computer in which processors are connected according to a hypercube graph. Recall from the discussion in section 13.2.1 that a hypercube of dimension d is built up recursively from 2^d hypercubes of dimension 0. This recursive structure of the hypercube is what makes it a particularly interesting architecture to study. In the next two sections, we show how the hypercube can emulate a variety of other architectures and how those virtual architectures are the basis for efficient data communication algorithms. Recall also that the nodes of a hypercube multiprocessor are numbered so that neighboring nodes have binary identifiers differing in exactly one bit.

13.3.2 Embeddings

In section 13.2.3, we noted that when messages contend for communication links, communication performance degrades. The special properties of the hypercube graph make it easy to design contention-free programs for hypercube multiprocessors. In fact, it is easy to carry out many communication operations efficiently using only nearest neighbor communication. In these communication schemes, messages pass along the links of a hypercube in a way that traces out a ring or tree or other simple graph. In this way, the graph is *embedded* into the hypercube. In this section, we review the basic embeddings that are relevant to this chapter. A more comprehensive description is provided in [Leighton 92].

The simplest graph that can be embedded into a hypercube using only nearest neighbor connections is a linear array or ring. This is easy to see in the case of the 2-cube. Sending a message around the four nodes in numerical order ($0 \rightarrow 1 \rightarrow 2 \rightarrow 3 \rightarrow 0$) involves communication between non-neighboring nodes (1 and 2), but sending the message along the route $0 \rightarrow 1 \rightarrow 3 \rightarrow 2 \rightarrow 0$ fixes the problem. The messages are passed between processors with binary identifiers differing in only one bit. If node 2 receives the message but does not return it to node 0, the 2-cube models a linear array of four processors. The node numbering shown in figure 13.4 is precisely the numbering required for embedding the nearest neighbor array or ring by bit manipulation. If we were to instead number the nodes counterclockwise in numerical order, flipping single bits to determine the route of the message would require us to send a message between two nodes at opposing corners of the cube.

This numbering scheme can be extended to hypercubes of any dimension by numbering the nodes of the ring according to the *binary reflected Gray code* [Leighton 92, Reingold et al 77]. To construct the Gray code, begin with the binary numbers

$$0$$
$$\underline{1}$$

Flip them over the underscore. Preface the numbers above the line with 0's and below the line with 1's as follows:

$$00$$
$$01$$
$$11$$
$$\underline{10}$$

At this point, we have constructed the Gray code ordering used in figure 13.4. Repeating the process a second time produces the Gray code ordering that embeds an 8-node ring into a 3-cube:

$$000 = \quad (0)_{10}$$
$$001 = \quad (1)_{10}$$
$$011 = \quad (3)_{10}$$
$$010 = \quad (2)_{10}$$
$$110 = \quad (6)_{10}$$
$$111 = \quad (7)_{10}$$

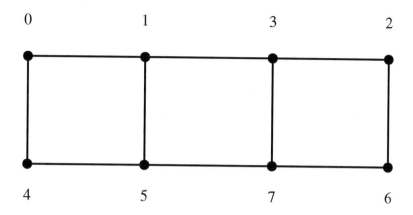

Figure 13.7: A 2×4 array embedded in a 3-cube.

$$101 = (5)_{10}$$
$$100 = (4)_{10}.$$

This Gray code ordering extends to any cube dimension and so permits the programmer to embed contention-free rings or one-dimensional arrays into any sized hypercube.

The Gray code ordering also permits the embedding of arrays or tori into the hypercube. For instance, to embed a 2×4 array into the 3-cube, we number the nodes in binary as in the following table:

	00	01	11	10
0	000	001	011	010
1	100	101	111	110

The top border of the table is the Gray code ordering of a 2-cube, and the left border is the Gray code ordering of a 1-cube. The identifiers of the nodes in the 2×4 array are constructed by taking the leftmost bits from the left border and the rightmost bits from the top border of the table. This ensures that the north, south, east, or west neighbor of a node in the array is a nearest neighbor of that node in the hypercube. The embedded array is shown in figure 13.7.

We can apply this same procedure to embed a 4×4 array into a 4-cube as follows:

	00	01	11	10
00	0000	0001	0011	0010
01	0100	0101	0111	0110
11	1100	1100	1100	1110
10	1000	1001	1011	1010

Notice that the nodes in positions $(1, i)$ and $(4, i)$ of this array have binary identifiers differing in the leftmost bit only. Thus, the square array can be closed into a cylinder by linking the corresponding nodes in its top and bottom rows. Similarly, the corresponding nodes on the left and right sides of the array have identifiers differing in exactly one bit. Linking those nodes closes the cylinder into a torus.

In this way, any array of size $2^{d_1} \times 2^{d_2}$ can be embedded into a cube with $p = 2^d$ processors as long as $d_1 + d_2 = d$. The array is formed into a torus by linking corresponding processors in the top and bottom rows of the array and in the left and right columns of the array. *Neighbors in the array or torus are nearest neighbors in the hypercube.*

By simple bit manipulation of the node identifiers, a tree of nodes can also be embedded in the hypercube. In particular, we can identify a *spanning tree* of the hypercube graph by linking only nearest neighbors in the hypercube. A spanning tree of a d-cube is simply a tree that includes all $p = 2^d$ nodes of the hypercube. One spanning tree of a 4-cube is shown in figure 13.8. Note that node 0 has children $1 = (0001)_2$, $2 = (0010)_2$, $4 = (0100)_2$, and $8 = (1000)_2$ in the tree. The identifiers of node 0's children thus differ from $0 = (0000)_2$ in exactly one bit. Similarly, node $1 = (0001)_2$ has children $3 = (0011)_2$, $5 = (0101)_2$, and $9 = (1001)_2$, and so on for all nodes in the tree.

In general, a spanning tree of height d is embedded into a d-cube by placing node 0 at the root of the tree. The remaining levels of the tree are identified in d steps where at step l, for $l = 1, \ldots, d$, each node with identifier $j < 2^l$ pairs with the node with identifier differing from j in bit l only. (Bits are numbered from right to left.) For each pair, the node with the smaller identifier is the parent node of its partner.

In the next section, we show how these embeddings can be used in communication routines for a hypercube multiprocessor. Embeddings can also be useful as a program development tool. Some application programs have

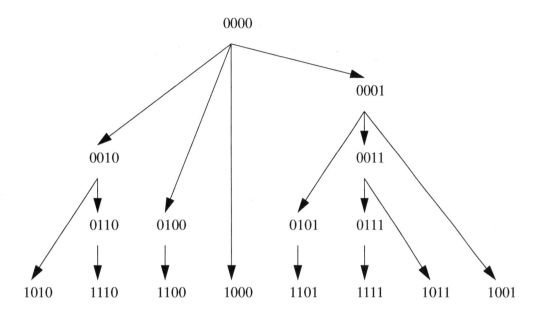

Figure 13.8: A spanning tree of a 4-cube. Source: [Saad & Schultz 89].

an obvious underlying data structure. For example, the bisection method for computing eigenvalues, divide and conquer methods, and branch-and-bound methods all are described by trees. One way to implement such methods on a hypercube multiprocessor is to embed the underlying tree data structure into the hypercube architecture. In this way, the tree serves as a *virtual architecture* for the computational problem. Embeddings can also permit the development of programs for one architecture on a different architecture. It is possible to develop a mesh-oriented program on a hypercube, for instance, by designing the communication scheme to embed a mesh into the hypercube. This practice has proven valuable for computational scientists using the mesh-connected Intel Touchstone Delta. Access to that machine is very limited, so it is often convenient to carry out program design and debugging for the Delta on a slower, but more accessible, hypercube multiprocessor.

13.3.3   Interprocessor communication schemes

A processor in a hypercube has direct access to only those data stored in its own local memory. However, all but the most embarrassingly parallel algorithm requires a processor to access data stored at other nodes. Because the cost of data transfer is typically high in comparison to the cost of computation, efficient hypercube programs rely on efficient communication routines. In this section, we discuss several fundamental algorithms for broadcasting data from one node of a hypercube to all others in that hypercube. These same algorithms are applied in reverse to gather data distributed among all nodes into a single node. We also present one more general algorithm to accumulate distributed data in all nodes. More information on data communication in hypercubes is provided in, for example, [Saad & Schultz 89].

The simplest broadcast algorithm is based on a linear array. Node 0 can broadcast data to all others simply by passing it along the links of a linear array embedded via a Gray code ordering of processors. Any node receiving a message passes it on to a neighbor unless that node is at the end of the array. A node identifies its neighbor in the array by finding its own position in the array, adding one to it, and determining the node at that position. Figure 13.9 shows the pseudocode for a node program to perform the linear array broadcast on a hypercube. In this pseudocode, MYID is the node identifier of the node executing the algorithm, and GRAY(j) is a function returning the identifier of the node at position j in a Gray code ordering of the p nodes. The positions and the node indices both range from 0 through $p - 1$. INVGRAY(k) is a function returning the position of the node with identifier k in the Gray code ordering.

To broadcast k bytes of data from node 0 to all others along the links of the array requires $p - 1$ communication steps. Using the message startup time β and the byte transfer time τ introduced in section 13.2.3 gives a total communication time of

$$T_{array} = (p - 1)(\beta + k\tau).$$

The linear array broadcast can be generalized to a two-dimensional array broadcast in an obvious way. Node 0 first broadcasts the data across the linear array of length $p_1 = 2^{d_1}$ defined by the top row of the array. The nodes of

In parallel,
do on all nodes j with binary labels β_j, $0 \leq j \leq 2^d - 1$:

```
    MYGRAYPOS = INVGRAY(MYID)
    MYNBR = GRAY(MYGRAYPOS + 1)
    If (MYID.eq.0) then
        Send data to MYNBR
    Else
        Receive data
        If (MYGRAYPOS.lt.2^d - 1) then
            Send data to MYNBR
        Endif
    Endif
```

Figure 13.9: The linear array broadcast algorithm for a hypercube.

that array then broadcast the data along the linear arrays of length $p_2 = 2^{d_2}$ defined by the columns of the array. The time for this broadcast is then

$$T_{2Darray} = ((p_1 - 1) + (p_2 - 1))(\beta + k\tau).$$

Thus, rearranging a linear array of length 8 into a 2×4 array cuts the cost of the data broadcast to $T_{2Darray} = \frac{4}{7}T_{array}$. Rearranging a linear array of length 16 into a 4×4 array cuts the cost of the data broadcast to $T_{2Darray} = \frac{2}{5}T_{array}$. We can combine this approach with the spanning tree broadcast we describe next to increase the parallelism in a two-dimensional array broadcast, but we defer discussion of that hybrid approach to section 13.4.2.

In section 13.3.2, we saw how to embed a spanning tree into a hypercube of dimension d. In this tree, the longest path from node 0 to any other node in the tree follows d links. This suggests that a spanning tree could be used to broadcast data from node 0 to all others in just d communication steps by distributing the data in parallel along the links of the tree. The pseudocode of figure 13.10 summarizes the steps in such a spanning tree broadcast (STB). Recall that for l steps with l varying from 0 through $d - 1$, processor j pairs with the processor having identifier differing from j in bit l only. In the implementation of the algorithm, the partner node's identifier is easily determined via the exclusive or operator (.xor.). Taking the exclusive or of j and 2^{l-1} returns the integer differing from j in bit l only.

The STB gives us a means for sending data from one node to all others in a d-cube in d time steps so that

$$T_{STB} = d(\beta + k\tau).$$

We can use the same mechanism to gather data distributed among the nodes into node 0. This operation is called a spanning tree gather (STG). Suppose that each processor has computed one element of a vector v so that v_j lies on processor j for $j = 0, \ldots, p - 1$. The STG proceeds for steps $l = 1, \ldots, d$. At each step, each node at level 2^{d-l+1} sends its data to its parent in the spanning tree. Thus, the sends originate from the leaves of the tree. When a node receives data, it appends it to the data it already holds and sends the accumulated data to its parent. Node 0 ultimately receives all elements of the vector v.

In parallel,
do on all nodes j with binary labels β_j, $0 \leq j \leq 2^d - 1$:

    For $l = 0, \ldots, d - 1$,
        MYPARTNER = MYID.xor.2^l
        MAXID = MAX(MYID,MYPARTNER)
        If (MAXID.lt.$2^l + 1$) then
            If (MYID.lt.MYPARTNER) then
                Send data to MYPARTNER
            Else
                Receive data
            Endif
        Endif
    Endfor

Figure 13.10: The STB algorithm for a hypercube.

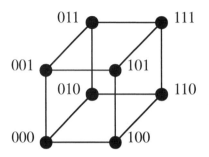

Figure 13.11: The binary identifiers of the nodes of a 3-cube.

The STG requires the same number of communication steps as the STB and, hence, the same number of communication startup charges. However, the length of the data doubles at each step, so the total communication cost for the STG is higher than for the STB. If each vector element is k bytes long, the vector v is pk bytes, and the cost of the STG is

$$T_{STG} = d\beta + \sum_{l=1}^{d} 2^{l-1}k\tau = d\beta + (p-1)k\tau.$$

The STB and the STG can be combined to accumulate and broadcast a distributed vector to all nodes in the hypercube. The distributed vector is gathered in node 0 using an STG; it is then distributed to all nodes by an STB. The final algorithm we present in this section serves the same purpose but can be carried out in less time than the combination of the STB and the STG. This algorithm is known as the alternate direction exchange (ADE).

The ADE relies on the recursive structure of the hypercube. Note that a d-cube can be split into a pair of $(d-1)$-cubes in any of d ways. And these $(d-1)$-cubes can be identified by the binary identifiers of their constituent nodes. As discussed in section 13.3.1, in a 3-cube, the three pairs of 2-cubes are the top and bottom, the front and back, and the two sides. As figure 13.11 shows, these pairs of 2-cubes are determined by the rightmost, middle, and leftmost bits, respectively, of the identifiers. That is, all nodes on the left side have zero as their least significant bit while all nodes on the right have that bit equal to one.

The ADE proceeds for d steps, using a different pair of $(d-1)$-cubes at each step. At each step, corresponding processors in the two $(d-1)$-cubes send their data to each other and accumulate the result. Figure 13.12 depicts the communication patterns at the three steps of an ADE on a 3-cube. At step l of the algorithm, for $l = 1, \ldots, d$, a node determines its partner by flipping bit l of its binary identifier. Each node begins with a single vector element (v_j in processor j, $j = 0, \ldots, 7$). It completes step one with two elements, step two with four elements, and ends with the full vector $v = (v_0, \ldots, v_7)^T$.

The pseudocode for an ADE for a d-cube is given in figure 13.13. Again, the partner node at each step is determined via the .xor. operator.

The ADE algorithm appears, like the STB or STG, to require d communication steps. Thus, it should take about half the time of the STB and STG combination. The actual time for the ADE, however, depends on some characteristics of hypercube on which it is implemented. Note that, at each step of the algorithm, the partner nodes simultaneously send messages to one another. When the startup costs for these head-to-head sends can be overlapped, the cost for such a pair of messages is much less than the cost of two messages sent in turn, and the cost for an ADE approaches

$$T_{ADE} = d\beta + 2(p - 1)k\tau.$$

The ADE thus requires the same data transfer time as the STG and STB combination but only half the number of startups.

Hypercube multiprocessors typically come equipped with communication library routines for basic data communication operations. For example, on an iPSC/2 or iPSC/860, a call to the library routine GCOLX with the data vector argument x accumulates distributed elements of x onto all nodes via an ADE. (See [Jessup 95] which is available via anonymous ftp in the /pub/HPSC directory at the cs.colorado.edu site.) It is not unusual, however, to need a routine that combines communication and computation in a way not provided by the libraries. Thus, it is important to understand the workings of the efficient communication algorithms in order to write efficient hypercube programs.

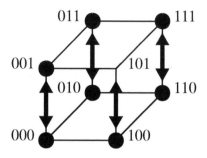

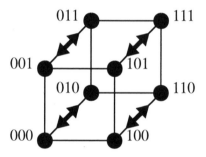

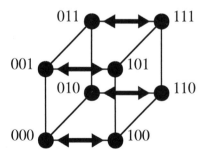

Figure 13.12: The steps of an alternate direction exchange (ADE) of the elements of vector $v = (v_0, v_1, \ldots, v_7)^T$ on a 3-cube.

In parallel,
do on all processors j with binary labels β_j, $0 \leq j \leq 2^d - 1$:

For $l = 1, \ldots, d$:
 MYPARTNER = MYID.xor.2^{l-1}.
 Send vector $v_{j+1}^{(l)}$ of length $2^{l-1}k$ to MYPARTNER.
 Receive vector $v_{j'+1}^{(l)}$ of length $2^{l-1}k$ from MYPARTNER.
 Insert newly received elements to form the vector $v_{j+1}^{(l+1)}$
  of length $2^l k$.
Endfor

Figure 13.13: The ADE algorithm for a hypercube.

13.4 The mesh multiprocessor

13.4.1 The mesh defined

A distributed-memory MIMD message-passing parallel computer may also be constructed with the processors connected in a two-dimensional array configuration. This type of computer is known as a *mesh* or *mesh-connected* multiprocessor. As for the hypercube, each node of the mesh consists of a processor (or separate communication and computation processors) and its own local memory. Transfer of data between nodes is accomplished by message passing across the interprocessor communication links.

The size of a mesh is defined by its number of rows p_1 and its number of columns p_2. The total number of processors is $p = p_1 p_2$. These numbers are not generally confined to powers of two and may be even or odd. A mesh node has two, three, or four nearest neighbors, depending on its position in the machine. If the mesh is connected further to form a torus, all nodes have four nearest neighbors.

The cost of the reduced number of neighbors is reduced versatility. However, many of the same concepts that apply to efficient hypercube program-

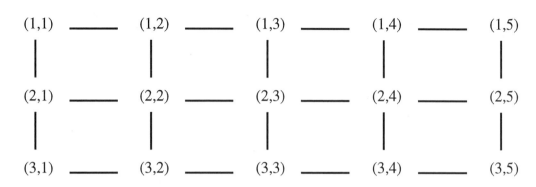

Figure 13.14: Numbering of the nodes of a 3×5 mesh.

ming extend to a mesh multiprocessor. Embeddings of alternate virtual architectures into the mesh and the communication algorithms dependent on them are the subjects of the next section. In describing these algorithms, we number the elements of the mesh according to their row and column positions. For example, the nodes of a 3×5 mesh would be labeled as in figure 13.14. In a one-dimensional mesh, the processors are taken to lie in a row so that their labels all begin with the index 1 as in the first row of the figure.

13.4.2 Embeddings and interprocessor communication schemes

As is the case for the hypercube multiprocessor, the fundamental communication operations on the mesh are the broadcasting of data from one node to all others, the gathering of distributed data into a single node, and the accumulation of distributed data in all nodes.

As is also true for the hypercube, the simplest broadcast algorithms are based on a linear array. When $p_1 = 1$ or $p_2 = 1$, the $p_1 \times p_2$ mesh reduces to a linear array, and data may only be broadcast from one node to all others via a linear array broadcast. When $p_1 = 1$, a linear array broadcast is carried out with node $(1, j)$ receiving data from node $(1, j - 1)$ and passing it to node $(1, j + 1)$, $j = 2, \ldots, p - 1$. The node in position (1,1) initiates the broadcast (and so sends but does not receive), and the node at $(1, p)$ ends it (and so receives but does not send). All communication occurs between nearest neighbors in the computer, meaning that the time for a linear array broadcast stays at

$$T_{array} = (p - 1)(\beta + k\tau).$$

We use the same communication model for the mesh as for the hypercube so that β is the communication startup time, and $k\tau$ is the time to transfer k bytes of data between neighboring nodes.

As discussed in section 13.3.2, when p_1 and p_2 are both greater than one, the mesh may be viewed as p_1 connected horizontal linear arrays of length p_2 or as p_2 connected vertical linear arrays of length p_1. A broadcast from the node at (1,1) to all others can be carried out by passing the data first along the top row of the array and then simultaneously along all columns of the array. Equivalently, the data may be passed down the first column and then along the rows. Again, all communication is between nearest neighbors in the mesh, and the communication time is

$$T_{2Darray} = ((p_1 - 1) + (p_2 - 1))(\beta + k\tau).$$

A comparison of T_{array} and $T_{2Darray}$ shows the advantage of greater connectivity: the time for a linear array broadcast is proportional to the product of p_1 and p_2 while the time for a two-dimensional mesh broadcast is proportional to their sum.

The cost for a one-dimensional (linear) array broadcast, and hence a two-dimensional array broadcast, may be reduced further by identifying a greater level of parallelism in those operations. To do this, we assume that both p_1 and p_2 are both powers of two. We also abandon the requirement of nearest neighbor communication while maintaining the requirement of a contention-free scheme. The wormhole routing used in modern message-passing architectures ensures that messages travel quickly between distant nodes as long as messages are passed on separate communication wires.

The key to the more efficient algorithm is to involve as many nodes in the broadcast as quickly as possible. In the linear array of eight nodes, we can do this by having the node at (1,1) send the message first to a node in the center position (1,5) of the array. The nodes at (1,1) and (1,5) may then distribute the message independently (and in parallel) in the two halves of the array. Applying this procedure recursively in the two halves increases the parallelism

further. In figure 13.15, we show how the messages are passed in a broadcast
from the node at (1,1) to all others in an array of length eight. The routes of
the messages are shown at each of the three time steps of the broadcast. Note
that no two messages ever travel over the same link simultaneously in this
linear array broadcast, so the time required for this contention-free broadcast
is

$$T_{1Dcf} = d(\beta + k\tau),$$

where $d = \log_2 p$. (This assumes that the time for a multihop message is about
the same as the time for a message sent between nearest neighbors.)

This one-dimensional algorithm is applied to a two-dimensional mesh by
broadcasting first along the first row of the mesh and then simultaneously down
all of the columns of the mesh. This ensures that increasing the dimension of
the mesh does not introduce contention for wires. Just like the STB broadcast
algorithm introduced for the hypercube in section 13.3.3, the contention-free
mesh algorithm works efficiently by embedding a spanning tree into the two-
dimensional mesh. Figure 13.16 shows the spanning tree created in a 4×4
mesh.

The steps of the spanning tree broadcast on a $p_1 \times p_2$ mesh are summarized
by the pseudocode in figure 13.17. The time to carry out this Mesh STB
broadcast for k bytes of data on a mesh with wormhole routing is about

$$T_{MeshSTB} = (d_1 + d_2)(\beta + k\tau),$$

where where $d_1 = \log_2 p_1$ and $d_2 = \log_2 p_2$. A comparison of $T_{2Darray}$ and
$T_{MeshSTB}$ shows that using the Mesh STB broadcast algorithm can lead to
a substantial savings in time when compared to the nearest neighbor one-
dimensional or two-dimensional array broadcast.

The Mesh STG is carried out by running the Mesh STB algorithm in
reverse. Data transmission begins from the leaves of the mesh spanning tree
and continues to its root. Data is accumulated along the way so that the
root node (1,1) completes the Mesh STG with a complete copy of all of the
data originally distributed among all nodes. On the hypercube, the alternate
direction exchange (ADE) presented in section 13.3.3 was the most efficient
means for such a gather operation. The ADE, however, relies completely on
the recursive structure of the hypercube and so does not extend to the mesh

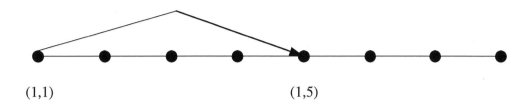

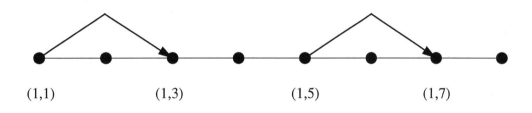

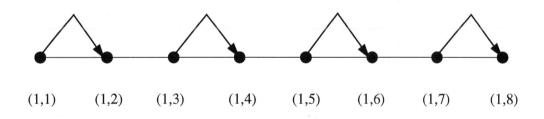

Figure 13.15: The messages passed at three communication steps of the contention-free linear array broadcast. Source: [Barnett et al 91].

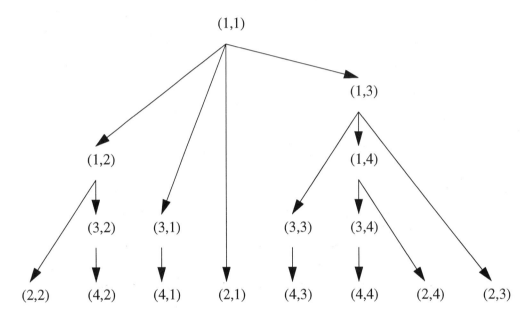

Figure 13.16: The spanning tree embedded into a 4×4 mesh by the efficient broadcast algorithm.

In parallel,
do on all nodes (i, j), with $1 \leq i \leq p_1$ and $1 \leq j \leq p_2$:

Get node identifier MYID = (MYROW, MYCOL).

If $MYROW = 1$ then
    For $l = 1, \ldots, d_2$,
        PARTNERCOL = MYCOL + $p_2/2^l$
        MYPARTNER = (1,PARTNERCOL)
        MAXCOL = max(1,PARTNERCOL)
        If (MAXCOL.le.$p_2/2^l$) then
            If (MYCOL.lt.PARTNERCOL) then
                Send data to MYPARTNER
            Else
                Receive data
            Endif
        Endif
    Endfor

    For $l = 1, \ldots, d_1$,
        PARTNERROW = MYROW + $p_2/2^l$
        MYPARTNER = (PARTNERROW,MYCOL)
        MAXROW = max(MYROW,PARTNERROW)
        If (MAXROW.le.$p_2/2^l$) then
            If (MYROW.lt.PARTNERROW) then
                Send data to MYPARTNER
            Else
                Receive data
            Endif
        Endif
    Endfor
Endif

Figure 13.17: The STB algorithm for a mesh.

architecture. Thus, a mesh spanning tree gather (MeshSTG) and broadcast must be combined to accumulate complete copies of distributed data on all nodes of the mesh. Node (1,1) then begins a Mesh STB to send that data back to all nodes. The exact cost of this combination depends on how much the data vector grows or shrinks during the exchange, but it is roughly bounded by $2T_{MeshSTB}$ where k is taken to be the length of the accumulated vector.

These mesh operations can also be applied if p_1 and p_2 are not powers of 2. For details of the modifications to arbitrary mesh size as well as for the optimizations available under wormhole routing, see [Barnett et al 91, Barnett et al 94].

As is true for the hypercube, mesh-connected multiprocessors also are equipped with library routines that perform the most basic communication operations. (See, for example, [Intel 93].) The user should also beware that when only a subset of the mesh processors are used, the subset may not actually represent a rectangular mesh of contiguous processors within the full mesh. On the Paragon, for example, a rectangular submesh is assigned only when specifically requested by the user. When the submesh is not rectangular, the communication algorithms presented in this section are no longer contention-free.

13.5   Some other multiprocessors

In the case of the tree multiprocessors, the graph and architectural issues are not easily separated. While the nodes in mesh or hypercube multiprocessors are typically identical, in tree multiprocessors, different positions in the tree are often allocated to different types of processors or controllers. For example, in the tree-based Teradata DBC/1012 Data Base Computer, two types of processors at the leaves process queries and access databases. Processors located above the leaves are dedicated to sorting and broadcasting operations [Almasi & Gottlieb 94].

The general-purpose Thinking Machines CM-5 architecture is based on a quaternary tree with SPARC processors at the leaves and routing chips at the other tree nodes. Each leaf processor has four parent routing chips. Each routing chip has two to four parent routing chips depending on its location in the tree. The idea is to increase the communication bandwidth between neighboring nodes higher up in the tree: these nodes occupy a larger subtree than those lower down in the tree and so are more likely to cause a communication bottleneck if their bandwidth is insufficient. Like a real tree, a tree of this type has "thicker" branches near the root than at the leaves. This architecture is termed a *fat tree* [Leiserson et al 92].

In an architecture like the CM-5, the concept of nearest neighbor no longer strictly applies. To route a message from one node to another, the message is sent up the tree to the least common ancestor of the two nodes, and then down to the destination node. As a message goes up the tree, it must choose which parent connection to take at each level. This choice is made randomly from among those links unobstructed by other messages. After the message has attained the height of the least common ancestor of the source and destination processors, it travels down the tree to its destination. The random choice at each level is designed to balance the load on the network. (The manufacturers of the CM-5 advertise a minimum of 5 Mbytes per second node to node data transfer regardless of other node communication loads. Near communications can attain up to 20 Mbytes per second because they may not have to be routed as far up the tree.)

The complexity of the routing scheme on the CM-5 means that it is difficult for the programmer to write communication routines. On this machine, all but the most expert programmers use library routines for all communication. For more information on the CM-5, see [Leiserson et al 92].

Other multiprocessor architectures have identical processors at all nodes but use a completely different interconnection mechanism. For example, the nodes of an Omega network are not connected directly by communication wires, but rather they are connected by means of switches [Hwang 93]. The IBM SP1 uses a high speed Omega switch by which a message may be sent quickly between any two nodes of the machine. Thus, the concept of nearest neighbor is not important in programming the IBM SP1.

On the SP1, however, the speed of communication is determined by the *transport layer* used. The transport layers represent different message passing protocols. In the layer of lowest bandwidth and highest latency (Ethernet/IP), messages are passed over ethernet connections between nodes without use of the high speed switch. In the layer of greatest bandwidth and lowest latency

(EUI-H), messages are passed using a low-overhead interface to the high speed switch. However, only the lowest bandwidth layers support multiple processes per node and multiple parallel jobs per node, so communication bandwidth may be sacrificed in the interest of greater parallelism.

Typically, the applications programmer does not have to be concerned with the details of transport layer programming. These are instead handled by various parallel programming languages. Nevertheless, the programmer may need to specify the layer to be used by the language and so must be aware of the benefits and disadvantages of the layer types. For more details on the SP1, see [Gropp et al 94a].

The details of programming a DM-MIMD multiprocessor differ from machine to machine. On some machines, the programmer has much control over such details as the routes traveled by messages, while, on others, such details are hidden in communication libraries. In either case, some knowledge of the inner workings of a particular computer is essential to writing efficient programs for it.

13.6   A sample program

In this section, we present a rough outline of how to construct a program for a generic DM-MIMD multiprocessor. This example is only intended to present some of the most basic issues one needs to consider in writing this type of program. The details of this process can vary substantially from program to program. They can also vary from machine to machine.

For our example, we develop a parallel program for computing $\| Ax \|_1$ on a DM-MIMD multiprocessor with p processors, given the $p \times p$ matrix A and the vector x of length p. Recall that if $y = Ax$, its 1-norm is given by $\| y \|_1 = \sum_{i=1}^p |y_i|$.

It is always best to begin with a carefully designed serial algorithm or code because it is hard to think in parallel when there are still errors in the serial ideas. Therefore, we begin by writing a MATLAB program for this computation:

```
y = A*x;
a = norm(y,1);
```

While we can be sure that this program is correct, it does not reveal the independent tasks comprising this computation. The next step in developing the parallel program is to identify those tasks. One way to break up the problem is to recognize that the ith element of the vector y comes from multiplying the vector x by row i of the matrix A. Therefore, the serial computation may be rewritten as

```
for  i = 1:p
        y(i) = A(i,:)*x;
end
a =   norm(y,1);
```

This reformulation of the problem clearly shows that the elements of the vector y are computed independently of one another. This suggests that the computation of $y = Ax$ can be implemented in parallel by assigning the computation of the element y_i to node $i - 1$, for $i = 1, \ldots, p$.

It then remains to implement the norm computation in parallel. Expanding the norm function call to show the operations performed now gives us the serial program

```
for  i = 1:p
        y(i) = A(i,:)*x;
end
a =   0;
for  i = 1:p
        a = a + abs(y(i));
end
```

Thus, computing the norm requires us to sum the absolute values of the distributed components of the vector y. Unlike the matrix-vector product, this sum cannot be implemented without communicating data between nodes.

Most DM-MIMD multiprocessors supply a library call to compute a sum of distributed data (called a *global sum*), but we can also write one ourselves by making simple modifications to a standard communication routine. For example, on a hypercube or mesh multiprocessor, we could perform a spanning tree gather but replace the vector accumulation with an addition at each step.

In this global sum routine, each leaf node of the spanning tree passes the absolute value of its vector element $|y_i|$ to its parent. Each parent node then adds the values received from its children to the absolute value of its own vector element $|y_i|$ and sends the result to its own parent. Continuing the process all the way up the spanning tree leaves the sum $\| y \|_1$ in the root node. If this norm is required by all nodes for subsequent computation, it can be broadcast to them from the root via a spanning tree broadcast.

This organization of the parallel algorithm means that every node $i - 1$, $i = 1, \ldots, p$, of the multiprocessor runs the following program:

```
y(i) = A(i,:)*x;
y(i) = abs(y(i));
call global_sum(a,y)
```

In this program, the routine global_sum leaves the sum of the distributed elements of y in A on one or all of the processors, depending on its implementation. To run this parallel program, we must first generate the matrix row $A(i,:)$ and the vector x on each node $i - 1$.

In this case, the computation done in the serial program takes p times as long as the parallel program run on p nodes. Moreover, because only one row of the matrix is used on a node, the serial program also takes p times the storage of the parallel program. The program thus appears to be efficient in terms of both time and storage. However, before we implement it, we need to estimate its overall efficiency. That is, we need to make sure that we have not designed a parallel program in which the cost of data communication outweighs the benefits of parallel computation.

Although the actual performance of a parallel program depends on many interacting factors, we can get a rough estimate of its cost by means of the analytical tools ω, β, and τ introduced in section 13.2.3 to model the costs of a floating-point operation, a message startup, and a byte transfer, respectively.

The serial computation of $\| Ax \|_1$ for a $p \times p$ matrix A requires p^2 multiplications and additions. The cost of this is roughly

$$T_1 = 2p^2\omega.$$

In comparison, the cost of executing the instructions y(i) = A(i,:)*x; and y(i) = abs(y(i)); in the parallel program is about $2p\omega$. If the computation is done in double precision, the numbers are eight bytes long, and the cost of the global sum on a hypercube or mesh is approximately $d(\omega + \beta + 8\tau)$. The total cost of the parallel algorithm is then

$$T_p = (2p + d)\omega + d\beta + 8d\tau.$$

Suppose that we wish to run our program on a DM-MIMD multiprocessor with $p = 2^d = 128$ where $\beta = 500$, $\omega = 10$, and $\tau = 1$. On this machine, $T_1 = 327,680$ and $T_p = 6256$ so that the theoretical speedup is $S = 52$. This represents an efficiency of 41% and so indicates that our algorithm may be worth implementing.

If, on the other hand, we want to run our program on a machine with only 16 nodes, the matrix size drops to 16×16. For this problem size, $T_1 = 5120$, $T_p = 243$, and $S = 2.1$. The small amount of computation performed is not enough to mask the communication cost, and the efficiency drops to only 13%. This poor efficiency may signal that we need to reconsider the design of the parallel implementation.

We have created our parallel algorithm by using one obvious division of the computation and communication steps. In general, it is not wise to proceed with the first thing that comes to mind without examining the alternatives. For example, we might want to rewrite the program to use a matrix of arbitrary order n. By using $p < n$, we could then assign more elements of y to each node and so increase the amount of computation performed relative to the number of messages sent.

We might even want to reconsider how we split up the matrix operations in the first place. For instance, it is often more efficient to split the matrix into blocks than into rows. (See [Golub & Van Loan 89] for more information on blocked algorithms.) We may also need to alter the program if a preceding computation distributes the matrix A among the nodes differently than we have required. If the program is to be used repeatedly or if the expected run time is very long, it is particularly important to develop an efficient implementation.

We now summarize the basic steps to remember when writing a program for a DM-MIMD multiprocessor.

1. Develop a good serial algorithm for the problem.

2. Identify the independent computational tasks in that algorithm.

3. Determine how to map those steps to the nodes.

4. Determine the data needs (initial and intermediate) of the parallel tasks.

5. Devise communication schemes to ensure that all data are in the proper place at the proper time and that the final result is easily accessible.

6. Assess the expected performance of the parallel algorithm. If the efficiency appears too small, return to step 2.

7. Write, test, debug, and run the parallel code.

14 SIMD Computing

According to the Flynn computer classification system [Flynn 72], a SIMD computer is a Single-Instruction, Multiple-Data machine. In other words, all the processors of a SIMD multiprocessor execute the same instruction at the same time, but each executes that instruction with different data.

The computers we discuss in this chapter are SIMD machines with distributed memories (DM-SIMD). They are sometimes referred to as *processor arrays* or as *massively-parallel* computers.

This chapter is divided into two main parts. In the first section, we discuss the general architecture of SIMD multiprocessors. Then we consider how these general features are embodied in two particular SIMD machines: the Thinking Machines CM-2 and the MasPar MP-2.

In the second section, we look into programming issues for SIMD multiprocessors, both architectural and language-oriented. In particular, we describe useful features of Fortran 90 and CM Fortran.

For detailed information on how to login and program specific SIMD computers such as CM-2 and the MasPar MP-1, refer to the documents in the /pub/HPSC directory at the cs.colorado.edu anonymous ftp site.

14.1   General architecture

Each of the processors in a distributed-memory SIMD machine has its own local memory to store the data it needs. Also each processor is connected to other processors in the computer and may send (or receive) data to (or from) any of them. In many respects, these computers are similar to distributed-memory MIMD (multiple instruction, multiple data) multiprocessors.

As stated above, the term SIMD implies that the same instruction is executed on multiple data. Hence the distinguishing feature of a SIMD machine is that all the processors act *in concert*. Each processor performs the same instruction at the same time as all the other processors, but each processor uses it own local data for this execution.

The array of processors is usually connected to the outside world by a sequential computer or workstation. The user accesses the processor array through this *front end* or *host* machine.

Using a SIMD computer for scientific computing means that many elements of an array can be computed simultaneously. Unlike vector processors,[1] the computation of these elements is not pipelined with different portions of neighboring elements being worked on at the same time. Instead large groups of elements go through the same computation in parallel.

In the following, we discuss the architectural features of SIMD multiprocessors concentrating on two computers in this class: the Connection Machine (CM-2) by Thinking Machines Corporation and the MasPar MP-2 by MasPar Computer Corporation. Similar computers include the Digital Equipment Corporation MPP series (technically the same as the MasPar machines), the Goodyear MPP, and the ICL DAP.

14.1.1   The Connection Machine (CM-2)

The CM-2 Connection Machine is a SIMD supercomputer manufactured by Thinking Machines Corporation (TMC). Data parallel programming is the natural paradigm for this machine allowing each processor to handle one data element or set of data elements at a time.

The initial concept of the machine was set forth in a Ph.D. dissertation by W. Daniel Hillis [Hillis 85]. The first commercial version of this computer was called the CM-1 and was manufactured in 1986. It contained up to 65,536 or 64K processors capable of executing the same instruction concurrently. As shown in figure 14.1, sixteen one-bit processors with 4K bits of memory apiece are on one chip of the machine. These chips are arranged in a hypercube[2] pattern. Thus the machine was available in units of 2^d processors where $d = 12$ through 16.

One of the original purposes of the computer was artificial intelligence; the eventual goal was a *thinking machine*. Each processor is only a one-

---

[1]See chapter 12 on vector computing for more information on vector processors.

[2]See chapter 13 on MIMD Computing for more information on a hypercube.

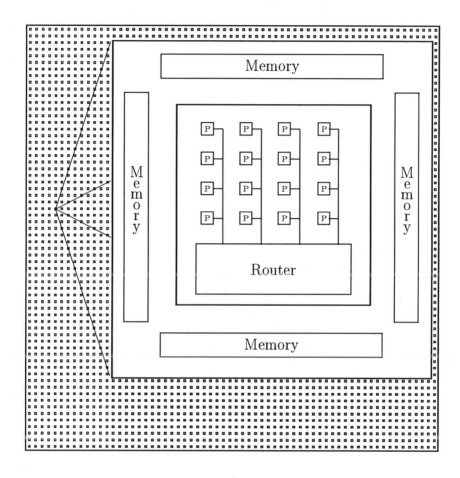

Figure 14.1: A representative blowup of one of the 64^2 processor chips in a Thinking Machines CM-1 or CM-2. Each CM-1/CM-2 chip contains 16 one-bit processors. The router connects the processor chips with other processor chips. Memory chips are associated with each CM-1/CM-2 chip.

bit processor. The idea was to provide one processor per pixel for image processing, one processor per transistor for VLSI simulation, or one processor per concept for semantic networks.

The first high-level language implemented for the machine is *Lisp, a parallel extension of Lisp. The design of portions of the *Lisp language are discussed in the Hillis dissertation.

As the first version of this supercomputer came onto the market, TMC discovered that there was also significant interest (and money) for supercomputers that can be used for numerical and scientific computing. Hence a faster version of the machine was produced in 1987; named the CM-2, this machine was the first of the CM-200 series of computers. It included floating-point hardware, a faster clock, and increased the memory to 64K bits per processor. These models emphasized the use of data-parallel programming. Both C* and CM Fortran were available on this machine in addition to *Lisp.

Announced in November 1991, a more recent machine is the CM-5. This is a MIMD machine that embodies many of the earlier Connection Machine concepts with more powerful processors, routing techniques, and I/O.

The following subsections discuss the characteristics and the performance of the CM-2. For further information, see the *Connection Machine CM-200 Series Technical Summary* [TMC 91d], *Parallel Supercomputing in SIMD Architectures* [Hord 90], chapter 7, or *Computer Architecture: Case Studies* [Baron & Higbie 92], chapter 18.

Characteristics

A Connection Machine (CM) may be considered as two main parts; these are depicted in figure 14.2. The *parallel processing unit* (PPU) is the SIMD portion of the machine and contains up to 64K single-bit processors; this part is the CM itself. The *front end* (FE) of the machine acts as a controller or host for the PPU and provides access to the rest of the world.

The FE is usually a small computer; for the CM, this may be a UNIX or a Symbolics Lisp workstation or a VAX. A CM may have up to four FE's; these do not need to all be the same type of machine. Programs are compiled, stored, and executed serially on the FE; any parallel operations in the program are recognized and pipelined to the PPU for execution there with the serial

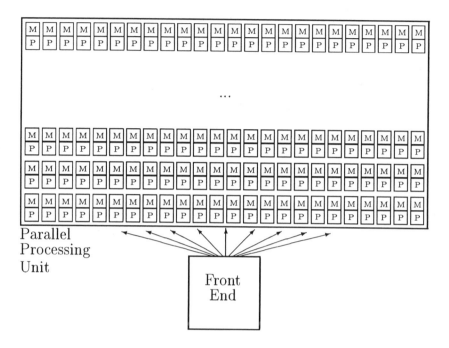

Figure 14.2:   The two main parts of a Thinking Machines Connection Machine system.

execution filling the pipeline and continuing until a response is needed back from the PPU. Thus CM programs have the familiar sequential control flow and do not require additional synchronization primitives as do programs for other multiprocessors.

Depending on the configuration of the machine, each one-bit PPU processor has 64K, 256K, or 1024K bits of memory, an arithmetic-logic unit (ALU), four one-bit registers, and interfaces for two forms of communication and I/O. These processors all work in parallel on simple instructions. In fact the PPU can be thought of as a large, synchronized drill team while the FE is the sergeant who yells out the commands. For example, all the PPU processors fetch something from their individual memories to their ALU's at one command; they all add something to that at the next command; and they all store their individual results into their own memories at the next command.

The language Paris (PARallel Instruction Set) is used to express the parallel operations that are to be run on the PPU. All *Lisp, CM Fortran, or C* parallel commands are compiled into Paris instructions. Such operations include parallel arithmetic operations (both floating-point and fixed), vector summation (and other reduction operations[3]), sorting, and matrix multiplication.

An alternate run-time system exists for machines with 64-bit floating-point accelerators. This is called the Slicewise model and provides a different viewpoint of the CM than the Paris model. This can be used only with CM Fortran programs but allows more efficient execution.

The PPU may be broken up into two or four sections as shown in figure 14.3. Each section has its own *sequencer* and can be used as a sub-PPU by itself or can be grouped with other sections. The nexus switch provides the pathway between a given FE and its current section(s) of the PPU.

A sequencer receives Paris instructions from the FE and breaks them down into a sequence of low-level instructions that can be handled by the one-bit processors. When that is done, the sequencer broadcasts these instructions to all the processors in its section. Each processor then executes the instructions in parallel with the other processors. When the execution of low-level instructions is completed, control is returned to the FE. Used independently, each section sets up its own grid layout for computation and communication for each array; if the sections are grouped together, one grid per array is laid over all the processors. These grids may be altered dynamically during the execution of the program.

Virtual processors are another feature of the CM. If the number of elements for the data of a given program is larger than the number of processors, the machine acts as if there were enough processors, providing virtual processors by assigning the data elements across the PPU in an efficient manner. In other words, each physical processor may act as one or more virtual processors.

Floating-point accelerators are optional and come in two types: 32-bit or 64-bit. A 32-bit FPA interprets the bits from two chips of sixteen one-bit processors as a single floating-point number allowing single precision arithmetic

---

[3]See chapter 12 on vector computing for more information on reduction operations.

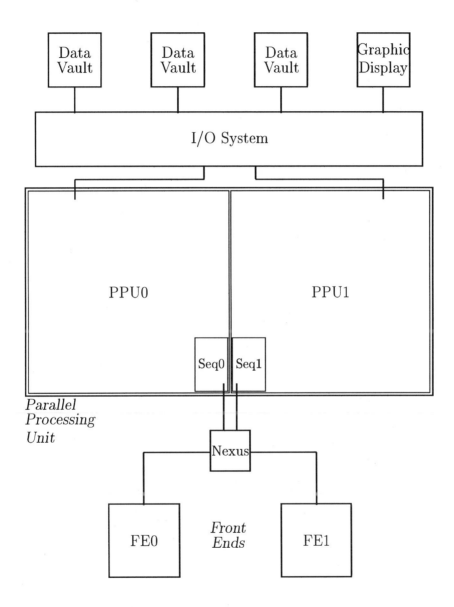

Figure 14.3: Breakdown of a Thinking Machines CM PPU into 2 sections with sequencers. Two front end machines, the nexus switch, and the I/O system are also shown.

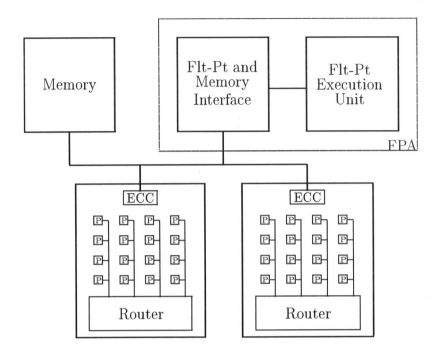

Figure 14.4: A pair of Thinking Machines CM-2 processor chips communicate with a single FPA.

computations. A 64-bit FPA also works with the contents of the processors contained on two chips and provides double precision arithmetic. For this reason, the processor chips are grouped in pairs with a floating-point and memory interface shared by each pair as shown in figure 14.4; it is common to think of each pair of chips as being equivalent to a floating-point processor. The addition of a floating-point unit to every 32 processors to form an FPA speeds up the processing of floating-point computations on the CM by more than a factor of twenty.

The Paris model of computation assigns full 32-bit or 64-bit words to the memory of each processor. These are passed to the FPA bit by bit when needed. The Slicewise model assigns different bits of a 32-bit or 64-bit word to each of the 32 processors on the two chips connected to a single FPA. When the FPA needs one of these words, each processor can send its bits concurrently with the other processors. In other words, one 32-bit word can be sent to the FPA in one load cycle as one bit from each of the 32 processors is sent to the FPA simultaneously. A 64-bit word requires only two load cycles since the data paths are 32-bits wide. Hence the amount of time spent passing data to and from the FPA is much less for the Slicewise model.

Data may be read and written serially through the FE to the PPU, but this is not a productive method for large amounts of data because of the cost of communication between the FE and PPU. An optional input/output system allows peak rates up to 320 Mbytes per second for transfers of data between the PPU processors and the I/O system buffers in parallel. One solution for speeding up peripherals is to attach multiple devices (i.e., disks) to the CM I/O system in parallel. Each device is connected to a separate CMIO bus and may transfer data in parallel with the other devices. Thus every 64 bits of data (the bandwidth of the bus) is transferred to a different device; this is called *file* or *disk striping*. Alternatively, these I/O buffers may attach to Thinking Machines *Data Vaults*; each Data Vault may contain 5 to 60 Gbytes of data. A maximum of eight data vaults can be on a given system, each transferring data at a peak rate of 40 Mbytes per second or an average rate of 20 Mbytes per second.

Another peripheral available through the I/O system is a *graphical display* for output generated by the PPU; this uses the Thinking Machines CM *framebuffer* managing up to one Gbyte of data per second. The high-resolution graphical display is a 19" color monitor. The framebuffer for the display attaches directly to the CM backplane, like an I/O controller, and holds raster image data. Alternatively, the image can be displayed in an X-window on a remote workstation or terminal. The size of the window determines the number of pixels; each PPU processor generates data for one pixel. Either 8-bit mode or 24-bit mode can be used.

In addition to the serial communication available between the FE and the PPU, the PPU processors may communicate with each other in different ways. The general form of communication is via the *router*; as shown in figure 14.1, a router is present on each chip of sixteen processors permitting each processor to communicate with any other processor on the PPU. The router also contains the logic to handle virtual processors.

Communication between processors on the same chip is called *local* or

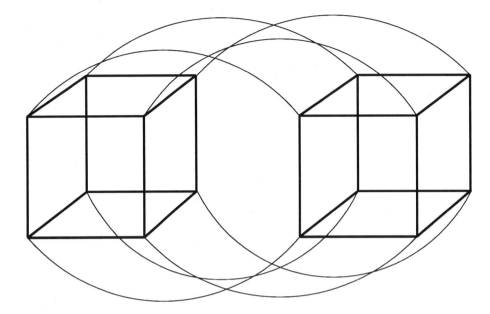

Figure 14.5: Part of a hypercube network.

on-chip communication. Clearly this is faster than any method of *off-chip* communication.

The interconnection of the processor chips across the PPU is based on a hypercube network, where each node of the hypercube is a processor chip containing sixteen processors. The operation of the router on a processor chip is broken up into twelve subcycles (where twelve is the maximum dimension of a CM-2 hypercube network). At subcycle k, the router is able to communicate with other processor chips (or nodes) along the kth dimension of the hypercube. Thus this interconnection pattern is sometimes called a *cube-connected cycle*. Figure 14.5 shows a 4-dimensional hypercube where the vertices of the network represent processor chips and the arcs represent connections to neighboring processor chips in the network.

When data can be treated as elements of a mesh or grid spread across the PPU, a second and faster method of off-chip communication may be used. This is called *NEWS* (*N*orth-*E*ast-*W*est-*S*outh) communication and allows the processors of each processor chip to communicate easily with the processors on the nearest neighbor chips on that mesh. However this type of communication

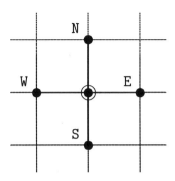

Figure 14.6: A two-dimensional NEWS grid.

is limited to nearest-neighbor messages and cannot be used for broadcasting information to all the processors. An example of a two-dimensional NEWS grid is shown in figure 14.6. The CM-1 only allowed a fixed, two-dimensional grid, but the CM-2 permits NEWS grids with up to 31 dimensions. Each chip on the PPU has an interface to the NEWS grid; processors with data elements required by or provided by their neighbors store pointers to those neighbors. When two or more sections of the PPU are being used independently, each section may have its own NEWS grid setup. When the sections are all ganged together; there is a single NEWS grid across the sections for each array or parallel structure.

Performance

A full CM-2 with 64K processors and double precision FPA's can perform arithmetic operations at the following speeds under the Paris model of execution[4] is shown in table 14.1.

The speed of memory reads and writes between each processor and its memory is greater than or equal to 5 Mbits/second. Each processor has 64K to 1024K bits or 8K to 128K bytes of memory and a full machine has 64K

---

[4]These figures were taken from the *Connection Machine Model CM-2 Technical Summary* [TMC 88], pp. 59-60.

Operation (Paris Model)	Single Precision Mflops
Fl. Pt. Addition	4,000
Fl. Pt. Multiplication	4,000
Fl. Pt. Division	1,500
Dot Product	10,000
4K×4K Matrix Mult	3,500

Table 14.1: CM-2 performance of single precision floating-point operations on a full 64K model, taken from [TMC 88].

processors providing a total memory of 512 Mbytes to 8 Gbytes. Hence given the minimum memory access speed of 5 Mbits/second for a single processor, the CM-2 can perform memory read/writes at about 300 Gbytes/second.

Other performance tests were done at the University of Colorado using the CM-2 at the National Center for Atmospheric Research (NCAR). This machine was just one-eighth the size of a full CM-2 (with merely 8192 processors) and contained only single precision FPA's; it also used the Paris model of execution. Standard CM Fortran routines were used exclusively for these tests. The results are summarized in table 14.2.

14.1.2   The MasPar MP-2

Manufactured by the MasPar Computer Corporation, the MasPar MP-1 and MP-2 are two other examples of DM-SIMD multiprocessors. Introduced in 1990, the MasPar MP-1 is the original MasPar machine. The newer MasPar MP-2 was brought out in 1992. While the MP-2 is larger and faster than the MP-1, the architectures of the two machines are similar.

Characteristics

Like the Thinking Machines CM, the MasPar MP-2 is broken up into two main parts: a front end (or *host*) machine and a processor array containing

Operation	Single Precision
(Paris Model)	Mflops
Fl. Pt. Addition	180
Fl. Pt. Multiplication	200
Fl. Pt. Division	70
Dot Product	56
Cosine	65
Exponential	60
Square Root	83

Table 14.2: Performance of NCAR CM-2 single precision floating-point operations.

the SIMD processors.

The front end of this machine may be a VAX or a DECstation 5000. Both types of these host machines run ULTRIX. The host machine permits communication of the SIMD processor array with the user. As with the CM-2, scalar processing is done on the front end.

The Data Parallel Unit (DPU) performs all the parallel computation and corresponds to the CM PPU; it is comprised of two parts. The first is the array of Processor Elements (PEs) incorporating 2^{10} (1024) to 2^{14} (16384) processors. Unlike the one-bit processors of the CM, a MasPar MP-2 processor (or PE) is a CPU capable of handling 32-bit values. Each of these processors contains 64 32-bit registers as well as an ALU and its own local memory. Thirty-two of the processors fit on a single chip in the PE. The older MasPar MP-1 contains only 4-bit processors but also fits 32 processors onto one chip.

The other part of the DPU is the Array Control Unit (ACU). This is a processor in itself with its own memory and instructions. The function of the ACU is to manage the operations of the PEs; it plays a similar role to that of the sequencers in the CM. An instruction is sent out by the ACU to all the processors in the PE array at the same time, and all the processors that are active simultaneously execute that instruction with their own data. The ACU is also the contact with the front end (host) machine.

Communication between the PEs is handled by the DPU in two ways. Ef-

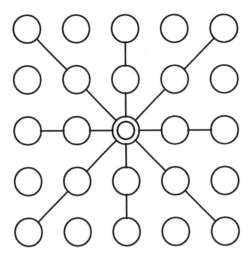

Figure 14.7: In a MasPar MP-2, each processing element (PE) is connected by the X-net to neighboring PE's along the north-south, east-west, northwest-southeast, and southwest-northeast axes.

ficient nearest-neighbor communication is done by the *X-net*; this is similar to the NEWS method of communication on the CM but goes in eight directions, as shown in figure 14.7. X-net communication should only be used for data communications in a regular pattern. The MasPar MP-2 processors are also connected by the *Global Router*. Like the CM router communication, communication via the Global Router is slower but permits message passing between any two processors in the DPU.

As with the CM architecture, there exist methods for allowing data on disks be read (or written), directly to (or from) the DPU. A framebuffer connection to the DPU is also present and permits animation on graphical displays generated directly from the DPU.

Two main programming languages are available for the MasPar MP-2: MPF (MasPar Fortran) and MPL (MasPar Programming Language). The first of these, MPF, is Fortran 90 with some extensions for the MasPar architecture; it includes special functions for passing data between the front end (host) machine and the DPU. The second language, MPL, is a version of C with parallel extensions; it is a low-level language and requires the programmer to have a good understanding of the machine architecture. The MasPar

MasPar	# Procs	R_{max}	N_{max}	$N_{1/2}$	R_{peak}
MP-2216	16384	1.6	11264	1920	2.4
MP-1216	16384	.473	11264	1280	.55
MP-1	16384	.44	5504	1180	.58
MP-2204	4096	.374	5632	896	.60
MP-1204	4096	.116	5632	640	.138
MP-2201	1024	.092	2816	448	.15
MP-1201	1024	.029	2816	320	.034
TMC CM-2	2048	10.4	33920	14000	28.

Table 14.3: Performance of MasPar MP-1 and MP-2, from [Dongarra 94]. The MasPar figures are for 64-bit numbers, while the Thinking Machines CM-2 values are for 32-bit numbers.

VAST-2 preprocessor allows the conversion of Fortran 77 programs to MPF programs for this machine. In addition, a programming environment called MPPE (MasPar Programming Environment) is installed on the host machine providing extra interactive support for the user to write, test, debug, and monitor parallel programs. Software libraries for mathematical computation, data display, and image processing are also available.

For more information on the MasPar SIMD computers, refer to [MasPar 92].

Performance

The table in table 14.3 compares the performance of various models of the MasPar MP-1 and MasPar MP-2 with differing numbers of processors. These figures are for double precision floating-point operations and are taken from [Dongarra 94]. Also included for comparison are Thinking Machines CM-2 figures for single precision floating-point operations. In this table, R_{max} represents the theoretical peak performance of a machine with the given number of processors (# **Procs**); R_{peak} is the best performance achieved for a problem of size N_{max}; and $N_{1/2}$ provides the size of the problem that executes at half the R_{max} performance.

14.2   Programming issues

Parallel programming is usually a greater challenge than programming for a sequential machine. In this section, we consider some of the problems and solutions associated with programming an SIMD multiprocessor. We also discuss some of the Fortran 90 features that are applicable to this type of architecture.

This discussion focuses on the CM-2 as a sample architecture using CM Fortran. Most of the concepts can easily be applied to the MasPar MP-2 and other DM-SIMD architectures. In particular, CM Fortran was a forerunner of Fortran 90 and contains most of the data-parallel constructs present in that language. Some additional constructs exist as well; those in that category are so indicated below. Some MPF constructs are also discussed. With a few exceptions (such as reduction operations), data-parallel operations behave in an elementwise fashion: each SIMD processor acts only on the array elements contained in its own memory.

14.2.1   Architectural organization considerations

When programming for the CM-2, the MP-2, or other SIMD architectures, it is best to think of the machine in the two parts shown in figure 14.2:

1. the front end to handle all the scalar operations and

2. the array of processors to execute all parallel operations.

For usual scientific computing applications, the front end (FE) is simply a sequential UNIX computer or workstation.[5] This is the machine to which the user logs in, the machine that compiles programs for the CM, and the machine that controls the execution of programs on the CM. All program variables used only in a sequential or scalar fashion are stored on the front end. The parallel processing unit (PPU) with its SIMD architecture handles all parallel or array operations, each processor taking care of one piece in unison with the other processors.

---

[5]Some installations use Symbolics Lisp machines as the FE to the CM.

```
ARRAYS
    Offset    Size  Type      Block/Class    Home    Name
         0    2048  REAL4     local          CM      A
      2048    2048  REAL4     local          CM      B
      4096    2048  REAL4     local          CM      C
```

Figure 14.8: **ARRAYS** Section of a CM Fortran Listing.

Since the editing and compiling of programs are done on the FE, programming on the CM is much like programming on any UNIX machine. In fact if a program does not use the CM parallel constructs, it executes entirely on the FE completely ignoring the PPU.

On the MasPar MP-2, the FE usually starts the program and sets up the initial data arrays; it also completes the program and collects the final results. Since communication is expensive between the FE and the DPU, it is best to contain most of the execution of a program to the DPU.

Homes

All program variables in CM Fortran programs are assigned a *home*. This is simply where the variable is stored. Since scalar variables can only be on the FE, that is their home. However arrays may be stored on the FE or on the PPU, depending on whether or not they are used in parallel operations. If they are used in both serial and parallel operations, they are stored on the PPU and copied to and from the FE for the serial operations. The exception to the rules above is for arrays of type **CHARACTER**; these are *always* stored on the FE.

To see where homes have been assigned for your variables, check the last part of your program listing. The two sections **VARIABLES** and **ARRAYS** provide the name, type, and size of the scalar variables and the arrays used in your program. The **ARRAYS** section also lists the Home for each array. Under this heading, the term **CM** refers to the PPU and **FE** to the FE. It is wise to double check this part of the program listing to make sure the arrays have been assigned as expected. A portion of a sample listing showing the ARRAYS section is given in figure 14.8.

Homes for variables in every program unit are assigned individually. In other words, the array Z in SUBROUTINE MYSUB may not be assigned the same home as Z in FUNCTION MYFTN. Each program module is treated as a unit.

Homes of actual and dummy arguments must match. If you pass an array to a function or argument, the dummy array argument must have the same home as the incoming array parameter. Otherwise unpredictable results are possible. There are a number of ways to force arrays to be assigned homes on the PPU:

- Declare the array in COMMON as all COMMON arrays are placed on the PPU;

- Put parallel operations in every program module for the appropriate arrays (even if they do nothing useful);

- Use the LAYOUT compiler directive:

$$\text{CMF\$ LAYOUT Z(:NEWS)}.$$

With :NEWS as an argument, Z *MUST* be in the PPU. (Other possible arguments are :SERIAL and :SEND. More on the LAYOUT compiler directive can be found in section 14.2.4.)

14.2.2   CM Fortran, MPF, and Fortran 90

The version of Fortran used on the CM is called CM Fortran. It is based on Fortran 77 and extended by parallel constructs; most of these constructs are contained in a subset of Fortran 90. On the MasPar MP-2, MPF (MasPar Fortran) is a version of Fortran 90 with extensions; many of the constructs are the same as those in CM Fortran.

On a CM, it is important to know that the control flow of a CM Fortran program is handled by the FE, as are all scalar statements like those found in Fortran 77. All data-parallel statements, including Fortran 90 statements, are executed on the PPU.

The following subsections introduce a few elements of CM Fortran, MPF, and Fortran 90 to get you started. For further reference, see the CM manuals ([TMC 91a], [TMC 91b], and [TMC 91c]), the MasPar manuals: ([MasPar 93b] and [MasPar 93a]), and some Fortran 90 references such as [Brainerd et al 90] and [Adams et al 92].

Arrays

An array on a SIMD multiprocessor may be considered a *data-parallel* object; this is true for CM Fortran arrays as well. In fact the only CM Fortran variables stored on the PPU are those arrays used in parallel operations; all scalars and all arrays not involved with parallel operations are stored on the FE of the CM.

The properties of an array are *rank*[6] and *shape*. The rank of an array is the number of its dimensions; e.g., the array declared as S(5,10) has rank 2. The shape of an array is its dimensions; so the shape of S(5,10) is 5×10. Two arrays with the same shape are said to be *conformable*. Most parallel operations require that the arrays involved be conformable.

Once an array has been declared, the use of the name of the array by itself (not subscripted) denotes the entire array with all its elements. Such usage implies a parallel operation is to be applied to the array. For instance the statement

$$S = 0.0$$

sets all the elements of S to zero in parallel on the PPU.

Subsections of an array can be specified by *triples*. The general form of a triple is

$$firstvalue : lastvalue : increment^7$$

For example if S is declared as above, then S(1:5:2,1:10) refers to the odd rows of S. The triple 1:5:2 specifies that rows 1, $1 + 2 = 3$, and $3 + 2 = 5$ are to be used; the triple 1:10 has an implied increment of 1 and so specifies

---

[6] In this context, *rank* has a different meaning than its customary mathematical definition.

[7] This differs from the triple form used by MATLAB: *firstvalue : increment : lastvalue*.

all ten columns. This second triple 1:10 could have been replaced by a single colon as in S(1:5:2,:) to imply that all the columns be used for the chosen rows.

As in Fortran 77 and Fortran 90, CM Fortran arrays may be declared by DIMENSION, COMMON, or *type* statements. In CM Fortran, they may also be declared using *array attribute* statements. For example assuming the array S is a *real* array, it could have been defined by the following array attribute statement:

REAL, ARRAY(5,10) :: S, T

In Fortran 90, the following statement would have the same effect:

REAL, DIMENSION(5,10) :: S, T

This simply says that both S and T are *real* arrays with 5 rows and 10 columns. Notice the comma after the type indicator REAL; this is how the array attribute statement is recognized by the compiler. The double colon :: is also a requirement; it must be placed between the array definition and the array names. You should recall that blank spaces in Fortran are traditionally ignored; hence any number of spaces can be added to this statement (even between the colons) or deleted from the statement.

Array constructors can be used to initialize the elements of an array in parallel. For instance if Z has been declared in CM Fortran by the statement

REAL, ARRAY(N) :: Z

then the statement

Z = REAL([1:N])

assigns 1.0 to Z(1), 2.0 to Z(2), and REAL(N) to Z(N). The corresponding Fortran 90 statements follow:

REAL, DIMENSION(N) :: Z

and

Z = (/ (REAL(I), I=1,N) /)

Array constructors can also be included in array attribute statements. In CM Fortran, this is done by adding a `DATA` parameter to the statement; thus the following CM Fortran statement has the effect of defining and initializing Z at once:

```
REAL, ARRAY(N), DATA ::  Z = [1:N]
```

The same effect can be achieved in Fortran 90 by the following statement:

```
REAL, DIMENSION(N) ::  Z = (/ (REAL(I), I=1,N) /)
```

This is an efficient way of assigning initial values to an array as it is done at load time. A limitation on the array constructor in CM Fortran is that is can only be used for one-dimensional arrays.

Array sections

Most of the parallel array facilities of Fortran 90 are part of CM Fortran. As mentioned in section 14.2.2, the ability to work with all the elements of an array or subsections of an array in parallel is provided; however in CM Fortran, all of the arrays involved in this type of parallel expression must be parallel arrays with homes on the PPU.

Using the name of an array implies using the entire array in parallel. For instance, the statement

```
Y = Z**2
```

causes `Y(1)` to be set to the value of `Z(1)**2`, `Y(2)` to `Z(2)**2`, etc. This also works for constant assignments; the statement

```
Y = -1.0
```

means that all the elements of Y are set to -1.0.

Subsections of arrays can be used in assignment statements as well. In the following statement

```
Y(1:10) = Z(11:20)
```

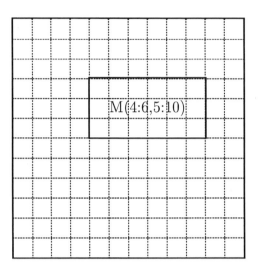

Figure 14.9: Subsection of array M(12,12).

the first ten elements of Y are set to the second ten elements of Z. Such statements execute in parallel.

An example of a subsection of a two-dimensional array is shown in figure 14.9. Here the array M has been declared as

REAL M(12,12)

and the 3×6 subsection is defined by

M(4:6,5:10)

Alternate DO loops

Additional control constructs exist in CM Fortran; these are similar to those in Fortran 90. The first of these are alternate forms of DO loops as demonstrated below:

```
N = 4096
DOWHILE (N .GT. 0)
   Z(1:N) = ...
   N = N/2
ENDDO

KK = 1
DO (N) TIMES
   KK = KK * K
ENDDO
```

The first of these loops assigns values to the first N elements of the array Z for N equal to decreasing powers of two. The second loop terminates when KK is equal to K^N where N is a non-negative integer. This form of the DO loop is useful when the loop index is not needed within the body of the loop. Note that the DO WHILE construct is a legal Fortran 90 construct; the second form of the DO loop, DO (N) TIMES, is not.

WHERE statements

The WHERE statements provide a means for working with a subset of a full array still as a parallel operation:

```
WHERE (Z .GT. 0.0)  Y = SQRT(Z)
```

Here all the CM processors actually compute Y = SQRT(Z). However only those processors with a value for Z greater than zero store the result. The intrinsic function SQRT is used on the entire array. If we wished to set the other (negative and zero) elements to zero at the same time, this operation could be programmed as follows:

```
WHERE (Z .GT. 0.0)
   Y = SQRT(Z)
ELSEWHERE
   Y = 0.0
ENDWHERE
```

In this set of statements, the elements of Y are set to zero when the corresponding element of Z is not greater than zero. Note that this construct acts in two steps. First all the processors compute Y = SQRT(Z), but only the values for elements of Y corresponding to non-zero elements of Z are stored. Then all the processors compute Y = 0, but only the values corresponding to zero or negative elements of Z are stored. In other words, the construct appears similar to the IF..THEN..ELSE..ENDIF statement but behaves a little differently.

FORALL statements

The FORALL construct is not a part of Fortran 90; however it is included in both CM Fortran and MPF. Such statements are very convenient for SIMD computers.

FORALL statements as in

```
FORALL (I=1:N)   Y(I) = I
```

can only contain one assignment. This statement is equivalent to the following DO loop:

```
DO I = 1,N
   Y(I) = I
ENDDO
```

Notice that there is a colon between the start and stop values of the FORALL index; this is in a *triple* format like the subscripts discussed earlier. An increment may be used as well after another colon, as the third element of the triple.

More than one index can be used within the FORALL statement; for instance the following statement:

```
FORALL (I=1:N, J=1:N)   S(I,J) = I
```

sets all the elements of the Ith row of S to I.

Often individual elements of an array need to be initialized to specific values. If the home of the array is on the PPU, it is best to use a FORALL statement for this purpose. For instance the statement

$$S(6,1) = S(6,2) + S(6,3)$$

is executed on the FE since it is essentially a scalar operation. However if we rewrite this statement as a **FORALL** statement,

$$\text{FORALL (I=6:6, J=1:1)} \quad S(I,J) = S(I,J+1) + S(I,J+2)$$

it is executed on the PPU. In effect, the sum of S(I,J+1) and S(I,J+2) for each I and J is computed by all the elements in the array, but only the processor containing the element in the first column of the sixth row of S stores this value into its array element.

A **FORALL** statement with dependencies that cannot be resolved is executed serially. Check for restrictions on the **FORALL** statement in the appropriate manual.

14.2.3 Built-in functions for CM Fortran and Fortran 90

The CM Fortran intrinsic functions are for the most part the same as the intrinsic functions described for Fortran 90. Additional functions for handling data-parallel data types are also present in both CM Fortran and Fortran 90. Some of these are described below.

To aid in the explanation of these built-in functions, assume the following arrays have the values given below:

$$A = \begin{pmatrix} 1 & 2 \\ 3 & 4 \\ 5 & 6 \end{pmatrix}$$

$$B = \begin{pmatrix} 2 & 4 & 5 \\ 3 & 8 & 5 \end{pmatrix}$$

$$C = \begin{pmatrix} 1, & -2, & 3, & -4, & 5, & -6 \end{pmatrix}$$

Intrinsic functions

The usual Fortran intrinsic functions are available in CM Fortran and Fortran 90. Moreover most of them can be used in a parallel fashion. For instance if A has been declared as above, then

$$\text{MOD(A,5)}$$

returns a matrix of the same type and shape as A containing the values of the elements of A *mod(5)*:

$$\text{MOD}(A, 5) = \begin{pmatrix} 1 & 2 \\ 3 & 4 \\ 0 & 1 \end{pmatrix}$$

Similarly the SQRT function can handle a whole array at once.

$$\text{SQRT}(A) = \begin{pmatrix} 1.000 & 1.414 \\ 1.732 & 2.000 \\ 2.236 & 2.449 \end{pmatrix}$$

Masks

Masks are logical arrays created by performing a relational operation on all the elements of a given array. Both the mask and the original array must conform. Consider the following examples.

$$\text{A.GT.0} = \begin{pmatrix} T & T \\ T & T \\ T & T \end{pmatrix}$$

$$\text{B.EQ.5} = \begin{pmatrix} F & F & T \\ F & F & T \end{pmatrix}$$

$$\text{C.LT.0} = \begin{pmatrix} F, & T, & F, & T, & F, & T \end{pmatrix}$$

Special functions

In addition to the normal Fortran intrinsic functions, CM Fortran provides several special functions to aid in the parallel operation of the machine. For the examples of the functions described below, assume the following: ARRAY is the name of any array of type real, integer, or logical; DIM is an integer denoting which particular dimension of the array (if any) the function is to be applied to; MASK is a logical array with the same shape as ARRAY telling which particular elements the function is to use; V, V1 and V2 are one-dimensional arrays or vectors; M1 and M2 are two-dimensional arrays or matrices; and SHIFT is an integer or integer array describing the shift to be made. For some of the following functions, DIM and MASK may be used as keyword parameters.

Reduction Operations: The following functions perform commonly used reduction operations. Except where noted, they are common to both CM Fortran and Fortran 90.

- SUM (ARRAY [, DIM] [, MASK]): This function computes the sum of all the elements of ARRAY, according to the values of DIM and MASK. (Note: in the last example, MASK is used as a keyword parameter, since the second parameter DIM is missing.)

$$
\begin{aligned}
\text{SUM(A)} &= 21 \\
\text{SUM(B, 1)} &= (5, \quad 12, \quad 10) \\
\text{SUM(B, 2)} &= (11, \quad 16) \\
\text{SUM(C, MASK=C.GT.0)} &= 9
\end{aligned}
$$

- PRODUCT (ARRAY [, DIM] [, MASK]): This function computes the product of all the elements of ARRAY according to the values of DIM and MASK. (Note: in the last example, MASK is used as a keyword parameter, since the second parameter DIM is missing.)

$$
\begin{aligned}
\text{PRODUCT(A)} &= 720 \\
\text{PRODUCT(B, 1)} &= (6, \quad 32, \quad 25) \\
\text{PRODUCT(B, 2)} &= (40, \quad 120) \\
\text{PRODUCT(C, MASK=C.GT.0)} &= 15
\end{aligned}
$$

- DOTPRODUCT (V1, V2): This function computes the dot product of the two vectors or one-dimensional arrays, V1 and V2.

$$
\begin{aligned}
\text{DOTPRODUCT(A(1,:),B(:,2))} &= 20 \\
\text{DOTPRODUCT(A(:,1),B(2,:))} &= 52 \\
\text{DOTPRODUCT(C,C)} &= 91
\end{aligned}
$$

- MAXVAL (ARRAY [, DIM] [, MASK]): This function finds the maximum value of all the elements of ARRAY according to the values of DIM and MASK.

$$
\begin{aligned}
\text{MAXVAL(A)} &= 6 \\
\text{MAXVAL(B(:,1))} &= 3 \\
\text{MAXVAL(C)} &= 5 \\
\text{MAXVAL(C,1,C.LT.0)} &= \text{-2}
\end{aligned}
$$

- MINVAL (ARRAY [, DIM] [, MASK]): This function finds the minimum value of all the elements of ARRAY according to the values of DIM and MASK.

$$
\begin{aligned}
\text{MINVAL(A)} &= 1 \\
\text{MINVAL(B(:,1))} &= 2 \\
\text{MINVAL(C)} &= \text{-6} \\
\text{MINVAL(C,1,C.GT.0)} &= 1
\end{aligned}
$$

- MAXLOC (ARRAY [, MASK]): This function returns an integer value or integer array representing the subscripts of the maximum values of all the elements of ARRAY according to the values of MASK. If more than one such location exists, which subscript is returned is non-deterministic.

$$
\begin{aligned}
\text{MAXLOC(A)} &= (3, \ 2) \\
\text{MAXLOC(B(:,3))} &= 1 \quad \text{(could also be 2)} \\
\text{MAXLOC(C)} &= 5 \\
\text{MAXLOC(C,C.LT.0)} &= 2
\end{aligned}
$$

- `MINLOC (ARRAY [, MASK])`: This function returns an integer value or integer array representing the subscripts of the minimum values of all the elements of `ARRAY` according to the values of `MASK`. If more than one such location exists, which subscript is returned is non-deterministic.

$$
\begin{aligned}
\texttt{MINLOC(A)} \quad &= (1, \ 1) \\
\texttt{MINLOC(B(:,3))} \quad &= 1 \quad \text{(could also be 2)} \\
\texttt{MINLOC(C)} \quad &= 6 \\
\texttt{MINLOC(C,C.GT.0)} &= 1
\end{aligned}
$$

- `COUNT (MASK [, DIM])`: This function returns the number of elements for which the `MASK` held true.

$$
\begin{aligned}
\texttt{COUNT(A.GT.0)} \quad &= 6 \\
\texttt{COUNT(A.GT.0,1)} &= (3, \ 3) \\
\texttt{COUNT(B.EQ.5)} \quad &= 2 \\
\texttt{COUNT(C.LE.0)} \quad &= 3
\end{aligned}
$$

- `ANY (MASK [, DIM])`: This function returns True if the `MASK` held true for any of the elements.

$$
\begin{aligned}
\texttt{ANY(A.GT.0)} \quad &= \text{T} \\
\texttt{ANY(A.GT.0,1)} &= (\text{T}, \ \text{T}) \\
\texttt{ANY(B.EQ.5)} \quad &= \text{T} \\
\texttt{ANY(C.LE.0)} \quad &= \text{T}
\end{aligned}
$$

- `ALL (MASK [, DIM])`: This function returns True if the `MASK` held true for all of the elements.

$$
\begin{aligned}
\texttt{ALL(A.GT.0)} \quad &= \text{T} \\
\texttt{ALL(A.GT.0,1)} &= (\text{T}, \ \text{T}) \\
\texttt{ALL(B.EQ.5)} \quad &= \text{F} \\
\texttt{ALL(C.LE.0)} \quad &= \text{F}
\end{aligned}
$$

Functions for matrices: The following built-in functions in CM Fortran
and Fortran 90 are used for manipulating matrices:

- TRANSPOSE (M1): This function returns the transpose of the matrix or
  two-dimensional array M1.

$$\text{TRANSPOSE(A)} = \begin{pmatrix} 1 & 3 & 5 \\ 2 & 4 & 6 \end{pmatrix}$$

- MATMUL (M1, M2): This function returns the result of the matrix multiplication of M1 by M2. (Note that the expression M1*M2 does *not* perform
  matrix multiplication. Instead it produces element-by-element multiplication; that is, for all i, j $(M1 * M2)_{i,j} = M1_{i,j} * M2_{i,j}$.)

$$\text{MATMUL(A, B)} = \begin{pmatrix} 8 & 20 & 15 \\ 18 & 44 & 35 \\ 28 & 68 & 55 \end{pmatrix}$$

$$\text{MATMUL(B, A)} = \begin{pmatrix} 39 & 50 \\ 52 & 68 \end{pmatrix}$$

- DIAGONAL (ARRAY [, FILL]): This function is only in CM Fortran. It
  creates a diagonal matrix from the vector ARRAY. The elements of the
  vector are placed on the diagonal and the value of FILL (if any) is placed
  in the other elements of the matrix. If there is no FILL value, the value
  of 0 (or .FALSE., if logical) is used.

$$\text{DIAGONAL(C)} = \begin{pmatrix} 1 & 0 & 0 & 0 & 0 & 0 \\ 0 & -2 & 0 & 0 & 0 & 0 \\ 0 & 0 & 3 & 0 & 0 & 0 \\ 0 & 0 & 0 & -4 & 0 & 0 \\ 0 & 0 & 0 & 0 & 5 & 0 \\ 0 & 0 & 0 & 0 & 0 & -6 \end{pmatrix}$$

$$\text{DIAGONAL}(\text{C}, 99) = \begin{pmatrix} 1 & 99 & 99 & 99 & 99 & 99 \\ 99 & -2 & 99 & 99 & 99 & 99 \\ 99 & 99 & 3 & 99 & 99 & 99 \\ 99 & 99 & 99 & -4 & 99 & 99 \\ 99 & 99 & 99 & 99 & 5 & 99 \\ 99 & 99 & 99 & 99 & 99 & -6 \end{pmatrix}$$

Other useful functions: In addition to the reduction operations listed above, both CM Fortran and Fortran 90 contain other functions directed toward handling data-parallel data objects. Some of these are given here:

- RANK (ARRAY): This CM Fortran function returns the rank of the given scalar or ARRAY. A similar Fortran 90 function is named SIZE.

$$\begin{aligned} \text{RANK(100)} &= 0 \\ \text{RANK(A)} \quad &= 2 \\ \text{RANK(B)} \quad &= 2 \\ \text{RANK(C)} \quad &= 1 \end{aligned}$$

- DSHAPE (ARRAY): This CM Fortran function returns the shape of the given scalar or ARRAY. In Fortran 90, this function is named SHAPE.

$$\begin{aligned} \text{DSHAPE(-1)} &= (\ ) \\ \text{DSHAPE(C)} \quad &= 6 \\ \text{DSHAPE(A)} \quad &= (\ 3, \quad 2\ ) \\ \text{DSHAPE(B)} \quad &= (\ 2, \quad 3\ ) \end{aligned}$$

- REPLICATE (ARRAY, DIM, NCOPIES): This CM Fortran function adds NCOPIES of the ARRAY along the given DIMension. The resultant array has the same rank as the original ARRAY, but the shape in greater in the given DIMENSION.

$$\text{REPLICATE}(A, 1, 2) \; = \; \begin{pmatrix} 1 & 2 \\ 3 & 4 \\ 5 & 6 \\ 1 & 2 \\ 3 & 4 \\ 5 & 6 \end{pmatrix}$$

$$\text{REPLICATE}(A, 2, 3) \; = \; \begin{pmatrix} 1 & 2 & 1 & 2 & 1 & 2 \\ 3 & 4 & 3 & 4 & 3 & 4 \\ 5 & 6 & 5 & 6 & 5 & 6 \end{pmatrix}$$

REPLICATE(A, 1, 0) = ( )
REPLICATE(C, 1, 2) =
$$(1, \quad -2, \quad 3, \quad -4, \quad 5, \quad -6, \quad 1, \quad -2, \quad 3, \quad -4, \quad 5, \quad -6)$$

- SPREAD (ARRAY, DIM, NCOPIES): This function is in both CM Fortran and Fortran 90. It produces NCOPIES of the ARRAY along DIM. The resultant array has rank one greater than that of the original ARRAY. This can also be used to make a vector from a scalar.

  SPREAD(-1, 1, 6) = $(-1, \quad -1, \quad -1, \quad -1, \quad -1, \quad -1)$
  SPREAD(-1, 1, 0) = ( )

$$\text{SPREAD}(A, 1, 2) \; = \; \left(\begin{pmatrix} 1 & 2 \\ 3 & 4 \\ 5 & 6 \\ 1 & 2 \\ 3 & 4 \\ 5 & 6 \end{pmatrix} \right)$$

$$\text{SPREAD}(A, 2, 3) \; = \; \left(\begin{pmatrix} 1 & 2 \\ 1 & 2 \\ 1 & 2 \end{pmatrix} \begin{pmatrix} 3 & 4 \\ 3 & 4 \\ 3 & 4 \end{pmatrix} \begin{pmatrix} 5 & 6 \\ 5 & 6 \\ 5 & 6 \end{pmatrix} \right)$$

$$\text{SPREAD}(C, 1, 2) \; = \; \begin{pmatrix} 1 & -2 & 3 & -4 & 5 & -6 \\ 1 & -2 & 3 & -4 & 5 & -6 \end{pmatrix}$$

$$\mathrm{SPREAD}(\mathtt{C}, 2, 3) \;=\; \begin{pmatrix} 1 & 1 & 1 \\ -2 & -2 & -2 \\ 3 & 3 & 3 \\ -4 & -4 & -4 \\ 5 & 5 & 5 \\ -6 & -6 & -6 \end{pmatrix}$$

- PACK (ARRAY, MASK [, V]): This function gathers elements from the ARRAY under the control of the MASK. If the vector V is specified, the result is placed on top of the values already there.

$$\begin{aligned}
\mathtt{PACK(B,\ B.GT.4)} &= \big(8,\ 5,\ 5\big) \\
\mathtt{PACK(A,\ A.GT.3,\ C)} &= \big(5,\ 4,\ 6,\ -4,\ 5,\ -6\big) \\
\mathtt{PACK(A,\ A.GT.0,\ C)} &= \big(1,\ 3,\ 5,\ 2,\ 4,\ 6\big)
\end{aligned}$$

- UNPACK (V, MASK, ARRAY): This function scatters elements from the vector V under the control of the MASK into the ARRAY.

$$\begin{aligned}
\mathtt{UNPACK(C,\ C.LT.0,\ [6[0.0]])} &= \big(0,\ 1,\ 0,\ -2,\ 0,\ 3\big) \\
\mathtt{UNPACK(C,\ C.LT.0,\ [6[-1.0]])} &= \big(-1,\ 1,\ -1,\ -2,\ -1,\ 3\big) \\
\mathtt{UNPACK(C,\ C.GT.0,\ C)} &= \big(1,\ -2,\ -2,\ -4,\ 3,\ -6\big)
\end{aligned}$$

- CSHIFT (ARRAY, DIM, SHIFT): This function does a *Circular SHIFT* on ARRAY returning a result that has the same type and shape as ARRAY.

$$\begin{aligned}
\mathtt{CSHIFT(C,\ 1,\ 1)} &= \big(-2,\ 3,\ -4,\ 5,\ -6,\ 1\big) \\
\mathtt{CSHIFT(C,\ 1,\ -1)} &= \big(-6,\ 1,\ -2,\ 3,\ -4,\ 5\big) \\
\mathtt{CSHIFT(C,\ 1,\ 2)} &= \big(3,\ -4,\ 5,\ -6,\ 1,\ -2\big) \\
\mathtt{CSHIFT(C,\ 1,\ -3)} &= \big(-4,\ 5,\ -6,\ 1,\ -2,\ 3\big)
\end{aligned}$$

$$\mathrm{CSHIFT}(\mathtt{A}, 1, 2) \;=\; \begin{pmatrix} 5 & 6 \\ 1 & 2 \\ 3 & 4 \end{pmatrix}$$

$$\mathrm{CSHIFT}(\mathtt{A}, 2, -1) \;=\; \begin{pmatrix} 2 & 1 \\ 4 & 3 \\ 6 & 5 \end{pmatrix}$$

$$\text{CSHIFT}(A, 1, [0, 1]) \;=\; \begin{pmatrix} 1 & 4 \\ 3 & 6 \\ 5 & 2 \end{pmatrix}$$

- EOSHIFT (ARRAY, DIM, SHIFT [, BOUNDARY]): This function does an *End-Off SHIFT* on ARRAY returning a result that has the same type and shape as ARRAY. The value of BOUNDARY (if any) is used to fill up the spaces made by shifting away from the edges; otherwise zero (or .FALSE.) is used.

$$
\begin{aligned}
\text{EOSHIFT(C, 1, 1)}\ \ &= \left(-2,\ \ 3,\ \ -4,\ \ 5,\ \ -6,\ \ 0\right) \\
\text{EOSHIFT(C, 1, -1)}\ \ &= \left(0,\ \ 1,\ \ -2,\ \ 3,\ \ -4,\ \ 5\right) \\
\text{EOSHIFT(C, 1, 3, 9)} &= \left(-4,\ \ 5,\ \ -6,\ \ 9,\ \ 9,\ \ 9\right)
\end{aligned}
$$

$$\text{EOSHIFT}(A, 2, -1) \;=\; \begin{pmatrix} 0 & 1 \\ 0 & 3 \\ 0 & 5 \end{pmatrix}$$

$$\text{EOSHIFT}(A, 1, [-1, 0]) \;=\; \begin{pmatrix} 0 & 2 \\ 1 & 4 \\ 3 & 6 \end{pmatrix}$$

$$\text{EOSHIFT}(A, 1, 2, 99) \;=\; \begin{pmatrix} 5 & 6 \\ 99 & 99 \\ 99 & 99 \end{pmatrix}$$

$$\text{EOSHIFT}(A, 1, 2, \text{REAL}([1:2])) \;=\; \begin{pmatrix} 5 & 6 \\ 1 & 2 \\ 1 & 2 \end{pmatrix}$$

There are many additional intrinsic functions that can be used to manipulate arrays in parallel, such a, RESHAPE in both CM Fortran and Fortran 90 and PROJECT available only in CM Fortran. Refer to the *CM Fortran Reference Manual* [TMC 91b] or the *MasPar Fortran Reference Manual* [MasPar 93a] for more information on these and other functions. Reference books on Fortran 90, such as the *Fortran 90 Handbook* by [Adams et al 92] and the *Programmers Guide to Fortran 90* by [Brainerd et al 90], may be helpful as well.

14.2.4 Compiler directives

Compilers for SIMD computers make use of compiler directives to define the assignment of array elements to different processors. The compiler directives in CM Fortran are used in two ways. The first is to specify the location of the elements of arrays with respect to each other; the second is to specify the use and homes of arrays in the common blocks of the program. The MasPar MPF language also provides compiler directives for mapping array elements to processors and for specifying the location of arrays and common blocks. Further MPF compiler directives assist with the calling of routines in other languages.

In this section, we discuss some of the compiler directives for both these machines. This is done to provide examples of how such directives are used; in many cases, these examples may be extended or modified for compilers on other SIMD machines. In the following, we briefly present the use and purpose of three CM Fortran compiler directives and three MPF compiler directives

$$
\begin{aligned}
&\texttt{CMF\$ LAYOUT } \textit{args} \\
&\texttt{CMPF MAP } \textit{args} \\
&\texttt{CMF\$ ALIGN } \textit{args} \\
&\texttt{CMF\$ COMMON } \textit{args} \\
&\texttt{CMPF ONDPU } \textit{args} \\
&\texttt{CMPF ONFE } \textit{args}
\end{aligned}
$$

as representative SIMD compiler directives.

All the CM Fortran compiler directives begin with the characters `CMF$`; they appear as comment lines to other Fortran compilers. The continuation of such lines if needed differs from the normal Fortran statement continuation; since the compiler directives look like comments, an ampersand (`&`) needs to be placed on the end of the line to be continued instead of the continuation character in column six. MPF directives begin with the characters `CMPF`.

CM Fortran `LAYOUT`

In this section and the next, a reference to a processor means a virtual processor. If an array has more elements than the number of physical processors, each processor acts as a virtual processor to more than one element. The

ordering of the assignment of elements to virtual processors depends on the
layout or ordering chosen for the given dimension of the array.

In the normal or default allocation of arrays on the PPU, each element is
placed on a single processor. When there are more elements than processors,
each element is mapped to a virtual processor. The intent is to allow each
processor to have its own piece of the action. Such a mapping for the array
V(0:N) is shown in figure 14.10.

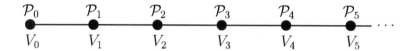

Figure 14.10: Normal layout for the array V(0:N).

However there are algorithms where it is more efficient to allow each processor to have a set of elements of a given array as those elements are used
together in the same computation. For instance in the case of arrays of two
or more dimensions, it may be best to have all the elements of one of the
dimensions be placed on the same processor. For instance suppose the array
Q is declared by the following DIMENSION statement

$$\text{DIMENSION} \quad \text{Q(3,0:N)}$$

Further suppose that

$$Q_{3,i} = \mathcal{F}(Q_{1,i}, Q_{2,i})$$

where $\mathcal{F}$ is some function of the two one-dimensional arrays that make up the
first two rows of Q. Then it might be desirable to have all three elements of
each column of the array on the same processor. This type of layout is shown
in figure 14.11. The compiler directive that assigns this layout is

$$\text{CMF\$ LAYOUT} \quad \text{Q(:SERIAL,:NEWS)}$$

Here the term SERIAL means that the elements on the array in this dimension
should all be on the same processor. The term NEWS implies that the elements
of the second dimension (each column) should be spread across the processors
in an order to allow efficient communication between nearest neighbors.

The general format of this directive is

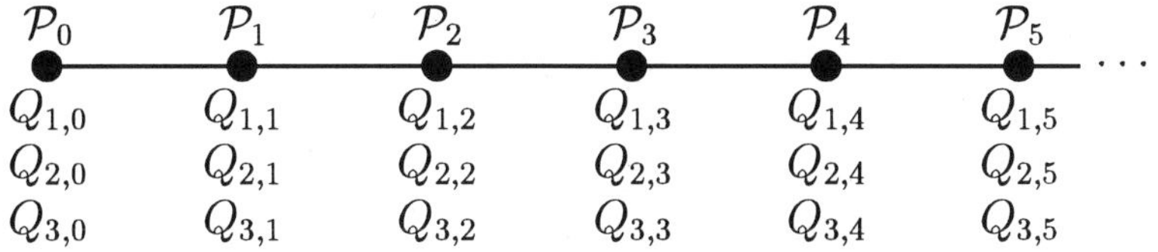

Figure 14.11: Layout for the array `Q(:SERIAL,:NEWS)`.

CMF$ LAYOUT ARRAY($weight1:order1, weight2:order2, \ldots,$
$weightN:orderN$)

This specifies an order and a weight to that order for each dimension of the given ARRAY. The order is as defined above; the weight is any constant expression that indicates the importance of that ordering for the given dimension in relation to the other dimensional orders. If the weight is missing, it is assumed to be 1; if the order is missing, it is assumed to be NEWS. The sample statement above

CMF$ LAYOUT Q(:SERIAL,:NEWS)

provides no weights. Weights have no meaning for serial ordering since the elements in that dimension should all be on the same processor.

There are three different types of *orders*. As mentioned in the last paragraph, the SERIAL order tells that compiler that all elements of the given dimension should be arranged sequentially on the same processor.

The NEWS order is the default ordering; it specifies that elements of the given dimension of the array should be stored on processors in such a way as to provide quick nearest neighbor communication. This is most often used in grid computations where each element of the array needs to be updated using the values of its neighboring elements as in the following two-dimensional expression:

$$X_{i,j} = \mathcal{F}(X_{i,j}, X_{i-1,j}, X_{i+1,j}, X_{i,j-1}, X_{i,j+1})$$

This two-dimensional form of a grid is shown in figure 14.12. So the default layout

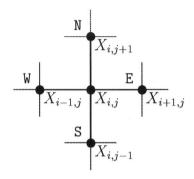

Figure 14.12: A two-dimensional NEWS grid.

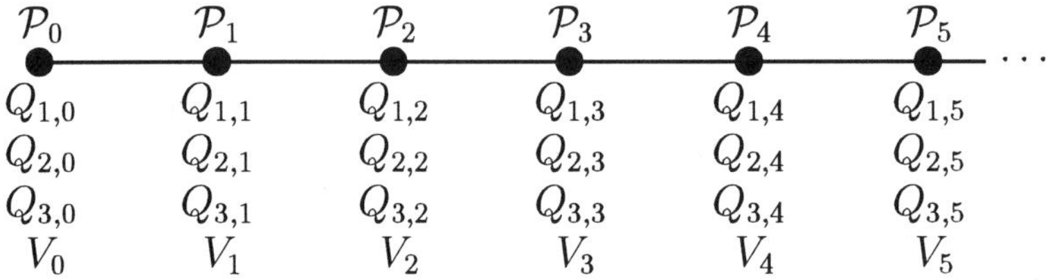

$$\mathcal{P}_0 \quad \mathcal{P}_1 \quad \mathcal{P}_2 \quad \mathcal{P}_3 \quad \mathcal{P}_4 \quad \mathcal{P}_5 \cdots$$

$$
\begin{array}{cccccc}
Q_{1,0} & Q_{1,1} & Q_{1,2} & Q_{1,3} & Q_{1,4} & Q_{1,5} \\
Q_{2,0} & Q_{2,1} & Q_{2,2} & Q_{2,3} & Q_{2,4} & Q_{2,5} \\
Q_{3,0} & Q_{3,1} & Q_{3,2} & Q_{3,3} & Q_{3,4} & Q_{3,5} \\
V_0 & V_1 & V_2 & V_3 & V_4 & V_5
\end{array}
$$

Figure 14.13: Layout for the arrays Q(:SERIAL,:NEWS) and V(:NEWS).

```
CMF$ LAYOUT  X(1.0:NEWS,1.0:NEWS)
```

would be best for this computation. Were the layout defined as

```
CMF$ LAYOUT  X(1000:NEWS,1:NEWS)
```

the compiler assumes that elements along the first axis of the array communicate far more often than those of the second axis. Thus it may assign elements of the first array on the same chip if possible to make the communication local.

The last ordering SEND arranges the elements of the array according to the unique addresses preassigned to each processor. This communication method uses the underlying hypercube of the CM, allowing each element to be quickly communicated to any other element in the array. This ordering is best if the communication is between local processors that are not also nearest neighbors. An example of a program that would benefit from this ordering is an FFT computation.

It is assumed that arrays declared in the same DIMENSION statement should have their first elements assigned to the same processor. Hence if the arrays Q and V have been declared together as in the statement

$$\text{DIMENSION} \quad \text{Q(3,0:N), V(0:N)}$$

and Q has the same LAYOUT directive as above, then the elements of both arrays are assigned to the processors in a manner similar to that shown in figure 14.13.

MasPar MPF MAP

The MAP compiler directive for the MasPar MP-2 is similar to the CM Fortran LAYOUT directive. It defines a mapping of the given array to the processor array elements. For example the statements

```
CMPF MAP Q(MEMORY, ALLBITS)
CMPF MAP V(ALLBITS)
```

also map the Q and V arrays as shown in figure 14.13. Here the parameter MEMORY has a meaning similar to that of :SERIAL for the CM; the elements in each column are to put in the same processor memory. The parameter ALLBITS corresponds to the CM Fortran :NEWS; along that dimension of the array, the elements are to be spread across the processor element array.

CM Fortran ALIGN

The ALIGN compiler directive is used to align the elements of two arrays with each other. For example suppose the following arrays are declared:

$$\text{DIMENSION} \quad \text{S(N,N), T(N)}$$

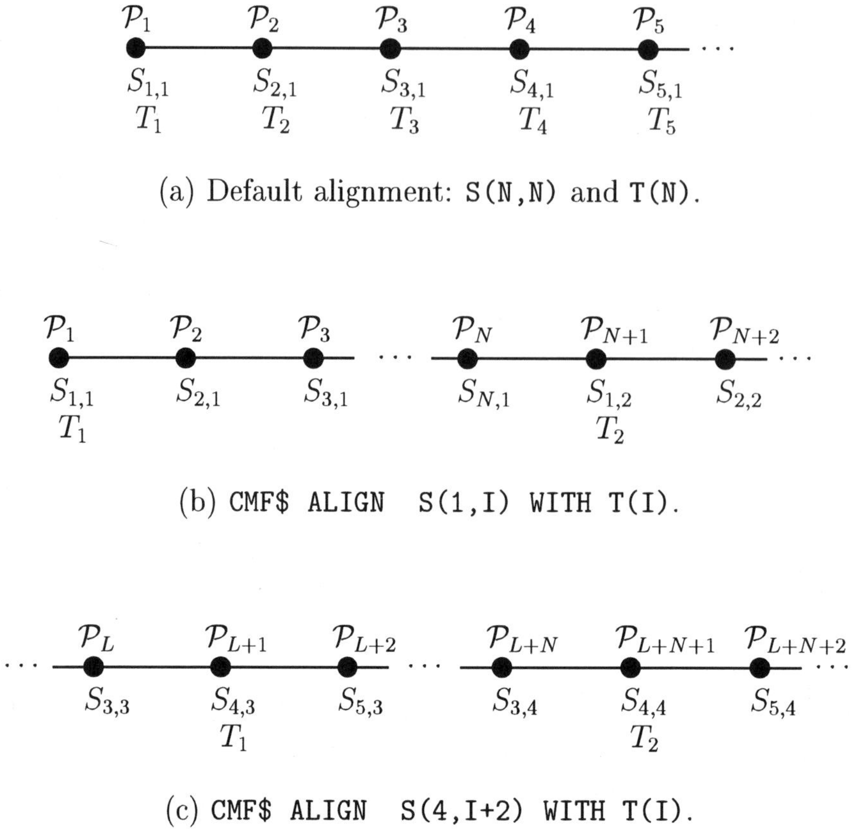

(a) Default alignment: S(N,N) and T(N).

(b) CMF$ ALIGN S(1,I) WITH T(I).

(c) CMF$ ALIGN S(4,I+2) WITH T(I).

Figure 14.14: Alignments of arrays S and T.

By default the elements of the arrays would be laid out so that S(1,1) and T(1) are on the same processor, S(2,1) and T(2) would be on the next processor, etc. This is the ordering shown in figure 14.14(a). However it might be desirable to have T(2) on the same processor as S(1,2) as shown in figure 14.14(b). This can be accomplished by the following ALIGN statement.

               CMF$ ALIGN   S(1,I) WITH T(I)

Similarly the statement

               CMF$ ALIGN   S(4,I+2) WITH T(I)

would produce a layout as shown in figure 14.14(c).

CM Fortran COMMON

The COMMON compiler directive is used to define a default home for the arrays in a given common block. The three possible forms of this directive are as follows:

```
CMF$ COMMON [, CMONLY] /blkname/
CMF$ COMMON FEONLY /blkname/
CMF$ COMMON INITIALIZE /blkname/
```

The first form of the directive tells the compiler to put the arrays contained in the given common block *blkname* on the PPU. The term CMONLY is optional and is used for clarity.

The middle or second form of the directive informs the compiler that the arrays should be placed on the FE. Otherwise the normal default for arrays in COMMON blocks would be on the PPU.

The third and final form of the directive also instructs the compiler to make the PPU be the home of the arrays in the common block *blkname*. But in addition, it allocates space on the FE for the common block as well to allow for the static initialization of the arrays. Such arrays may be initialized by DATA statements.

MasPar MPF ONDPU

To force an array or common block in MPF to be on the DPU, use the ONDPU compiler directive:

```
CMPF ONDPU   A, DPUCBLK
```

This command asserts that the array A and the common block DPUCBLK must reside on the DPU. This happens whether or not the array or common block is used in any parallel operations.

MasPar MPF ONFE

The ONFE compiler directive performs the reverse of the ONDPU directive. The command

```
        CMPF ONFE  B, FECBLK
```

will make sure that the array B and the common block FECBLK are placed on
the front end.

V Applications

15 Molecular Dynamics

15.1 Introduction

Molecular dynamics is concerned with simulating the motion of molecules to gain a deeper understanding of chemical reactions, fluid flow, phase transitions, droplet formation, and other physical phenomena that derive from molecular interactions. These studies include not only the motion of many molecules as in a fluid, but also the motion of a single large molecule consisting of hundreds or thousands of atoms, as in a protein. Much of this work uses simple classical Newtonian mechanics. This chapter is concerned only with these classical systems and uses mathematical concepts that should be familiar to most upper division undergraduates in a physical science or engineering curriculum.

Computers are a critically important tool for these studies because there simply is no other way to trace the motion of a large number of interacting particles. The earliest of these computations were done in the 1950s by Berni Alder and Tom Wainwright at Lawrence Livermore National Laboratory. They studied the distribution of molecules in a liquid, using a model in which the molecules are represented as *hard spheres* which interact like billiard balls [Alder & Wainright 59, Alder & Wainright 60]. Using the fastest computer at that time, an IBM 704, they were able to simulate the motions of 32 and 108 molecules in computations requiring 10 to 30 hours. Now it is possible to perform hard sphere computations on systems of over a million particles. A hard sphere model, with millions of molecules, has been used at NASA Ames Research Center by Leonardo Dagum to simulate hypersonic flow conditions encountered by flight vehicles at high altitudes [Dagum 88]. Computations using a more realistic molecular model known as *Lennard-Jones* have been performed at IBM Kingston by Lawrence Hannon, George Lie, and Enrico Climenti to study the flow of fluids [Hannon et al 86]. In these computations the fluid was represented by $\sim 10^4$ interacting molecules. Even though this is miniscule compared with the number of molecules in a gram of water, the behavior of the flow was like that in a real fluid.

Another class of molecular dynamics computations is concerned with the internal motion of molecules especially proteins and nucleic acids, vital components of biological systems. The goal is to gain a better understanding of the function of these molecules in biochemical reactions. Interestingly, it has been found that quantum effects do not have much influence on the overall dynamics of proteins except at low temperatures. Thus classical mechanics is sufficient to model the motions, but still the computational power required for following the motion of a large molecule is enormous. For example, simulating the motion of a 1,500-atom molecule, a small protein, for a time interval of 10^{-10} seconds is a six hour computation on a Cray X-MP.

Martin Karplus and Andrew McCammon, in an interesting article "The Molecular Dynamics of Proteins" [Karplus & McCammon 86], describe a discovery concerning the molecule myoglobin that could only have been made through molecular dynamics. The interest in myoglobin comes about because it stores oxygen in biological systems. Whales, for example, are able to stay under water for long periods of time because of a large amount of myoglobin in their bodies. It was known that the oxygen molecule binds to a particular site in the myoglobin molecule but it was not understood how the binding could take place. X-ray crystallography work shows that large protein molecules tend to be folded up into compact three-dimensional structures, and in the case of myoglobin the oxygen sites were known to lie in the interior of such a structure. There did not seem to be any way that an oxygen atom could penetrate the structure to reach the binding site. Molecular dynamics provided the answer to this puzzle. The picture of a molecule provided by X-ray crystallography shows the average positions of the atoms in the molecule. In fact these atoms are in continuous motion, vibrating about positions of equilibrium. Molecular dynamics simulation of the internal motion of the molecule showed that a path to the site, wide enough for an oxygen atom, could open up for a short period of time.

Scientific visualization is particularly important for understanding the results of a molecular dynamics simulation. The millions of numbers representing a history of the positions and velocities of the particles is not a very revealing picture of the motion. How is one to recognize the formation of a vortex in this mass of data, or the nature of the bending and stretching of a large molecule?

How is one to gain new insights? Pictures and animations enable the scientist to literally see the formation of vortices, to view protein bending, and thus to gain new insights into the details of these phenomena.

Computations that involve following the motion of a large number of interacting particles, whether the particles are atoms in a molecule, or molecules of a fluid or solid, or even particles in the discrete model of a vibrating string or membrane are similar in the following respect. They involve a long series of time steps, at each of which Newton's laws are used to determine the new positions and velocities from the old positions and velocities and the forces. The computations are quite simple but there are many of them. To achieve accuracy, the time steps must be quite small, and therefore many are required to simulate a significantly long *real* time interval. In computational chemistry the time step is typically about a femtosecond (10^{-15} seconds), and the total computation may represent $\sim 10^{-10}$ seconds which could cost about 100 hours of computer time. The amount of computation at each step can be quite extensive. In a system consisting of n particles the force computations at each step may involve $\mathcal{O}(n^2)$ operations. Thus it is easy to see that these computations can consume a large number of machine cycles. In addition, animation of the motion of the system can make large demands on memory.

The objective of this chapter is to help you gain some understanding of the nature of these computations. We use the term *particle* to refer to the interacting objects: atoms or molecules. We concentrate on three different models: Hooke's Law, Lennard-Jones, and hard sphere. In the Hooke's Law model the force acts as if the particles were connected to their neighbors by springs. Lennard-Jones is a model with forces that are strongly repulsive at very short interparticle distances, attractive at larger distances, and extremely weak attractive at very large distances. In the hard sphere model, already mentioned, the particles interact as if they were billiard balls – they bounce off each other when they are a certain distance apart, otherwise they do not interact. Normally the Lennard-Jones model is used for three-dimensional systems, but it is instructive to use it also for one-dimensional and two-dimensional systems.

After describing these models we discuss the equations of motion for each model, and then we consider numerical methods for solving these equations. Solving the equations of motion for the Hooke's Law model and the Lennard-Jones model is intrinsically different from solving them for the hard sphere

model. Solving the equations of motion for the hard sphere model requires
solving some simple problems in geometry; in particular, we must determine
when and where two spheres moving at constant velocity will collide. In one
dimension this is extremely simple, but in two and three dimensions you need
to draw on a knowledge of vector analysis. With Hooke's Law and Lennard-
Jones models we must solve a system of differential equations. For this we
use two numerical methods: Euler's method and Verlet's method. We use
Euler's method because it is the simplest of any method we could reasonably
use and thus it provides the easiest introduction to some of the basic ideas of
numerically solving the equations of motion. Verlet's method is only slightly
more complex but more accurate. It is the simplest of the numerical methods
used in serious molecular dynamics computations.

Finally, we consider the exact solution of the equations of motion for the
Hooke's Law model. This is the only model of the three considered here
that admits an exact solution. Exact solutions are important for us because
they provide a means for testing the accuracy of our numerical methods. An
understanding of this part of the chapter requires some elementary knowledge
of matrix eigenvalues and eigenvectors.

15.2   Models

Models of particle systems are characterized by the nature of the interactions
between the particles. Generally it is assumed that the forces between the
particles are conservative, two-body forces; that is, energy is conserved and
the total force acting on a particle due to the other particles is the sum of the
forces between pairs of particles. Thus the force acting on particle i is given
by an expression of the form

$$f_i = \sum_{\substack{j=1 \\ j \neq i}}^{n} f_{i,j}, \tag{15.1}$$

where f_i is the total force on particle i due to the other particles, $f_{i,j}$ is the force
on particle i due to particle j, and n is the number of particles in the system.
Force is a vector quantity, so the sum in equation (15.1) is a vector sum. The
order of the indices is important: the first index identifies the particle acted

on, the second index identifies the particle causing the action. Newton's third law tells us that

$$f_{i,j} + f_{j,i} = 0.$$

There is an important relation between potential energy and force in a conservative system. If r is the position of a particle, $f(r)$ the force acting on it, and $\phi(r)$ its potential energy, then

$$f(r) = -\nabla \phi(r). \tag{15.2}$$

Thus we can describe a model in terms of the force or the potential energy. For example, if

$$\phi(r) = \|r\|^2, \quad \text{where } r = \begin{pmatrix} x \\ y \\ z \end{pmatrix},$$

then the three components of the force are

$$
\begin{aligned}
f_x &= -\frac{\partial \phi}{\partial x} = -2x, \\
f_y &= -\frac{\partial \phi}{\partial y} = -2y, \\
f_z &= -\frac{\partial \phi}{\partial z} = -2z.
\end{aligned}
$$

Since potential energy is a scalar quantity it is often more convenient to describe the model in terms of its potential energy function, ϕ.

At a point of minimum potential energy the partial derivatives of the potential energy are zero, and thus it is a point at which all of the forces are zero. Accordingly we call this point an *equilibrium point*.

We now consider three models, referred to as the *Hooke's Law* model, *HL* for short, the *Lennard-Jones* model, *LJ* for short, and the *hard sphere* model, *HS* for short. Of these three, the LJ model comes closest to representing real molecular systems. On the other hand, the LJ model presents the most difficult computational challenge. The HL model is an approximation to the LJ model when the particles have low kinetic energy, thus remaining close to their equilibrium positions; similarly, the HS model is an approximation to the LJ model when the particles have high kinetic energy, or when attractive forces are very weak.

15.2.1 Hooke's Law model

In the HL model the potential energy of a particle is proportional to the square of its displacement from its equilibrium position. Figure 15.1 shows the potential energy function for a particle in a one-dimensional system; and figure 15.2 shows the force on the particle that, according to equation (15.2), must be proportional to the displacement of the particle from its equilibrium position, and directed towards it. The equations for the potential energy, and force are:

$$\phi(x) = \frac{k}{2}(x - x^{eq})^2 + \phi^{min}, \tag{15.3}$$
$$f(x) = -k(x - x^{eq}),$$

where k is a constant, sometimes referred to as the *force constant*; ϕ^{min} is a constant, the minimum potential energy; and x^{eq} is the equilibrium position of the particle. Notice that

$$f(x) = -\frac{d\phi(x)}{dx}$$

as required by equation (15.2). Thus the force is proportional to the displacement of the particle from its equilibrium position and it is directed towards the equilibrium position. The most familiar example of a system subject to a HL force is a small mass suspended by a spring: it moves up and down under the influence of a HL force imposed by the spring, figure 15.3.

In two dimensions this model is described by

$$\phi(x, y) = \frac{k}{2}((x - x^{eq})^2 + (y - y^{eq})^2) + \phi^{min},$$

which we may write more compactly as

$$\phi(r) = \frac{k}{2}\|r - r^{eq}\|^2 + \phi^{min}.$$

We now consider the more interesting case of HL models for systems of more than one particle, starting with the two-particle case. It helps to think of a physical system of two masses connected by a spring as illustrated in

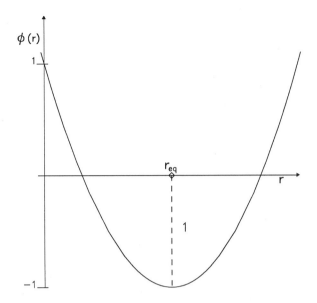

Figure 15.1: Potential energy of a particle in the HL model varies as the square of its displacement from equilibrium.

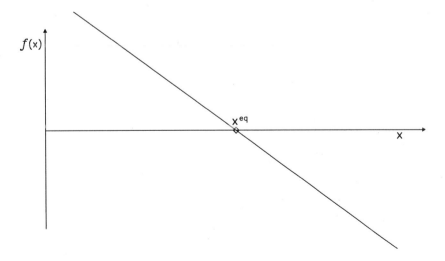

Figure 15.2: Force on a particle in the HL model is proportional to its displacement from equilibrium and in the direction of the equilibrium point.

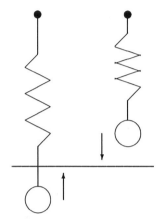

Figure 15.3: The motion of a small mass suspended from a spring typifies the motion of a particle subject to an HL force.

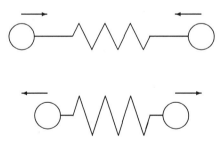

Figure 15.4: The motion of two small masses connected by a spring illustrates the nature of the motion of a two-particle system in the HL model.

figure 15.4. The force of the spring acts along a line joining the particles that we take to be the x-axis. We assume that when the particles are separated by a distance d the spring is neither stretched or compressed, so the system is in equilibrium. When the distance between the particles is less than d the spring is compressed and the force acts to drive the particles apart; when the spring is stretched the force acts to bring the particles closer together. The potential energy function is

$$\phi(x_1, x_2) = \frac{k}{2}(x_1 - x_2 + d)^2 + \phi^{min}.$$

The forces can be obtained by taking the appropriate derivatives of the potential energy giving

$$
\begin{aligned}
f_1(x_1, x_2) &= -k(x_1 - x_2 + d), \\
f_2(x_1, x_2) &= k(x_1 - x_2 + d).
\end{aligned}
$$

A one-dimensional, four-particle system is shown in figure 15.5. An illustration of the motion of this system is shown in figure 15.6. The potential energy of this system is

$$\phi(x_1, x_2, x_3, x_4) = \frac{k}{2}((x_1 - x_2 + d)^2 + (x_2 - x_3 + d)^2 + (x_3 - x_4 + d)^2) + \phi^{min}.$$

This system is in equilibrium when the particles are ordered from left to right, each a distance d from its neighbors. The forces are

$$
\begin{aligned}
f_1(x_1, x_2) &= -k(x_1 - x_2 + d), \\
f_2(x_1, x_2, x_3) &= -k(2x_2 - x_1 - x_3), \\
f_3(x_2, x_3, x_4) &= -k(2x_3 - x_2 - x_4), \\
f_4(x_3, x_4) &= -k(x_4 - x_3 - d).
\end{aligned}
$$

Extension of these equations to n-particle systems should be obvious.

The equations are simplified if we fix the equilibrium position of the first particle to be at the origin and we define new variables q_i as follows:

$$x_i = (i - 1)d + q_i. \tag{15.4}$$

Figure 15.5: A one-dimensional, four-particle HL model. Motion is restricted to the x dimension and is oscillatory.

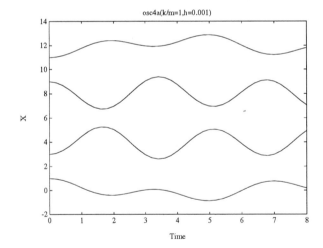

Figure 15.6: Illustration of the motion of a four-particle system. The curves show the position of the four particles as functions of time. These results were for a system with $k/m = 1$, k is the force constant, and m is the mass of a particle. The equilibrium positions of the particles are 0, 4, 8, 12. The parameter h is the time step used in integrating the equations of motion.

Thus q_i denotes the displacement from the equilibrium position of the ith particle. In the new variables the potential energy is

$$\phi(q_1, q_2, q_3, q_4) = \frac{k}{2}((q_1 - q_2)^2 + (q_2 - q_3)^2 + (q_3 - q_4)^2) + \phi^{min},$$

and the forces are

$$\begin{aligned}
f_1(q_1, q_2) &= -k(q_1 - q_2), \\
f_2(q_1, q_2, q_3) &= -k(2q_2 - q_1 - q_3), \\
f_3(q_2, q_3, q_4) &= -k(2q_3 - q_2 - q_4), \\
f_4(q_3, q_4) &= -k(q_4 - q_3).
\end{aligned}$$

15.2.2 Lennard-Jones model

We consider a three-dimensional system. The potential energy function for a pair of particles, 1 and 2, in the LJ model is given by

$$\phi(r_1, r_2) = \left(\frac{1}{\|r_1 - r_2\|^{12}} - \frac{2}{\|r_1 - r_2\|^6} \right). \tag{15.5}$$

The units have been chosen to locate the minimum of the potential energy at $\|r_1 - r_2\| = 1$, and the value of the minimum equal to -1. This potential function is illustrated in figure 15.7. Considering the slope of this function we see that the force is strongly repulsive at small distances, and is attractive at large distances, becoming extremely weak at very large distances. The crossover between the repulsive region and the attractive region occurs where $\|r_1 - r_2\| = 1$, the point of minimum potential energy. Note that at this point the force of the interaction is zero since the derivative of ϕ is zero.

The forces on particle 1 can be determined from the basic formula, equation (15.2):

$$\begin{aligned}
f_{1,x} &= 12 \left(\frac{1}{\|r_1 - r_2\|^{14}} - \frac{1}{\|r_1 - r_2\|^8} \right)(x_1 - x_2), \\
f_{1,y} &= 12 \left(\frac{1}{\|r_1 - r_2\|^{14}} - \frac{1}{\|r_1 - r_2\|^8} \right)(y_1 - y_2), \\
f_{1,z} &= 12 \left(\frac{1}{\|r_1 - r_2\|^{14}} - \frac{1}{\|r_1 - r_2\|^8} \right)(z_1 - z_2).
\end{aligned}$$

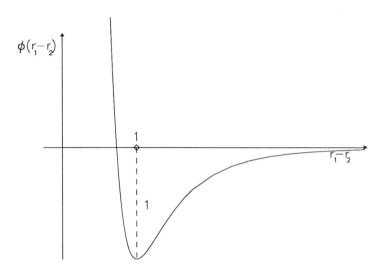

Figure 15.7: Potential energy of a particle in the LJ model.

In a many-particle LJ system the force on each particle is determined by summing over the pairwise interactions, using the above formulas. Since the force between widely separated pairs is very weak it is sometimes neglected: a cutoff distance is chosen, and the force between particles separated by more than the cutoff is ignored.

The equilibrium configuration is not easily determined. If there are only four particles, then the particles are at the corners of a regular tetrahedron, as shown in figure 15.8. But what about larger systems?

Particle coordinates for equilibrium configurations of five and six particles are shown in table 15.1. Pictures of these configurations are shown in figures 15.9 and 15.10.

15.2.3   Hard sphere model

This model is best visualized as a collection of hard, perfectly elastic, balls — like ball bearings, or billiard balls. The interactions between particles are like collisions between these balls. Two-dimensional and one-dimensional versions

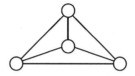

Figure 15.8: Minimum energy configuration for four particles in a three-dimensional LJ model: particles are located at the corners of a regular tetrahedron.

5 Particles		
x	y	z
0.751308	0.888623	1.293299
0.906831	1.164906	0.343358
1.619112	0.680194	0.846874
0.418779	1.771369	0.967718
1.398694	1.623815	1.085234

6 Particles		
x	y	z
5.951021	6.316232	5.356796
6.260952	5.532089	4.827508
6.271016	5.496511	5.822353
5.294356	5.769662	4.845783
5.614350	4.949941	5.311340
5.304420	5.734084	5.840627

Table 15.1: x-, y-, z-coordinates of particles in a five-particle LJ model, and a six-particle LJ model: these results were obtained by Elizabeth Eskow using minimization software developed by Robert Schnabel.

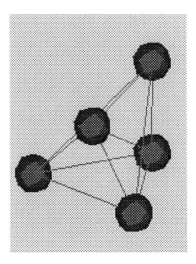

Figure 15.9: Minimum energy configuration for five particles.

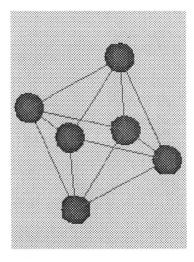

Figure 15.10: Minimum energy configuration for six particles.

Atom	Length (cm.)	Time (sec.)
He	2.87×10^{-8}	3.38×10^{-14}
Ne	3.10×10^{-8}	6.80×10^{-14}
Ar	3.83×10^{-8}	1.54×10^{-13}
Kr	4.11×10^{-8}	1.95×10^{-13}
Xe	4.43×10^{-8}	2.42×10^{-13}

Table 15.2: Length and time scale factors for the LJ model.

of this model, as well as the three-dimensional model, are studied: in two dimensions it is called the *hard disk model*, and in one dimension it is called the *hard rod model*.

The potential energy for a pair of particles is

$$\phi(r_1, r_2) = \begin{cases} 0, & \|r_1 - r_2\| > \sigma \\ \infty, & \|r_1 - r_2\| \leq \sigma \end{cases}.$$

Thus there is no force acting on the particles except at the instant when they are a distance σ apart. At that point an instantaneous force is applied, causing a change in velocities. We discuss this further in the next section when we consider the equations of motion. For the present it is sufficient to think of the collision as if it were between two billiard balls of radius σ.

The HS model can be viewed as an approximation to the LJ model with high-velocity particles. When particles in the LJ model are moving at high velocities the effect of the attractive force is quite small. The particles move at high speed in straight lines (approximately) until they get close enough for the repulsive force to come into play, at which time they collide as in the hard sphere model. The repulsive force rises so steeply in the LJ model it has almost the same effect as a collision between hard spheres of diameter slightly less than 1.

15.2.4 Units and the connection with real systems

The models all have potentials that depend only on distance; i.e., they are spherically symmetric. Therefore they serve best as models for systems composed of atoms of helium (He), neon (Ne), argon (Ar), Krypton (Kr), or Xenon (Xe). Our computations with the LJ model apply to any of these systems by appropriate choice of a scale factor for length and time. In table 15.2 we show the scale factors for length and time for these elements. The interpretation of the numbers in this table can be illustrated for the case of argon: the *length* entry means that a distance of 1 unit, $\|r\| = 1$, in the formula for the LJ potential equation (15.5) represents 3.83×10^{-8} cm.; similarly, the *time* entry means that a time unit of 1 in the equations of motion (next section) represents 1.54×10^{-13} seconds, assuming the mass we use in the equations of motion is 1.

15.3 Equations of motion

Newton's second law gives us the equation for the motion of a particle:

$$ma = f,$$

where m is the particle's mass, a is its acceleration, and f is the force acting on it. From this equation and a knowledge of the initial position and the initial velocity of the particle we can, in principle, determine its position and velocity at future times. In a system of interacting particles their motion is determined by solving many of these equations, one for each particle. The equations are interdependent because the force on a particle is a function of the position of some or all of the other particles. The solution of these equations is our major concern in the next section. In this section we look at the form of these equations for the different models in order to gain an understanding of the nature of the problems we are trying to solve.

15.3.1   One-dimensional systems

The equations of motion for a one-dimensional system of two interacting particles are:

$$m_1 \ddot{x}_1 = f_1,$$
$$m_2 \ddot{x}_2 = f_2,$$

where the acceleration is represented by $\ddot{x}$; that is,

$$\ddot{x}_i = \frac{d^2 x_i}{dt^2}.$$

We assume that all particles have the same mass: $m = m_1 = m_2$.

If we write the equations of motion in matrix form we have

$$m \begin{pmatrix} \ddot{x}_1 \\ \ddot{x}_2 \end{pmatrix} = \begin{pmatrix} f_1 \\ f_2 \end{pmatrix}. \tag{15.6}$$

These equations can be written more compactly as

$$m\ddot{x} = f, \tag{15.7}$$

with the understanding that $\ddot{x}$ and f are the vectors in equation (15.6). We could express the equations of motion for a system of n particles by exactly the same simple equation, with $\ddot{x}$ denoting a vector of n accelerations, and $f(x)$ denoting a vector of n forces. For example, the explicit matrix equation of a four-particle system, figure 15.5, is

$$m \begin{pmatrix} \ddot{x}_1 \\ \ddot{x}_2 \\ \ddot{x}_3 \\ \ddot{x}_4 \end{pmatrix} = \begin{pmatrix} f_1 \\ f_2 \\ f_3 \\ f_4 \end{pmatrix}$$

The equations of motion for four particles in the HL model are

$$\begin{aligned} m\ddot{q}_1 &= -k(q_1 - q_2), \tag{15.8} \\ m\ddot{q}_2 &= -k(2q_2 - q_1 - q_3), \\ m\ddot{q}_3 &= -k(2q_3 - q_2 - q_4), \\ m\ddot{q}_4 &= -k(q_4 - q_3), \end{aligned}$$

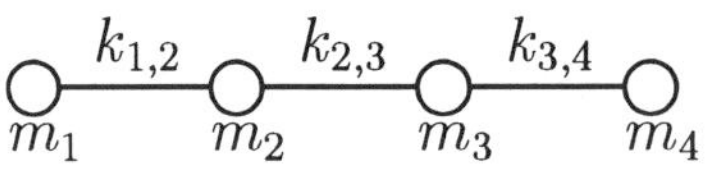

Figure 15.11: A four-particle HL model with unequal masses and unequal force constants.

where the q_is were defined in the last section, equation (15.4).

We assumed that the particles are identical so that a common force constant, and common masses are used throughout. You might check your understanding of these equations by deriving the equations of motion for a nonhomogeneous HL model consisting of four particles with unequal masses, and unequal force constants, as illustrated in figure 15.11.

The equations of motion for two particles with LJ forces, acting in just one dimension, are

$$m\ddot{x}_1 = 12\left(\frac{1}{(x_1 - x_2)^{13}} - \frac{1}{(x_1 - x_2)^7}\right),$$

$$m\ddot{x}_2 = 12\left(\frac{1}{(x_2 - x_1)^{13}} - \frac{1}{(x_2 - x_1)^7}\right).$$

In a four-particle system they are

$$m\ddot{x}_1 = 12\sum_{j=2}^{4}\left(\frac{1}{(x_1 - x_j)^{13}} - \frac{1}{(x_1 - x_j)^7}\right),$$

$$m\ddot{x}_2 = 12\sum_{\substack{j=1\\j\neq2}}^{4}\left(\frac{1}{(x_2 - x_j)^{13}} - \frac{1}{(x_2 - x_j)^7}\right),$$

$$m\ddot{x}_3 = 12\sum_{\substack{j=1\\j\neq3}}^{4}\left(\frac{1}{(x_3 - x_j)^{13}} - \frac{1}{(x_3 - x_j)^7}\right),$$

$$m\ddot{x}_4 = 12\sum_{\substack{j=1\\j\neq4}}^{4}\left(\frac{1}{(x_4 - x_j)^{13}} - \frac{1}{(x_4 - x_j)^7}\right).$$

In the two-particle example the particles are in equilibrium when they are unit distance apart, but the four-particle case is a little different. When the particles are unit distance apart, then the force on each is *almost*, but not exactly, zero: although the force from particles unit distance away is zero, particles 2 and 3 units away exert a small attractive force. In fact the forces for unit separation are:

$$
\begin{aligned}
f_1 &= +0.0978, \\
f_2 &= +0.0923, \\
f_3 &= -0.0923, \\
f_4 &= -0.0978.
\end{aligned}
$$

Therefore if we placed the particles at locations 0, 1, 2, 3 along the x-axis we would expect the first two to start moving to the right, and the second two to start moving to the left.

The equations of motion for the HS model must be expressed a little differently. Consider two particles moving along the x-axis. Their positions are given by

$$
\begin{aligned}
x_1 &= x_1^{(0)} + \dot{x}_1^{(0)}(t - t^{(0)}), \\
x_2 &= x_2^{(0)} + \dot{x}_2^{(0)}(t - t^{(0)}),
\end{aligned}
$$

where $x_i^{(0)}$ denotes position at time $t^{(0)}$, and the velocities, $\dot{x}_i$, are the velocities at $t^{(0)}$. If the particles are moving towards each other, then they will collide at some time, say $t^{(1)}$. At this instant they change their velocities and the new positions are given by

$$
\begin{aligned}
x_1 &= x_1^{(1)} + \dot{x}_1^{(1)}(t - t^{(1)}), \\
x_2 &= x_2^{(1)} + \dot{x}_2^{(1)}(t - t^{(1)}).
\end{aligned}
$$

Of course if the particles are moving away from each other, then there is no collision. Thus solving the equations of motion in this case amounts to determining the time of the next collision; moving the particles to their positions at that time; and then determining the new velocities of the colliding particles. This process is repeated over and over. Thus we compute the motion from collision to collision.

A one-dimensional HS model like this is not very interesting because the particles gradually move farther and farther apart, going off to $+\infty$ or $-\infty$. But we can make it interesting if we put "walls" on left and right constraining the particles to remain in some interval. When a particle hits the wall then we can assume it bounces back, i.e., reverses its velocity. Or we can assume that the particles are confined to a circle, as if we joined left and right ends of an interval of the x-axis — this kind of assumption is referred to as a *periodic boundary condition*.

15.3.2   Two-dimensional systems

We now assume

$$r = \left(\begin{array}{c} x \\ y \end{array} \right).$$

The position of the ith particle is denoted r_i. With this understanding we can write the equations of motion for an n-particle system exactly as before, equation (15.7), but with r in place of x:

$$m\ddot{r} = f.$$

Consider a two-particle system with

$$r \;=\; \left(\begin{array}{c} r_1 \\ r_2 \end{array} \right),$$

$$\;=\; \left(\begin{array}{c} x_1 \\ y_1 \\ x_2 \\ y_2 \end{array} \right).$$

These coordinates are illustrated in figure 15.12. Thus we can write the equations of motion as follows:

$$m \left(\begin{array}{c} \ddot{x}_1 \\ \ddot{y}_1 \\ \ddot{x}_2 \\ \ddot{y}_2 \end{array} \right) \;=\; \left(\begin{array}{c} f_{1,x} \\ f_{1,y} \\ f_{2,x} \\ f_{2,y} \end{array} \right) \tag{15.9}$$

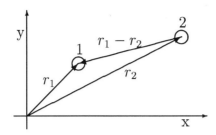

Figure 15.12: A two-particle, two-dimensional system.

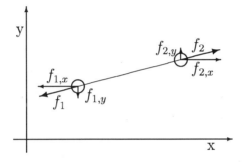

Figure 15.13: Force vectors in a two-particle, two-dimensional system.

where $f_{i,x}$ and $f_{i,y}$ are the x- and y-components of the force on particle i.

The forces are directed along the line through the centers of the two particles, as illustrated in figure 15.13.

Now we can write the equations of motion for two-particle HL and LJ models using equation (15.9) and the forces given in the last section. The equations of motion for a two-particle HL model are

$$
m \begin{pmatrix} \ddot{x}_1 \\ \ddot{y}_1 \\ \ddot{x}_2 \\ \ddot{y}_2 \end{pmatrix} = k \left(1 - \frac{d}{\|r_\|\|} \right) \begin{pmatrix} x_1 - x_2 \\ y_1 - y_2 \\ x_2 - x_1 \\ y_2 - y_1 \end{pmatrix},
$$

and for a two-particle LJ model they are

$$
m \begin{pmatrix} \ddot{x}_1 \\ \ddot{y}_1 \\ \ddot{x}_2 \\ \ddot{y}_2 \end{pmatrix} = 12 \left(\frac{1}{\|r_1 - r_2\|^{14}} - \frac{1}{\|r_1 - r_2\|^8} \right) \begin{pmatrix} x_1 - x_2 \\ y_1 - y_2 \\ x_2 - x_1 \\ y_2 - y_1 \end{pmatrix}.
$$

The equations for an n-particle system have the same form, the only difference being that when $n > 2$, a pairwise sum over interactions must be made to determine $f_{i,x}, f_{i,y}$; e.g., for the LJ model

$$
f_{i,x} = 12 \sum_{\substack{j=1 \\ j \neq i}}^{n} \left(\frac{1}{\|r_i - r_j\|^{14}} - \frac{1}{\|r_i - r_j\|^8} \right) (x_i - x_j),
$$

$$
f_{i,y} = 12 \sum_{\substack{j=1 \\ j \neq i}}^{n} \left(\frac{1}{\|r_i - r_j\|^{14}} - \frac{1}{\|r_i - r_j\|^8} \right) (y_i - y_j).
$$

Thus in an n-particle system the equations of motion are:

$$
m \begin{pmatrix} \ddot{r}_1 \\ \ddot{r}_2 \\ \vdots \\ \ddot{r}_n \end{pmatrix} = \begin{pmatrix} f_1 \\ f_2 \\ \vdots \\ f_n \end{pmatrix},
$$

where $\ddot{r}_i$ and f_i are two-element column vectors:

$$
\ddot{r}_i = \begin{pmatrix} \ddot{x}_i \\ \ddot{y}_i \end{pmatrix}, \quad f_i = \begin{pmatrix} f_{i,x} \\ f_{i,y} \end{pmatrix}.
$$

The equations of motion for a two-dimensional HS system are the obvious extension of the one-dimensional equations. We can write the equation for the ith particle in vector form as follows:

$$
r_i = r_i^{(1)} + \dot{r}_i^{(1)} (t - t^{(1)}).
$$

The work of the computation is determining when the next collision will occur. Here there is an essential difference from the one-dimensional computation. In one dimension only the particles on the left and right of a given particle are collision candidates; furthermore, these particles remain candidates for the entire calculation — in one dimension neighbors remain neighbors. Not so in two dimensions. Now the number of possibilities we must examine is much larger. Naively, we might consider every pair of particles but we can do better than this. One can, for example, divide space into bins of a certain size. If the bin size is chosen appropriately then for a given particle its collision candidates are the other particles in the same bin or in neighboring bins.

Another scheme uses a timetable of predicted collision times. Suppose that at some point we determine for every particle the time and partner for its next collision, *assuming no other collisions take place*. Thus we produce a timetable for collisions. The entry with the earliest collision time in the timetable is the next collision. Once we process that collision, we need to update the timetable. This updating process takes some work, but it may lead to less work overall than the naive approach. The updating process involves looking for a new collision partner for each of the two particles that just collided. When those collision partners have been found some other entries in the timetable may need to be updated, any that had one member of the colliding pair as a collision partner must be updated.

There is no need to consider three-dimensional systems separately. The formulas and issues are the same as for the two-dimensional systems just discussed. The only difference being that a z-component must be added to the vectors.

15.4 Numerical solution of the equations of motion

In general the equations of motion do not admit an analytic solution so they must be solved numerically. The numerical method is normally a *time stepping* algorithm; that is, the solution is generated incrementally in time starting from a set of initial conditions. This solution is simply a list of numbers, usually the particle positions and velocities at the time steps. The positions and velocities at any one time represent the *state* of the system at that time.

The process of generating the numerical solution is easy to describe in

broad outline, though the details can be rather difficult. The broad outline
is this. Starting from an initial state at a given time, the state at a slightly
later time is computed, then the next state, and so on. The time interval
between states is either constant or variable; in order to control the error, it
must be kept small enough so that values of important variables undergo little
change, but if the interval is too small then the computation becomes too slow.
More complex algorithms usually accumulate less error per step and so bigger
steps are possible, but they also take more time per step. Obviously these
tradeoffs are an important consideration in designing a program for solving
a molecular dynamics problem. Normally the state is not recorded at every
time step, they are too close together. Instead, a different time interval that
is a multiple of the time step is used for recording results. Important physical
parameters such as energy, momentum, mean separation of particles, and so
forth may be recorded during the computation or generated later from the
state information.

Accuracy of the numerical solution is an important consideration. As a
practical matter this usually must be estimated by indirect methods. Running
the computation for different values of the time step is one indirect technique,
and running with different numerical precision is another. Changing from
single to double precision is relatively easy, and there are software tools for still
higher precision [Bailey 91]. Special cases in which an exact solution is possible
to test an algorithm are also used. The HL model is one of these special cases.
It is possible to express the solution of the equations of motion for this system
in terms of the eigenvalues and eigenvectors of a certain matrix. Although the
eigenvalues and eigenvectors may have to be computed numerically, the error
from computation is negligible compared with the error from a time stepping
algorithm. Therefore the HL model admits, for all practical purposes, an exact
solution against which numerical solutions can be compared.

The efficiency of these computations on vector and parallel computers depends on the models and the algorithms. The HL model, for example, can be
run very efficiently on a vector computer but the HS model cannot. In most
particle systems the neighbors of a particle change with time, thus the set of
interacting pairs change and because of this an efficient decomposition of the
computation for a parallel computer can be difficult to find.

For HL and LJ systems we use algorithms for solving second order differential equations. We consider two of these algorithms: Euler's method and Verlet's method. The HS system is essentially different since no differential equation needs to be solved. The work of the computation consists mainly in determining when the next collision occurs. For this we use only the naive algorithm mentioned: consideration of the more efficient algorithms would carry us beyond the scope of this chapter.

15.4.1 Euler's method

A discussion of the basic ideas of Euler's method can be found in chapter 2 on Numerical Review and any elementary numerical analysis book, for example [Conte & de Boor 80, Burden & Faires 85]. We describe it briefly in terms of the equations we are interested in solving. We begin with

$$m\ddot{x} = f(x), \tag{15.10}$$

with initial conditions

$$x(0) = x^{(0)}, \quad \dot{x}(0) = \dot{x}^{(0)}.$$

The first step is to write the second order differential equation, equation (15.10), as a pair of coupled first order differential equations:

$$\begin{aligned} m\dot{u} &= f(x), \\ \dot{x} &= u, \end{aligned}$$

where u is just a new name for $\dot{x}$. The next step is to use Euler's method on these equations to get the following formulas for generating the solution at times h, $2h$, $3h$, ...:

$$\begin{aligned} x(t+h) &\approx x(t) + hu(t), \\ u(t+h) &\approx u(t) + \frac{h}{m}f(x(t)). \end{aligned}$$

The code segment for this computation is:

```
DO T = 1,NSTEP
  CALL FORCE(X,Y,Z,FX,FY,FZ)
  DO I = 1,NPART
    X(I) = X(I) + H*U(I)
    Y(I) = Y(I) + H*V(I)
    Z(I) = Z(I) + H*W(I)
    U(I) = U(I) + (H/M)*FX(I)
    V(I) = V(I) + (H/M)*FY(I)
    W(I) = W(I) + (H/M)*FZ(I)
  END DO
       .
       .
       .
  (Write positions and velocities)
       .
       .
       .
END DO
```

Figure 15.14: Code segment for a three-dimensional computation on an n-particle system.

```
DO T = 1,NSTEP
    X(T+1) = X(T) + H*U(T)
    U(T+1) = U(T) + (H/M)*F(X(T))
END DO
```

The extension of these formulas to multiparticle, multidimensional problems is straightforward. The code segment for a three-dimensional computation on an n-particle system, assuming all particles have the same mass, is given in figure 15.14. The procedure FORCE evaluates the the forces on the particles from their current positions that are stored in the arrays X, Y, Z and returns the x-, y-, and z-components of the forces in the arrays FX, FY, FZ. The remark after the inner loop indicates a block of code that would write position and velocity information. Generally this information is not written at every time step because H is so small; rather it is written at a larger interval that is an integer multiple of H. Therefore, this block of code would include a test to determine if writing should take place at the current time.

If $f(x)$ is *well-behaved* it can be shown that the error in the computed solution is $\mathcal{O}(h)$. As an illustration of the error in solving the equations of motion with Euler's method we show the error as a function of time for a one-dimensional, four-particle HL system in figure 15.15. This result was obtained with a stepsize of $h = 0.01$. If we reduce the stepsize by a factor of ten, that is $h = 0.001$, then we get the results shown in figure 15.16. Comparison of the error for the two different values of h shows that the peaks in the error have been reduced by about a factor of 10 as we would expect because h is reduced by reduced by this factor and the error should be $\mathcal{O}(h)$.

While Euler's method has the virtue of simplicity it is far less accurate than other methods we might use. Verlet's method is still a relatively simple method but gives much better accuracy.

15.4.2 Verlet's method

The name we give to this method is commonly used in the molecular dynamics literature, but it is known in mathematics as *Störmer's method*. Actually there is a class of Störmer methods, of which this is the simplest. Henrici shows that the error in this method is $\mathcal{O}(h^2)$ [Henrici 62].

The basic idea is to approximate the second derivative with a finite difference,

$$\ddot{x} \approx \frac{x(t+h) - 2x(t) + x(t-h)}{h^2}.$$

The error in this approximation of the second derivative is $\mathcal{O}(h^2)$, as shown in chapter 2. If we use this approximation in equation (15.10) we obtain

$$x(t+h) \approx 2x(t) - x(t-h) + h^2 f(x(t)).$$

This, then, is the basis of the algorithm. We can generate the solution with the following code segment:

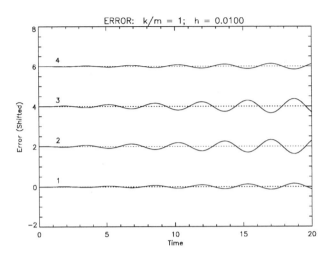

Figure 15.15: Error in numerical solution by Euler's method of one-dimensional, four-particle HL model: $h = 0.01$, $k/m = 1$. The curves have been shifted vertically: the point of zero error for a particular curve is its position at t=0.

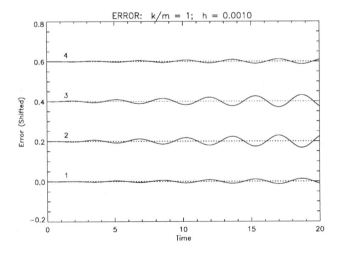

Figure 15.16: Error in numerical solution by Euler's method of one-dimensional, four-particle HL model: $h = 0.001$, $k/m = 1$. The curves have been shifted vertically: the point of zero error for a particular curve is its position at t=0.

```
DO T = 1,NSTEP
   X(T+1) = 2*X(T) - X(T-1) + H**2*F(X(T))
END DO
```

There are important differences between this code segment and the corresponding code for Euler's method. Notice that two previous values of X are required at each time step, unlike Euler's method which required just one. Notice also that the force term has a factor H**2, not H as in Euler's method. And, finally, notice that the velocity does not appear.

Something special must be done to start the iteration because the initial values for the problem are usually position and velocity, not two position values. A Taylor series expansion can be used to compute $x(h)$ given $x(0)$ and $\dot{x}(0)$:

$$x(h) \approx x(0) + h\dot{x}(0) + \frac{h^2}{2} f(x(0)).$$

The fact that the force term is $\mathcal{O}(h^2)$ implies that we are adding a very small number to a much larger number at every step, resulting in a loss in accuracy. This can be mitigated by using a different form of the algorithm, called the *summed form*:

```
DO T = 1,NSTEP
   DX(T) = DX(T-1) + H*F(X(T))
   X(T+1) = X(T) + H*DX(T)
END DO
```

It is easy to verify that this is mathematically equivalent to the original algorithm: DX(T) is simply the name of (X(T+1) - X(T))/H. In other words, if all computations were exact (no roundoff error) then this code would produce the same result as the original. But real computations are not exact and the summed form gives a more accurate result.

The velocity can be computed from the position using a central difference approximation:

$$\dot{x}(t) \approx \frac{x(t+h) - x(t-h)}{2h}.$$

Alternatively, it can be computed within the algorithm as follows:

```
DO T = 1,NSTEP
  X(T+1) = X(T) + H*(U(T) + H*F(X(T))/2)
  U(T+1) = U(T) + H*(F(X(T+1))+F(X(T)))/2
END DO
```

An efficient implementation of this only requires one computation of the force at each time step, and saving it for use in the next time step. This form has the numerical accuracy of the summed form, but it requires more computations per time step. If velocities are not needed then the summed form should be used. If double precision arithmetic is used then the original unsummed form of the algorithm may give acceptable accuracy.

An idea of the difference in accuracy between Euler's method and Verlet's method is illustrated in figures 15.17 and 15.18, which correspond to figures 15.15 and 15.16: the same computation except Verlet's method was used. Comparison of the peaks in these error curves with the corresponding curves for Euler's method shows that Verlet's method is far more accurate than Euler's method: for $h = 0.01$ the error in Verlet's method is smaller than the error in Euler's method by a factor of 10^{-3}. Comparison of the error in Verlet's method for $h = 0.01$ with that for $h = 0.001$ shows that the error is reduced by about a factor of 100, confirming the $\mathcal{O}(h^2)$ behavior of the error.

Verlet's algorithm belongs to the class of *symplectic* algorithms. These are important for particle dynamics because they keep the total energy almost constant. The article by Sanz-Serna [Sanz-Serna 92] describes many of these algorithms and has an extensive list of references.

Extension of this algorithm to three-dimensional, multiparticle systems should be evident from the discussion of Euler's method.

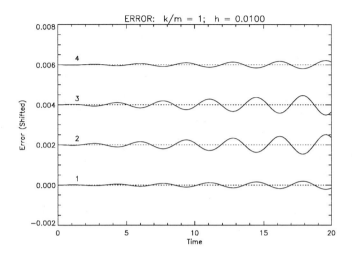

Figure 15.17: Error in numerical solution by Verlet's method of one-dimensional, four-particle HL model: $h = 0.01$, $k/m = 1$. The curves have been shifted vertically: the point of zero error for a particular curve is its position at t=0.

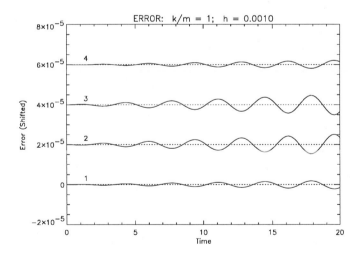

Figure 15.18: Error in numerical solution by Verlet's method of one-dimensional, four-particle HL model: $h = 0.001$, $k/m = 1$. The curves have been shifted vertically: the point of zero error for a particular curve is its position at t=0.

15.4.3   Hard sphere collisions

Determining the motion of a system of particles that are modeled as hard spheres does not require solving a differential equation, so the methods described above do not apply. Since hard sphere particles travel in straight lines with constant speed between collisions we only need to know when collisions occur and the velocities after collision in order to follow the motions of the particles. It sounds simple, but determining the sequence of collisions is a computationally intensive task. Bear in mind that we must determine which pair of particles will collide next after each collision, a task that must take into consideration all pairs of particles. We consider the easy case of a one-dimensional system first.

One-dimensional system

In one dimension the particles collide *head-on* as illustrated in figure 15.19. The conservation laws require that the total momentum and total energy do not change. Therefore we have:

$$u_{1,old} + u_{2,old} = u_{1,new} + u_{2,new} \quad (conservation\ of\ momentum),$$
$$u_{1,old}^2 + u_{2,old}^2 = u_{1,new}^2 + u_{2,new}^2 \quad (conservation\ of\ energy),$$

where u denotes velocity, and it is assumed that both particles have the same mass. These equations can be solved easily for $u_{1,new}$ and $u_{2,new}$:

$$u_{1,new} = u_{2,old}, \quad u_{2,new} = u_{1,old} \tag{15.11}$$

Thus the particles simply exchange velocities when they collide.

Assume we have a system of hard-sphere particles ordered along the x-axis so that

$$x_1 < x_2 < \ldots < x_n.$$

as illustrated in figure 15.20 for $n = 6$.

To determine the time of the next collision me must consider all pairs, executing a segment of code that looks the code in figure 15.21. After execution of this segment we know that the time of the next collision is COLLTIME and that the collision partners are particles I and I+1, provided that COLLTIME $\neq$

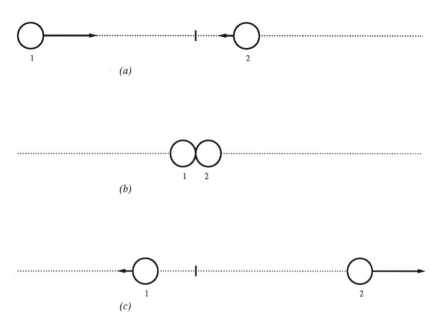

Figure 15.19: Collision of hard sphere particles in one dimension: (a) one time unit before collision; (b) instant of collision; (c) one time unit after collision. In this illustration particle 1 is travelling at four times the speed of particle 2 before the collision. At collision the particles exchange velocities so after the collision particle 2 is travelling at four times the speed of particle 1.

Figure 15.20: A six-particle, one-dimensional system of hard spheres.

INFINITY. The code segment we execute at a collision updates the positions of all particles, and the velocities of the colliding pair. It looks like that in figure 15.22.

Two-, three-dimensional systems

In two- and three-dimensional systems collisions are not necessarily head on, they may be oblique, as illustrated in figure 15.23.

In an oblique collision the interaction or impact is along the line drawn between the centers of the particles at the instant of collision: there is no force exerted on the particles in the plane tangent to the two particles at the point of impact — our HS model assumes that the particles are perfectly smooth. An analysis like that used for the one dimensional case shows that in an oblique collision the particles exchange the components of their velocities *along the line between the centers of the particles*, no other velocity components are changed. Thus the effect of an oblique collision is as illustrated in figure 15.24.

The critical part of the computation for updating velocities is the determination of which particles will collide next. In its simplest form the steps in this computation are as follows. We let r_i and $\dot{r}_i$ denote the position vector and velocity vector of the ith particle, and we let σ denote the diameter of a particle. It is convenient to define new position and velocity vectors

$$r_{i,j} = r_i - r_j, \quad \dot{r}_{i,j} = \dot{r}_i - \dot{r}_j$$

that represent the position and velocity of the ith particle relative to the jth particle. If we want to be explicit about the relative position at time t then we write $r_{i,j}(t)$. The test to determine if two particles collide can be broken into two parts: determine if they are approaching each other; if they are approaching each other then determine their distance of closest approach.

The particles are approaching each other if the component of their relative velocity in the direction of $r_{i,j}$ is negative; i.e.,

$$r_{i,j} \cdot \dot{r}_{i,j} < 0,$$

where the product on the left is a scalar (dot) product: this notation is also used below to denote scalar product. The idea is illustrated in figure 15.25.

```
COLLTIME = INFINITY
DO I = 1,N-1
  IF ((U(I+1) - U(I) .LT. 0)
    COLLTIMENEW = (X(I+1) - X(I))/(U(I) - U(I+1))
    IF(COLLTIMENEW .LT. COLLTIME)
      COLLPART = I
      COLLTIME = COLLTIMENEW
    END IF
  END IF
END DO
```

Figure 15.21: Code segment to determine time of next collision.

```
DO I = 1,N
  X(I) = X(I) + U(I)*COLLTIME
END DO
TMP = U(COLLPART)
U(COLLPART) = U(COLLPART + 1)
U(COLLPART + 1) = TMP
```

Figure 15.22: Code segment for collision updates.

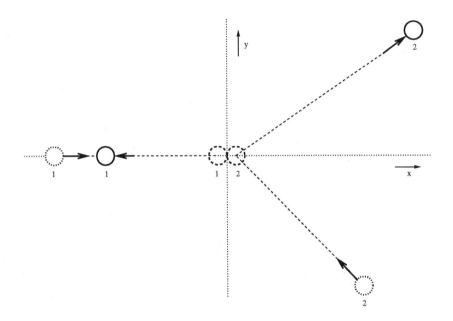

Figure 15.23: Collision of hard spheres in two dimensions: positions at one time unit before collision (dotted circles); at collision (dashed circle); and one time unit after collision (solid circles) are shown. Particles 1 and 2 are travelling at the same speed, s, before the collision. Particle 2 is travelling along a line that is 45° from the x-axis. At collision they exchange x-components of velocity; their y-components of velocity are unchanged. Therefore, after collision particle 1 is travelling in the x-direction at speed $s/\sqrt{2}$, while particle 2 is travelling upward and to the right, at an angle of $\arctan(1/\sqrt{2})$ to the x-axis, with a speed equal to $s\sqrt{3/2}$.

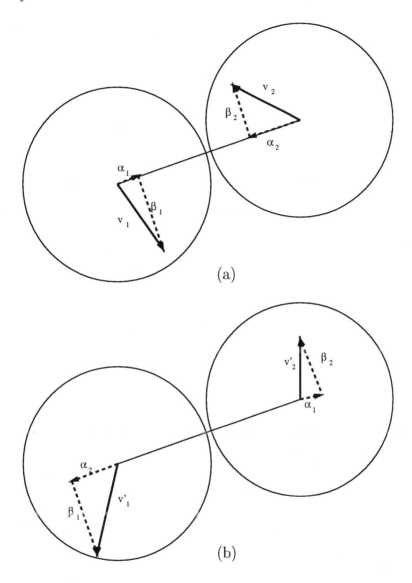

Figure 15.24: Detail of velocity components at instant before collision (a), and instant after collision (b). At collision the force is exerted only along the line joining the centers of the particles; no force is exerted in the plane perpendicular to this line because the particles are assumed to be *smooth*. Thus velocity components along this line are exchanged at the instant of collision, but velocity components in the plane perpendicular to this line are unchanged.

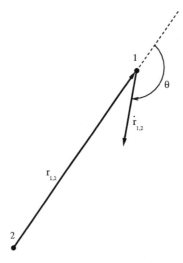

Figure 15.25: Illustration of necessary condition for a collision.

They actually collide if the following condition is true:

$$\|b_{i,j}\| \le \sigma, \quad \|b_{i,j}\|^2 = \|r_{i,j}\|^2 - \left(\frac{r_{i,j} \cdot \dot{r}_{i,j}}{\|\dot{r}_{i,j}\|}\right)^2 .$$

Some simple geometrical considerations, illustrated in figure 15.26 show that $b_{i,j}$ is the distance of closest approach if each particle had diameter zero.

If the particles do collide, then the time of the collision is given by the formula

$$t = -\frac{1}{\|\dot{r}_i - \dot{r}_j\|} \left(\frac{(r_i - r_j) \cdot (\dot{r}_i - \dot{r}_j)}{\|\dot{r}_i - \dot{r}_j\|} + (\sigma^2 - \|b_{i,j}\|^2)^{\frac{1}{2}}\right) .$$

This equation can be understood by observing that the second factor on the right, namely

$$-\left(\frac{r_{i,j} \cdot \dot{r}_{i,j}}{\|\dot{r}_{i,j}\|} + (\sigma^2 - \|b_{i,j}\|^2)^{\frac{1}{2}}\right) ,$$

is the distance to be traversed in the direction of $\dot{r}_{i,j}$ before collision: refer to figure 15.26 and do some elementary geometry. It is evident from these formulas that determining the time of the next collision is a good deal more complicated than in the one-dimensional case.

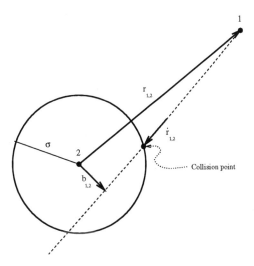

Figure 15.26: Illustration of collision parameters. Here $r_{1,2}$ is the position of the center of particle 1 relative to the center of particle 2; and $\dot{r}_{1,2}$ is the velocity of particle 1 relative to the velocity of particle 2. Thus you can think of particle 2 as fixed at the origin, and particle 1 at $r_{1,2}$ moving with velocity $\dot{r}_{1,2}$. For a collision to occur particle 1 must be moving towards the origin, and must come within a distance σ (the particle's diameter) of the origin. Particle 1 is moving towards the origin if the dot product $r_{1,2} \cdot \dot{r}_{1,2} < 0$ as in this picture. In this picture $\|b_{1,2}\|$ is the distance of closest approach by particle 1 to the origin: $b_{1,2}$ is a vector perpendicular to $\dot{r}_{1,2}$. The point at which the collision occurs is the point where the trajectory of particle 1 crosses the circle of radius σ around the origin, and the time of collision is the time it takes to move from $r_{1,2}$ to this point travelling at velocity $\dot{r}_{1,2}$.

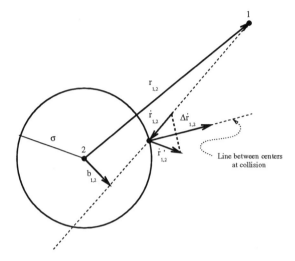

Figure 15.27: Change in velocity at collision of hard spheres. Parameters σ, $r_{1,2}$, $\dot{r}_{1,2}$, $b_{1,2}$ are as defined in figure 15.25; $\dot{r}'_{1,2}$ is the relative velocity after collision and $\Delta\dot{r}_{1,2}$ is the change in relative velocity due to the collision. Note that $\Delta\dot{r}_{1,2}$ is perpendicular to the line between centers at the time of collision.

When the particles do collide then the new velocities are easily computed. There is no change in velocity in the plane tangent to the spheres at the collision point; the velocity changes only in the direction of the line between the centers of the particle, as illustrated in figure 15.27.

The change in velocity takes place, as noted earlier, in the direction of the vector $(r_i - r_j)$ at the time of collision. In particular,

$$\dot{r}_{i,new} = \dot{r}_{i,old} + \Delta\dot{r}_{i,old}$$
$$\dot{r}_{j,new} = \dot{r}_{j,old} - \Delta\dot{r}_{i,old}$$

where

$$\Delta\dot{r}_{i,old} = -\frac{(\dot{r}_{i,old} - \dot{r}_{j,old}) \cdot (r_{i,old} - r_{j,old})(r_{i,old} - r_{j,old})}{\|r_{i,old} - r_{j,old}\|^2}.$$

You should be able to verify that this formula gives the same result for the one-dimensional case as obtained earlier, equation (15.11).

At this point you have all of the necessary formulas for constructing a program to solve the equations of motion by Euler's method or Verlet's method for any of the models — HL, LJ, or HS.

15.5   Exact solution of the equations of motion for HL model

We noted earlier that the Hooke's Law model is special because, unlike the other models, it admits an exact solution. Here we outline the main ideas leading to the exact solution. In order to understand the material in this section, you need to know about eigenvalues and eigenvectors of matrices at an elementary level; for example, see section 2.5.4 of chapter 2.

Consider first the simple case of a single particle moving under the influence of an HL force, illustrated earlier in figure 15.3. The equation of motion for this particle is

$$m\ddot{q}(t) = -kq(t),\tag{15.12}$$

where m is the mass of the particle, k is the force constant, and $q(t)$ is the displacement of the particle from its equilibrium position at time t.

It is easy to verify that

$$q(t) = Q\cos(\omega t + \delta)$$

satisfies equation (15.12). If we substitute this expression for $q(t)$ into equation (15.12) we obtain

$$-mQ\omega^2\cos(\omega t + \delta) = -kQ\cos(\omega t + \delta).$$

The left side agrees with the right side provided that

$$\omega = \sqrt{\frac{k}{m}}.$$

Therefore

$$q(t) = Q\cos\left(\sqrt{k/m}\,t + \delta\right)\tag{15.13}$$

is a solution.

What about the parameters Q and δ appearing in this solution? We show here that they are determined by the initial conditions for the motion. Notice that when $t = 0$ we have

$$q(0) = Q\cos(\delta), \quad \dot{q}(0) = -\omega Q \sin(\delta).$$

Therefore, if we impose the initial conditions

$$q(0) = 1, \quad \dot{q}(0) = 0, \tag{15.14}$$

we easily find that

$$Q = 1, \quad \delta = 0.$$

Thus for the initial conditions given in equation (15.14) the solution to the equations of motion is

$$q(t) = \cos\left(\sqrt{k/m}\,t\right) \tag{15.15}$$

Thus the solution is a periodic function of time, with frequency $(2\pi)^{-1}\sqrt{k/m}$.

Now we turn to the more complicated case of a chain of four particles considered earlier and illustrated in figure 15.5. Following the one-particle example above, we first "guess" a solution and then show that this guess satisfies the equations of motion provided certain parameters related to the frequency of the motion are satisfied. We will find that there are three distinct solutions, known as *modes*. These modes correspond to the general solution, equation (15.13), for the one-particle case. The initial conditions determine a linear combination of the modes that is the particular solution, corresponding to equation (15.15).

The equations of motion for four particles in the HL model were given in equation (15.8). It is convenient to express them in matrix form:

$$m\ddot{q} = -kMq \tag{15.16}$$

where q is a vector and M is a matrix:

$$q = \begin{pmatrix} q_1 \\ q_2 \\ q_3 \\ q_4 \end{pmatrix}, \quad M = \begin{pmatrix} 1 & -1 & 0 & 0 \\ -1 & 2 & -1 & 0 \\ 0 & -1 & 2 & -1 \\ 0 & 0 & -1 & 1 \end{pmatrix}.$$

Of course q is a function of time but for simplicity we often write q rather than $q(t)$.

We proceed by analogy with the one-particle case just considered, guessing that the solution to these four equations has the form:

$$q = Q\cos(\omega t + \delta), \quad Q = \begin{pmatrix} Q_1 \\ Q_2 \\ Q_3 \\ Q_4 \end{pmatrix}.$$

This looks like the solution for the one-particle case except q and Q are now vectors, q_i representing the displacement of the ith particle from its equilibrium position:

$$q_i = Q_i\cos(\omega t + \delta).$$

We can verify that this guess is indeed a solution by the same process as before, that is by substituting it into the equations of motion, equation (15.16). Substitution and a little algebraic manipulation produces the result

$$MQ = \frac{m\omega^2}{k}Q. \tag{15.17}$$

You may recognize this as the usual form of a matrix eigenvalue equation where Q is an eigenvector and $m\omega^2/k$ is an eigenvalue of the matrix M; see equation (2.14) of chapter 2. Thus our guessed solution satisfies the equations of motion provided that $m\omega^2/k$ and Q are an eigenvalue-eigenvector pair of the matrix M.

Since M is a 4×4 matrix it has four eigenvalues; these are

$$\lambda_0 = 0, \ \lambda_1 = 2 - \sqrt{2}, \ \lambda_2 = 2, \ \lambda_3 = 2 + \sqrt{2}$$

and the corresponding eigenvectors are

$$\psi_0 = \begin{pmatrix} 1 \\ 1 \\ 1 \\ 1 \end{pmatrix}, \ \psi_1 = \begin{pmatrix} 1 \\ \sqrt{2} - 1 \\ -\sqrt{2} + 1 \\ -1 \end{pmatrix}, \ \psi_2 = \begin{pmatrix} 1 \\ -1 \\ -1 \\ 1 \end{pmatrix}, \ \psi_3 = \begin{pmatrix} \sqrt{2} - 1 \\ -1 \\ 1 \\ -\sqrt{2} + 1 \end{pmatrix}.$$

An explanation of where these results came from would take us too far afield, but you can easily verify them using equation (15.17), or you could use MATLAB to compute the eigenvalues and eigenvectors numerically. Remember that an eigenvector can be normalized in various ways. Here we normalized them so that $\|\psi\|_\infty = 1$; on the other hand MATLAB normalizes them so that $\|\psi\|_2 = 1$.

Each eigenvalue-eigenvector pair represents a particular motion of the four-particle system, a mode. Consider the mode represented by λ_3, ψ_3. The motion for this mode is given by

$$q = \begin{pmatrix} \sqrt{2} - 1 \\ -1 \\ +1 \\ -\sqrt{2} + 1 \end{pmatrix} \cos\left(\sqrt{(2 + \sqrt{2})k/m}\, t + \delta \right).$$

Notice that each particle oscillates with a frequency

$$\frac{1}{2\pi}\sqrt{(2 + \sqrt{2})k/m}.$$

Thus the frequency of the oscillation is proportional to the square root of the eigenvalue of the mode, and the relative amplitudes of the motion are determined by the eigenvector of the mode. Figure 15.28 illustrates this mode. Since it has the largest eigenvalue, and therefore the highest frequency, we call it the *high-frequency mode*.

The mode associated with λ_1 is called the *low frequency mode* since λ_1 is the smallest eigenvalue excepting zero (see below). The motion associated with this mode is illustrated in figure 15.29.

The eigenvalue $\lambda_0 = 0$ represents the state of no relative motion; i.e., no motion of the particles relative to each other. Notice that zero for the eigenvalue implies $\omega = 0$, which implies that the solution is independent of time. This situation would occur if we started all the particles in their equilibrium positions with no initial velocity: they would remain motionless. Here we are concerned only with states of relative motion; so we focus on the modes represented by the nonzero eigenvalues.

The importance of the modes comes from the fact that any motion of the four-particle system can be expressed as a linear combination of the modes.

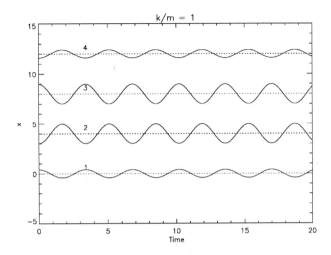

Figure 15.28: Illustration of the high-frequency mode of a four-particle system with $k/m = 1$, $\delta = 0$. Equilibrium positions of the four particles are assumed to be 0, 4, 8, 12.

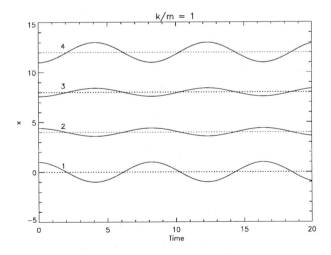

Figure 15.29: Illustration of the low-frequency mode of a four-particle system with $k/m = 1$, $\delta = 0$. Equilibrium positions of the four particles are assumed to be 0, 4, 8, 12.

We illustrate this for a particular case. Consider the motion determined by initial conditions

$$q(0) = \begin{pmatrix} 1 \\ 0 \\ 0 \\ -1 \end{pmatrix}, \quad \dot{q}(0) = 0. \tag{15.18}$$

We express the general solution as a linear combination of the modes:

$$q(t) = \sum_{i=1}^{3} c_i \psi_i \cos\left(\sqrt{(k/m)\lambda_i}\, t + \delta_i\right), \tag{15.19}$$

where c_1, c_2, c_3 are arbitrary constants. We must determine c_1, c_2, c_3 and $\delta_1, \delta_2, \delta_3$ so that $q(t)$ defined by equation (15.19) satisfies the initial conditions specified in equation (15.18).

To satisfy the condition on $q(0)$, set t equal to zero on the right side of equation (15.19) and substitute the result for $q(0)$ into the initial condition equation. This gives:

$$\sum_{i=1}^{3} c_i \cos(\delta_i)\psi_i = \begin{pmatrix} 1 \\ 0 \\ 0 \\ -1 \end{pmatrix}. \tag{15.20}$$

Next differentiate both sides of equation (15.19) with respect to t to obtain the following expression for $\dot{q}(t)$:

$$\dot{q}(t) = -\sum_{i=1}^{i=3} c_i \psi_i \sqrt{k/m\lambda_i} \sin\left(\sqrt{(k/m)\lambda_i}\, t + \delta_i\right).$$

Then set t equal to zero in this equation and substitute the result into the initial condition equation for $\dot{q}(0)$, giving the result

$$\sum_{i=1}^{3} c_i \sqrt{\frac{k}{m}\lambda_i} \sin(\delta_i)\psi_i = \begin{pmatrix} 0 \\ 0 \\ 0 \\ 0 \end{pmatrix}. \tag{15.21}$$

The four equations in equation (15.20) together with the four equations in equation (15.21) give us a system of eight linear equations in six unknowns d_1, d_2, ..., d_6 where

$$d_i = c_i \cos(\delta_i), \quad d_{i+3} = c_i \sin(\delta_i) \quad (i = 1, 2, 3).$$

It might seem that we cannot solve these equations because the number of equations exceeds, by 2, the number of unknowns. However two of the equations are redundant. Note that the sum of the components of each eigenvector is zero, and that the same is true for the sum of the components of $q(0)$ and $\dot{q}(0)$. Therefore there are at most six independent equations, not eight. These equations can be solved, for example with MATLAB, to obtain the result

$$d_1 = -8.535533906e - 01, \ d_2 = 0, \ d_3 = -3.535533906e - 01,$$
$$d_4 = d_5 = d_6 = 0.$$

From this is follows that

$$c_1 = -8.535533906e - 01, \ c_2 = 0, \ c_3 = -3.535533906e - 01,$$
$$\delta_1 = \delta_2 = \delta_3 = 0.$$

To confirm your understanding of this you might verify that these coefficients and phase angles produce a solution to the equations of motion that satisfies the initial conditions.

Similarly it is possible to obtain a solution to the equations of motion for any valid set of initial conditions (i.e., the sums of the components of $q(0)$ and of $\dot{q}(0)$ are both zero.)

The procedure for solving the problem with a chain of n atoms is the same; the matrix M has order n and has the form

$$M = \begin{pmatrix} 1 & -1 & 0 & 0 & \cdots & 0 \\ -1 & 2 & -1 & 0 & \cdots & 0 \\ 0 & -1 & 2 & -1 & \cdots & 0 \\ \vdots & & & & & \vdots \\ 0 & 0 & \cdots & -1 & 2 & -1 \\ 0 & 0 & \cdots & & -1 & 1 \end{pmatrix}.$$

The eigenvalues of M are given by

$$\lambda_p = 2\left(1 - \cos(\frac{p\pi}{n})\right)$$

and the eigenvectors are given by

$$\psi_j^{(p)} = A \cos\left(\frac{p\pi}{n}(j - \frac{1}{2})\right)$$

where $\psi_j^{(p)}$ is the jth component of the pth eigenvector. The index p takes values $0, 1, \ldots, n-1$; and the index j takes values $1, 2, \ldots, n$. The coefficient A, the normalization factor, is arbitrary. In the four-particle example we chose it so that the element of maximum magnitude in $\psi^{(p)}$ has magnitude 1; i.e.,

$$\|\psi^{(p)}\|_\infty = 1.$$

It is worth noting that for every n there is one eigenvalue equal to zero. The trivial mode corresponding to this eigenvalue would be ignored, just as we did in the case for the four-particle chain.

In order to check your understanding of the above discussion you might try to solve the following problems.

1. What is the high frequency mode for a five-particle system?

2. What is the upper bound on the frequency for any one-dimensional chain, assuming $k/m = 1$?

3. What is the exact solution of the equations of motion for a five-particle system with the following initial conditions?

$$q(0) = \begin{pmatrix} 1.0 \\ 0.5 \\ -0.5 \\ -0.5 \\ -0.5 \end{pmatrix}, \quad \dot{q}(0) = 0.$$

Assume $k/m = 1$.

4. What is the exact solution of the equations of motion for a five-particle system with the following initial conditions?

$$q(0) = 0, \quad \dot{q}(0) = \begin{pmatrix} 1 \\ 0 \\ 0 \\ 0 \\ -1 \end{pmatrix}.$$

Assume $k/m = 1$.

5. Assume a two-dimensional HL model with 8 particles in each dimension. What is the matrix form of the equations of motion?

6. Assume a three-dimensional HL model with 8 particles in each dimension. What is the matrix form of the equations of motion?

16 Advection

16.1 Introduction

When the first automatic computers were built in the middle of this century, weather forecasting was identified as a major problem to be attacked by these new machines. It remains so today. Meteorology, the study of the atmosphere for the purpose of weather forecasting, now has advanced to the state where reasonably accurate five-day forecasts are possible. Climatology, the study of the earth's climate, is closely related. It requires predicting effects over relatively large time intervals and these predictions are not nearly as reliable as the five-day weather forecast. Indeed, the results of computations predicting global warming have been under dispute. Thus the need for improving the speed and accuracy of these computations continues.

Computations in climatology and meteorology require models of the atmosphere and those aspects of the earth's surface — mountains, sea, ice, and so forth – that affect the weather and climate. The mathematical description of these models generally takes the form of a complex system of partial differential equations.

Among these equations is one relatively simple equation, called the *advection equation*, that describes the transport of a substance in the atmosphere. In meteorology the word *convection* refers to the vertical transport of matter due to rising air, and *advection* refers to the horizontal transport of matter due to wind. Thus the advection equation might describe the transport of a moist air mass, a dust cloud, or chemical substances by wind. Studies of large-scale air pollution, as might occur over southern California or northern Europe, represent an important application of this equation.

The advection equation and its solution is the principal subject of this chapter. There are two good reasons for including it in a course on high-performance scientific computing. In the first place, advection is a part of important applications of high-performance computing: meteorology, climatology, and air pollution. It would go beyond the level of this tutorial to treat the full system of equations for a climate model or an air pollution model, but we can learn something about the computational issues by studying the advection component of these models. Second, many of the computations done on high-performance computers involve the solution of partial differential equations. The advection equation is a relatively simple partial differential equation, so it provides a natural introduction to a broad class of computations performed on supercomputers.

This chapter consists of a detailed discussion of the one-dimensional advection equation and its solution by several algorithms. These algorithms are illustrated by a number of examples, and a careful analysis of the algorithms is made in order to explain the results. Two algorithms are the focus of most of the discussion, the Leapfrog algorithm and the Trapezoid algorithm. They have been chosen because they are relatively simple and because they serve to illustrate two fundamentally different kinds of numerical methods. The discussion closes with a short section on the two-dimensional advection equation. Here enough material is given to provide the reader the tools necessary to construct and analyze the two-dimensional forms of the Leapfrog and Trapezoid algorithms.

Advection implies the existence of a fluid flow, wind, that is responsible for the transport. Of course the velocity of this flow is not constant but it may be almost constant over relatively long periods. Most of this chapter. assumes constant wind velocity, mainly to keep the introduction to this subject reasonably easy to follow. In the section on the two-dimensional advection equation, a variable velocity is introduced, illustrating the added complexity it causes.

It is assumed that you have had a conventional undergraduate course in numerical analysis. It does not assume a course in partial differential equations, but it does assume that you are familiar with partial derivatives and finite difference approximations of derivatives. The mathematical analysis of the algorithms we discuss requires an elementary knowledge of Fourier series. You may wish to read the chapter 2 on Numerical Analysis Review to refresh your knowledge. If you are interested in background reading in the areas of meteorology and climatology you might consider the books *Meteorology* [Miller & Anthes 80], *An Introduction to Three-Dimensional Climate Modeling*

[Washington & Parkinson 86], and *Numerical Prediction and Dynamic Meteorology* [Haltiner & Williams 80]. The first of these should be easily understood by any undergraduate student, the other two are considerably more advanced, the latter has a good discussion of the advection equation and of numerical methods for solving it. Another good source for numerical methods in this area is *Numerical Methods Used in Atmospheric Models*[Mesinger & Arakawa 76]. The technical report *Running Air Pollution Models on the Connection Machine* [Zlatev & Waśniewski 92] has an easily understood discussion of air pollution models and the use of a massively parallel computer to study them.

16.2   One-dimensional advection equation

Imagine a small particle suspended in a fluid flowing with constant velocity u. After an interval of time t' the particle is carried downstream a distance ut'. Similarly, a group of particles suspended in the fluid are carried along with the flow, each moving a distance ut', thus their positions relative to each other do not change. Now imagine a very large number of tiny particles suspended in the fluid and let $\phi(x, t)$ represent their density (the number of particles per unit interval). Figure 16.1 illustrates this density at time $t = 0$ and at a slightly later time $t = 0.2$ when $u = 1$: the density distribution is simply translated a distance 0.2 in the direction of the flow.

In this model a group of particles in the fluid is carried along by the flow, maintaining their positions relative to each other, therefore it is a trivial matter to determine where every particle is, or what the density distribution of particles is, at a later time. However, when the velocity of the flow is not constant and when the flow is not one-dimensional, this determination cannot be made so easily. Then we must use mathematical equations to represent the physical model and numerical methods to solve them. Nevertheless it is instructive to start our study of the mathematical issues with the simple one-dimensional, constant velocity model. The equation that describes it is called the *one-dimensional advection equation*, expressed as follows:

$$\frac{\partial \phi(x, t)}{\partial t} + u \frac{\partial \phi(x, t)}{\partial x} = 0. \tag{16.1}$$

The independent variables x and t are position and time, respectively; $\phi(x, t)$

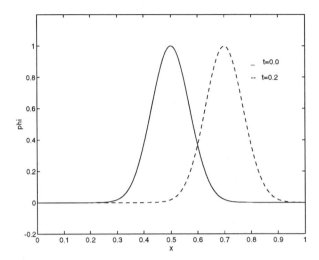

Figure 16.1: Example of a density distribution of particles in a one-dimensional flow at time $t = 0$ and at a later time $t = 0.2$ when $u = 1$.

is the density of particles; and u is the velocity of the flow, which we assume to be constant and in the positive x direction. In more realistic, more complex models, the right-hand side of this equation is not zero: it may contain terms that represent the effects of diffusion, sources of substances injected into the flow (for example, automobile exhaust), and reactions between chemicals in the flow.

It is customary to use the notation ϕ_t and ϕ_x to denote partial derivatives with respect to t and x, then the advection equation has the form

$$\phi_t + u\phi_x = 0.$$

We can verify that this equation models a distribution of particles carried along by a flow moving at velocity u. Suppose that the initial density of particles is given by $f(x)$; i.e., $\phi(x, 0) = f(x)$. As time advances the particles are carried along, downstream, so that all of them have moved a distance ut' by time t', as described already. Therefore

$$\phi(x, t') = \phi(x - ut', 0) = f(x - ut').$$

Thus, if the solution to the advection equation is

$$\phi(x, t) = f(x - ut), \tag{16.2}$$

we know that it correctly represents the particle flow in our simple model. So let us verify that equation (16.2) is a solution to equation (16.1) by substituting it into equation (16.1). For the first term on the left of equation (16.1) we have, with $z = x - ut$,

$$\frac{\partial \phi(x,t)}{\partial t} = \frac{\partial f(z)}{\partial t} = \frac{df(z)}{dz}\frac{\partial z}{\partial t} = \frac{df(z)}{dz}(-u);$$

and for the second term on the left of equation (16.1) we have

$$u\frac{\partial \phi(x,t)}{\partial x} = u\frac{\partial f(z)}{\partial x} = u\frac{df(z)}{dz}\frac{\partial z}{\partial x} = u\frac{df(z)}{dz},$$

therefore

$$\frac{\partial \phi(x,t)}{\partial t} + u\frac{\partial \phi(x,t)}{\partial x} = \frac{df(z)}{dz}(-u + u) = 0,$$

thus equation (16.1), the advection equation, is satisfied.

16.2.1 Advection equation is $d\phi/dt = 0$

Here is another view of what the advection equation represents. Imagine that you could shrink to the size of a particle and move along with the flow. The change you would see in ϕ with time is given by the total derivative $d\phi/dt$ and from elementary calculus we know that

$$\frac{d\phi}{dt} = \phi_t + \frac{dx}{dt}\phi_x.$$

Comparing this with the left-hand side of the advection equation and noting that $dx/dt = u$ we see that the advection equation is equivalent to the equation

$$\frac{d\phi}{dt} = 0.$$

It is important to recognize that this result for $d\phi/dt$ holds *in a frame of reference that is moving with the flow*. In fluid dynamics this frame of reference is called *Lagrangian*, while a frame of reference that is fixed is called *Eulerian*.

16.2.2   Advection equation as a conservation law

Assume a collection of particles with density $\phi(x,t)$ is flowing with velocity $u(x,t)$. Now consider an infinitesimal interval of length $2dx$, centered at $x = a$. If particles are not spontaneously created or destroyed, then the number of particles entering the interval minus the number of particles leaving it must equal the net change in the number of particles in the interval. The number entering and leaving per unit time is

$$\begin{aligned} N_{\text{enter}} &= u(a - dx, t)\phi(a - dx, t) \\ N_{\text{leave}} &= u(a + dx, t)\phi(a + dx, t) \end{aligned}$$

and the net increase in the number of particles per unit time in the interval is

$$N_{\text{increase}} = \left.\frac{\partial\phi(a,t)}{\partial t}\right|_{x=a} dx \, .$$

Conservation of particles requires

$$N_{\text{enter}} - N_{\text{leave}} = N_{\text{increase}} \, ,$$

hence

$$u(a - dx, t)\phi(a - dx, t) - u(a + dx, t)\phi(a + dx, t) = \frac{\partial\phi(a,t)}{\partial t} 2dx \, ,$$

or

$$\frac{u(a - dx, t)\phi(a - dx, t) - u(a + dx, t)\phi(a + dx, t)}{2dx} = \frac{\partial\phi(a,t)}{\partial t} \, .$$

If we now take the limit $dx \to 0$ this becomes

$$-\left.\frac{\partial u(x,t)\phi(x,t)}{\partial x}\right|_{x=a} = \frac{\partial\phi(a,t)}{\partial t} \, .$$

Now use the fact that a can be any point on the x-axis and rearrange terms, then

$$\frac{\partial\phi(x,t)}{\partial t} + \frac{\partial u(x,t)\phi(x,t)}{\partial x} = 0 \, , \tag{16.3}$$

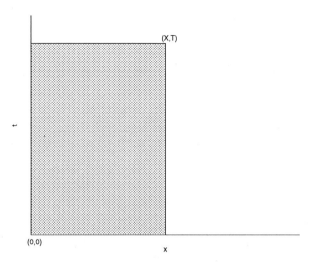

Figure 16.2: Solution of the advection equation is to be determined in the rectangle $R = [0, X] \times [0, T]$.

which is the advection equation when we assume that the velocity can depend on x and t. If the velocity is constant then this becomes the advection equation as already given, equation (16.1). Equation (16.3) is important to remember because it indicates how formulas for the constant velocity case, appearing later, must be modified in order to account for a variable velocity.

This derivation is also important because it suggests how the advection equation must be modified if particles are being created, as would be necessary in a model that takes into account the continual injection of a polluting substance. A nonzero term on the right of this equation would be required to represent this source of particles.

16.2.3   Boundary conditions

Usually we are interested in knowing the solution to the advection equation in a finite region of the x-t plane. Here we assume that this region is the rectangle $R = [0, X] \times [0, T]$, as illustrated in figure 16.2.

Restricting our attention to the rectangle R suggests that we specify the initial condition

$$\phi(x,0) = f(x)$$

on the interval $[0, X]$. However, this does not determine the solution everywhere in R. We can see this in the following way. In our earlier discussion we verified that the solution is given by equation (16.2). It follows from this equation that $\phi(x,t)$ is constant on the straight line $x = ut + x_0$ and on this line $\phi(x,t) = f(x_0)$, where x_0 can be any point in $[0, X]$: this line is called a *characteristic*. Three characteristics are illustrated in figure 16.3, corresponding to $x_0 = 0.2$, 0.4, and 0.6. Thus for every point on the x-axis where $f(x)$ is defined we have such a characteristic and if $f(x)$ is known on an interval, say $[0, X]$, then the associated characteristics sweep out a region as indicated in figure 16.4. Thus it is evident that there is a region of R, bounded on the left by the t-axis and on the right by the characteristic through the origin, in which the solution is completely undetermined.

We could resolve this problem by specifying the value of $\phi(x,0)$ on a larger interval. Our previous discussion should make it evident that if $\phi(x,0)$ is specified on the interval $[\text{-}uT, X]$ then the solution is determined everywhere in R. But this is not very satisfactory because we must use points outside the region R.

Here is a better way to deal with the problem. If we specify the value of $\phi(0,t)$ on the t-interval $[0, T]$ and on the x- interval $[0, X]$ then the solution is determined everywhere in R. To illustrate, suppose that we know $\phi(0, t')$ where $t' \in [0, T]$, then $\phi(x,t)$ is determined along the line

$$t = \frac{1}{u}x + t'.$$

This follows from the fact that the solution to the advection equation is constant along the characteristics, straight lines with slope $1/u$; therefore if we specify the solution at any point on one of these lines it is known everywhere else on that line. Thus if we specify the value of $\phi(0,t)$ on $[0, T]$ and $\phi(x,0)$ on $[0, X]$ then the solution is determined at every point in R.

How should we choose $\phi(0,t)$? A common choice, and the one we adopt here, is periodic boundary conditions; that is, we require

$$\phi(0,t) = \phi(X,t), \quad t \in [0, T].$$

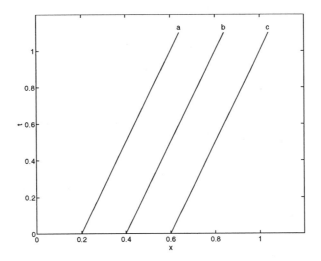

Figure 16.3: Three characteristics: (a) $\phi(x,t) = f(0.2)$, (b) $\phi(x,t) = f(0.4)$, (c) $\phi(x,t) = f(0.6)$.

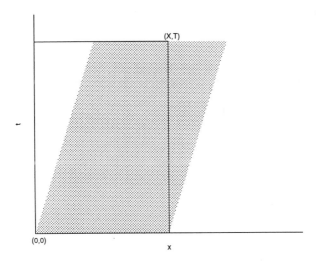

Figure 16.4: The shaded region is swept out by the characteristics and in this region the solution is determined. In the unshaded region, bounded on the left by the t-axis and on the right by the characteristic through the origin, the solution is completely undetermined.

We can see that periodic boundary conditions determine $\phi(0, t)$ for $t \in [0, T]$ in the following way. From the characteristics it is seen that the value of $\phi(X, t)$ for $t \in [0, X/u]$ is completely determined by $\phi(x, 0)$; similarly, the value of $\phi(X, t))$ for $t \in [X/u, 2X/u]$ is completely determined by $\phi(x, X/u)$, etc. The use of periodic boundary conditions is equivalent to assuming that the region R is wrapped on the surface of a cylinder, with the t-axis parallel to the axis of the cylinder; or, equivalently, to extending the region R periodically along the x-axis from $-\infty$ to $+\infty$.

Before closing this section we must say some additional words about characteristics. In general they are not straight lines as they are here, nor is the solution constant along them. They do have the following important property: if initial data are provided along a curve $\mathcal{C}$ in the problem space, and this curve is not a characteristic, then that data determines the solution along any characteristic intersecting $\mathcal{C}$; on the other hand, if $\mathcal{C}$ is a characteristic then the initial data is not sufficient to determine the solution elsewhere in the problem space. Note, for example, that if we specified the value of ϕ along a characteristic, its value elsewhere in R is completely undetermined.

Characteristics are important not only to the theory of partial differential equations, they also provide a means for solving partial differential equations — the solution along a characteristic can be determined by integrating an ordinary differential equation (in our case it is $d\phi/dt = 0$). Not all partial differential equations have characteristics; there are three broad classes of partial differential equations – hyperbolic, parabolic, and elliptic – only the first two have characteristics. In fact the notion of characteristics is intimately connected with this classification. For more on this subject see the books *Numerical Solution of Partial Differential Equations: Finite Difference Methods* [Smith 78] and *Partial Differential Equations* [Carrier & Pearson 76].

16.3   Numerical methods: first order

Here we turn our attention to some simple numerical methods for solving the advection equation. We focus on *finite difference algorithms*, so named because they are based on approximating the partial derivatives in the advection equation by finite differences. Here we consider algorithms based on first-order

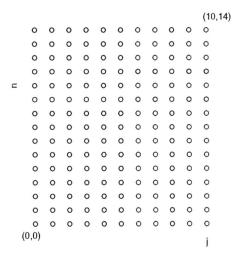

Figure 16.5: A point in R_g is identified by the ordered pair (j, n). Two corner points are labeled here for illustration: $(0,0)$ at the lower left corner; $(10,14)$ at the upper right corner.

approximations of the partial derivatives. While these algorithms are not accurate enough for serious computational work, they provide a convenient tool for introducing ideas we use later in discussing more accurate finite difference methods.

16.3.1   Discretization of R

Finite-difference methods generate the solution on the points of a regularly spaced grid denoted by R_g, figure 16.5. The values of x and t at the grid point (j, n) are x_j and t_n:

$$x_j = j\Delta x, \text{ and } t_n = n\Delta t\,;$$

in particular,

$$x_0 = 0, \ x_J = X, \ t_0 = 0, \text{ and } t_N = T\,.$$

Thus the continuous region R is discretized and replaced by the region R_g, also referred to as "the grid".

   The solution ϕ at (j, n) is denoted by $\phi(x_j, t_n)$ or equivalently by $\phi_{j,n}$. For brevity, we often use $\phi_{j,n}$. The notation $\phi_{0:J,n}$ denotes the set of values

$\phi_{0,n}, \phi_{1,n}, \ldots, \phi_{J,n}$, the solution at all values of x at time t_n; similarly, $\phi_{0:J,0:N}$ denotes the solution on the entire region R_g.

In R_g periodic boundary conditions are expressed with

$$\phi_{0,0:N} = \phi_{J,0:N}\,,$$

and the initial condition is expressed with

$$\phi_{0:J-1,0} = f(x_{0:J-1})\,.$$

Periodic boundary conditions make it unnecessary to specify $\phi_{J,0}$.

16.3.2   First-order difference methods

Recall from Taylor's series that

$$\begin{aligned}
\phi(x + \Delta x, t) &= \phi(x,t) + \phi_x(x,t)\Delta x + \mathcal{O}((\Delta x)^2)\,, \\
\phi(x, t + \Delta t) &= \phi(x,t) + \phi_t(x,t)\Delta t + \mathcal{O}((\Delta t)^2)\,.
\end{aligned}$$

Therefore,

$$\begin{aligned}
\phi_x(x,t) &= \frac{\phi(x + \Delta x, t) - \phi(x,t)}{\Delta x} + \mathcal{O}(\Delta x)\,, \\
\phi_t(x,t) &= \frac{\phi(x, t + \Delta t) - \phi(x,t)}{\Delta t} + \mathcal{O}(\Delta t)\,.
\end{aligned} \qquad (16.4)$$

The expressions $\phi(x + \Delta x, t) - \phi(x,t)$ and $\phi(x, t + \Delta t) - \phi(x,t)$ appearing here are called " forward differences".

A forward difference algorithm, Algorithm 1

If we use equations (16.4, 16.4) to replace the partial derivatives in the advection equation we obtain

$$\frac{\phi_{j,n+1} - \phi_{j,n}}{\Delta t} + u\frac{\phi_{j+1,n} - \phi_{j,n}}{\Delta x} + \mathcal{O}(\Delta x) + \mathcal{O}(\Delta t) = 0\,.$$

Then if we drop the terms $\mathcal{O}(\Delta x)$ and $\mathcal{O}(\Delta t)$ we obtain a finite difference approximation of the advection equation:

$$\frac{\phi_{j,n+1} - \phi_{j,n}}{\Delta t} + u\frac{\phi_{j+1,n} - \phi_{j,n}}{\Delta x} = 0. \qquad (16.5)$$

The numerical method we discuss now computes a solution to this equation.

Rearrange the terms in equation (16.5) to obtain

$$\phi_{j,n+1} = \phi_{j,n} - \alpha(\phi_{j+1,n} - \phi_{j,n}), \qquad (16.6)$$

where

$$\alpha = u\frac{\Delta t}{\Delta x}.$$

Here you can see that we have the basis of an algorithm for computing the solution at time t_{n+1} in terms of the solution at time t_n. Since the initial condition provides the value of $\phi_{0:J-1,0}$ we can can compute the value of $\phi_{0:J-1,1}$ with equation (16.6), then the value of $\phi_{0:J-1,2}$, and so on. Thus we have Algorithm 1.

[Algorithm 1]

$\phi_{0:J-1,0} = f(x_{0:J-1})$
for $n = 0 : N - 1$
    $\phi_{J,n} = \phi_{0,n}$
    $\phi_{0:J-1,n+1} = \phi_{0:J-1,n} - \alpha(\phi_{1:J,n} - \phi_{0:J-1,n})$

One detail of this algorithm deserves special comment. The update of the value of ϕ at the right boundary, $j = J - 1$, is special because of the periodic boundary conditions; for example,

$$\phi_{J-1,n+1} = \phi_{J-1,n} - \alpha(\phi_{0,n} - \phi_{J-1,n}).$$

Notice that the value of $\phi_{0,n}$ is used on the right side of this expression, so this update does not have the same pattern as the update at the other points,

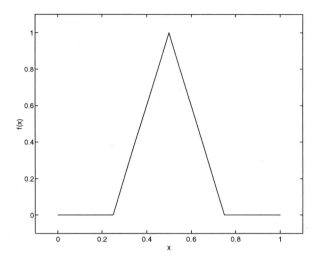

Figure 16.6: The triangular pulse used to specify the initial condition for Algorithm 1.

$j = 0, 1, \ldots J - 2$. However, by copying the value $\phi_{0,n}$ to the point (J, n) (see the first statement in the loop) the update of the right boundary point is done within the single vector statement of the time-stepping loop. Not only does this make the code simpler in appearance, it has the potential for producing a faster computation on a machine with vector arithmetic.

Test of the forward-difference algorithm, Algorithm 1

Unfortunately Algorithm 1 is far from satisfactory as we now see with an example. We use a triangular pulse for the initial condition:

$$f(x) = \begin{cases} 0 & \text{if } 0 \le x \le 0.25 \,; \\ 4x - 1 & \text{if } 0.25 < x \le 0.5 \,; \\ 3 - 4x & \text{if } 0.5 < x \le 0.75 \,; \\ 0 & \text{if } 0.75 < x \le 1 \,. \end{cases} \tag{16.7}$$

This function is illustrated in figure 16.6.

The computed and exact solutions at $t = 0.1$ for $u = 0.4$ and $\Delta t = \Delta x = 0.01$ are shown in figure 16.7. Since $\Delta t = 0.01$ this is the result after ten iterations, $N = 10$ in Algorithm 1. The computed solution at $t = 0.2$ is shown

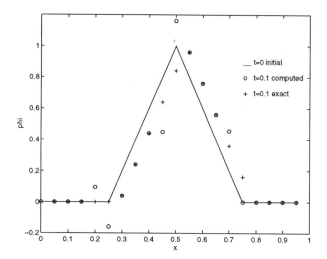

Figure 16.7: The solution computed by Algorithm 1, marked by "o", and the exact solution, marked by "+", at $t = 0.1$.

in figure 16.8. In this second picture the computed solution has grown by two orders of magnitude, so large that the exact solution appears to be on the x-axis in the scale of this figure. Clearly this algorithm is producing completely unsatisfactory results after just twenty time steps. The reason for this behavior is explained shortly, but first we look at a slightly different algorithm that gives much better results.

A backward difference algorithm, Algorithm 2

Again from Taylor's series we have

$$\phi(x - \Delta x, t) = \phi(x, t) - \phi_x(x, t)\Delta x + \mathcal{O}((\Delta x)^2),$$

from which we obtain

$$\phi_x(x, t) = \frac{\phi(x, t) - \phi(x - \Delta x, t)}{\Delta x} + \mathcal{O}(\Delta x). \tag{16.8}$$

The expression $\phi(x, t) - \phi(x - \Delta x, t)$ appearing here is called a "backward difference".

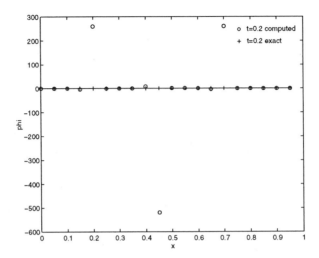

Figure 16.8: The solution computed by Algorithm 1, marked by "o", and the exact solution, marked by "+", at $t = 0.2$.

If we use this new formula, equation (16.8), for $\phi_x(x, t)$ and the old formula, equation (16.4), for $\phi_t(x, t)$ in the advection equation we obtain

$$\frac{\phi_{j,n+1} - \phi_{j,n}}{\Delta t} + u\frac{\phi_{j,n} - \phi_{j-1,n}}{\Delta x} + \mathcal{O}(\Delta x) + \mathcal{O}(\Delta t) = 0 \, .$$

As before, we drop the terms $\mathcal{O}(\Delta x)$ and $\mathcal{O}(\Delta t)$ and arrive at a new formula for computing $\phi_{j,n+1}$:

$$\phi_{j,n+1} = \phi_{j,n} - \alpha(\phi_{j,n} - \phi_{j-1,n}) \, . \tag{16.9}$$

Thus we have:

[Algorithm 2]

$\phi_{0:J-1,0} = f(x_{0:J-1})$
for $n = 0 : N - 1$
$\quad \phi_{-1,n} = \phi_{J-1,n}$
$\quad \phi_{0:J-1,n+1} = \phi_{0:J-1,n} - \alpha(\phi_{0:J-1,n} - \phi_{-1:J-2,n})$

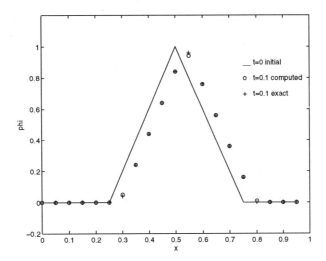

Figure 16.9: The solution computed by Algorithm 2, marked by "o", and the exact solution, marked by "+", at $t = 0.1$.

Notice that we have again used the bordering idea: in this case for the update of $\phi_{0,n}$.

Test of the backward difference algorithm, Algorithm 2

To illustrate the performance of this algorithm we again choose the triangular pulse, figure 16.6, for the initial condition. The computed solution at $t = 0.1$ and $t = 0.2$ for $u = 0.4$ and $\Delta t = \Delta x = 0.01$ is shown in figures 16.9 and 16.10. It is evident that this algorithm produces much better results. Figure 16.11 shows the computed solution after a much longer time, $t = 10$. Notice that the shape of the pulse is damped and spread out quite a bit but the peak is in about the right position: the peak of the exact solution is at $x = 0.5$ because the velocity of the pulse is 0.4 so it will have moved a distance of 4 at $t = 10$, thus returning to its initial position because of the periodic boundary conditions.

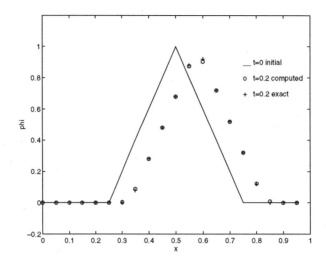

Figure 16.10: The solution computed by Algorithm 2, marked by "o", and the exact solution, marked by "+", at $t = 0.2$.

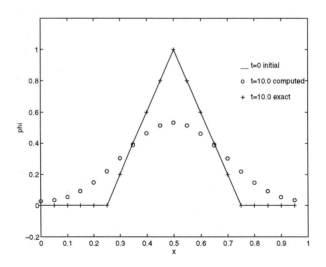

Figure 16.11:   The solution computed by Algorithm 2, marked by "o", and the exact solution, marked by "+", at $t = 10.0$.

Analysis of Algorithms 1 and 2

We use a well-known technique due to von Neumann for analyzing the two methods just described. It assumes that the solution is represented as a Fourier series

$$\phi(x,t) = c_0 + \sum_{\nu=1}^{\infty} (c_\nu \cos(2\pi\nu(x - ut)) + s_\nu \sin(2\pi\nu(x - ut))) \,. \qquad (16.10)$$

This analysis shows how the algorithms act on individual terms in the Fourier series and from this we are able to understand their effect on the triangular pulse and other functions. We should remind you that in this context u is referred to as the *phase velocity* of the wave represented by the sine or cosine term in which it appears.

The triangular pulse we have been using is represented by a Fourier series with coefficients:

$$c_\nu = \left\{ \begin{array}{ll} 1/4 & \text{if } \nu = 0; \\ -4/(\nu\pi)^2 & \text{if } \nu \text{ is odd}; \\ 0 & \text{if } \nu \text{ is divisible by 4}; \\ 8/(\nu\pi)^2 & \text{otherwise} \end{array} \right\} ; \quad s_\nu = 0. \qquad (16.11)$$

For example, taking terms up to $\nu = 7$ we have the approximation

$$\begin{aligned} \phi(x,t) \approx \ & \frac{1}{4} + \frac{1}{\pi^2} \left(-4\cos(2\pi(x - ut)) + \frac{8}{4}\cos(4\pi(x - ut)) \right. \\ & \left. - \frac{4}{9}\cos(6\pi(x - ut)) - \frac{4}{25}\cos(10\pi(x - ut)) \right. \\ & \left. + \frac{8}{36}\cos(12\pi(x - ut)) - \frac{4}{49}\cos(14\pi(x - ut)) \right). \end{aligned}$$

Figure 16.12 shows a plot of this Fourier series approximation to the triangular pulse using terms up to $\nu = 15$, when $t = 0$.

For the analysis that follows it is more convenient to put the Fourier series into exponential form:

$$\phi(x,t) = \sum_{\nu=-\infty}^{+\infty} a_\nu e^{i\omega(x - ut)} \,.$$

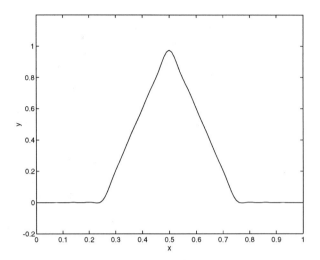

Figure 16.12: Fourier series approximation of the triangle function, using terms up to $\nu = 15$.

For brevity we use $\omega = 2\pi\nu$.

The analysis proceeds by examining the effect of the two algorithms on a single Fourier component, $e^{i\omega(x-ut)}$. The real part of this component represents a cosine wave with frequency ν and phase velocity u; and the imaginary part a sine wave with frequency ν and phase velocity u. By determining the effect of the algorithm on a single Fourier component we are able to infer its effect on the entire Fourier series, and thus its effect on the solution to the advection equation. The following analysis could deal explicitly with the cosine or sine waves but the algebra is simpler if we use the exponential $e^{i\omega(x-ut)}$.

We choose $X = 1$ so that the region R is the rectangle $[0, 1] \times [0, T]$; thus in R_g we have $J\Delta x = 1$ and $N\Delta t = T$. It should be evident from the earlier discussion that if

$$\phi(x, 0) = e^{i\omega x}, \tag{16.12}$$

then the solution to the advection equation is

$$\phi(x, t) = e^{i\omega(x-ut)},$$

the single Fourier component we will consider. Notice that it does satisfy the initial condition equation (16.12) and the periodic boundary conditions.

Now compare the solution generated by Algorithm 1 with this exact solution. For the initial condition we have

$$\phi_{0:J-1,0} = e^{i\omega x_{0:J-1}} , \qquad (16.13)$$

and after one time step

$$\phi_{0:J-1,1} = \phi_{0:J-1,0} - \alpha(e^{i\omega\Delta x}\phi_{0:J-1,0} - \phi_{0:J-1,0}) ,$$

which can be written

$$\phi_{0:J-1,1} = A\phi_{0:J-1,0} ,$$

where

$$A = 1 - \alpha(e^{i\omega\Delta x} - 1) . \qquad (16.14)$$

Thus the effect of a time step, one iteration of Algorithm 1, is to multiply the solution at the present time step by A. Therefore, after n time steps the computed solution is

$$\phi_{0:J-1,n} = A^n \phi_{0:J-1,0} . \qquad (16.15)$$

Notice that the exact solution can be expressed in exactly the same form:

$$\phi_{0:J-1,n}^{(\text{exact})} = B^n \phi_{0:J-1,0} ,$$

where

$$B = e^{-i\omega u\Delta t} .$$

Thus the difference between the computed and exact solutions is contained in the difference between A and B.

In order to see this difference clearly we put the complex number A into polar form, that is

$$A = \rho e^{i\theta} . \qquad (16.16)$$

If the computed solution were equal to the exact solution then $\rho = 1$ and $\theta = -\omega u\Delta t$. Notice that ρ determines whether the amplitude of the computed solution increases with time or decreases with time: $\rho > 1$ implies increasing amplitude, $\rho < 1$ implies decreasing amplitude, and $\rho = 1$ implies that the amplitude does not change.

Substitution of A, given by equation (16.16), and $\phi_{0:J-1,0}$, given by equation (16.13), into equation (16.15) gives

$$\phi_{0:J-1,n} = \rho^n e^{i\omega(x_{0:J-1} + \frac{\theta}{\omega\Delta t} n\Delta t)} .$$

From this formula it is evident that the phase velocity of a wave in the computed solution is

$$u^{(\text{computed})} = -\frac{\theta}{\omega\Delta t} . \tag{16.17}$$

Considering the real and imaginary parts of A in equation (16.14) and some simple algebra we can find ρ and θ of A in equation (16.16):

$$\rho^2 = 1 + 2\alpha(1+\alpha)(1 - \cos(\omega\Delta x)) , \tag{16.18}$$

and

$$\theta = \arctan\left(-\frac{\alpha\sin(\omega\Delta x)}{1 - \alpha(\cos(\omega\Delta x) - 1)} \right) .$$

Therefore

$$\rho \geq 1$$

because $\alpha > 0$ and $(1 - \cos(\omega\Delta x)) \geq 0$; and

$$u^{(\text{computed})} = \arctan\left(\frac{\alpha\sin(\omega\Delta x)}{1 - \alpha(\cos(\omega\Delta x) - 1)} \right) \frac{1}{\omega\Delta t} . \tag{16.19}$$

We see here that the amplitude of every Fourier component increases with time (excepting the special case where $1 - \cos(\omega\Delta x) = 0$). This explains the very large values of the computed solution after just 20 time steps shown in figure 16.8. Figure 16.13 shows the amplification factor ρ as a function of frequency when $u = 0.4$ and $\Delta t = \Delta x = 0.01$. Notice that the higher frequency components are amplified the most; since these are least important so far as the general shape of the exact solution is concerned, their amplification causes a serious distortion of the computed solution.

The dependence of the phase velocity on frequency, equation (16.19) is made evident by a graph, shown in figure 16.14. We see here that the phase velocity is less than the correct value (0.4) at all frequencies, with the low

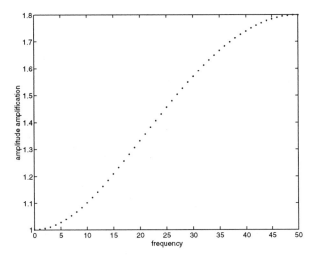

Figure 16.13: The amplification factor ρ of Fourier components for a solution computed by Algorithm 1 shown as a function of frequency, ν, when $u = 0.4$ and $\Delta t = \Delta x = 0.01$.

frequency waves having velocities closest to the correct value. The word "dispersion" is used to describe the fact that waves of different frequency travel at different speeds — the word derives from the fact that a travelling wave form such as the triangular pulse spreads out (disperses) when its Fourier components are travelling at different speeds.

The important conclusion we draw from this analysis is that the error in the computed solution is influenced by two factors — growth in amplitude of the Fourier components and dispersion. This error should decrease as $\Delta x \to 0$ and $\Delta t \to 0$. To confirm this expectation we look at the values of ρ and θ in this limit. In going to the limit it is useful to hold α constant, which means that we must keep the ratio $\Delta t / \Delta x$ constant. From equation (16.18) and a Taylor series expansion of $\cos(\omega \Delta x)$ we have

$$\rho^2 = 1 + 2\alpha(1 + \alpha)(\omega \Delta x)^2/2 + \mathcal{O}((\Delta x)^4) \,;$$

and from equation (16.19) we have

$$
\begin{aligned}
u^{\text{(computed)}} &= \alpha \omega \Delta x \frac{1}{\omega \Delta t} + \mathcal{O}((\Delta x)^2) \,, \\
&= u + \mathcal{O}((\Delta x)^2).
\end{aligned}
$$

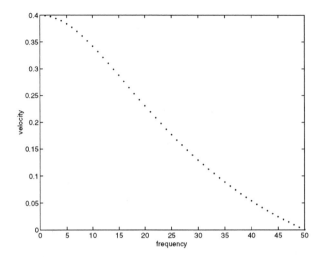

Figure 16.14: The phase velocity for the solution computed by Algorithm 1 shown as a function of frequency, ν, when $u = 0.4$ and $\Delta t = \Delta x = 0.01$.

Thus we can conclude that in the limit

$$\Delta x \to 0, \quad \Delta t \to 0, \quad \text{with } \frac{\Delta t}{\Delta x} \text{ constant,}$$

we have

$$\rho \to 1, \quad u^{(\text{computed})} \to u \,.$$

Let us now turn our attention to Algorithm 2. Having introduced the basic ideas of the analysis already we go immediately to the results. In this case we find

$$\phi_{0:J-1,n} = A^n \phi_{0:J-1,0}, \quad \text{where now } A = 1 - \alpha(1 - e^{-i\omega\Delta x}) \,.$$

Hence

$$\rho^2 = 1 - 2\alpha(1 - \alpha)(1 - \cos(\omega\Delta x)) \,. \tag{16.20}$$

We can distinguish three cases:

$$\rho^2 \begin{cases} < 1 & \text{if } \alpha < 1, \ \omega\Delta x \neq 0; \\ = 1 & \text{if } \alpha = 1; \\ > 1 & \text{if } \alpha > 1. \end{cases}$$

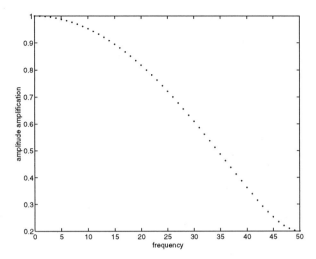

Figure 16.15: The amplification factor ρ of Fourier components for the solution computed by Algorithm 2 shown as a function of frequency, ν, when $u = 0.4$ and $\Delta t = \Delta x = 0.01$.

Therefore the amplitude of the Fourier components (excepting the zero frequency component) decreases with time if $\alpha < 1$; remains constant if $\alpha = 1$; and increases with time if $\alpha > 1$. In fact when $\alpha = 1$ the computed solution *is* the exact solution, as can be seen by replacing α by 1 in equation (16.9).

The phase velocity in the computed solution is given by equation (16.17) where now

$$\theta = \arctan\left(\frac{-\alpha \sin \omega \Delta x}{1 - \alpha(1 - \cos \omega \Delta x)}\right). \tag{16.21}$$

To verify your understanding of this show that the computed velocity is equal to the exact velocity when $\alpha = 1$ using equation (16.17) and equation (16.21).

We again illustrate the dependence of ρ and the phase velocity on frequency: ρ is shown as a function of frequency in figure 16.15; the phase velocity is shown as a function of frequency in figure 16.16. In both of these figures $u = 0.4$ and $\Delta t = \Delta x = 0.01$.

Notice that the decrease in amplitude is most significant for the higher frequencies which are least important for the general shape of the solution. Figure 16.12 shows, for example, that the shape of the triangle pulse is well

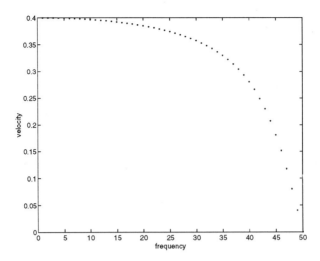

Figure 16.16: The phase velocity for the solution computed by Algorithm 2 shown as a function of frequency, ν, when $u = 0.4$ and $\Delta t = \Delta x = 0.01$.

represented by the first fifteen frequencies. Also, the effect of dispersion is reduced because it is the high frequency components, which are damped, that are travelling significantly more slowly. Thus the general shape of the solution is better preserved in Algorithm 2 than in Algorithm 1.

An analysis like that done for Algorithm 1 shows that Algorithm 2 has the same limiting behavior; that is, $\rho \to 1$ and $u^{(\text{computed})} \to u$ when $\Delta x \to 0$ and $\Delta t \to 0$ with $\Delta t/\Delta x$ constant. Again, you might confirm your understanding of this analysis by confirming this claim.

We might have expected Algorithm 2 would give more accurate results than Algorithm 1 by a loose qualitative argument based on characteristics. Figure 16.17 shows the grid points used to compute a value of the solution at the next time level. The computation in Algorithm 1 uses a weighted average of ϕ at points b and c to determine the value of ϕ at point d, and Algorithm 2 uses a weighted average of the values of ϕ at points a and b to determine the value of ϕ at point d. The line passing through point d in the figure represents a characteristic, therefore we know that the value of ϕ all along this line is the value of ϕ at point d. In particular, at the point where the characteristic passes between a and b, ϕ has the same value that it has at d. Therefore, since

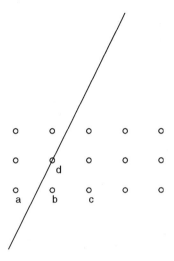

Figure 16.17: Segment of R_g. In Algorithm 1 the values of ϕ at points b and c are used to compute the value of ϕ at point d. In Algorithm 2 the values of ϕ at points a and b are used to compute the value of ϕ at point d. The line going through point d is a characteristic.

a and b lie closer to this point than the pair b and c it is reasonable to expect that a weighted average of values of ϕ at a and b (Algorithm 2) would give the more accurate estimate of ϕ at d.

The technique employed in Algorithm 2, wherein a backward difference is used so that the grid point at $(i - 1, n)$ is used in approximating the partial derivative with respect to x at the grid point (i, n), is called *upwinding*. This name derives from the fact that the point $(i - 1, n)$ is upwind, with respect to the flow velocity, from the point (i, n): note that $u > 0$ implies that the fluid is flowing from $(i - 1, n)$ to (i, n).

An algorithm is called *unstable* when it causes growth in the amplitude of the Fourier components of a solution. Thus we say that Algorithm 1 is unstable, and Algorithm 2 is unstable if $\alpha > 1$, otherwise it is stable. An algorithm is said to be *dissipative* if it causes a decrease in the amplitude of the Fourier coefficients. Thus we say that Algorithm 2 is dissipative if $\alpha < 1$.

It is quite evident that the dimensionless parameter α plays an important role in determining the characteristics of a solution. It is sometimes called the *Courant number* in recognition of the work by Courant, Friedrichs, and

Lewy [Courant et al 67] in which the concept of stability was introduced. The condition $\alpha \leq 1$ for stability is called the "von Neumann condition".

16.3.3   Confirmation of analysis of Algorithm 2

Having analyzed in detail the effect of Algorithm 2 on a Fourier component we now confirm this analysis with some numerical experiments. Because of periodic boundary conditions for $x \in [0, 1]$, our given domain, we know that the exact solution of the advection equation satisfies the equation

$$\phi(x, n/u) = \phi(x, 0), \quad n = 0, 1, 2, \ldots; \qquad (16.22)$$

that is, ϕ returns to its initial value after a time interval of $1/u$. However we know from the analysis above that the amplitude of a Fourier component changes by a factor of ρ at each time step and the velocity of a Fourier component is less than u. Therefore, the numerical solution will not satisfy equation (16.22) and our analysis allows us to predict the difference exactly. It is instructive to run a few tests in which we compare the results of a numerical solution with the predicted results at times $1/u$, $2/u$, and $3/u$.

In these tests we use a single Fourier component and two different frequencies for the initial condition; namely,

$$\phi_{0:J-1,0} = \sin(2\pi\nu(0 : J - 1)\Delta x), \quad \nu = 1, 5. \qquad (16.23)$$

We also use $u = 0.4$, and $\Delta x = \Delta t = 0.01$, hence $\alpha = 0.4$. Therefore the times of interest are 2.5, 5.0, and 7.5.

To make our prediction we must first determine ρ and the wave velocity from equations (16.17, 16.20, 16.21). A little arithmetic produces the results shown in table 16.1.

Now it is easy to predict the amplitude of the solution at each of the three times. For example, since $\Delta t = 0.01$ it takes 250 steps to reach $t = 2.5$; hence the amplitude at this time is ρ^{250}. Thus for $\nu = 1$ the amplitude is 0.8882941 at $t = 2.5$. A summary of the predicted amplitudes is given in table 16.2.

We can also predict the distance the wave will have traveled at $t = 2.5$. For example, if $\nu = 1$ then its velocity is 0.3999684 and so it will have traveled a distance 0.9999210; or, to put it another way, it will *lag* behind its correct

ν	ρ	*phase velocity*
1	0.999526302628	0.399968405541
5	0.988183752053	0.399203038709

Table 16.1: Values of ρ and phase velocity at two different frequencies, ν, as predicted by analysis of Algorithm 2. Parameter values are $u = 0.4$, $\Delta t = \Delta x = 0.01$.

	Amplitude		
ν	$t = 2.5$	$t = 5.0$	$t = 7.5$
1	0.888294097037e+00	0.789066402830e+00	0.700923027804e+00
5	0.512185418526e-01	0.262333902951e-02	0.134363599877e-03

Table 16.2: Predicted amplitude of solution (a sine wave) at times $t = 1/u$, $t = 2/u$, and $t = 3/u$ and frequencies $\nu = 1$ and $\nu = 5$ for Algorithm 2. Parameter values are $u = 0.4$, $\Delta t = \Delta x = 0.01$.

	Distance		
ν	$t = 2.5$	$t = 5.0$	$t = 7.5$
1	0.999921013853	1.999842027706	2.999763041560
5	0.998007596772	1.996015193545	2.994022790318

Table 16.3: Predicted distance traveled by the solution at times $t = 1/u$, $t = 2/u$, and $t = 3/u$ and frequencies $\nu = 1$ and $\nu = 5$ for Algorithm 2. Parameter values are $u = 0.4$, $\Delta t = \Delta x = 0.01$.

position by 0.0000790. A summary of the predicted distances travelled is shown in table 16.3.

Next we compare these predictions with the numerical results. Values of the solution obtained from the execution of Algorithm 2 using the initial condition given above in equation (16.23), with $\nu = 1$, are shown in table 16.4. The values are shown in the neighborhood of the peak and the zero in the solution: note that the exact solution would have a peak at $x = 0.25$ and would be zero at $x = 0.5$ when $t = 2.5$, $t = 5.0$, and $t = 7.5$. It is apparent from simply scanning this table that the peak and zero are approximately where they should be.

To confirm the agreement between the results in table 16.4 with our prediction we can compare the peak value of the solution with the predicted amplitude. This comparison is complicated somewhat because the peak value does not occur exactly at $x = 0.5$ but at a point slightly less than this — our prediction in table 16.3 would put the peak at 0.499921. Since the slope of the solution is zero at the peak the difference in its value at 0.5 from its value at 0.499921 is very small. Therefore we will ignore this difference, and indeed the computed solution (0.88829398764) at $t = 2.5$ and $x = 0.25$ agrees with the predicted amplitude (table 16.2) to six significant figures. The same close agreement occurs at $t = 5.0$; the agreement is to five significant figures at $t = 7.5$, the effect of the lag in position becoming slightly more noticeable at this later time.

We can locate the zero by interpolation. For example, using linear interpolation on the values at $x = 0.49$ and $x = 0.50$, $t = 2.5$, we estimate that the zero is at 0.4999210, indicating a lag in position of 0.0000790 as predicted. We can compare the computed and analytical position of the zero for the other times similarly, however there is a more precise way to make all of these comparisons.

We can simply compute the value of $A \sin(2\pi\nu(x - u't))$ at $x = 0.25$, 0.5 and $t = 2.5$, 5.0, 7.5 using the predicted amplitude for A and the predicted phase velocity for u'. The results of this computation are in table 16.5. Comparison of these results with those shown in table 16.4 show complete agreement to 11 significant figures excepting one case where the agreement is to 10 significant figures a difference that can be attributed to roundoff error.

Results obtained for Algorithm 2 when $\nu = 5$ are shown in table 16.6. Here

t	x	ϕ
2.50E+00	2.40E-01	8.8656882336E-01
2.50E+00	2.50E-01	8.8829398764E-01
2.50E+00	2.60E-01	8.8651346138E-01
⋮	⋮	⋮
2.50E+00	4.90E-01	5.5336464230E-02
2.50E+00	5.00E-01	-4.4084665887E-04
2.50E+00	5.10E-01	-5.6216417728E-02
⋮	⋮	⋮
5.00E+00	2.40E-01	7.8755815032E-01
5.00E+00	2.50E-01	7.8906601414E-01
5.00E+00	2.60E-01	7.8745979489E-01
⋮	⋮	⋮
5.00E+00	4.90E-01	4.8764207570E-02
5.00E+00	5.00E-01	-7.8320287310E-04
5.00E+00	5.10E-01	-5.0327522372E-02
⋮	⋮	⋮
7.50E+00	2.40E-01	6.9960466739E-01
7.50E+00	2.50E-01	7.0092225094E-01
7.50E+00	2.60E-01	6.9947361459E-01
⋮	⋮	⋮
7.50E+00	4.90E-01	4.2969760017E-02
7.50E+00	5.00E-01	-1.0435715192E-03
7.50E+00	5.10E-01	-4.5052784556E-02

Table 16.4: Results obtained from the execution of Algorithm 2 with initial condition $\phi(x,0) = \sin(2\pi x)$: note $\nu = 1$. Column 1 is time (t), column 2 is position (x), column 3 is computed solution $(\phi(x,t))$. Parameter values are $u = 0.4$, $\Delta t = \Delta x = 0.01$.

$A\sin(x - u't)$			
x	$t = 2.5$	$t = 5.0$	$t = 7.5$
0.25	8.8829398764E-01	7.8906601414E-01	7.0092225094E-01
0.50	-4.4084665887E-04	-7.8320287309E-04	-1.0435715192E-03

Table 16.5: Predicted value of the solution at times $t = 2.5$, $t = 5.0$, and $t = 7.5$ using predicted amplitudes A' from table 16.2 and predicted velocity u' from table 16.1.

the results are tabulated only in the neighborhood of the first peak ($x = 0.05$) and the first zero ($x = 0.10$). The predicted values of the solution are shown at the points of interest in table 16.7. Again we see that the the results agree to eleven significant figures.

16.4 Numerical methods: higher order

Here we consider more accurate finite difference algorithms for solving the advection equation. The first of these is a popular scheme known as the *Leapfrog algorithm*, the second is sometimes called the *Trapezoid algorithm*. An important difference between these two algorithms is that Leapfrog is an *explicit method*, meaning the solution at a new time point is obtained by simply evaluating an expression involving values of the solution at an earlier time, or times, just as in the first-order algorithms introduced in the last section. On the other hand the Trapezoid algorithm is an *implicit method*, meaning that it is necessary to solve a system of equations in order to determine the solution at a new time point. Thus an implicit method requires more work at each time step. However, as we see later, an implicit method may be competitive with an explicit method.

16.4.1 Leapfrog algorithm

This algorithm uses second-order approximations to the derivatives rather than the first-order approximations used in Algorithm 1 and Algorithm 2 of the last

t	x	ϕ
2.50E+00	4.00E-02	4.9606375828E-02
2.50E+00	5.00E-02	5.1118239843E-02
2.50E+00	6.00E-02	4.7626294381E-02
$\vdots$	$\vdots$	$\vdots$
2.50E+00	9.00E-02	1.2749372798E-02
2.50E+00	1.00E-01	-3.2038390821E-03
2.50E+00	1.10E-01	-1.8843436870E-02
$\vdots$	$\vdots$	$\vdots$
5.00E+00	4.00E-02	2.5766375562E-03
5.00E+00	5.00E-02	2.6028098598E-03
5.00E+00	6.00E-02	2.3742009995E-03
$\vdots$	$\vdots$	$\vdots$
5.00E+00	9.00E-02	4.9279465093E-04
5.00E+00	1.00E-01	-3.2754922924E-04
5.00E+00	1.10E-01	-1.1158303087E-03
$\vdots$	$\vdots$	$\vdots$
7.50E+00	4.00E-02	1.3329201135E-04
7.50E+00	5.00E-02	1.3200164366E-04
7.50E+00	6.00E-02	1.1779003538E-04
$\vdots$	$\vdots$	$\vdots$
7.50E+00	9.00E-02	1.6935663057E-05
7.50E+00	1.00E-01	-2.5082724013E-05
7.50E+00	1.10E-01	-6.4645839294E-05

Table 16.6: Results obtained from the execution of Algorithm 2 with initial condition $\phi(x,0) = \sin(10\pi x)$: note $\nu = 5$. Column 1 is time (t), column 2 is position (x), column 3 is computed solution $(\phi(x,t))$. Parameter values are $u = 0.4$, $\Delta t = \Delta x = 0.01$.

x	$A\sin(x - u't)$		
	$t = 2.5$	$t = 5.0$	$t = 7.5$
0.05	5.1118239842E-02	2.6028098597E-03	1.3200164366E-04
0.10	-3.2038390821E-03	-3.2754922924E-04	-2.5082724013E-03

Table 16.7: Predicted value of the solution near the first peak and near the first zero at times $t = 2.5$, $t = 5.0$, and $t = 7.5$, using predicted amplitudes A' from table 16.2 and predicted velocity u' from table 16.1.

section. Proceeding as before from Taylor's series we have

$$\phi(x + \Delta x, t) = \phi(x, t) + \phi_x(x, t)\Delta x + \frac{1}{2}\phi_{xx}(x, t)(\Delta x)^2 + \mathcal{O}((\Delta x)^3),$$

$$\phi(x - \Delta x, t) = \phi(x, t) - \phi_x(x, t)\Delta x + \frac{1}{2}\phi_{xx}(x, t)(\Delta x)^2 - \mathcal{O}((\Delta x)^3).$$

Subtracting the second equation from the first we have

$$\phi_x(x, t) = \frac{\phi(x + \Delta x, t) - \phi(x - \Delta x, t)}{2\Delta x} + \mathcal{O}((\Delta x)^2).$$

If the term $\mathcal{O}((\Delta x)^2)$ is deleted we have the central difference approximation of $\phi_x(x, t)$.

In a similar way we can obtain the following expression for the partial derivative with respect to t:

$$\phi_t(x, t) = \frac{\phi(x, t + \Delta t) - \phi(x, t - \Delta t)}{2\Delta t} + \mathcal{O}((\Delta t)^2),$$

and if the term $\mathcal{O}((\Delta t)^2)$ is deleted, we have the central difference approximation of $\phi_t(x, t)$.

Substitution of these expressions for the partial derivatives into the advection equation gives

$$\frac{\phi(x, t + \Delta t) - \phi(x, t - \Delta t)}{2\Delta t} + u\frac{\phi(x + \Delta x, t) - \phi(x - \Delta x, t)}{2\Delta x}$$
$$+ \mathcal{O}((\Delta x)^2) + \mathcal{O}((\Delta t)^2) = 0.$$

Now if we drop the terms $\mathcal{O}((\Delta x)^2)$ and $\mathcal{O}((\Delta t)^2)$ and confine x and t values to the grid space R_g we obtain this finite difference approximation to the advection equation:

$$\frac{\phi_{j,n+1} - \phi_{j,n-1}}{2\Delta t} + u\frac{\phi_{j+1,n} - \phi_{j-1,n}}{2\Delta x} = 0. \qquad (16.24)$$

Rearranging terms in this equation to give

$$\phi_{j,n+1} = \phi_{j,n-1} + \alpha(\phi_{j+1,n} - \phi_{j-1,n}), \qquad (16.25)$$

we have the basic formula for the Leapfrog algorithm. Its name derives from the fact that in computing $\phi_{j,n+1}$ the algorithm ignores (leaps over) the grid point (j, n), the point at which the partial derivatives are being approximated in equation (16.24). The template used in the update is shown in figure 16.18. To verify your understanding of this formula you might try deriving it under the assumption that the velocity is not constant, using equation (16.3).

Since equation (16.25) is based on a finite difference approximation to the advection equation having an error that is second order in Δt we say that the accuracy of equation (16.25), and the Leapfrog algorithm, is *second order*. See the book by Richtmyer and Morton [Richtmyer & Morton 67, pages 67,68] for a more detailed explanation of "order of accuracy" of finite difference approximations to partial differential equations.

Equation (16.25) requires values at two time levels, n and $n-1$, in order to compute the solution at the next time level, $n+1$. It is apparent that something special is required to start an algorithm based on this equation because the initial condition specifies the value of ϕ at only one time level, $n = 0$. A simple solution to this problem uses a forward difference approximation for ϕ_t in the first step, keeping the central difference approximation for ϕ_x. This leads to the following formula for the first time step:

$$\phi_{j,1} = \phi_{j,0} - \frac{\alpha}{2}(\phi_{j+1,0} - \phi_{j-1,0}). \qquad (16.26)$$

The penalty for using this formula is a slightly larger error in the first step, and it is reasonable to expect that the overall accuracy of the computation is not significantly affected by this.

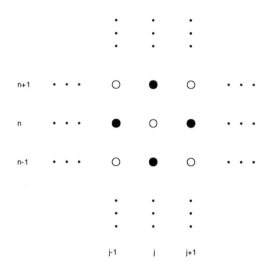

Figure 16.18: The set of points (template) used in a Leapfrog computation. The value of the solution at the point $(j, n+1)$ is computed from values of the solution at the points $(j-1, n)$, $(j+1, n)$, $(j, n-1)$.

[Algorithm 3: Leapfrog]

$\phi_{0:J-1,0} = f(x_{0:J-1})$

Insert border values for periodic boundary conditions:
$\phi_{-1,0} = \phi_{J-1,0}$, $\phi_{J,0} = \phi_{0,0}$

First time step:
$\phi_{0:J-1,1} = \phi_{0:J-1,0} - \frac{\alpha}{2}(\phi_{1:J,0} - \phi_{-1:J-2,0})$

Main time stepping loop:
for $n = 1:N-1$
    $\phi_{-1,n} = \phi_{J-1,n}$, $\phi_{J,n} = \phi_{0,n}$
    $\phi_{0:J-1,n+1} = \phi_{0:J-1,n-1} - \alpha(\phi_{1:J,n} - \phi_{-1:J-2,n})$

Figure 16.19: Leapfrog: Algorithm 3.

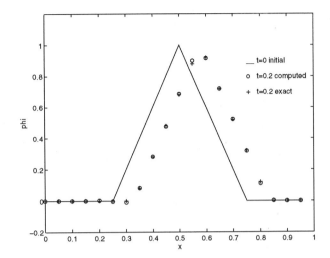

Figure 16.20: The solution computed by the Leapfrog algorithm, marked by "o", and the exact solution, marked by "+", at $t = 0.2$; with $\Delta t = \Delta x = 0.01$ and $u = 0.4$.

Thus we arrive at the Leapfrog algorithm shown in figure 16.19 that is based on equations (16.25, 16.26). Note that the amount of arithmetic required in each time step of this algorithm is the same as for Algorithms 1 and 2.

Figures 16.20 and 16.21 show the computed solution and the exact solution at $t = 0.2$ and $t = 10$, using the same triangular pulse, figure 16.6, for the initial values as in the earlier examples. Comparison of these results with those obtained earlier for Algorithm 2, figures 16.10 and 16.11, shows a slight improvement in accuracy at $t = 0.2$ and a large improvement at $t = 10$. Notice especially, for the Leapfrog algorithm, that the peak at $t = 10$ has not decreased nearly so much, and the shape of the curve is better preserved.

16.4.2 Analysis of the Leapfrog algorithm

We analyze the Leapfrog algorithm almost like we did for Algorithms 1 and 2: we again consider the effect of the algorithm on a single Fourier component, but the computation of A is a little different because the formula for the Leapfrog algorithm involves three time levels, rather than only the two that were required in Algorithms 1 and 2.

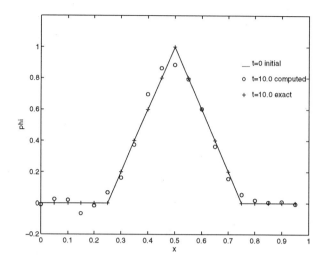

Figure 16.21: The solution computed by the Leapfrog algorithm, marked by "o", and the exact solution, marked by "+", at $t = 10.0$; with $\Delta t = \Delta x = 0.01$ and $u = 0.4$.

Skipping a few of the initial steps, which should now be familiar to you, we have the following expression for the computed solution:

$$\phi_{0:J-1,n} = A^n \phi_{0:J-1,0}\,.$$

Next we determine A by substituting this solution into equation (16.25), the Leapfrog equation. This yields

$$A^{n+1}\phi_{0:J-1,0} = A^{n-1}\phi_{0:J-1,0} - \alpha A^n \phi_{0:J-1,0}\left(e^{i\omega\Delta x} - e^{-i\omega\Delta x}\right).$$

Move the terms to the left side, do a little more algebra, and this becomes

$$A^2 + 2i\alpha A \sin(\omega\Delta x) - 1 = 0\,.$$

The solutions of this quadratic equation are:

$$\begin{aligned}
A_+ &= -i\alpha \sin(\omega\Delta x) + (1 - \alpha^2 \sin^2(\omega\Delta x))^{1/2}\,, & (16.27)\\
A_- &= -i\alpha \sin(\omega\Delta x) - (1 - \alpha^2 \sin^2(\omega\Delta x))^{1/2}\,. & (16.28)
\end{aligned}$$

Therefore we have two solutions of equation (16.25), the "+" solution

$$\phi_{0:J-1,n} = A_+^n \phi_{0:J-1,0}\,,$$

and the "-" solution

$$\phi_{0:J-1,n} = A_-^n \phi_{0:J-1,0}.$$

Furthermore, since equation (16.25) is linear, any linear combination of these two solutions is also a solution of this equation. Thus the general solution of equation (16.25) is

$$\phi_{0:J-1,n} = c_1 A_+^n \phi_{0:J-1,0} + c_2 A_-^n \phi_{0:J-1,0},$$

where c_1 and c_2 are constants determined by the initial conditions. We return to this point later but now let us look closely at the "+" and "-" solutions.

Following the path taken earlier we put A_+ and A_- into polar form:

$$A_+ = \rho_+ e^{i\theta_+}, \quad A_- = \rho_- e^{i\theta_-}. \tag{16.29}$$

Then we determine ρ_+, θ_+, ρ_-, and θ_- from equations (16.27) and (16.28). Notice that the square root term in these equations,

$$(1 - \alpha^2 \sin^2(\omega \Delta x))^{1/2},$$

can be real or imaginary according to the magnitude of α, therefore we distinguish two cases: $\alpha \leq 1$, and $\alpha > 1$. If $\alpha \leq 1$ it is real for all values of $\omega \Delta x$; if $\alpha > 1$ it is imaginary for some values of $\omega \Delta x$. We see that this distinction is crucial for stability: the case $\alpha \leq 1$ giving a stable algorithm; the case $\alpha > 1$ giving an unstable algorithm.

The case $\alpha \leq 1$

If $\alpha \leq 1$ then with some straightforward algebra

$$\rho_+ = 1, \quad \theta_+ = -\arctan\left(\frac{\alpha \sin(\omega \Delta x)}{(1 - \alpha^2 \sin^2(\omega \Delta x))^{1/2}}\right),$$

and

$$\rho_- = 1, \quad \theta_- = \arctan\left(\frac{\alpha \sin(\omega \Delta x)}{(1 - \alpha^2 \sin^2(\omega \Delta x))^{1/2}}\right) + \pi.$$

Since ρ_+ and ρ_- are both unity, the solutions corresponding to A_+ and A_- are stable and nondissipative.

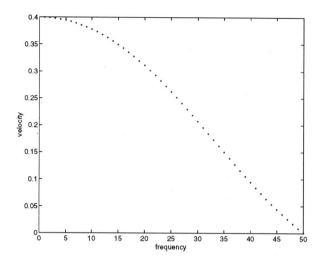

Figure 16.22: The phase velocity for the solution computed by the Leapfrog algorithm, "+" solution, shown as a function of frequency, ν, when $u = 0.4$ and $\Delta t = \Delta x = 0.01$.

Following our earlier consideration of θ in Algorithms 1 and 2, the phase velocity is given by the the following equation:

$$
\begin{aligned}
u_+^{(\text{computed})} &= -\frac{\theta_+}{\omega \Delta t}, \\
&= \arctan\left(\frac{\alpha \sin(\omega \Delta x)}{(1 - \alpha^2 \sin^2(\omega \Delta x))^{1/2}}\right) \frac{1}{\omega \Delta t}.
\end{aligned}
\tag{16.30}
$$

Figure 16.22 shows the phase velocity, $u_+^{(\text{computed})}$, as a function of frequency for $\alpha = 0.4$. This should be compared with the corresponding figures, figures 16.14 and 16.16, for Algorithms 1 and 2. Notice that the dispersion in the low frequency waves is comparable to Algorithm 1 and significantly worse than for Algorithm 2.

The consideration of θ_- is a little different because of the term π. If we split it off so that we have

$$
e^{i\theta_-} = e^{i\pi} e^{i(\theta_- - \pi)} = -e^{i(\theta_- - \pi)}
\tag{16.31}
$$

then the phase velocity for the "-" solution is given by

$$
\begin{aligned}
u_-^{(\text{computed})} &= -\frac{\theta_- - \pi}{\omega \Delta t} \\
&= -\arctan\left(\frac{\alpha \sin(\omega \Delta x)}{(1 - \alpha^2 \sin^2(\omega \Delta x))^{1/2}}\right) \frac{1}{\omega \Delta t} \\
&= -u_+^{(\text{computed})}.
\end{aligned}
$$

Thus the "-" solution can be seen as a wave moving with the same speed as the "+" solution but in the opposite direction. This wave is somewhat peculiar because it changes sign at every step due to the factor $e^{i\pi}(= -1)$ which we had split off: note that from equations (16.29, 16.31)

$$
A_- = -\rho_- e^{i(\theta_- - \pi)}.
$$

If we observe the "-" solution only at every other time step, then it looks like a normal wave travelling in the direction of $-x$.

Considering the limiting behavior with $\Delta x \to 0$ and $\Delta t \to 0$, as before, we find

$$
u_+^{(\text{computed})} = u + \mathcal{O}((\omega \Delta x)^2).
$$

Thus the "+" solution converges to the exact solution as the grid spacing goes to zero. The "-" solution is an artifact of the numerical method and is not physically meaningful.

In the special case that $\alpha = 1$, the "+" solution and the exact solution are identical, for we have

$$
\begin{aligned}
u_+^{(\text{computed})} &= -\frac{\theta_+}{\omega \Delta t}, \\
&= \arctan\left(\frac{\sin(\omega \Delta x)}{(1 - \sin^2(\omega \Delta x))^{1/2}}\right) \frac{1}{\omega \Delta t}, \\
&= \arctan\left(\frac{\sin(\omega \Delta x)}{\cos(\omega \Delta x)}\right) \frac{1}{\omega \Delta t}, \\
&= \frac{\omega \Delta x}{\omega \Delta t}, \\
&= u.
\end{aligned}
$$

The final line in this sequence of equalities may need clarification: recall that $\alpha = u\Delta t/\Delta x$, consequently $\alpha = 1$ implies $u = \Delta x/\Delta t$. Notice that in this case a particle moving with the fluid moves from one grid point to the next in each time step – thus we might have expected something special in this case.

The case $\alpha > 1$

If $\alpha > 1$ then it is possible that $\alpha^2 \sin^2(\omega\Delta x) > 1$ and in this case A_+ and A_- are pure imaginary (see equations 16.27 and 16.28). In particular we have

$$
\begin{aligned}
A_+ &= i\left(-\alpha\sin(\omega\Delta x) + (\alpha^2\sin^2(\omega\Delta x) - 1)^{1/2}\right), \\
A_- &= i\left(-\alpha\sin(\omega\Delta x) - (\alpha^2\sin^2(\omega\Delta x) - 1)^{1/2}\right).
\end{aligned}
$$

hence

$$
\rho_+ = \alpha\sin(\omega\Delta x) - (\alpha^2\sin^2(\omega\Delta x) - 1)^{1/2}, \quad \theta_+ = -\frac{\pi}{2},
$$

and

$$
\rho_- = \alpha\sin(\omega\Delta x) + (\alpha^2\sin^2(\omega\Delta x) - 1)^{1/2}, \quad \theta_- = -\frac{\pi}{2}.
$$

It is evident that $\rho_- > 1$, hence *the Leapfrog algorithm is unstable when $\alpha > 1$.*

The general solution

We have seen that the numerical solution provided by the Leapfrog algorithm is a linear combination of the "+" solution and the "-" solution:

$$
\phi_{0:J-1,n} = c_1 A_+^n \phi_{0:J-1,0} + c_2 A_-^n \phi_{0:J-1,0} . \tag{16.32}
$$

Also, we have seen that if $\alpha \le 1$ then the "+" solution is a reasonably good approximation to the solution, and the "-" solution is a bad approximation. Therefore c_2 must be small relative to c_1 if the Leapfrog algorithm is to be accurate. We turn our attention now to determining these two constants.

From equation (16.32) and the values for $\phi_{0:J-1,0}$ and $\phi_{0:J-1,1}$ we solve the following system of equations for c_1 and c_2:

$$
\begin{aligned}
\phi_{0:J-1,0} &= c_1 A_+^0 \phi_{0:J-1,0} + c_2 A_-^0 \phi_{0:J-1,0}, \\
\phi_{0:J-1,1} &= c_1 A_+^1 \phi_{0:J-1,0} + c_2 A_-^1 \phi_{0:J-1,0} .
\end{aligned}
$$

The first of these equations easily reduces to

$$c_1 + c_2 = 1 .$$

The second equation, making use of the special first step, becomes

$$\phi_{0:J-1,0} - \frac{\alpha}{2}(\phi_{1:J,0} - \phi_{-1:J-2,0}) = c_1 A_+ \phi_{0:J-1,0} + c_2 A_- \phi_{0:J-1,0} .$$

Then, using a single Fourier component for the initial condition,

$$\phi_{0:J-1,0} = e^{i\omega x_{0:J-1}} ,$$

the second equation becomes

$$c_1 A_+ + c_2 A_- = 1 - i\alpha \sin(\omega \Delta x) .$$

Substitution of the values for A_+ and $A-$ and then solving the two equations for c_1 and c_2 gives

$$c_1 = \frac{1}{2}(1 + (1 - \alpha^2 \sin^2(\omega \Delta x))^{-1/2}) ,$$

$$c_2 = \frac{1}{2}(1 - (1 - \alpha^2 \sin^2(\omega \Delta x))^{-1/2}) .$$

Therefore as $\Delta x \to 0$, keeping α constant, $c_1 \to 1$ and $c_2 \to 0$, and we see that the "+" solution dominates in this limit. More specifically, we have

$$c_1 = 1 + \left(\frac{\alpha \omega \Delta x}{2}\right)^2 + \mathcal{O}((\omega \Delta x)^4) ,$$

$$c_2 = -\left(\frac{\alpha \omega \Delta x}{2}\right)^2 + \mathcal{O}((\omega \Delta x)^4) .$$

Thus we see that the presence of high-frequency components not only causes dispersion but also causes the unwanted "-" solution to become a significant part of the solution.

16.4.3 Numerical experiments with the Leapfrog algorithm

Here we illustrate some additional results obtained from using the Leapfrog algorithm with different initial conditions, longer execution times, different velocities and different sets of values for Δx and Δt.

Sine wave

Here we use

$$\phi_{0:J-1,0} = \sin(\omega x_{0:J-1})$$

for the initial condition. Figures 16.23 and 16.24 show results for the velocities $u = 0.4$ and $u = 0.8$, respectively. In each figure the initial condition and the computed solution at $t = 20$ and $t = 40$ are shown. Note that, because of the periodic boundary conditions, the exact solution $\phi(x, t)$ is identical to the initial condition $\phi(x, 0)$ at any time t for which the distance ut is an integer. Therefore the three curves representing the initial condition, the exact solution at $t = 20$, and the exact solution at $t = 40$ should coincide for both $u = 0.4$ and $u = 0.8$. Consequently the separation of the curve for the computed solution from that for the initial condition represents the error in the computed solution.

The effect of dispersion is apparent in these figures: at $t = 20$ the sine wave is lagging behind the initial sine wave and at $t = 40$ it is lagging still further. Reducing the size of Δx and Δt reduces the lag as we would expect (recall equation (16.30)). We do not see any change in the amplitude of the wave with time which is also expected since the amplification factor is 1.

Comparison of results for the two different velocities shows a slight decrease in the dispersion for the higher velocity. You can use equation (16.30) to determine what this decrease should be.

Dispersion increases as the frequency of the sine wave increases (recall figure 16.22). This effect is illustrated in the next set of results, figures 16.25 and 16.26, where the frequency of the sine wave is 3.

Triangular pulse

Now we present a similar set of results for the case when $\phi(x, 0)$ is the triangular pulse, equation (16.7). They are shown in figures 16.27 and 16.28. The same parameter sets as for the sine function are used. As we have seen already dispersion distorts the triangular shape, though not nearly so much as it did for Algorithms 1 and 2.

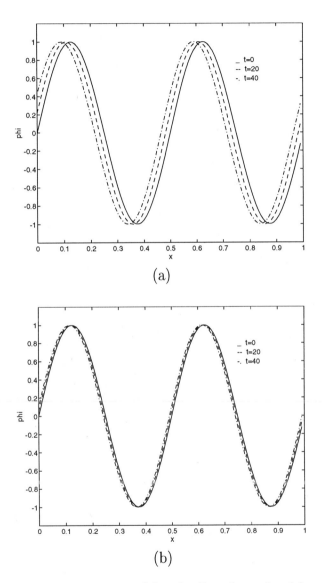

(a)

(b)

Figure 16.23: The solution computed by the Leapfrog algorithm at $t = 20$ and $t = 40$ with $u = 0.4$; with $\Delta x = \Delta t = 0.01$ (a), and $\Delta x = \Delta t = 0.005$ (b). The initial condition is $\phi(x, 0) = \sin(4\pi x)$.

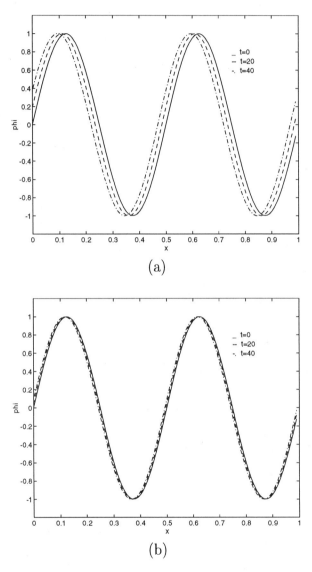

Figure 16.24: The solution computed by the Leapfrog algorithm at $t = 20$ and $t = 40$ with $u = 0.8$; with $\Delta x = \Delta t = 0.01$ (a), and $\Delta x = \Delta t = 0.005$ (b). The initial condition is $\phi(x, 0) = \sin(4\pi x)$.

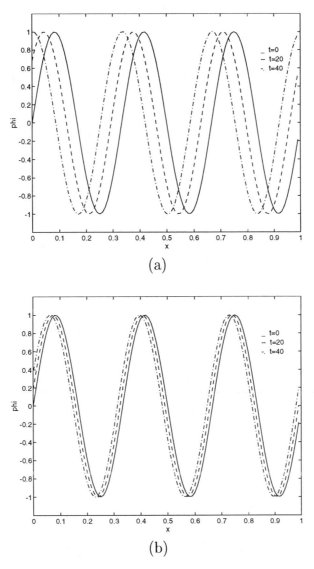

Figure 16.25: The solution computed by the Leapfrog algorithm at $t = 20$ and $t = 40$ with $u = 0.4$; with $\Delta x = \Delta t = 0.01$ (a), and $\Delta x = \Delta t = 0.005$ (b). The initial condition is $\phi(x, 0) = \sin(6\pi x)$.

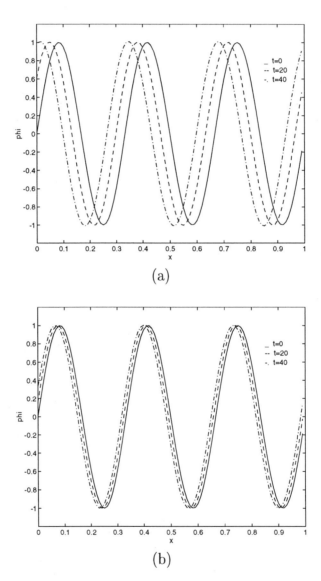

Figure 16.26: The solution computed by the Leapfrog algorithm at $t = 20$ and $t = 40$ with $u = 0.8$; with $\Delta x = \Delta t = 0.01$ (a), and $\Delta x = \Delta t = 0.005$ (b). The initial condition is $\phi(x, 0) = \sin(6\pi x)$.

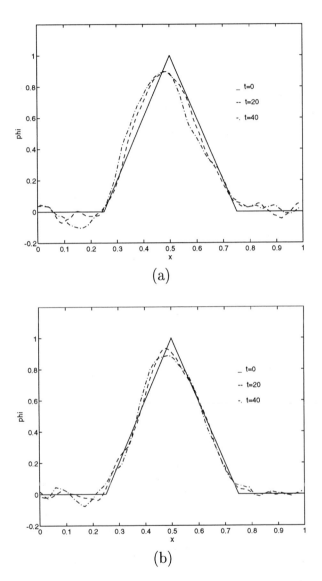

Figure 16.27: The solution computed by the Leapfrog algorithm at $t = 20$ and $t = 40$ with $u = 0.4$; with $\Delta x = \Delta t = 0.01$ (a), and $\Delta x = \Delta t = 0.005$ (b). The initial condition for ϕ is the triangular pulse.

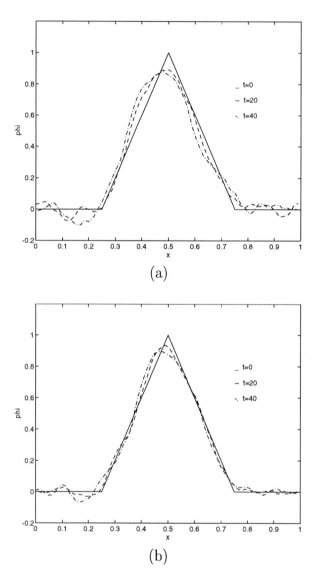

Figure 16.28: The solution computed by the Leapfrog algorithm at $t = 20$ and $t = 40$ with $u = 0.8$; with $\Delta x = \Delta t = 0.01$ (a), and $\Delta x = \Delta t = 0.005$ (b). The initial condition for ϕ is the triangular pulse.

Square pulse

The square pulse is zero on the interval [0,1] except on a subinterval in the center where it is equal to 1; specifically,

$$f(x) = \begin{cases} 0 & \text{if } 0 \le x \le 0.25; \\ 1 & \text{if } 0.25 < x < 0.75; \\ 0 & \text{if } 0.75 \le x \le 1. \end{cases}$$

The results displayed in figures 16.29 and 16.30 use this square pulse for the initial condition. The first figure uses velocity $u = 0.4$, the second uses velocity $u = 0.8$.

It is evident that there is a much greater distortion of the square pulse than the triangular pulse. Notice that even when the smaller values of Δx and Δt are used, figures 16.29b and 16.30b, the distortion remains rather large. The distortion is caused by dispersion. The reason the dispersion is worse for the square pulse than for the triangular pulse is that the amplitude of the high frequency components of the square pulse are larger than for the triangular pulse, hence the effect of dispersion is more noticeable.

We can make this more explicit by looking at the two sets of Fourier coefficients. We saw earlier, equations (16.10) and (16.11), that the Fourier coefficients for the triangular pulse are

$$c_\nu = \begin{cases} 1/4 & \text{if } \nu = 0; \\ -4/(\nu\pi)^2 & \text{if } \nu \text{ is odd}; \\ 0 & \text{if } \nu \text{ is divisible by 4}; \\ 8/(\nu\pi)^2 & \text{otherwise} \end{cases}, \quad s_\nu = 0.$$

The Fourier coefficients for the square pulse are

$$c_\nu = \begin{cases} 1/2 & \text{if } \nu = 0; \\ -2/\nu\pi & \text{if } \nu \text{ is } 1, 5, 9, \ldots; \\ 2/\nu\pi & \text{if } \nu \text{ is } 3, 7, 11, \ldots; \\ 0 & \text{if } \nu \text{ is even} \end{cases}, \quad s_\nu = 0.$$

The important thing to notice in comparing these two sets of coefficients is that the coefficients for the triangular pulse decrease proportional to $(1/\nu)^2$, while those for the square pulse decrease proportional to $1/\nu$. Thus high frequencies are much stronger in the square pulse.

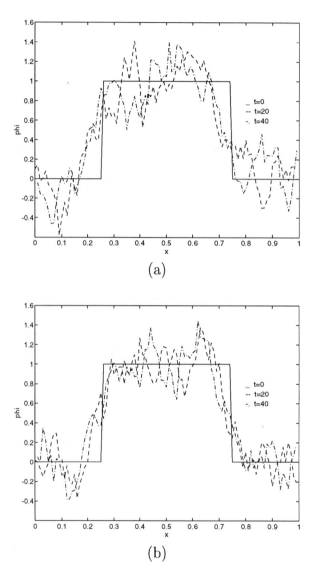

(a)

(b)

Figure 16.29: The solution computed by the Leapfrog algorithm at $t = 20$ and $t = 40$ with $u = 0.4$; with $\Delta x = \Delta t = 0.01$ (a), and $\Delta x = \Delta t = 0.005$ (b). The initial condition for ϕ is the square pulse.

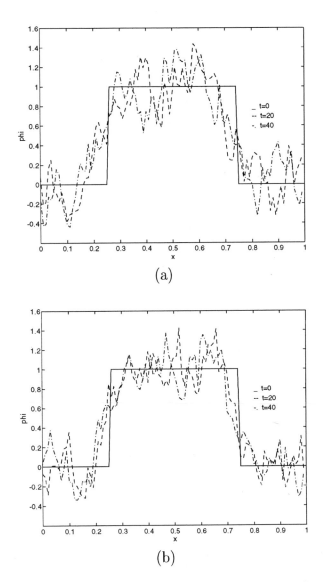

Figure 16.30: The solution computed by the Leapfrog algorithm at $t = 20$ and $t = 40$ with $u = 0.8$; with $\Delta x = \Delta t = 0.01$ (a), and $\Delta x = \Delta t = 0.005$ (b). The initial condition for ϕ is the square pulse.

Gaussian pulse

Our final example uses a Gaussian pulse for the initial condition, defined by

$$f(x) = e^{-250(x-1/2)^2}.$$

The results obtained from using this function as the initial condition are displayed in figures 16.31 and 16.32; the first is for velocity $u = 0.4$, the second is for velocity $u = 0.8$.

Here we can see a significant improvement in the accuracy of the results when Δx and Δt are decreased to 0.005. However, the peak lags behind its correct position, a result of the slower speed of the Fourier components and we see substantial error on the upstream side (left side of the peak) caused by dispersion

16.4.4 Trapezoid algorithm

The Trapezoid algorithm is based on finite difference approximations of the partial derivatives ϕ_x and ϕ_t at intermediate points in the grid. In particular, to obtain the solution at time $n + 1$ we use approximations of the partial derivatives at $(j, n+1/2)$, $j = 0, 1, 2, \ldots, J-1$. The points used in the approximation of ϕ_x and ϕ_t at $(j, n + 1/2)$ are shown as filled circles in figure 16.33.

These approximations are

$$\phi_t(j\Delta x, (n + 1/2)\Delta t) \approx \frac{\phi_{j,n+1} - \phi_{j,n}}{\Delta t},$$

$$\phi_x(j\Delta x, (n + 1/2)\Delta t) \approx \frac{1}{2}\left(\frac{\phi_{j+1,n} - \phi_{j-1,n}}{2\Delta x} + \frac{\phi_{j+1,n+1} - \phi_{j-1,n+1}}{2\Delta x}\right).$$

The origin of the first of these should be evident from the earlier discussions. The second is easily seen to be the arithmetic mean of two approximations to ϕ_x, one at time level n, the other at time level $n + 1$. Substitution of these approximations into the advection equation gives

$$\frac{\phi_{j,n+1} - \phi_{j,n}}{\Delta t} + \frac{u}{2}\left(\frac{\phi_{j+1,n} - \phi_{j-1,n}}{2\Delta x} + \frac{\phi_{j+1,n+1} - \phi_{j-1,n+1}}{2\Delta x}\right) = 0. \quad (16.33)$$

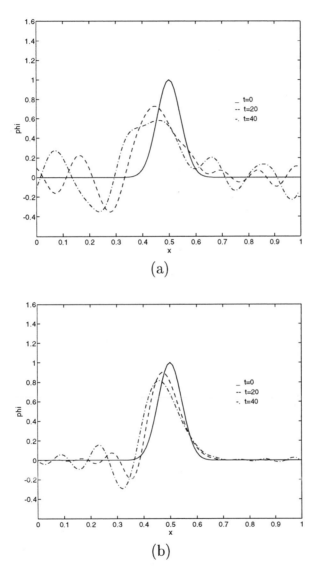

Figure 16.31: The solution computed by the Leapfrog algorithm at $t = 20$ and $t = 40$ with $u = 0.4$; with $\Delta x = \Delta t = 0.01$ (a), and $\Delta x = \Delta t = 0.005$ (b). The initial condition for ϕ is the Gaussian pulse.

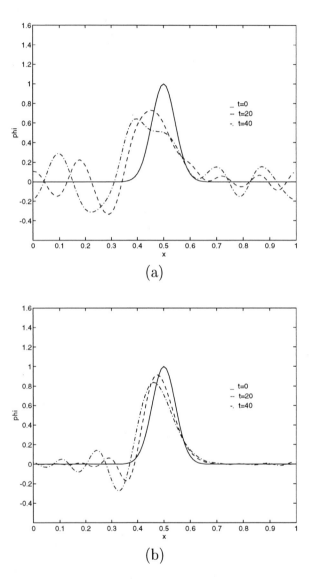

Figure 16.32: The solution computed by the Leapfrog algorithm at $t = 20$ and $t = 40$ with $u = 0.8$; with $\Delta x = \Delta t = 0.01$ (a), and $\Delta x = \Delta t = 0.005$ (b). The initial condition for ϕ is the Gaussian pulse.

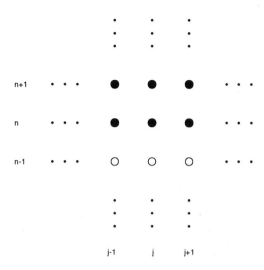

Figure 16.33: The Trapezoid algorithm is based on finite difference approximations of the partial derivatives ϕ_x and ϕ_t at $(j, n + 1/2)$ in this figure. The template for the Trapezoid algorithm is indicated by the filled circles.

This equation is the basis of the Trapezoid algorithm. Using techniques from our earlier discussion you can verify that this equation differs from the advection equation by an error term that is $\mathcal{O}((\Delta t)^2)$. Thus it follows that the accuracy of the Trapezoid algorithm is second order, the same as for the Leapfrog algorithm.

This equation involves values of ϕ at only one time level, n, to determine the solution at the next time level, $n + 1$. In order to see the essence of the Trapezoid algorithm let us put all of the terms involving time level $n + 1$ on the left side and the terms involving time level n on the right side, thus

$$-\alpha\phi_{j-1,n+1} + 4\phi_{j,n+1} + \alpha\phi_{j+1,n+1} = \alpha\phi_{j-1,n} + 4\phi_{j,n} - \alpha\phi_{j+1,n}, \quad (16.34)$$

where $\alpha = u\Delta t/\Delta x$ as before. Clearly, we cannot use this equation by itself to determine a value of ϕ at time level $n + 1$ from values of ϕ we already know at time level n. We have to solve a system of equations, obtained by writing equation (16.34) for $j = 0, 1, 2, \ldots, J - 1$.

These equations are illustrated with a small example, $J = 5$. For $j = 0$ we have

$$-\alpha\phi_{4,n+1} + 4\phi_{0,n+1} + \alpha\phi_{1,n+1} = \alpha\phi_{4,n} + 4\phi_{0,n} - \alpha\phi_{1,n}.$$

Here the periodic boundary conditions were used to replace $\phi_{-1,n+1}$ with $\phi_{4,n+1}$ and $\phi_{-1,n}$ with $\phi_{4,n}$. For $j = 1$ we have

$$-\alpha\phi_{0,n+1} + 4\phi_{1,n+1} + \alpha\phi_{2,n+1} = \alpha\phi_{0,n} + 4\phi_{1,n} - \alpha\phi_{2,n} \, ;$$

and so on. Thus the system of equations we obtain is (in matrix form)

$$\begin{pmatrix} 4 & \alpha & 0 & 0 & -\alpha \\ -\alpha & 4 & \alpha & 0 & 0 \\ 0 & -\alpha & 4 & \alpha & 0 \\ 0 & 0 & -\alpha & 4 & \alpha \\ \alpha & 0 & 0 & -\alpha & 4 \end{pmatrix} \begin{pmatrix} \phi_{0,n+1} \\ \phi_{1,n+1} \\ \phi_{2,n+1} \\ \phi_{3,n+1} \\ \phi_{4,n+1} \end{pmatrix} = \begin{pmatrix} \alpha\phi_{4,n} + 4\phi_{0,n} - \alpha\phi_{1,n} \\ \alpha\phi_{0,n} + 4\phi_{1,n} - \alpha\phi_{2,n} \\ \alpha\phi_{1,n} + 4\phi_{2,n} - \alpha\phi_{3,n} \\ \alpha\phi_{2,n} + 4\phi_{3,n} - \alpha\phi_{4,n} \\ \alpha\phi_{3,n} + 4\phi_{4,n} - \alpha\phi_{0,n} \end{pmatrix}.$$

$$(16.35)$$

This system must be solved in order to advance the solution one time step. Notice that nothing special needs to be done at the start. The initial conditions are used to evaluate the right-hand side of equation (16.35) for $n = 0$, then the equations can be solved to obtain the solution for $n = 1$. This solution can be substituted into the right-hand side and the equations solved again to obtain the solution at $n = 2$, and so on.

The right-hand side vector can be expressed as a matrix vector product very similar to the one on the left-hand side, namely

$$\begin{pmatrix} \alpha\phi_{4,n} + 4\phi_{0,n} - \alpha\phi_{1,n} \\ \alpha\phi_{0,n} + 4\phi_{1,n} - \alpha\phi_{2,n} \\ \alpha\phi_{1,n} + 4\phi_{2,n} - \alpha\phi_{3,n} \\ \alpha\phi_{2,n} + 4\phi_{3,n} - \alpha\phi_{4,n} \\ \alpha\phi_{3,n} + 4\phi_{4,n} - \alpha\phi_{0,n} \end{pmatrix} = \begin{pmatrix} 4 & -\alpha & 0 & 0 & \alpha \\ \alpha & 4 & -\alpha & 0 & 0 \\ 0 & \alpha & 4 & -\alpha & 0 \\ 0 & 0 & \alpha & 4 & -\alpha \\ -\alpha & 0 & 0 & \alpha & 4 \end{pmatrix} \begin{pmatrix} \phi_{0,n} \\ \phi_{1,n} \\ \phi_{2,n} \\ \phi_{3,n} \\ \phi_{4,n} \end{pmatrix}.$$

Notice that this matrix is the same as the coefficient matrix, equation (16.35), with the sign of α reversed everywhere.

The form of the system of equations for any J should be evident from this example. The coefficient matrix is a $J \times J$ array that is almost tridiagonal, having 4 on the main diagonal, $-\alpha$ on the lower codiagonal, α on the upper codiagonal, α in the lower left corner, $-\alpha$ in the upper right corner, and zeros elsewhere. The nonzero values in the lower left corner and the upper right corner are a consequence of the periodic boundary conditions. The form of the right-hand side for any J should also be evident. In the subsequent discussion we use M to denote the coefficient matrix and $b_{0:J-1,n}$ to denote the right-hand side, so we write the system to be solved

$$M\phi_{0:J-1,n+1} = b_{0:J-1,n}. \tag{16.36}$$

It is important to notice that M is constant, a consequence of the fact that we are assuming u is constant. To verify your understanding of this derivation of the equations to be solved, you can now try deriving them under the assumption that the velocity is not constant, using equation (16.3).

Thus we arrive at the algorithm shown in figure 16.34.

For the moment we do not discuss the solution of the system of equations except to make a few observations. Since M is constant we need to perform an LU factorization only once, not at every time step. Also, the simple sparse structure of M makes it unnecessary to store the full matrix. Nevertheless, the amount of arithmetic required at each time step and the memory requirements are going to be larger than for any of the preceding algorithms. We will see what we get in return for this extra cost.

Figures 16.35 and 16.36 show the computed solution and the exact solution at $t = 0.2$ and $t = 10$, using the triangular pulse again for the initial conditions. Comparison of these figures with the corresponding ones for the Leapfrog algorithm, figures 16.20 and 16.21, show only very small differences. Thus it seems that the cost of the Trapezoid algorithm has not paid a dividend. The next example shows that it has.

We repeat the first example above but with $u = 1.03$, instead of $u = 0.4$. Leapfrog produces the results shown in figure 16.37. On the other hand Trapezoid produces the far more accurate results shown in figure 16.38. The large errors in the Leapfrog results should not be surprising because Leapfrog is unstable when $\alpha > 1$ and in this computation $\alpha = 1.03$.

As this example shows, the Trapezoid algorithm allows us to obtain accurate results in the region where the Leapfrog algorithm is unstable. Notice that this implies that we can use the Trapezoid algorithm to reduce the number of time steps in a computation. For example, suppose that the solution at $t = 0.2$ when $u = 0.4$ is computed as before, but with $\Delta x = 0.01$ and $\Delta t = 0.04$, making $\alpha = 1.6$. The computational effort is one-fourth of its previous value (we now take 5 steps instead of 20). The Trapezoid algorithm gives

[Algorithm 4: Trapezoid]

```
Set initial values:
```
$\phi_{0:J-1,0} = f(x_{0:J-1})$

```
Construct coefficient matrix M:
for j = 0 : J − 1
```
$\quad M_{j,j} = 4$
```
for j = 0 : J − 2
```
$\quad M_{j,j+1} = \alpha$
```
for j = 1 : J − 1
```
$\quad M_{j-1,j} = -\alpha$
$M_{0,J-1} = -\alpha$
$M_{J-1,0} = \alpha$

```
Main time stepping loop:
for n = 1 : N
    Insert border values:
```
$\quad \phi_{-1,n-1} = \phi_{J-1,n-1}; \;\; \phi_{J,n-1} = \phi_{0,n-1}$
```
    Construct right-hand side vector b:
    for j = 0 : J − 1
```
$\quad\quad b_j = \alpha\phi_{j-1,n-1} + 4\phi_{j,n-1} - \alpha\phi_{j+1,n-1}$
```
    Solve
```
$M\phi_{0:J-1,n} = b$

Figure 16.34: Trapezoid: Algorithm 4.

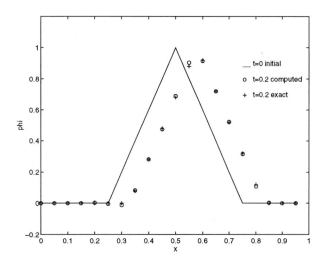

Figure 16.35: The solution computed by the Trapezoid algorithm, marked by "o", and the exact solution, marked by "+", at $t = 0.2$: with $\Delta t = \Delta x = 0.01$ and $u = 0.4$.

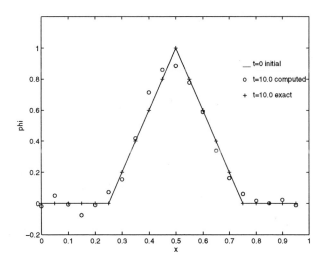

Figure 16.36: The solution computed by the Trapezoid algorithm, marked by "o", and the exact solution, marked by "+", at $t = 10.0$: with $\Delta t = \Delta x = 0.01$ and $u = 0.4$.

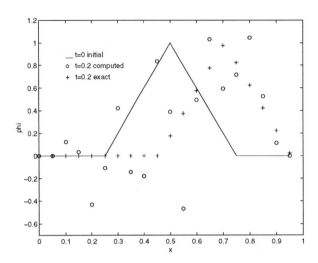

Figure 16.37: The solution computed by the Leapfrog algorithm, marked by "o", and the exact solution, marked by "+", at $t = 0.2$: with $\Delta t = \Delta x = 0.01$ and $u = 1.03$.

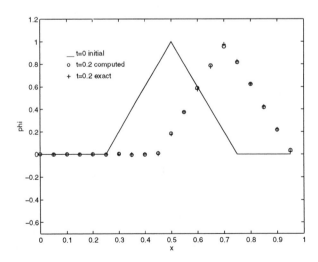

Figure 16.38: The solution computed by the Trapezoid algorithm, marked by "o", and the exact solution, marked by "+", at $t = 0.2$: with $\Delta t = \Delta x = 0.01$ and $u = 1.03$.

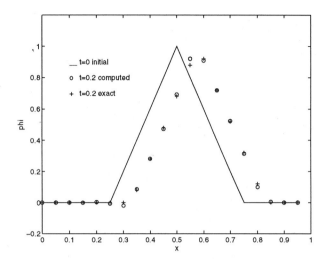

Figure 16.39: The solution computed by the Trapezoid algorithm, marked by "o", and the exact solution, marked by "+", at $t = 0.2$: with $\Delta t = 0.04$, $\Delta x = 0.01$ and $u = 0.4$. Note that in this case $\alpha = 1.6$, a value for which the Leapfrog algorithm is unstable.

the results shown in figure 16.39 which are nearly the same as those obtained earlier when $\Delta t = 0.01$. The Trapezoid algorithm is about three times slower that Leapfrog, therefore a reduction of the computational effort by a factor of four means that the actual time taken by Trapezoid in this computation is about 3/4 of the time taken by Leapfrog.

16.4.5   Analysis of the Trapezoid algorithm

Since the procedure should now be familiar we leave the details to the reader. Assume the numerical solution has the form

$$\phi_{0:J-1,n} = A^n \phi_{0:J-1,0} \, .$$

Then it follows from equation (16.33) that

$$A = \frac{2 - i\,\alpha \sin(\omega \Delta x)}{2 + i\,\alpha \sin(\omega \Delta x)} \, .$$

Using the polar form

$$A = \rho e^{i\theta},$$

we find

$$
\begin{aligned}
\rho &= 1; \\
\theta &= -\arctan\left(\frac{\alpha \sin(\omega \Delta x)}{1 - (\alpha^2/4)\sin^2(\omega \Delta x)}\right).
\end{aligned}
$$

Therefore the phase velocity of the computed solution is

$$u^{(\text{computed})} = \arctan\left(\frac{\alpha \sin(\omega \Delta x)}{1 - (\alpha^2/4)\sin^2(\omega \Delta x)}\right)\frac{1}{\omega \Delta t}.$$

Thus the Trapezoid algorithm is stable and not dissipative for all values of α, and its dispersion is comparable to the Leapfrog algorithm, equation (16.30).

16.4.6   Solution of the linear equations

We turn our attention now to the system of linear equations that must be solved at each iteration, equation (16.36). Consider the coefficient matrix M. Is this matrix well-conditioned? Figure 16.40 shows the 2-norm condition number as a function of α for $J = 100, 200$. Since $\Delta x = 1/J$, the condition numbers shown apply to $\Delta x = 0.01$ and $\Delta x = 0.005$, values used in our computations. Thus we see that M is well conditioned for any α we might reasonably use, and so we can expect good accuracy in the numerical solution.

A straightforward and efficient way to solve these linear equations is to use LU decomposition and backsolving. For a full matrix the number of operations required to compute L and U is proportional to the cube of the order, but for this matrix it is just proportional to the order (i.e., to J). We can see this with a small example, a coefficient matrix of order 5. The form of this matrix is

$$
\begin{pmatrix}
x & x & 0 & 0 & x \\
x & x & x & 0 & 0 \\
0 & x & x & x & 0 \\
0 & 0 & x & x & x \\
x & 0 & 0 & x & x
\end{pmatrix},
$$

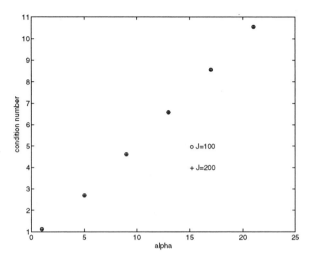

Figure 16.40: The 2-norm condition number of the coefficient matrix M as a function of α for $J = 100$ and $J = 200$.

where x denotes a nonzero element. Assume there is no pivoting. After the first stage of Gauss elimination, in which we put zeros into the first column below the diagonal, we have the form

$$
\begin{pmatrix}
x & x & 0 & 0 & x \\
0 & x & x & 0 & x \\
0 & x & x & x & 0 \\
0 & 0 & x & x & x \\
0 & x & 0 & x & x
\end{pmatrix} .
$$

Notice that two new nonzero elements have been introduced: one in the last row and one in the last column. After the second stage in which zeros are put in the second column below the main diagonal we have the form

$$
\begin{pmatrix}
x & x & 0 & 0 & x \\
0 & x & x & 0 & x \\
0 & 0 & x & x & x \\
0 & 0 & x & x & x \\
0 & 0 & x & x & x
\end{pmatrix} .
$$

It should now be evident that at the end of this process we have an upper triangular matrix with the form

$$
\begin{pmatrix}
x & x & 0 & 0 & x \\
0 & x & x & 0 & x \\
0 & 0 & x & x & x \\
0 & 0 & 0 & x & x \\
0 & 0 & 0 & 0 & x
\end{pmatrix}.
$$

Thus the final matrix only has nonzero elements on the diagonal, the upper codiagonal and the last column. This is the final form of a coefficient matrix of any order. Also, at every stage the number of arithmetic operations performed is a constant: two divides, four multiplies, four adds (the last stage only requires half of these). So we see that the work required to perform the LU decomposition is proportional to J, and remember that this only needs to be done once if u is independent of t.

We have assumed that no pivoting is required, now we verify that this is reasonable. Pivoting is done to prevent growth of the matrix elements during the elimination process.[1] However, it is possible to show that there is virtually no growth for these matrix elements when pivoting is ignored. The argument is a little tedious so we skip it, only suggesting how it goes. It is easy to see that after the first stage the diagonal element in the second row has the value $4 + \alpha^2/4$. Generalizing this it can be seen after the kth stage the diagonal element in row $k + 1$, d_{k+1}, is given by the simple recurrence relation

$$
d_{k+1} = 4 + \left(\frac{\alpha^2}{d_k} \right).
$$

From this it follows that no diagonal element exceeds $4 + \alpha^2$, excepting the diagonal element in the lower right corner that grows somewhat. Similar arguments can be applied to the off-diagonal elements.

What about storage? Clearly $3J$ memory cells are sufficient to contain M and they are sufficient to contain M at any stage of the reduction process since each fill-in is matched by a new zero element. Of course keeping track of the

---

[1]The reason for this is that the bound on the error in the solution is proportional to the magnitude of the largest element of the matrix in any stage of the elimination process.

elements requires a bit of cleverness. Besides the elements of M we need to save the multipliers, and this requires another $2J - 1$ memory cells. In short, we can handle all of the storage required in a rectangular array of J rows and 5 columns.

Other methods could have been used to solve these equations. The matrix M has a special form: each row differs from its predecessor by a cyclic shift of one position to the right. This is called a *Toeplitz* matrix. There are special algorithms for solving linear systems of equations in which the coefficient matrix is a Toeplitz matrix. However, they provide no real advantage for this problem. Furthermore, if u depends on x, then the Toeplitz nature of M is destroyed.

16.4.7 Numerical experiments with the Trapezoid algorithm

These experiments are like those conducted with the Leapfrog algorithm; however we take advantage of the fact that we can use values of $\alpha > 1$.

Triangular pulse

In the first group of figures, figure 16.41 and figure 16.42, the initial condition is the triangular pulse. Figure 16.41 is to be compared with the corresponding results for Leapfrog, figure 16.27. There are small, not very significant, differences between these two sets of results as we might expect. Figure 16.42 shows results for similar computations but with the velocity, u, equal to 1.2. While the accuracy deteriorates, it is not too bad when $\Delta x = \Delta t = 0.005$. Of course Leapfrog would be totally inaccurate in this case because $\alpha = 1.2$.

Square pulse

The second group of figures, figure 16.43 and figure 16.44, show the results obtained when the initial condition is the square pulse. The results are less accurate than for the triangular pulse because, as already explained, dispersion plays a stronger role. Comparison of figure 16.43 with corresponding results

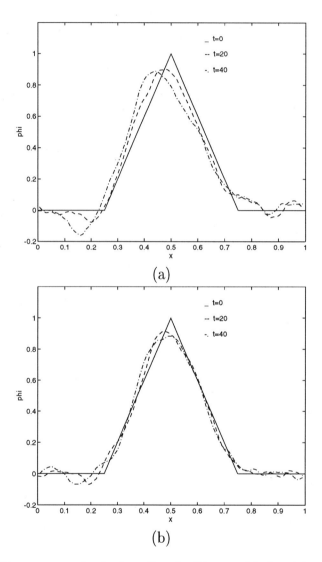

Figure 16.41: The computed solution, for the Trapezoid algorithm, at $t = 20$ and $t = 40$ with $u = 0.4$: with $\Delta x = \Delta t = 0.01$ (a), and $\Delta x = \Delta t = 0.005$ (b). The initial condition for ϕ is the triangular pulse.

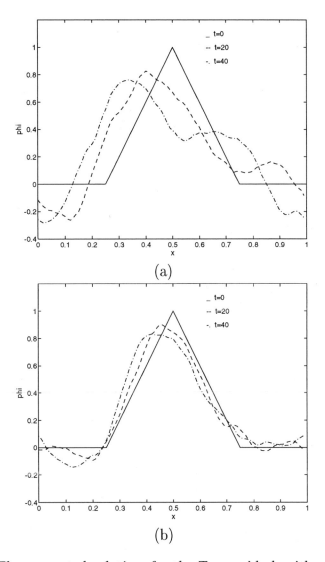

Figure 16.42: The computed solution, for the Trapezoid algorithm, at $t = 20$ and $t = 40$ with $u = 1.2$: with $\Delta x = \Delta t = 0.01$ (a), and $\Delta x = \Delta t = 0.005$ (b). The initial condition for ϕ is the triangular pulse.

for the triangular pulse, figure 16.29, shows again that both algorithms are producing results of about the same accuracy.

Gaussian pulse

The last group of figures, figure 16.45 and figure 16.46, show results obtained when the initial condition is the Gaussian pulse. Figure 16.45 can be compared with figure 16.31 for the Leapfrog algorithm. The results in this group show the same characteristics of those seen in the other groups of figures.

Comparing the cost of Leapfrog and Trapezoid algorithms

The one advantage that the Trapezoid algorithm seems to enjoy over Leapfrog is that it can handle problems with larger values of velocity wherein $\alpha > 1$. But we can certainly apply the Leapfrog algorithm to problems in which $u = 1.2$ by simply reducing the ratio $\Delta t/\Delta x$ so that $\alpha < 1$. For example, if we use the parameter values $\Delta x = 0.005$, $\Delta t = \Delta x \times 2/3$ and $u = 1.2$ we have $\alpha = 0.8$ and we can expect to obtain results like those obtained earlier for the Trapezoid algorithm when $u = 0.8$. The results are be identical because more iterations, time steps, are required thus producing a greater accumulation of roundoff error, and the values used for ϕ are slightly different since a different grid, R_G, is used. Figure 16.47 shows results for the Leapfrog algorithm with these parameter values and the triangular pulse for the initial condition. They are very close to those shown in figure 16.28b for the Trapezoid algorithm with $\Delta x = \Delta t = 0.005$ and $u = 0.8$, though not identical, and they are also very close to those obtained from the Trapezoid algorithm, figure 16.42b (note scale is slightly different).

Thus similar results can be obtained from the two algorithms even at larger velocities. What about costs — how does the computation time for one compare with that for the other. The added complexity of the Trapezoid algorithm means a larger computation time, and timing experiments show that the Trapezoid algorithm is about three times slower than the Leapfrog algorithm per time step. Since the total computation time is proportional to the number of time steps, we would have to have a reduction of Δt by more than one-third before the Leapfrog algorithm would cost more than the Trapezoid algorithm. Thus if $u > 3$ the Trapezoid algorithm would have the advantage.

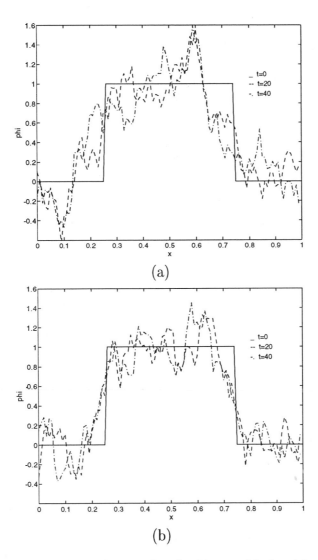

(a)

(b)

Figure 16.43: The computed solution, for the Trapezoid algorithm, at $t = 20$ and $t = 40$ with $u = 0.4$: with $\Delta x = \Delta t = 0.01$ (a), and $\Delta x = \Delta t = 0.005$ (b). The initial condition for ϕ is the square pulse.

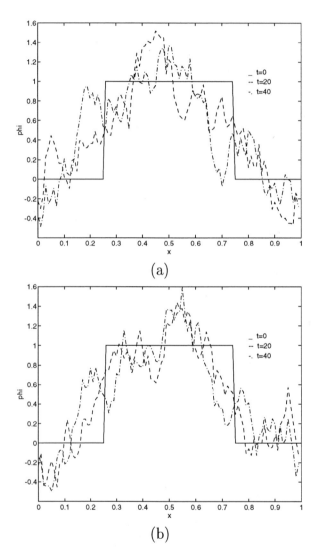

Figure 16.44: The computed solution, for the Trapezoid algorithm, at $t = 20$ and $t = 40$ with $u = 1.2$: with $\Delta x = \Delta t = 0.01$ (a), and $\Delta x = \Delta t = 0.005$ (b). The initial condition for ϕ is the square pulse.

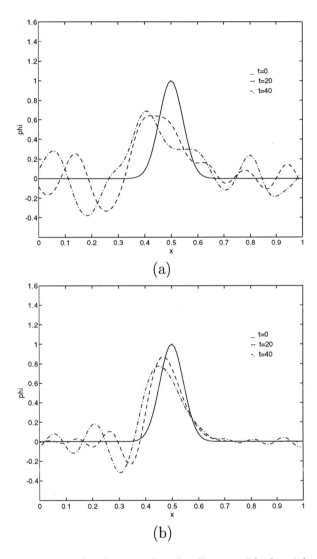

(a)

(b)

Figure 16.45: The computed solution, for the Trapezoid algorithm, at $t = 20$ and $t = 40$ with $u = 0.4$: with $\Delta x = \Delta t = 0.01$ (a), and $\Delta x = \Delta t = 0.005$ (b). The initial condition for ϕ is the Gaussian pulse.

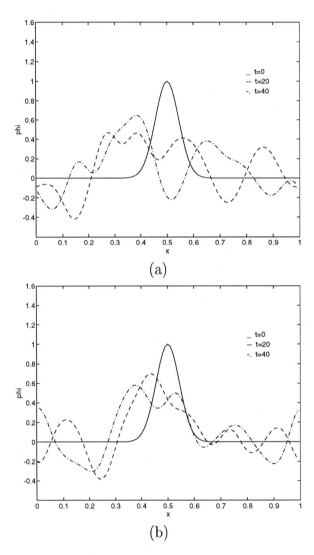

(a)

(b)

Figure 16.46: The computed solution, for the Trapezoid algorithm, at $t = 20$ and $t = 40$ with $u = 1.2$: with $\Delta x = \Delta t = 0.01$ (a), and $\Delta x = \Delta t = 0.005$ (b). The initial condition for ϕ is the Gaussian pulse.

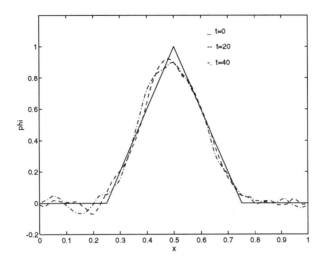

Figure 16.47: The computed solution, for the Leapfrog algorithm, at $t = 20$ and $t = 40$ with $u = 1.2$, $\Delta x = 0.005$, $\Delta t = 2/3\Delta x$ (a). The initial condition for ϕ is the triangular pulse.

We could also find the Trapezoid algorithm advantageous when T is very large. Then we may want to increase Δt in order to reduce the number of time steps. If this causes $\alpha > 1$, we cannot use the Leapfrog algorithm.

Here we have considered a constant velocity field. In cases where the velocity field is not constant, in time and space, situations can arise where the Leapfrog algorithm could become unstable because of an increase of the velocity in some part of the domain. This could be treated by dynamically changing Δt but this would increase the complexity, hence increasing the cost of the algorithm; alternatively a value for Δt could be chosen that is so small as to guarantee stability throughout the time and space domain of the computation and this would also increase the cost of the Leapfrog algorithm. On the other hand, this concern does not arise for the Trapezoid algorithm; it remains stable.

16.5   Two-dimensional problems

In these problems ϕ is a function of two space variables, x and y, and time, t; that is, $\phi = \phi(x, y, t)$. Thus ϕ now represents the density of a substance on a surface, on a particular level of the atmosphere for example.

The two-dimensional advection equation is

$$\frac{\partial \phi}{\partial t} + \frac{\partial u \phi}{\partial x} + \frac{\partial v \phi}{\partial y} = 0\,,$$

where u and v are the x and y components of the velocity which may be functions of x, y, t; thus

$$\phi = \phi(x, y, t), \quad u = u(x, y, t), \quad v = v(x, y, t)\,.$$

The equation is sometimes written in the more compact, vector form, as

$$\frac{\partial \phi}{\partial t} + \nabla \cdot c\phi = 0\,,$$

where c is the velocity vector.

The mathematical space of interest, R, is three-dimensional: $R = [0, X] \times [0, Y] \times [0, T]$. The grid, R_g, is a three-dimensional structure with spacing between points given by Δx, Δy, and Δt. A point in R_g is specified by (j, k, n): $j = 0, 1, 2, \ldots, J$; $k = 0, 1, 2, \ldots, K$; $n = 0, 1, 2, \ldots, N$. It represents the point $(j\Delta x, k\Delta y, n\Delta t)$ in R.

It is customary to use the vertical axis in R to represent values of t (time) and the horizontal plane to representing the physical space, with coordinate axes x and y, as illustrated in figure 16.48. Here a single plane of R_g is shown at $t = 0$ $(n = 0)$ with $\Delta x = \Delta t = 0.1$ The initial conditions for a computation consist of specifying ϕ on this plane.

To display a set of values for ϕ at a particular time we use the vertical axis to represent the values of ϕ. Figure 16.49 is an illustration.

Periodic boundary conditions are expressed with the equations:

$$
\begin{aligned}
\phi(0, y, t) &= \phi(X, y, t), \quad y \in [0, Y], \quad t \in [0, T]\,, \\
\phi(x, 0, t) &= \phi(x, Y, t), \quad x \in [0, X], \quad t \in [0, T]\,.
\end{aligned}
$$

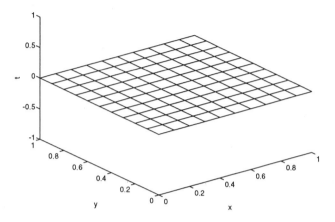

Figure 16.48: Illustration of the space R with a plane of R_g superposed at $t = 0$. Grid spacing is $\Delta x = \Delta y = 0.1$.

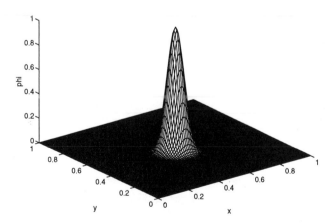

Figure 16.49: Representation of ϕ for a Gaussian density centered at $(0.5, 0.5)$.

16.5.1 Leapfrog and Trapezoid algorithms in two dimensions

It should be unnecessary to repeat the details of the finite difference approximations for these algorithms. The only change is that we must use a finite difference approximation for ϕ_y in addition to one for ϕ_x. We also allow u and v to be functions of x and y.

Thus one obtains the following Leapfrog formula for determining the solution at the the point $(j, k, n + 1)$;

$$
\begin{aligned}
\phi_{j,k,n+1} &= \phi_{j,k,n-1} + \frac{\Delta t}{\Delta x}(u_{j+1,k}\phi_{j+1,k,n} - u_{j-1,k}\phi_{j-1,k,n}) \\
&+ \frac{\Delta t}{\Delta y}(v_{j,k+1}\phi_{j,k+1,n} - v_{j,k-1}\phi_{j,k-1,n}).
\end{aligned}
$$

The computation is very much like the one-dimensional case: it is explicit, and it requires a special step to get started. From this information you should now be able to construct a Leapfrog algorithm for this two-dimensional problem.

The Trapezoid equations are a little more complicated to construct. Following the pattern of our earlier derivation of the Trapezoid algorithm we arrive at the 2-D Trapezoid discretization of the advection equation, the analog of equation 16.33:

$$
\begin{aligned}
&\frac{\phi_{j,k,n+1} - \phi_{j,k,n}}{\Delta t} \\
&+ \frac{1}{4\Delta x}(u_{j+1,k}\phi_{j+1,k,n} - u_{j-1,k}\phi_{j-1,k,n} + u_{j+1,k}\phi_{j+1,k,n+1} - u_{j-1,k}\phi_{j-1,k,n+1}) \\
&+ \frac{1}{4\Delta y}(v_{j,k+1}\phi_{j,k+1,n} - v_{j,k-1}\phi_{j,k-1,n} + v_{j,k+1}\phi_{j,k+1,n+1} - v_{j,k-1}\phi_{j,k-1,n+1}) \\
&= 0.
\end{aligned}
$$

If we rearrange the terms so that those at time $n + 1$ are on the left and those at time n are on the right, then we have

$$
\begin{aligned}
-\alpha_{j-1,k}\phi_{j-1,k,n+1} - \beta_{j,k-1}\phi_{j,k-1,n+1} + 4\phi_{j,k,n+1} \\
+\alpha_{j+1,k}\phi_{j+1,k,n+1} + \beta_{j,k+1}\phi_{j,k+1,n+1} =
\end{aligned}
$$

$$\alpha_{j-1,k}\phi_{j-1,k,n} + \beta_{j,k-1}\phi_{j,k-1,n} + 4\phi_{j,k,n}$$
$$-\alpha_{j+1,k,n}\phi_{j+1,k,n} - \beta_{j,k+1}\phi_{j,k+1,n}, \qquad (16.37)$$

where, for any (j,k) pair,

$$\alpha_{j,k} = u_{j,k}\frac{\Delta t}{\Delta x}, \quad \beta_{j,k} = v_{j,k}\frac{\Delta t}{\Delta y}.$$

As in the one-dimensional problem we again must solve a system of linear equations to advance the solution from time n to time $n+1$. Equation (16.37) is one member of that system, which now consists of $J \times K$ equations.

To put the system into matrix vector form we must establish an ordering of the equations. Each equation corresponds to a grid point, that is, a (j,k) pair. We put the equations into the following order:

$$(0,0),\ (1,0),\ldots,(J-1,0),\ (0,1),\ (1,1),\ldots,(J-1,K-1)$$

With this ordering the system of equations to be solved at each time step is

$$M\Phi^{(n+1)} = M^{(-)}\Phi^{(n)},$$

where M and $M^{(-)}$ are square matrices with $J \times K$ rows; and $\Phi^{(n)}$ is a column vector with $J \times K$ elements. These quantities are best described with a small example.

Let $J = K = 4$, then the matrices are 16×16 and the vector has 16 elements. It is easiest if we look at the matrices and vector in block form. The solution vector, $\Phi^{(n)}$ is

$$\Phi^{(n)} = \begin{pmatrix} \Phi_0^{(n)} \\ \Phi_1^{(n)} \\ \Phi_2^{(n)} \\ \Phi_3^{(n)} \end{pmatrix}, \text{ where } \Phi_i^{(n)} = \begin{pmatrix} \phi_{0,i,n} \\ \phi_{1,i,n} \\ \phi_{2,i,n} \\ \phi_{3,i,n} \end{pmatrix}.$$

Notice that the order of the elements is the order we established above: the first element is $\phi_{0,0,n}$, the second is $\phi_{1,0,n}$ and so forth, the last element being $\phi_{3,3,n}$. The coefficient matrix M is

$$M = \begin{pmatrix} A_0 & B_1 & O & -B_3 \\ -B_0 & A_1 & B_2 & O \\ O & -B_1 & A_2 & B_3 \\ B_0 & O & -B_2 & A_3 \end{pmatrix},$$

where

$$A_i = \begin{pmatrix} 4 & \alpha_{1,i} & 0 & -\alpha_{3,i} \\ -\alpha_{0,i} & 4 & \alpha_{2,i} & 0 \\ 0 & -\alpha_{1,i} & 4 & \alpha_{3,i} \\ \alpha_{0,i} & 0 & -\alpha_{2,i} & 4 \end{pmatrix},$$

and

$$B_i = \begin{pmatrix} \beta_{0,i} & 0 & 0 & 0 \\ 0 & \beta_{1,i} & 0 & 0 \\ 0 & 0 & \beta_{2,i} & 0 \\ 0 & 0 & 0 & \beta_{3,i} \end{pmatrix}$$

and, finally, O is a 4×4 matrix of zeros. The matrix $M^{(-)}$ on the right hand side is

$$M^{(-)} = \begin{pmatrix} A_0^{(-)} & -B_1 & O & B_3 \\ B_0 & A_1^{(-)} & -B_2 & O \\ O & B_1 & A_2^{(-)} & -B_3 \\ -B_0 & O & B_2 & A_3^{(-)} \end{pmatrix},$$

where

$$A_i^{(-)} = \begin{pmatrix} 4 & -\alpha_{1,i} & 0 & \alpha_{3,i} \\ \alpha_{0,i} & 4 & -\alpha_{2,i} & 0 \\ 0 & \alpha_{1,i} & 4 & -\alpha_{3,i} \\ -\alpha_{0,i} & 0 & \alpha_{2,i} & 4 \end{pmatrix}.$$

Thus $M^{(-)}$ is identical to M except for reversing the sign on every off-diagonal element.

With this information you should be able to construct and analyze algorithms for the two-dimensional advection equation. As a start you might first return to the easy case in which the velocity is constant and repeat some of the one-dimensional experiments described earlier. For this you only need to make u or v zero. Next try a velocity directed diagonally across R_g, for example using $u = 0.4/\sqrt{2}$ and $v = 0.4/\sqrt{2}$. Finally, you can begin computations in which the velocities are not constant.

The fact that α and β are variables complicates the analysis, and you will probably find it helpful to assign bounds on these numbers in order to conclude something useful about stability and dispersion. Following ideas in this chapter, you can study dispersion by following sine waves in the two-dimensional system, observing their velocity as a function of frequency.

16.6   Conclusion

In an introductory chapter such as this one it is not an easy matter to decide what to put in and what to leave out. Our decision was to focus on just a few methods and study them in detail with the idea that you would then have the background to study other methods on your own. You will find descriptions of other methods and pointers to the literature in some of the references already given, especially [Haltiner & Williams 80] and [Mesinger & Arakawa 76]. Another source with some results on amplification of amplitude and phase speed for different algorithms is given in a paper by Wen-Yih Sun [Sun 93].

A considerable amount of time was spent on the analysis of the algorithms because we believe the reader should understand not only what an algorithm does but why. Although you can use an algorithm without knowing the "why", you cannot do a good job of developing new and better algorithms without this knowledge. Furthermore, even the user can benefit because he is often faced with the possibility that his own errors may have caused some peculiar results in a computation. He has an easier time recognizing whether or not this is true if he understands the "why".

The particular technique we used for the analysis of these algorithms was based on the Fourier series representation of functions. It is a powerful technique and it demonstrates an important application of Fourier series that you might find useful in other contexts. However, while it was useful in this case its use is restricted to linear systems. An analysis of stability of a non-linear partial differential equation requires other methods.

Periodic boundary conditions have been used throughout this chapter, primarily because they simplify the computations but they also are used in practice. Another boundary condition you might want to explore is, in one-dimension, $\phi(0, t) = 0$ for all t. Thus ϕ is held constant and equal to zero at all grid points along the t-axis. There are some problems here that you have to deal with: note that a larger range of the x-axis is needed. A comparison of the computations done here with the same computations using this boundary condition would show the effects of periodic boundary conditions.

In meteorology and climatology serious computations require enormous computing power. This means "supercomputers" and implies some form of parallel computation. We have said little about this important subject here but our use of vector notation in many of the formulas indicates, at least, the easily parallelizable parts of these computations.

17 Computerized Tomography

17.1 An introduction to computerized tomography

The word *tomography* derives from the Greek word *tomos* meaning "a piece cut off" [Webster]. The method of tomography examines the inside of a three-dimensional object by creating two-dimensional images of cross-sections of the object. Each image is created by passing radiation through one plane of the object, measuring its attenuation, and using that attenuation to map the density of the object in that plane. Because computers are used to create the image from the measured data, tomography is often called *computerized tomography* (CT). In this chapter, we examine some uses of computerized tomography in various scientific fields and review the filtered backprojection algorithm used to recreate images from measured data.

Perhaps the most famous form of tomography is the medical diagnostic tool known as the CT scan. For a CT scan, a patient sits or lies inside a ring mounted with an X-ray source directly opposite an X-ray detector. Figure 17.1 sketches a possible setup for a CT scan. A set of parallel rays of X-ray photons is directed through the patient's body. When a ray passes through a body part, some X-ray photons are absorbed, with dense materials such as bone and tumors absorbing more than soft muscles and skin. The detector measures the number of photons passed through the body and so determines how much the ray was attenuated by absorption. The average density of the body along the path of each ray can then be determined by comparing the incident and transmitted intensities of the ray.

By rotating the ring around the patient, rays can be sent along any number of paths in one plane of the body. Using the attenuation readings for all of the rays, it is possible to create an approximate map of the density of the object within the plane. Translating the ring along the body and repeating

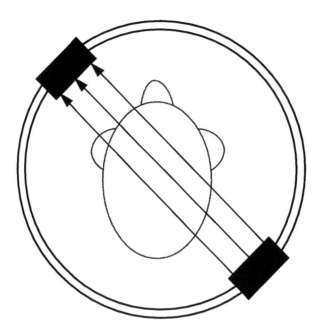

Figure 17.1: CT scan apparatus

the procedure builds a sequence of two-dimensional images that together form a rough three-dimensional image of the interior of the object. In this chapter, we concentrate on forming a single two-dimensional image.

Because we pass only a finite number of rays through the body, we cannot find an exact or continuous map of the density throughout the slice. Instead, we overlay the slice with a two-dimensional grid as shown in figure 17.2. We then use the measured attenuations to approximate the density of the body in each box of the grid. If r rays are passed through a $\sqrt{n} \times \sqrt{n}$ grid with $r > n$, we are required to solve an overdetermined linear system with an $r \times n$ coefficient matrix to produce a digitized image of the object in one plane. An overdetermined system is one with more equations than unknowns. Such a system typically does not have an exact solution, so the image is reconstructed from the best approximate solution. The mathematical details of how to reconstruct the image from the data are covered in sections 17.3–17.8 of this chapter.

The first published description of CT was authored by Sir Godfrey

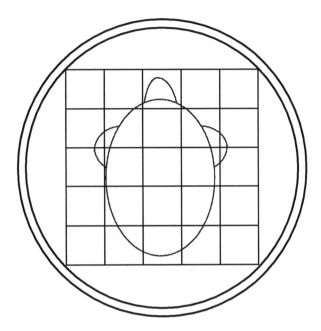

Figure 17.2: A two-dimensional grid overlaid on the body slice

Hounsfield of EMI Ltd. in London and appeared in 1973 [Hounsfield 73]. Hounsfield's scans used X-rays of very low intensity, and it took many hours of exposure to gather the data. An eight by eight grid was superimposed on the object, and the attenuations of sixty four rays were measured. The resulting 64×64 system took hours to solve on EMI's then state-of-the art ICL computer [Spencer 89].

Present day problems are much larger but take less time to solve. The resolution desired for modern CT scans demands that the grid boxes be at most 1–3 mm on a side. This means that a 148×148 or greater grid is used for a typical brain scan. The radiation source is moved to pass X-rays through each row of the grid at many different angles of incidence. When 148 rays are passed at each of 180 different angles, a total of 26,640 rays are passed. This translates into a 26,640 by 21,904 linear system. Recorded intensities from each ray are sent directly to the computer where the image is reconstructed [Hall 79]. A CT scan using even more data is shown in section 17.8.3 of this chapter.

Present day backprojection algorithms are based on the first practical algorithm presented by by A.M. Cormack of Tufts University in the early 1960's [Cormack 63, Cormack 64]. Hounsfield and Cormack shared a Nobel prize in 1979 for their pivotal work on CT [Russ 92].

17.2   Some other types and applications of tomography

Tomography is used in a variety of medical and other scientific applications. In most cases, a form of tomography other than CT is used. The different types of tomography vary in the sources and types of radiation used, but the same image reconstruction algorithms can be used with all forms of tomography. In this section, we review some different methods and applications of tomography.

17.2.1   Positron Emission Tomography (PET)

Particles of matter that cannot be divided into smaller particles are known as *elementary particles*. An atom is not an elementary particle, but its constituent particles, the proton, the neutron, and the electron, are. For most elementary particles, there exists an *antiparticle*, a piece of matter having the same charge as the elementary particle but of opposite sign. The positively charged antiparticle corresponding to the electron is known as the *positron*. When an elementary particle and its corresponding antiparticle meet, they annihilate one another and produce radiation. In particular, the collision of an electron and a positron results in two high energy gamma-ray photons travelling in opposite directions. (See, for example, [Sears et al 80].)

The annihilation of the electron and the positron is the basis of a medical or physiological diagnostic technique known as positron emission tomography (PET). In response to Hounsfield's work on CT [Hounsfield 73], the first PET apparatus capable of rendering tomographic images was completed in the early 1970's by Edward Hoffman, Michael E. Phelps, and Michael M. Ter-Pogossian of the Mallinckrodt Institute of Radiology at Washington University of School of Medicine. Their apparatus resembles that for a CT scan in that the patient sits inside a ring of gamma-ray detectors, but the radiation used is generated and detected in a different way. Specifically, CT is a form of *transmission* tomography: the image is constructed by measuring X-rays passed through the patient. In contrast, PET is a form of *emission* tomography: the patient takes a drug containing a positron-emitting chemical, and the positrons annihilate the electrons in the patient's body to produce gamma-ray photons to be detected by the PET apparatus [Hall 79, Tilyou 91].

PET is a well-established research tool and is beginning to be used in imaging psychological and neurological disorders of the brain such as seizures and in diagnosing heart disease and cancer [Tilyou 91].

17.2.2 Heavy ion tomography

A third medical imaging technique involves accelerating heavy charged particles or *nuclei* such as protons or helium ions through the patient. Like CT and unlike PET, heavy ion tomography is a form of transmission tomography in which heavy ions are directed at the patient. The image is formed by measuring either the stopping point of the ion inside the patient or its residual energy upon exit. Because of their mass, heavy ions are unlikely to scatter, and the heavy ion tomographic image gives an accurate high contrast map of the thickness and electron density of the material. T.F. Budinger and his colleagues first described a setup for heavy ion tomography in 1975. In their experiments, the heavy ions were accelerated in the 184-inch cyclotron at Berkeley. Because of the unwieldy source of the particles, heavy ion tomography is not as widely used as CT scans [Hall 79].

17.2.3 Some nonmedical applications

Some applications of tomography differ from the CT scan in that the "patient" cannot be brought into the lab. In particular, tomographic techniques can be used to model the interior structure of the earth. In seismic tomography, shock waves from earthquakes replace X-rays, and seismic stations located around the globe replace the X-ray detectors. A tomographic image is produced by determining the average density over paths through the patient by measuring the average velocity of the seismic waves. The velocity of a wave depends on the temperature of the earth (waves travel faster in colder regions) and on the alignment of crystallographic axes. Geologists measure the velocities of many crisscrossing rays and solve the resulting overdetermined system to get detailed information about variations of density and temperature in the earth's interior. In this century, research by standard seismic techniques has established that the earth has a crust, an upper and a lower mantle, and a core [Anderson & Dziewonski 84].

Similar techniques are applied to study ocean currents and variations in such properties of the ocean as temperature, salinity, and density. In the ocean, low-frequency sound waves are generated and detected by a set of transceivers mounted on long cables rooted in the sea bed. The transceivers emit waves at known intervals and determine how fast the waves travel between the different pairs of transceivers. The variations in travel time translate into maps of the temperature and pressure inside the circle of transceivers: sound travels faster in warmer water (like near the surface) and when under higher pressure (like near the bottom) [Spindel & Worcester 90].

Smaller scale tomography problems arise in various industrial applications. For example, as part of the oil refining process, oil mixes with catalyst pellets inside of a steel cylinder. The mathematical properties of this mixture are not understood, meaning that measurements of density must be done by experiment rather than by modelling. In this case, gamma rays, rather than X-rays, are projected through the pipe and their attenuations measured. The method of image reconstruction is identical to that for the CT scan [Friedman 88]. Some other applications of the various forms of tomography include iceberg detection on ocean-going vessels [Herman 80], airport luggage scanning [Faber 93], viewing inside the wrappings of mummies [Pickering 93], and the resolution of astronomical features [Hall 79].

17.3   The model

For the remainder of this chapter, we concentrate on reconstructing an image via a CT scan as presented in, for example, [Friedman 88, Herman 80, Rosenfeld & Kak 82]. Excellent sources of additional information at a basic level are [Kak & Slaney 88, Russ 92]. Recall that for a CT scan, X-rays are passed through one plane of an object from various angles. The intensity of each ray is measured before and after it passes through the object. In this section, we review the mathematical fundamentals of the process of mapping the density of the object from the measured ray attenuations.

If we introduce a variable s that measures the distance from the source along a ray, we can write down an expression for how the intensity of a ray changes as it passes through an object assuming that the ray travels in the xy-plane. Specifically, the intensity I changes with respect to the distance s according to

$$\frac{dI}{ds} = -\mu(x, y)\ I\ , \tag{17.1}$$

where $\mu(x, y)$ is the density of the object. Because the density and intensity must always be nonnegative, the negative sign in equation (17.1) shows that, if the intensity changes, it decreases with increasing distance s.

To relate the initial and transmitted intensities of the ray to the density, we must group all terms involving the intensity and integrate the resulting equation

$$\frac{dI}{I} = -\mu(x, y)\ ds. \tag{17.2}$$

The lefthand side is a definite integral in terms of I. If I_0 is the initial intensity of the ray and I_T is its final intensity,

$$\int_{I_0}^{I_T} \frac{dI}{I} = \ln\ (I_T/I_0) = -\ln\ (I_0/I_T).$$

To integrate the righthand side of equation (17.1), we must integrate a function of x and y with respect to the the variable s. This is not inconsistent because the distance s along the ray is itself a function of x and y. If the beam originates at the point (x_0, y_0) and distance s_0 from the origin, the length from that point to any other point (x, y) on the ray is

$$s - s_0 = \sqrt{(x - x_0)^2 + (y - y_0)^2}.$$

To integrate the righthand side we must use a special sort of integral known as a *line* or *path integral*. If σ denotes the line in the xy-plane followed by the ray, the line integral is written

$$\int_\sigma \mu(x,y) \ ds \ .$$

The usual definite integral $\int_a^b f(x)dx$ measures the area beneath the curve of the integrand between $x = a$ and $x = b$ by summing infinitesimally small increments of area between those points. In contrast, a line integral measures the "weight" of the curve itself. For example, if the object is of constant density $\mu(x,y) = \gamma$ and the ray is of length s_T, the line integral is just the length of the line times the density of the object

$$\int_\sigma \mu(x,y) \ ds \ = \gamma s_T.$$

Integrating equation (17.2)

$$\ln (I_0/I_T) = \int_\sigma \mu(x,y) \ ds \tag{17.3}$$

then tells us how the ratio of the initial and transmitted intensities of a ray is related to the amount of material through which the ray passes. When the density of the object depends explicitly on x and y, we must rewrite the integral to remove the dependence on s before we can evaluate the integral. Details of this procedure are presented in most calculus books (see, for example, [Leithold 76]). Line integrals arise in many problems in mathematics and physics, and the path over which one integrates need not be a straight line. In the next section, we explain how to use these integrals in the context of the image reconstruction problem.

17.4   The discretized problem

When recreating an image by tomography we do not evaluate the line integral of equation (17.3). Indeed, the density $\mu(x,y)$ is the unknown quantity we are trying to find. In 1917, Radon showed how to extract the density from the righthand side of equation (17.3) by transform methods, but his formulas are based on continuous projection data instead of the finite set of measurements produced in an actual scan. They are inaccurate when applied to finite data sets, especially those subject to some experimental error. In addition, his formulas do not lend themselves to an efficient computational algorithm [Herman 80]. Hence, subsequent research has focused on developing a good computational algorithm. In this section, we develop a discretized formulation of the image reconstruction problem that can be solved on a computer. We show how the problem translates into an overdetermined system. In sections 17.5–17.8, we derive the method of filtered backprojection typically used to solve the discretized problem in medical and other image reconstruction applications.

Because we use a finite number of rays, we cannot obtain a continuous approximation for the density $\mu = \mu(x, y)$ over the entire grid, so we instead approximate the density in all boxes of the grid. If we are studying a $\sqrt{n} \times \sqrt{n}$ grid, we assume that the density is constant within each box. Denoting the density in box B_i by μ_i, for $i = 1, \ldots, n$, we then compute an approximation to μ of the following form:

$$\mu \approx \mu_1 \text{ in } B_1 \text{ and } \mu_2 \text{ in } B_2 \text{ and } \ldots \text{ and } \mu_n \text{ in } B_n.$$

To make this approximation easier to work with, we introduce a new function δ_i which equals 1 inside box i but equals 0 in any other box. For example, if we number the boxes of a 3×3 grid as shown in figure 17.3, δ_8 has the value shown in figure 17.4.

Rewriting the grid itself as a matrix allows us to represent the discrete problem as a system of linear equations. For the 3×3 example, we can write the approximate density in the grid in matrix form as follows

$$
\begin{aligned}
\mu &\approx \begin{pmatrix} \mu_1 & \mu_2 & \mu_3 \\ \mu_4 & \mu_5 & \mu_6 \\ \mu_7 & \mu_8 & \mu_9 \end{pmatrix} \\
&= \mu_1 \begin{pmatrix} 1 & 0 & 0 \\ 0 & 0 & 0 \\ 0 & 0 & 0 \end{pmatrix} + \mu_2 \begin{pmatrix} 0 & 1 & 0 \\ 0 & 0 & 0 \\ 0 & 0 & 0 \end{pmatrix} + \ldots + \mu_9 \begin{pmatrix} 0 & 0 & 0 \\ 0 & 0 & 0 \\ 0 & 0 & 1 \end{pmatrix} \\
&= \mu_1 \delta_1 + \mu_2 \delta_2 + \ldots + \mu_9 \delta_9.
\end{aligned}
$$

For a grid with n boxes, we can write

$$\mu \approx \mu_1 \delta_1 + \mu_2 \delta_2 + \ldots + \mu_n \delta_n$$

Figure 17.3: Numbering of grid boxes.

Figure 17.4: The matrix function δ_8.

$$= \sum_{i=1}^{n} \mu_i \delta_i.$$

For ray σ_j, the path integral from equation (17.3) then becomes

$$\int_{\sigma_j} \mu(x,y) \, ds \approx \int_{\sigma_j} \sum_{i=1}^{n} \mu_i \delta_i. \qquad (17.4)$$

If the measured attenuation for ray σ_j is denoted

$$p_j = \ln \left(I_0 / I_T \right), \qquad (17.5)$$

equations (17.3) through (17.5) define the relations

$$p_j = \int_{\sigma_j} \mu(x,y) \, ds \approx \int_{\sigma_j} \sum_{i=1}^{n} \mu_i \delta_i = \sum_{i=1}^{n} \mu_i \int_{\sigma_j} \delta_i, \quad j = 1, \ldots, n. \qquad (17.6)$$

If we make r attenuation measurements p_j, $j = 1, \ldots, r$, equation (17.6) defines a rectangular system of equations

$$M\mu \approx p, \qquad (17.7)$$

where the $r \times n$ coefficient matrix has elements $M_{ji} = \int_{\sigma_j} \delta_i$ for $j = 1, \ldots, r$ and $i = 1, \ldots, n$. The vector p on the righthand side has r elements $p(j) = p_j$, and the solution vector μ has n elements $\mu(i) = \mu_i$.

In practice, there are more measurements than grid boxes so that $r > n$. Thus, the system in equation (17.7) is overdetermined, and we cannot expect to find an exact solution of it in general, no matter how we try to solve it. One approach to this problem is to compute the solution $\hat{\mu}$ that minimizes the residual error $\| M\hat{\mu} - p \|_2$. Such a solution is said to satisfy the system $M\mu = p$ in the *least squares sense*. (See, for example, [Golub & Van Loan 89]). Note, however, that the solution vector μ must have all nonnegative components to make physical sense as the component μ_i represents the density of the object in box B_i. Thus, to obtain a realistic approximation to μ we must actually solve a *linear least squares problem with nonnegativity constraints*. Methods for solving the constrained least squares problems involve advanced techniques in matrix algebra and are discussed in [Lawson & Hanson 74], and the application of such least squares techniques to tomography is the subject of continuing research [Kaufman 93]. In this chapter, we do not study methods for constrained least squares problems but rather concentrate on the method of filtered backprojection commonly used in practical tomography problems.

17.5 An introduction to backprojection

In its simplest form, backprojection is essentially a mechanism for deducing the density of an object from measured ray attenuations. In the Mathematical Recreations column in the September 1990 issue of Scientific American, A.K. Dewdney gives the basic idea of how to recreate an image from measured data [Dewdney 90]. He uses a two-dimensional grid with its boxes colored either black or white and attempts to determine the coloring of each box by passing rays vertically or horizontally through the grid. At its origin, a ray is assigned a value of zero. When it emerges from the grid, it has an integer value equal to the number of black boxes through which it has passed. Note that this is opposite to the situation described in section 17.3 where the assigned value or intensity decreases with increasing distance from the ray's origin. In this section, we examine Dewdney's method for 3 × 3 grids.

Figure 17.5 shows the values that would be measured in Dewdney's approach by passing three rays horizontally through the given 3 × 3 grid. (To make the source of the measurements clearer, the black and white boxes are also shown in figure 17.5.) The rays passed through the first and last rows of the grid have the value 3. Thus, all three boxes in those rows must be black. The ray passed through the second row has value 2 and tells us only that two of the three boxes in that row are black. To determine its exact structure, we must pass more rays through the grid. Figure 17.6 shows the measured values for three vertical rays. These data tell us that the first and third columns of the grid are colored black. Because the first and third rows are also black, the center vertical measurement 2 means that the center square must be white.

It is not difficult to devise an example for which this simple deductive algorithm fails. Figure 17.7 shows horizontal and vertical measurements for a new and unknown (and ambiguous) 3 × 3 grid. From these measurements, we can be certain only that the center row of the grid has three white boxes. Either of the two configurations in figure 17.8 would give the same readings. In this 3 × 3 case, we can discern the correct pattern by sending one more ray diagonally through the grid from upper left to lower right. This diagonal ray passed through the first grid would pass through two black boxes, while the diagonal ray passed through the second grid would pass through only one.

An alternative to this precise deductive process is to produce a gray scale

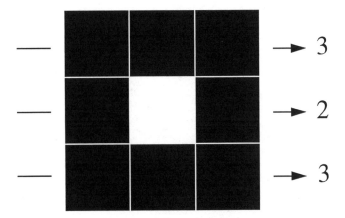

Figure 17.5: The result of passing horizontal rays through the grid.

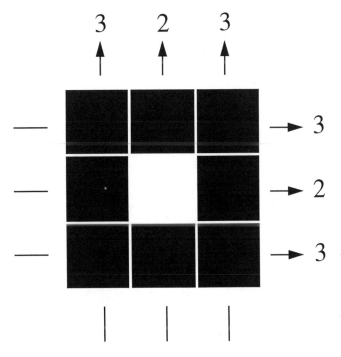

Figure 17.6: The result of passing horizontal and vertical rays through the grid.

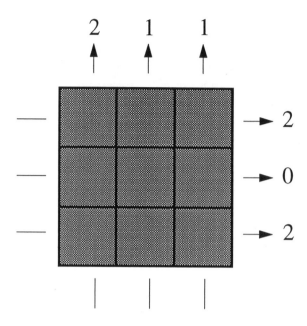

Figure 17.7: The result of passing horizontal and vertical rays through the unknown grid.

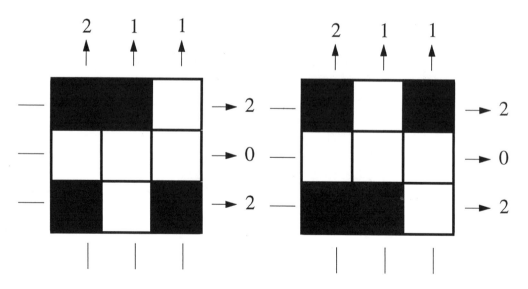

Figure 17.8: The two possible colorings of the grid.

1	5/6	1
5/6	4/6	5/6
1	5/6	1

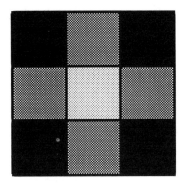

Figure 17.9: A gray scale rendering of figure 17.6.

coloring of the grid. In this case, a ray's value upon exit is divided by the number of boxes through which it passes. All boxes in the ray's path are then colored the same shade of gray defined by that value. In this way, the measured value of the ray is *backprojected* along its path. Repeating this process for many rays produces a rough approximation to the black and white image in varying shades of gray.

The gray scale coloring of the grid of figure 17.6 is shown in figure 17.9. In this case, the horizontal rays assign values of 1, 2/3, and 1 to the boxes in the first, second, and third rows, respectively. The vertical rays give values of 1, 2/3, and 1 to the boxes in the first, second, and third columns. Summing these values and dividing by the number of rays per box gives the average value per ray marked on the left grid of figure 17.6. The shade of gray corresponding to each value appears on the right grid. While the black bordering rows and columns of the grid are not exactly resolved by this process, the center box correctly appears lighter than the surrounding ones. A better representation could be obtained by passing more rays through the grid or by combining the gray scale and deductive algorithms to recognize such features as a fully blackened row or column.

These procedures give the most fundamental idea behind the process of backprojection. In order to accurately reproduce larger and more complicated images, however, it is necessary to turn to the more sophisticated procedure that is the subject of the rest of this chapter.

17.6   The filtered backprojection method

To present a mathematical formulation of filtered backprojection, we first assign unique angle and distance parameters to each ray as shown in figure 17.10 The origin is located at the center of the grid overlaying the object. The line L runs through the origin in the same direction as the rays. The ray σ_j is identified by its perpendicular distance t_j from L and the angle θ that the perpendicular to the ray makes with the x-axis. The measured attenuation of σ_j after it passes through the object is denoted

$$p_j = p(\theta, t_j) = \int_{\sigma_j} \mu(x, y) ds. \qquad (17.8)$$

Our goal is to devise a mathematical relationship between the measured attenuations p_j and the density $\mu(x, y)$ that we are trying to determine.

To reproduce the image, we need to measure the attenuations of a large number of rays. In this chapter, we organize the attenuations by passing K equally spaced parallel rays $\sigma_0, \ldots, \sigma_{K-1}$ through the object at each of q equally spaced angles $\theta_0, \ldots, \theta_{q-1}$ for a total of $r = qK$ attenuation measurements. For example, at each angle θ, we can send K rays at distances from the line L equal to $t_0, t_1, \ldots, t_{K-1}$ with $t_j = t_0 + j\Delta t$.

We can then measure and plot the K attenuations $p_0, p_1, \ldots, p_{K-1}$ as in figure 17.11. (The circles show the measured values of the continuous function $p(t)$.) In this figure, $p_0 = p_1 = p_{10} = 0$ because rays 0, 1, and 10 do not pass through the patient. This means that $I_0 = I_T$ and $\ln I_0/I_T = 0$ in those three cases.

The collection of attenuation values measured for one set of the parallel lines comprises a *parallel projection*. An alternate way of organizing these measurements uses a set of rays emanating at several angles from a single starting point. The set of attenuations of this collection of rays is termed a *fan-beam projection*. We discuss only parallel projections in this chapter. The backprojection method using fan beam projections is discussed in [Kak & Slaney 88, Rosenfeld & Kak 82].

The projection data and the density are related via *Fourier transforms*. We develop this relationship in three steps. In this section, we present the continuous Fourier transform and its inverse then derive a continuous back-

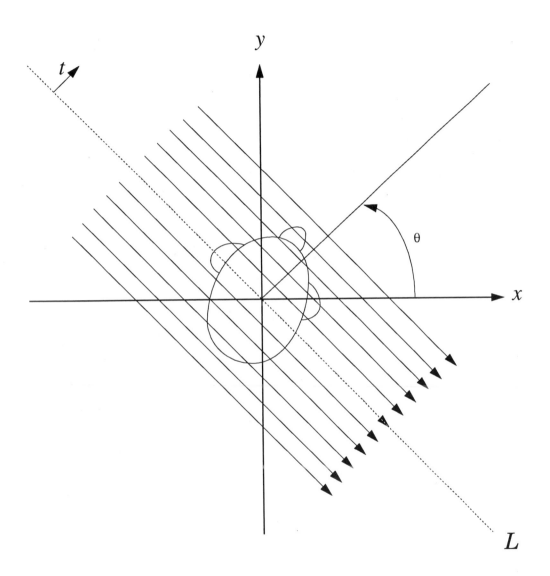

Figure 17.10: The geometry.

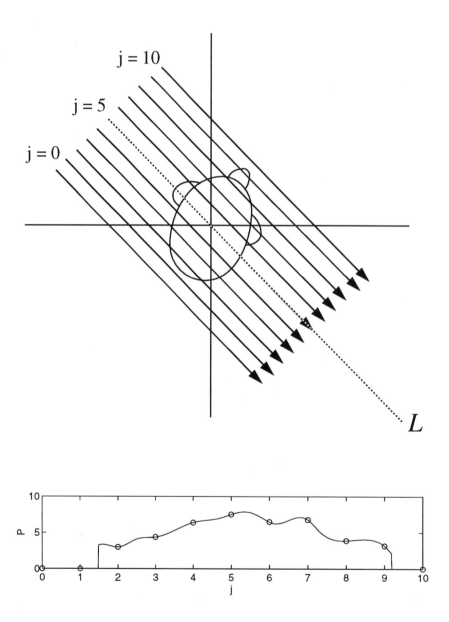

Figure 17.11: The projections $p_0, \ldots, p_{10}$ measured for rays 0 through 10 are plotted as circles on a plot of the continuous projection $p(\theta, t)$.

projection algorithm. This algorithm cannot be applied to the discrete data set $p_0, \ldots, p_{K-1}$ but serves to show the general backprojection approach. In section 17.7, we discretize both the transforms and the backprojection algorithm and introduce the concept of image filtering. In section 17.8, we show how to alter the discretized filtered backprojection algorithm to make it computationally correct and efficient.

17.6.1 The Fourier transform and its inverse

The Fourier transform is a way to convert a continuous function of one variable to a continuous function of frequency of that variable. For example, a function of space is transformed to a function of spatial frequency, and a function of time is transformed to a function of temporal frequency. The Fourier transform is generally applied when it is more convenient to do a computation in the frequency domain, and we will see that this is indeed the case for backprojection. In this section, we present only those aspects of the Fourier transform that are relevant to backprojection. For more information, see, for example, [Kahaner et al 77, Loan 92]. The Fourier transform is also introduced in chapter 2.

If $f(x)$ is a continuous one-dimensional function of distance x and

$$\int_{-\infty}^{+\infty} |f(x)| dx < +\infty,$$

the Fourier transform of $f(x)$ is defined by

$$F(u) = \int_{-\infty}^{+\infty} f(x) e^{-i2\pi ux} dx, \tag{17.9}$$

where $F(u)$ is a one-dimensional function of spatial frequency. We can extract the function $f(x)$ from its Fourier transform by means of its inverse Fourier transform

$$f(x) = \int_{-\infty}^{+\infty} F(u) e^{i2\pi ux} du. \tag{17.10}$$

The Fourier transform can also be applied to higher-dimensional functions. The two-dimensional Fourier transform of the function $f(x, y)$ is

$$F(u, v) = \int_{-\infty}^{+\infty} \int_{-\infty}^{+\infty} f(x, y) e^{-i2\pi(ux+vy)} \, dx \, dy,$$

and the inverse Fourier transform of $F(u, v)$ is

$$f(x, y) = \int_{-\infty}^{+\infty} \int_{-\infty}^{+\infty} F(u, v) e^{i2\pi(ux+vy)} \ du \ dv.$$

A two-dimensional function can also be transformed in only one of its variables. For example, the two-dimensional function $f(x, y)$ can be transformed in the x dimension alone as

$$F(u, y) = \int_{-\infty}^{+\infty} f(x, y) e^{-i2\pi ux} \ dx$$

or in the y dimension alone as

$$F(x, v) = \int_{-\infty}^{+\infty} f(x, y) e^{-i2\pi vy} \ dy.$$

This property implies that we can actually replace a two-dimensional Fourier transform by a pair of one-dimensional Fourier transforms taken in turn. One possible organization is as follows:

$$\begin{aligned} F(u, v) &= \int_{-\infty}^{+\infty} \int_{-\infty}^{+\infty} f(x, y) e^{-i2\pi(ux+vy)} \ dx \ dy \\ &= \int_{-\infty}^{+\infty} [\int_{-\infty}^{+\infty} f(x, y) e^{-i2\pi ux} \ dx] e^{-i2\pi vy} \ dy \\ &= \int_{-\infty}^{+\infty} F(u, y) e^{-i2\pi vy} \ dy. \end{aligned}$$

Alternatively, we can transform the function $f(x, y)$ first in the variable y and then in x to form

$$F(u, v) = \int_{-\infty}^{+\infty} F(x, v) e^{-i2\pi ux} \ dx.$$

In addition, the Fourier transform is not confined to the Cartesian coordinate system. For instance, a function $g(\theta, t)$ expressed in polar coordinates can be transformed in the angular variable θ or the radial variable t or in both by integrating over the full ranges of those variables:

$$G(\eta, \rho) = \int_{0}^{2\pi} \int_{0}^{+\infty} g(\theta, t) e^{-i2\pi(\eta\theta+\rho t)} \ dt \ d\theta.$$

In some cases, we need to take the Fourier transform not of one function but rather of the special integral of the product of two functions called a *convolution* and defined in one dimension by

$$f * g \equiv \int_{-\infty}^{+\infty} f(\alpha)g(x - \alpha)d\alpha. \tag{17.11}$$

The Fourier transform of a convolution is the product of the Fourier transforms of the functions used in that convolution. Thus, if the Fourier transforms of $f(x)$ and $g(x)$ are available, the process of taking the Fourier transform of $f * g$ reduces to a simple multiplication in frequency space.

$$\int_{-\infty}^{+\infty} (f * g)e^{-i2\pi ux} \, dx \; = \; F(u)G(u) \tag{17.12}$$

$$= \; (\int_{-\infty}^{+\infty} f(x)e^{-i2\pi ux} \, dx)(\int_{-\infty}^{+\infty} g(x)e^{-i2\pi ux} \, dx).$$

This result is a statement of the *Convolution Theorem*. Taking the inverse Fourier transform of both sides of equation (17.13) gives an alternative definition of the convolution

$$f * g \equiv \int_{-\infty}^{+\infty} F(u)G(u)e^{i2\pi ux}du.$$

In the case of filtered backprojection, the functions with which we work represent an image or its Fourier transform. The image $\mu(\theta, t)$ is a map of the density in the xy-plane expressed in terms of polar coordinates. Its Fourier transform $U(\theta, \rho)$ is thus a function of spatial frequency, also expressed in polar coordinates. An image for which $U(\theta, \rho)$ is large when ρ is large is one with rapid variation in density across the xy-plane. If this variation is not actually a property of the depicted object, such an image is termed *noisy*. The Convolution Theorem gives an easy way of improving the quality of noisy images. For example, a function $g(x - \alpha)$ can be constructed so that the convolution of $U(\theta, \rho)$ and g either removes or enhances the contribution to the image $\mu(x, y)$ of certain frequencies. Thus, the Fourier transform $G(\rho)$ of the filter function g acts as a filter in frequency space. Details of how to improve an image by filtering out noise are presented in, for instance, [Russ 92]. As we shall see in section 17.6.2, filters are also a necessary part of the most basic backprojection algorithm.

17.6.2   A continuous formulation of backprojection

Now that the basic tools of the backprojection algorithm have been defined, we can return to the problem of how to produce a discrete approximation of the density $\mu \approx \sum_{j=1}^{n} \mu_j \delta_j$ from the measured projection data $p_0, \ldots, p_{K-1}$. While these data are clearly discrete, it simplifies the derivation of the backprojection method to assume first that we instead have a continuous parallel projection for a given angle θ. This corresponds to passing an infinity $(K \to +\infty)$ of rays through the object at the angle θ and collecting their attenuations $p_j, \, j = 0, \ldots, +\infty$, to form a continuous function $p(\theta, t)$. In this case, t is a continuous variable measuring the perpendicular distance from the line L through the origin to the ray. Although t varies continuously from $-\infty$ to $+\infty$, the projection can have nonzero values only for those values of t within the confines of the object. The full range of t is included only for convenience in the derivation.

We construct the relation between the continuous projection and the density by first considering the special case of $\theta = 0$ and then generalizing that result to hold for any value of θ. Both the special and general cases rely upon the Fourier transform and inverse Fourier transform. We conclude this section by showing the efficient rearrangement of the general relation that defines the continuous filtered backprojection algorithm.

The special case $\theta = 0$

We first relate the density and the projection for a continuous projection taken parallel to the y-axis. The angle of this projection is $\theta = 0$, and the line L running through the origin at angle $\theta = 0$ is the y-axis itself. This means that the perpendicular distance t from the ray to the line L is just the x-coordinate of that ray. That is, $t = x$, and

$$p(\theta = 0, t) = p(\theta = 0, x) = \int_{\sigma} \mu(x, y) ds.$$

The distance along the ray from its starting point is

$$s - s_0 = \sqrt{(x - x_0)^2 + (y - y_0)^2}.$$

However, when $t = x$, this distance varies with y alone so that $ds = dy$. Thus, when $\theta = 0$, the line integral along the path of the ray can be written as the definite integral in y

$$p(\theta = 0, t) = p(\theta = 0, x) = \int_\sigma \mu(x, y)dy = \int_{-\infty}^{+\infty} \mu(x, y)dy. \qquad (17.13)$$

Again, we include the full range of y for convenience even though only those y values between the emitter and detector can actually contribute to the value of the integral.

It is not immediately clear how to obtain an expression for the density from equation (17.13), but the Fourier transform provides the key. To see this, we first write the two-dimensional Fourier transform of the density

$$U(u, v) = \int_{-\infty}^{+\infty} \int_{-\infty}^{+\infty} \mu(x, y)e^{-i2\pi(ux+vy)} \, dx \, dy.$$

If we then consider the case $v = 0$, we are left with the two-dimensional Fourier transform of the density along the u-axis in frequency space

$$\begin{aligned} U(u, 0) &= \int_{-\infty}^{+\infty} \int_{-\infty}^{+\infty} \mu(x, y)e^{-i2\pi ux} \, dx \, dy \\ &= \int_{-\infty}^{+\infty} \left[\int_{-\infty}^{+\infty} \mu(x, y) \, dy \right] e^{-i2\pi ux} \, dx. \end{aligned} \qquad (17.14)$$

Notice that the expression in square brackets in this equation is just $p(\theta = 0, x)$ so we actually have the important result

$$U(u, 0) = \int_{-\infty}^{+\infty} p(\theta = 0, x)e^{-i2\pi ux} \, dx. \qquad (17.15)$$

That is, taking the one-dimensional Fourier transform of the projection $p(\theta = 0, x)$ along the y-axis gives us one line in frequency space (namely, the u-axis) of the two-dimensional Fourier transform of the density.

Thus, we can extract one line ($t = x$) of the density $\mu(x, y)$ by taking the one-dimensional Fourier transform of $U(u, 0)$. This operation *backprojects* $U(u, 0)$ from frequency space onto the line $t = x$ in the spatial domain. If we could generalize this result to produce any line of $U(u, v)$, we could use it to find $U(u, v)$ in all of frequency space. By taking the inverse Fourier transform of that representation of $U(u, v)$ for all u and v, we could determine the density $\mu(x, y)$ for all x and y. We consider the more general case next.

Using any old θ

As it turns out, equation (17.15) is easily modified to apply to any angle θ. With the simple matrix operation known as the *Jacobi rotation* or *Givens rotation*, we can get other lines of the Fourier transform of the density $\mu(x, y)$ from projections taken at arbitrary angles θ.

A Jacobi rotation $\mathcal{J}$ is defined by a 2×2 matrix function of an angle θ. Applying this matrix to a vector in the xy-plane rotates the vector about the angle θ in that plane as follows:

$$\mathcal{J}\begin{pmatrix} x \\ y \end{pmatrix} = \begin{pmatrix} \cos\theta & \sin\theta \\ -\sin\theta & \cos\theta \end{pmatrix}\begin{pmatrix} x \\ y \end{pmatrix} = \begin{pmatrix} x\cos\theta + y\sin\theta \\ -x\sin\theta + y\cos\theta \end{pmatrix} = \begin{pmatrix} t \\ s \end{pmatrix}.$$

The Jacobi rotation lets us rotate the xy-coordinate system into the ts-coordinate system as shown in figure 17.12. This means that we can treat a projection taken at an angle θ in the xy-plane as a projection taken at an angle 0 in the ts-plane. The ts-coordinate system is a natural one to use for our problem as the variable t represents the distance of the ray from the line L through the origin, and the variable s represents the distance travelled along the ray from its source. For example, the point at location (x_0, y_0) in this figure has s- and t-coordinates s_0 and t_0, where $s_0 = -x_0\sin\theta + y_0\cos\theta$ and and $t_0 = x_0\cos\theta + y_0\sin\theta$.

Following the same steps as we used in the xy-coordinate system to derive equation (17.13), we can write down the equation for a parallel projection in the ts-plane running parallel to the s-axis:

$$p(\theta, t) = \int_{-\infty}^{+\infty} \mu(t, s) \, ds. \tag{17.16}$$

Its one-dimensional Fourier transform is

$$\begin{aligned} P(\theta, \rho) &= \int_{-\infty}^{+\infty} p(\theta, t)e^{-i2\pi\rho t} \, dt \\ &= \int_{-\infty}^{+\infty} \int_{-\infty}^{+\infty} \mu(t, s)e^{-i2\pi\rho t} \, ds \, dt. \end{aligned}$$

Rewriting this in xy-coordinates gives us

$$P(\theta, \rho) = \int_{-\infty}^{+\infty} \int_{-\infty}^{+\infty} \mu(x, y)e^{-i2\pi\rho(x\cos\theta + y\sin\theta)} \, dx \, dy = U(\theta, \rho). \tag{17.17}$$

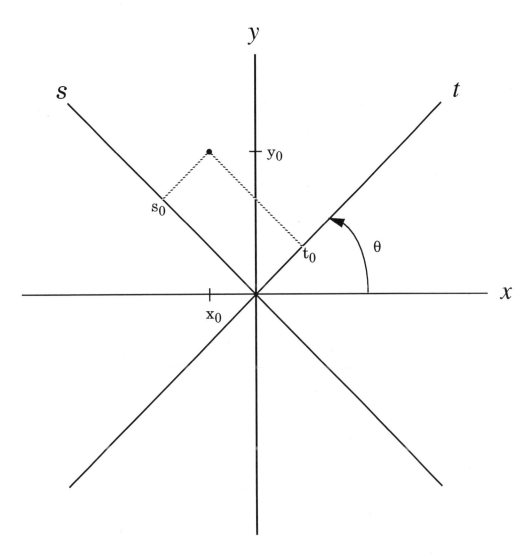

Figure 17.12: Rotating the xy-coordinate system into the ts-coordinate system.

Notice that the righthand side of equation (17.17) is now the two-dimensional Fourier transform of the density.

Thus, taking the one-dimensional Fourier transform of the continuous parallel projection at angle θ gives the two-dimensional Fourier transform of the density along the line in frequency space defined by the polar coordinates (θ, ρ). We can extract the density $\mu(x, y)$ along the line $t = x \cos \theta + y \sin \theta$ for the given angle θ by taking the two-dimensional inverse Fourier transform of $P(\theta, \rho)$ for that angle.

This result is known as the *Fourier Slice Theorem*, and it gives a mechanism for determining the density. That is, if we take a continuous parallel projection at every angle θ and evaluate the Fourier transform of each projection at every radial coordinate ρ, we produce the Fourier transform of $\mu(x, y)$ at every angle θ and every radial coordinate ρ in frequency space. This gives us a complete, continuous representation of the Fourier transform of the density. Taking the inverse Fourier transform of that transform, in turn, gives us a continuous representation of the density in the xy-plane.

Filtered backprojection

The Fourier Slice Theorem leads us to take the two-dimensional inverse Fourier transform of $P(\theta, \rho)$ in polar coordinates:

$$\mu(x, y) \;=\; \int_0^{2\pi} \int_0^{+\infty} P(\theta, \rho) e^{i2\pi\rho(x \cos \theta + y \sin \theta)} \rho \, d\rho \, d\theta. \qquad (17.18)$$

A little thought, however, reveals that using the full angle range of 0 to 2π in equation (17.18) causes redundant calculations. Because the value of the projection depends only on the density of the object along the path of the ray, the value of the projection does not change when the positions of the x-ray source and detector are swapped. Mathematically, this means that $p(\theta, -\rho) = p(\theta + \pi, \rho)$.

Because the projections are the same for θ and $\theta + \pi$, the Fourier transforms of those projections are also equal. That is, $P(\theta, -\rho) = P(\theta + \pi, \rho)$. These observations allow us to reduce the angle interval of equation (17.18) in the following way.

$$\mu(x, y) \;=\; \int_0^{\pi} \int_0^{+\infty} P(\theta, \rho) e^{i2\pi\rho(x \cos \theta + y \sin \theta)} \rho \, d\rho \, d\theta$$

$$+ \int_\pi^{2\pi} \int_0^{+\infty} P(\theta, \rho) e^{i2\pi\rho(x \cos\theta + y \sin\theta)} \rho \, d\rho \, d\theta$$

$$= \int_0^\pi \int_0^{+\infty} P(\theta, \rho) e^{i2\pi\rho(x \cos\theta + y \sin\theta)} \rho \, d\rho \, d\theta$$

$$+ \int_0^\pi \int_0^{+\infty} P(\rho, \theta + \pi) e^{i2\pi\rho(x \cos(\theta+\pi) + y \sin(\theta+\pi))} \rho \, d\rho \, d\theta. \quad (17.19)$$

With the substitution $P(\theta, -\rho) = P(\theta + \pi, \rho)$, we then get

$$
\begin{aligned}
\mu(x, y) &= \int_0^\pi \int_0^{+\infty} P(\theta, \rho) e^{i2\pi\rho(x \cos\theta + y \sin\theta)} \rho \, d\rho \, d\theta \\
&+ \int_0^\pi \int_0^{+\infty} P(\theta, -\rho) e^{-i2\pi\rho(x \cos\theta + y \sin\theta)} \rho \, d\rho \, d\theta \\
&= \int_0^\pi \int_0^{+\infty} P(\theta, \rho) e^{i2\pi\rho(x \cos\theta + y \sin\theta)} \rho \, d\rho \, d\theta \\
&+ \int_0^\pi \int_{-\infty}^0 P(\theta, \rho) e^{i2\pi\rho(x \cos\theta + y \sin\theta)} |\rho| \, d\rho \, d\theta,
\end{aligned}
$$

and we can recombine these terms into the single integral

$$\mu(x, y) = \int_0^\pi \int_{-\infty}^{+\infty} P(\theta, \rho) e^{i2\pi\rho(x \cos\theta + y \sin\theta)} |\rho| \, d\rho \, d\theta. \quad (17.20)$$

Equation (17.20) shows that projection data for angles 0 to π are sufficient to construct the density $\mu(x, y)$.

Using $t = x \cos\theta + y \sin\theta$ in equation (17.20) gives us

$$\mu(x, y) = \int_0^\pi \left[\int_{-\infty}^{+\infty} P(\theta, \rho) |\rho| e^{i2\pi\rho t} \, d\rho \right] d\theta. \quad (17.21)$$

The interior integral

$$C(\theta, t) = \int_{-\infty}^{+\infty} P(\theta, \rho) |\rho| e^{i2\pi\rho t} \, d\rho \quad (17.22)$$

is the basis of the filtered backprojection algorithm. In particular, the function $|\rho|$ is an *inverse filter*: multiplying $P(\theta, \rho)$ by $|\rho|$ increases the relative influence of $P(\theta, \rho)$ at high frequencies. Hence, $C(\theta, t)$ is known as a *filtered projection*.

The integral in equation (17.21) forms a map of the density in the xy-plane by accumulating all of the filtered projections for all angles θ from 0 to π. Each filtered projection $C(\theta, t)$ contributes to the density along the line of constant $t = x\cos\theta + y\sin\theta$ for its particular value of θ. In this way, each filtered projection is backprojected into the xy-plane.

17.7   A discrete filtered backprojection algorithm

Equations (17.21) and (17.22) define the process of filtered backprojection given continuous projection data. While these equations do establish a relationship between the projections and the density of the object, they are impractical because experimental data cannot be continuous. The projection measured at a given angle is a collection of values denoted

$$p = (p_0, p_1, \ldots, p_{K-1})^T. \tag{17.23}$$

We repeat this measurement for q different angles $\theta_0, \ldots, \theta_{q-1}$. In this section, we show how to discretize both the Fourier transform and the filtered backprojection algorithm to operate on the discrete data.

We begin by introducing the discrete Fourier transform and the discrete inverse Fourier transform. Just as the continuous filtered backprojection algorithm is based on the continuous Fourier transform, the discrete algorithm is based on the discrete Fourier transform. Discretization of the backprojection algorithm proceeds in three steps. First, we show how the continuous backprojection algorithm is modified to work in a finite frequency or spatial domain. Second, we consider the case of a finite number of angles. Finally, we move to a finite number of rays.

17.7.1   The discrete Fourier transform and its inverse

The discrete Fourier transform is a mechanism for transforming a set of measurements of the spatial function $f(x)$ to a set of values of a function of spatial frequency $F(u)$. (In this section, we work with spatial variables only. However, just as is the case for the continuous Fourier transform, the discrete Fourier transform can be applied to convert a function of any variable to a function of the frequency of that variable.) An introduction to the Fourier transform is provided in chapter 2.

We assume that we have an even number of measurements of the continuous function $f(x)$ taken at the equally spaced x-values so that $x_0, \ldots, x_{K-1}$. That is, $f(x_j) = f_j$ for $j = 0, \ldots, K - 1$. We also assume that the function $f(x)$ is zero outside of the range $[x_0, x_{K-1}]$ so that our samples f_j represent all important parts of $f(x)$. The sum

$$F_l = F(u_l)/\Delta x = \sum_{j=0}^{K-1} f_j e^{-i2\pi lj/K} \tag{17.24}$$

defines the *discrete Fourier transform* of these data at frequency u_l.

Note that the terms of the sum in equation (17.24) form a periodic series in l. That is, because $e^{-i2\pi j} = 1$ for all integer values of j, $F_{-l} = F_{K-l}$. In particular, $F_{-K/2} = F_{K/2}$, so we need only compute the K values $F_{-K/2}, \ldots, F_{K/2-1}$ to have all information about F. The K data values $f_0, \ldots, f_{K-1}$ thus actually lead to K distinct discrete Fourier transform values $F_{-K/2}, \ldots, F_{K/2-1}$.

Similarly, for $j = 0, \ldots, K - 1$,

$$f_j = (1/K) \sum_{l=-K/2}^{K/2-1} F_l e^{i2\pi lj/K} \tag{17.25}$$

defines the discrete inverse Fourier transform at x_j of values $F_0, \ldots, F_K$ in frequency space.

The Convolution Theorem still applies in the discrete case. The convolution $f * g$ has the K elements

$$h_l = (f * g)_l = \sum_{j=0}^{K-1} f(j)g(l - j). \tag{17.26}$$

This convolution is well-defined for $l = 0, \ldots, K - 1$ when the data f and g are assumed periodic with period K. In that case, the elements of g with negative indices are evaluated by the relation $g(-m) = g(K - m)$. This sort of convolution is termed a *periodic* or *circular* convolution.

Applying a periodic convolution directly to aperiodic data results in an incorrect result as the terms involving elements $g_{-(K-1):1}$ contribute incorrectly to the result. This *interperiod interference* is remedied by affixing zero elements to the data vectors [Rosenfeld & Kak 82, Kak & Slaney 88]. We illustrate this "zero padding" in context in section 17.7.3.

The convolution is related to the Fourier transforms of f and g by

$$H_l = \sum_{j=-K/2}^{K/2-1} (f * g)_j e^{-i2\pi lj/K} = F_l G_l,$$

for $l = -K/2, \ldots, K/2 - 1$.

The discrete Fourier transform can also be applied in two dimensions. The two-dimensional discrete Fourier transform of the function $f(x,y)$ for samples taken at $f_{jk} = f(x_j, y_k)$ for $j, k = 0, \ldots, K - 1$ is

$$F_{lm} = F(u_l, v_m)/\Delta x \Delta y = \sum_{j=0}^{K-1} \sum_{p=0}^{K-1} f_{jk} e^{-i2\pi(lj/K + mk/K)}. \qquad (17.27)$$

The two-dimensional discrete inverse Fourier transform is

$$f_{jk} = (1/K) \sum_{l=-K/2}^{K/2-1} \sum_{m=-K/2}^{K/2-1} F_{lm} e^{-i2\pi(lj/K + mk/K)}. \qquad (17.28)$$

As in the continuous case, the 2D discrete Fourier transform can be written as a pair of one-dimensional discrete Fourier transforms:

$$F_{lm} = \sum_{j=0}^{K-1} [\sum_{k=0}^{K-1} f_{jk} e^{-i2\pi mk/K}] e^{-i2\pi lj/K}.$$

17.7.2 Continuous backprojection in a finite frequency domain

In section 17.6, we developed a continuous formulation of filtered backprojection by relating the density $\mu(x, y)$ of the object to the projection data as follows:

$$\mu(x, y) \;=\; \int_0^\pi C(\theta, t) \, d\theta, \qquad (17.29)$$

where the transform

$$C(\theta, t) = \int_{-\infty}^{+\infty} P(\theta, \rho)|\rho|e^{i2\pi\rho t}\, d\rho \qquad (17.30)$$

defines a filtered projection. The first thing to note is that our CT scan cannot provide the full range of frequencies required by equation (17.30). In particular, we saw in section 17.7.1 that the Fourier transform maps a function of the discrete spatial data $t_{-K/2}, \ldots, t_{K/2-1}$ onto a function of the discrete frequencies $u_{-K/2}, \ldots, u_{K/2}$ with $-u_{-K/2} = 1/(2\Delta t) = u_{K/2}$. Even our continuous formulation of backprojection should account for this finite frequency domain. The easiest way to zero out the contribution to $C(\theta, t)$ of frequencies outside of the range $[-1/(2\Delta t), 1/(2\Delta t)]$ is to introduce the new filter function

$$B(\rho) = \begin{cases} |\rho|, & |\rho| \leq 1/(2\Delta t) \\ 0, & \text{otherwise.} \end{cases} \qquad (17.31)$$

The filtered projection of equation (17.30) may then be rewritten in the equivalent form

$$C(\theta, t) = \int_{-\infty}^{+\infty} P(\theta, \rho)B(\rho)e^{i2\pi\rho t}\, d\rho. \qquad (17.32)$$

This representation is valid in the finite frequency domain imposed by the measured data.

The use of the new filter function $B(\rho)$ also allows us to write the filtered projection in terms of spatial variables. The integral in equation (17.30) is the inverse Fourier transform of the product of the function $B(\rho)$ and the one-dimensional Fourier transform $P(\theta, \rho)$ of the projection $p(\theta, t)$. The Convolution Theorem therefore says that $C(\theta, t)$ is the convolution of the projection $p(\theta, t)$ and the inverse Fourier transform of $B(\rho)$, where the inverse Fourier transform of $B(\rho)$ is given by

$$
\begin{aligned}
b(t) &= \int_{-\infty}^{+\infty} B(\rho)e^{i2\pi\rho t}dt \\
&= \frac{1}{2(\Delta t)^2}\frac{\sin 2\pi t/(2\Delta t)}{2\pi t/(2\Delta t)} - \frac{1}{4(\Delta t)^2}\left(\frac{\sin \pi t/(2\Delta t)}{\pi t/(2\Delta t)}\right)^2. \quad (17.33)
\end{aligned}
$$

We can thus redefine the filtered projection of equation (17.32) by the convolution

$$C(\theta, t) \;=\; \int_{-\infty}^{+\infty} p(\theta, \alpha) b(t - \alpha) \, d\alpha. \tag{17.34}$$

In this case, the variable α measures position. Note that we could not have used the Convolution Theorem in this way with the original filter $|\rho|$. Because the integral $\int_{-\infty}^{+\infty} |\rho| d\rho$ is not bounded, the inverse Fourier transform of $|\rho|$ does not exist.

The filtered projections defined by either equation (17.32) or equation (17.34) may be accumulated as in equation (17.29) to give the density for all x and y.

17.7.3   A discrete formulation of filtered backprojection

Equations (17.29) and (17.32) or (17.34) fully define the continuous backprojection algorithm on a finite frequency or spatial domain. Our projection data, however, is not continuous but rather is measured at the equally $K - 1$ equally spaced positions $t_j = t_0 + j\Delta t$, $j = 0, \ldots, K - 1$, at the q equally spaced angles $\theta_0, \ldots, \theta_{q-1}$ with $\theta_l = l\frac{\pi}{q}$. We now describe the changes necessary to make the algorithm operate on the discrete data.

Our first step toward the discrete algorithm is to replace the integral in equation (17.29) with the sum over angles

$$\mu(x, y) \approx \frac{\pi}{q} \sum_{l=0}^{q-1} C(\theta_l, t). \tag{17.35}$$

The angle size $\frac{\pi}{q}$ replaces the angle differential $d\theta$, and $C(\theta_l, t)$ is the result of evaluating equation (17.32) or equation (17.34) at angle θ_l. Equation (17.35) gives us the density only at those values of x and y satisfying $t = x \cos \theta_l + y \sin \theta_l$ for $l = 0, \ldots, q - 1$.

Discretizing with respect to the angle in this way lends new significance to the filter function $|\rho|$ of equation (17.30). In confining our data to the discrete angles $\theta_0, \ldots, \theta_{q-1}$, we are replacing the continuous data domain with one made up of q lines through the origin in frequency space as depicted in figure 17.13 for the case $K = 3$. In this figure, we can see that when $|\rho| = \sqrt{u^2 + v^2}$ is small, the points in the frequency domain are denser than when $|\rho|$ is large. Thus, a better approximation to the continuous frequency domain is obtained when $|\rho|$ is small. Filtering $U(\theta_l, \rho)$ with $|\rho|$ is essentially a mechanism for weighting the data to compensate for the sparser fill of the frequency domain at larger $|\rho|$ [Kak & Slaney 88]. The filter $B(\rho)$ of equation (17.32) operates in exactly the same way. The filter $b(t)$ of equation (17.34) is equivalent although it operates in the spatial domain.

The last step in devising our discrete algorithm is to discretize $C(\theta_l, t)$. We do this first for the spatial domain representation of equation (17.34). In this case, we must discretize with respect to both position variables t and α. The projections $p = (p_0, \ldots, p_{K-1})^T$ defined in equation (17.23) are taken at positions $t_j = t_0 + j\Delta t$, $j = 0, \ldots, K - 1$, and we first take $\alpha_k = k\Delta t$ for $k = -\infty, \ldots, +\infty$. Equation (17.34) then becomes

$$C(\theta_l, t_j) = \sum_{k=-\infty}^{+\infty} p(\theta_l, \alpha_k) b(t_j - \alpha_k)\, \Delta t,$$

for $j = 0, \ldots, K-1$. If we take $t_m = t_0 + m\Delta t = t_0 + (j-k)\Delta t$, the convolution can be rewritten as

$$C(\theta_l, t_j) = \Delta t \sum_{m=-(K-1)}^{K-1} p(\theta_l, t_j - t_m) b(t_m). \tag{17.36}$$

Equation (17.36), however, is a discrete periodic convolution applied to the aperiodic measured projection data. To avoid interperiod interference, we must affix $K - 1$ zero elements $p_{-(K-1)}, \ldots, p_{-1}$ to the vector of projections $p = (p_0, \ldots, p_{K-1})^T$. This is an example of zero padding. The discrete values of b are derived from equation (17.33), where

$$b_m = b(t_m) = \begin{cases} 1/(2\Delta t)^2, & m = 0 \\ 0, & \text{even } m \\ -1/(m\pi\Delta t)^2, & \text{odd } m \end{cases} \tag{17.37}$$

for $m = -(K-1), \ldots, K-1$.

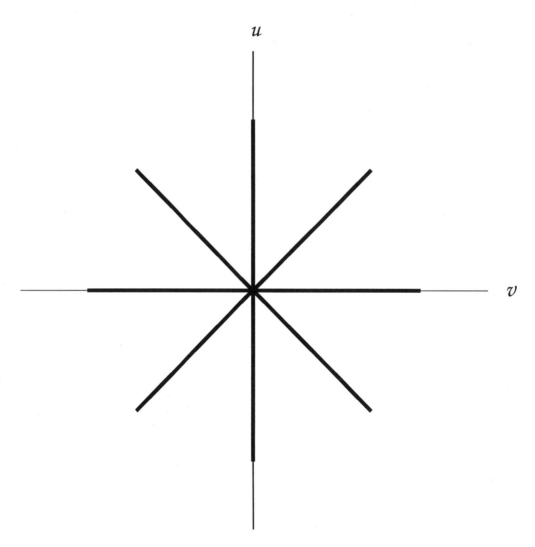

Figure 17.13: The data domain in frequency space.

We can determine the filtered projection via equation (17.36) for the positions $\hat{t} = (t_0, \ldots, t_{K-1})^T$. The result is a vector quantity of length K that we can denote $C(\theta_l, \hat{t})$. To compute $C(\theta_l, t_j)$ we find the dot product of two vectors of length $2K - 1$. If we don't actually perform the arithmetic operations involving the zero elements $p_{-(K-1)}, \ldots, p_{-1}$, the dot product uses $K - 1$ floating-point additions and K floating-point multiplications. Computing the vector quantity $C(\theta_l, \hat{t})$ requires K such dot products, and so it takes $\mathcal{O}(K^2)$ floating-point operations.

Another way to formulate the method is to use the frequency domain convolution from equation (17.32) instead of the spatial domain convolution of equation (17.34). Discretizing the frequency domain convolution in equation (17.32) at first appears to be a simple matter of replacing the continuous inverse Fourier transform in that equation with its discrete form. However, equation (17.32) is just a different representation of the periodic convolution in equation (17.36) applied to the aperiodic projection data. Thus, it too is subject to interperiod interference. To ensure that the frequency domain convolution is correctly evaluated, we must still construct it using the zero-padded projection data. Thus, the discretization of equation (17.32) proceeds by the following steps at each angle θ_l:

- Pad the projection data $p = (p_0, \ldots, p_{K-1})^T$ with zeros to form the new projection vector $\tilde{p} = (p_{-(K-1)}, \ldots, p_{K-1})^T$. This zero padding ensures that we do not introduce interperiod interference.

- Take the discrete Fourier transform of $\tilde{p}$ to form $P(\theta_l, \tilde{\rho})$ and so bring the computation into the frequency domain. Both $P(\theta_l, \tilde{\rho})$ and the frequency vector $\tilde{\rho}$ are of length $2K - 1$.

- Take the discrete Fourier transform of the spatial domain filter $b = (b_{-(K-1)}, \ldots, b_{(K-1)})^T$ to produce the frequency domain filter $B(\tilde{\rho})$.

- Discretize equation (17.32) by convolving $P(\theta_l, \tilde{\rho})$ in the frequency domain and taking the discrete inverse Fourier transform of the result to produce $C(\theta_l, \tilde{t})$. Like $P(\theta_l, \tilde{\rho})$ and $B(\tilde{\rho})$, $C(\theta_l, \tilde{t})$ is a vector of length $2K - 1$.

These steps are summarized by

$$C(\theta_l, \tilde{t}) = \text{DIFT}\left(\text{DFT}(p(-(K{-}1) : K{-}1)) .* \text{DFT}(b(-(K{-}1) : K{-}1))\right). \quad (17.38)$$

In this equation, DFT and DIFT stand for the operations of taking the discrete Fourier transform or discrete inverse Fourier transform of a vector. As in MATLAB, .∗ is the elementwise vector product operator. We have also used MATLAB vector element notation to emphasize that the vectors are now of length $2K - 1$.

While the padded projection vector $\tilde{p}$ has at least $K - 1$ zero elements, its discrete Fourier transform generally has $2K - 1$ nonzero elements. The computed projection $C(\theta_l, \hat{t})$ also has $2K - 1$ nonzero elements. Padding the data vector as we have thus spreads the filtered projection out to nearly double the length of the original projection data when the computation is done in the frequency domain. We see in section 17.8.1, however, that the discrete transforms of vectors of length $2K - 1$ can be implemented in $\mathcal{O}((2K - 1)\log_2(2K - 1))$ floating-point operations. For the large values of K typical of CT applications, it is therefore more efficient to evaluate the convolution in the zero padded frequency domain representation of equation (17.38) than in the spatial representation of equation (17.36).

It remains to compute the frequency domain convolutions $C(\theta_l, \tilde{t})$ for each of the q angles and sum them via equation (17.29) to get the density. Note that each convolution $C(\theta_l, \tilde{t})$ can contribute density values along each of the $q(2K - 1)$ lines in the xy-plane defined by $t_j = x\cos\theta_l + y\sin\theta_l$. Recall that we originally discretized the image by overlaying it with a $\sqrt{n} \times \sqrt{n}$ grid. The gridsize n is generally unrelated to q and K and, furthermore, is typically much smaller than $q(K - 1)$. The process of coloring the discretized image thus amounts to compressing the available data onto the grid and summing appropriately.

For a given angle θ_l, the vector $C(\theta_l, \tilde{t})$ gives the values of the filtered projections only at the values of t given by the elements of $\tilde{t}$. These filtered projection values are methodically mapped to the grid by mapping each $C(\theta_l, \tilde{t}_j)$ onto its closest grid box. The values of t to which they are mapped are determined by evaluating $\tau_i = x_i \cos\theta_l + y_i \sin\theta_l$ at the center (x_i, y_i) of each box B_i, $i = 1, \ldots, n$, of the grid for each angle θ_l. It is not always the case that τ_i corresponds exactly to an element of the vector $\tilde{t}$. Thus, the element of $\tilde{t}$

closest to τ_i and within box B_i is used to approximate τ_i. All values $C(\theta_l, \tau_i)$, $l = 1, \ldots, q-1$, mapped to a box B_i are summed to produce the density within that box.

This summing process and equation (17.38) define the discrete filtered backprojection algorithm. The final section of this chapter is devoted to the computer implementation of this algorithm.

17.8   A computational filtered backprojection algorithm

17.8.1   The fast Fourier transform

The filtered backprojection algorithm uses the discrete Fourier transform and the discrete inverse Fourier transform as given in equations (17.24) and (17.25), respectively. Devising a good computational backprojection algorithm thus requires devising a good computational algorithm for evaluating these transforms. The first step in creating this algorithm is to determine the number of floating-point operations needed for it.

One approach to the problem derives from a straightforward rearrangement of equation (17.24). If we apply equation (17.24) to each data element $f_0, \ldots, f_{K-1}$ in turn, we are left with the K equations

$$F_{-K/2} = \sum_{j=0}^{K-1} e^{-i2\pi(-\frac{K}{2})\frac{j}{K}} f_j$$

$$F_{-K/2+1} = \sum_{j=0}^{K-1} e^{-i2\pi(-\frac{K}{2}+1)\frac{j}{K}} f_j$$

$$\vdots$$

$$F_{K/2-1} = \sum_{j=0}^{K-1} e^{-i2\pi(\frac{K}{2}-1)\frac{j}{K}} f_j.$$

If we set $w = e^{i2\pi/K}$, these equations become

$$F_{-K/2} = \sum_{j=0}^{K-1} e^{(-\frac{K}{2})j} f_j$$

$$F_{-K/2+1} = \sum_{j=0}^{K-1} e^{(-\frac{K}{2}+1)j} f_j$$

$$\vdots$$

$$F_{K/2-1} = \sum_{j=0}^{K-1} e^{(\frac{K}{2}-1)j} f_j.$$

This system of K linear equations in K unknowns defines the discrete Fourier transform $F = (F_{-K/2}, \ldots, F_{K/2-1})^T$ of the discrete data $f = (f_0, \ldots, f_{K-1})^T$.

If we define the $K \times K$ matrix W with elements $W_{j,l+\frac{K}{2}} = w^{-jl}$, for $j = 0, \ldots, K-1$ and $l = -\frac{K}{2}, \ldots, \frac{K}{2} - 1$, we can rewrite this transform as the matrix-vector product

$$F = W^T f. \tag{17.39}$$

The corresponding discrete inverse Fourier transform is $f = \frac{1}{K} W F$.

These relations imply that $\mathcal{O}(K^2)$ floating-point operations are required to evaluate a discrete Fourier transform or its inverse. A different arrangement of these computations shows that, in fact, they can each be done in $\mathcal{O}(K \log K)$ operations. The more efficient algorithms for computing the discrete Fourier and discrete inverse Fourier transforms are known as the *Fast Fourier Transform (FFT)* and the *Inverse FFT (IFFT)*. Variants of these algorithms were discovered independently by several authors [Kahaner et al 77], but they were first made famous by Cooley and Tukey at IBM [Cooley & Tukey 65].

The FFT relies on the recursive structure of the discrete Fourier transform to reduce the operation count. At its highest level, the FFT splits the sum of equation (17.24) into sums of even and odd numbered terms as follows:

$$
\begin{aligned}
F_l &= \sum_{j=0}^{K-1} f_j e^{-i2\pi l j/K} \\
&= \sum_{j=0}^{K/2-1} f_j e^{-i2\pi l(2j)/K} + \sum_{j=0}^{K/2-1} f_{2j+1} e^{-i2\pi l(2j+1)/K} \\
&= \sum_{j=0}^{K/2-1} f_j e^{-i2\pi l j/(K/2)} + w^{-l} \sum_{j=0}^{K/2-1} f_{2j+1} e^{-i2\pi l j/(K/2)} \\
&= F_l^e + w^{-l} F_l^o.
\end{aligned}
$$

That is, we can express the discrete Fourier transform F_l of length K as the weighted sum of the two discrete Fourier transforms F_l^e and F_l^o, each of length $K/2$. The superscript e means that the sum is formed from the even-indexed terms of F_l, and the superscript o means that the sum is formed from the odd-indexed terms of F_l. The result is a statement of the *Danielson-Lanczos Lemma*.

Suppose that K is a power of 2. Applying the lemma a second time allows us to split F_l into the sum of four discrete Fourier transforms of length $K/4$

$$\begin{aligned} F_l &= F_l^{ee} + w^{-l}F^{eo} + w^{-l}(F_l^{oe} + w^{-l}F_l^{oo}) \\ &= F_l^{ee} + w^{-l}[(F_l^{eo} + F_l^{oe}) + w^{-l}F_l^{oo}]. \end{aligned}$$

Dividing a third time gives us a sum of eight discrete Fourier transforms of length $K/8$

$$\begin{aligned} F_l = \ F_l^{eee} &+ w^{-l}\{(F_l^{eeo} + F_l^{eoe} + F_l^{oee}) \\ &+ w^{-l}[(F_l^{eoo} + F_l^{oeo} + F_l^{ooe}) + w^{-l}F_l^{ooo}]\}. \end{aligned}$$

Note that the number of times that w^{-l} multiplies a term is equal to the number of o's in the superscript of that term.

We can continue this recursive division until the resulting discrete Fourier transforms are of length one. This process takes $\log_2 K$ divisions, and the resulting length-one transformations F_l^x have binary superscripts x that are $\log_2 K$ bits long. We can then copy the discrete Fourier transforms of length one directly from the data vector f because

$$F_l^x = \sum_{j=0}^{0} f_j^x w^{-lj} = f_0^x,$$

where f_0^x is just one element of the vector f.

To determine which element of f gives each f_0^x we follow these steps:

- Reverse the superscript x of each length-one transform (i.e., eeo becomes oee.)

- Substitute 0 for e and 1 for o.

The decimal value of each superscript then gives the index of the corresponding element of the vector f. This process is illustrated in figure 17.14 for $K = 8$. The formation of the length-one transforms is the result of traversing a binary tree from its root to its leaves.

The beauty of this approach is that once we have identified which element of the vector f corresponds to which of the length-one transforms, we can easily determine F_l by recursively summing transformations of lengths 1 through $K/2$ with multiplication by w^{-l} in the appropriate places. The nodes of the tree in figure 17.14 represent the series formed at each step of the division process. That is, the nodes at level i of the tree each represent series of length $2^{\log_2 K - i}$, for $i = 0, \ldots, \log_2 K$. The source of the superscripts at the leaves becomes clearer if we look at the how the data elements are actually scattered among the nodes of the tree. These are shown in figure 17.15. The number of o's in a leaf subscript and, hence, the number of 1's its binary identifier shows the number of branches to the right there are in the path from the root to that leaf.

The w^{-l} multipliers are placed correctly by applying the Danielson-Lanczos Lemma while traversing the tree from its leaves to its root. For the example of figures 17.14 and 17.15, the summation would follow the $\log_2 8 = 3$ levels of the tree:

$$\text{level 2:} \quad \begin{aligned} f^{ee} &= f^{eee} + w^{-l} f^{eeo}, \\ f^{eo} &= f^{eoe} + w^{-l} f^{eoo}, \\ f^{oe} &= f^{oee} + w^{-l} f^{oeo}, \\ f^{oo} &= f^{ooe} + w^{-l} f^{ooo} \end{aligned}$$

$$\text{level 1:} \quad \begin{aligned} f^{e} &= f^{ee} + w^{-l} f^{eo}, \\ f^{o} &= f^{oe} + w^{-l} f^{oo} \end{aligned}$$

$$\text{level 0:} \quad f = f^{e} + w^{-l} f^{o}.$$

At level i, we combine two pairs of series of length $2^{\log_2 K - i - 1}$ and so perform $2^{\log_2 K - i - 1}$ floating-point multiplications and $2^{\log_2 K - i - 1}$ floating-point additions. Summing these operation counts over the levels $i = 0, \ldots, \log_2 K - 1$ of the tree shows that the FFT does indeed require $\mathcal{O}(K \log_2 K)$ floating-point operations.

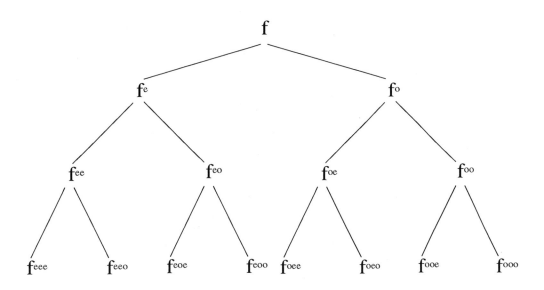

superscripts at leaves:	*eee*	*eeo*	*eoe*	*eoo*	*oee*	*oeo*	*ooe*	*ooo*
reverse pattern :	*eee*	*oee*	*eoe*	*ooe*	*eeo*	*oeo*	*eoo*	*ooo*
$e = 0, o = 1$:	000	100	010	110	001	101	011	111
decimal value :	0	4	2	6	1	5	3	7

Figure 17.14: The tree underlying an FFT of length eight.

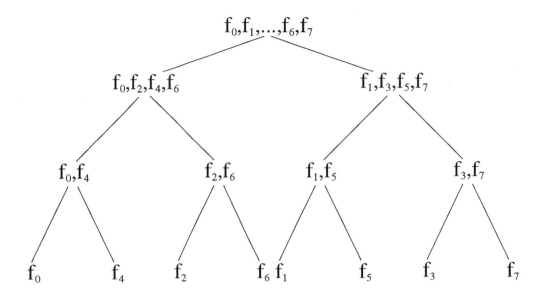

Figure 17.15: The data elements at the nodes of the FFT tree.

We use the notation $FFT(f)$ for the operation of taking the FFT of the data vector f and the notation $IFFT(F)$ for taking the IFFT of the vector F. Then the periodic convolution $f * g$ is defined elementwise by

$$f * g = IFFT(FFT(f). * FFT(\text{shift}(g))), \qquad (17.40)$$

where the elements of g are shifted as in the definition of the discrete convolution in equation (17.26). Remember that when this periodic convolution is applied to aperiodic data, that data must be padded with enough zeros to prevent interperiod interference. As in MATLAB, the notation .* denotes the elementwise product of two vectors.

In our derivation of the FFT, we assumed that all vectors involved were of a length K equal to a power of two. In practice, we can force this assumption to be satisfied in a very easy way by padding the vector with an appropriate number of zeros. Even if these zero elements are actually used in the computation, the loss of efficiency is not great. For example, doubling the length of a vector from K to $2K$ raises the cost of the FFT from an $\mathcal{O}(K \log_2 K)$ operation to an $\mathcal{O}(2K \log_2 K)$ operation. For large K, the FFT with the length

$2K$ vector is still faster than the $\mathcal{O}(K^2)$ matrix-vector product approach of equation (17.39). The advantage of the FFT, even on a vector padded with many zeros, is most pronounced for the very long vectors typical in an image reconstruction problem. The efficient FFT algorithm is the basis of the efficient implementation of backprojection.

17.8.2 A computational example

In this section, we work through an example of the discrete filtered backprojection algorithm based on equations (17.38) and (17.35) and on the data mapping procedure described at the end of section 17.7. For this example, we take a CT scan of a beam of constant density with an elliptical cross-section. The beam is loaded into the scanner so that it runs perpendicular to the plane crossed by the rays. The beam is located slightly below and to the left of the scanner's center.

If we could create a continuous and exact representation of the beam's cross-section, it would appear as a solid white ellipse. Figure 17.16 shows how that two-dimensional cross-section appears when the image is discretized. In this figure, the full field of view has been overlaid with a 128×128 grid.

For this example, we carry out the CT scan by passing $K = 128$ parallel rays through the beam at each of $q = 90$ angles and recording the attenuations I_0/I_T of each. According to equation (17.5), taking the logarithm of these measurements defines the projection $p = (p_0, \ldots, p_{K-1})^T$ at each angle. We store the projection for the ray at angle l and position j as element (l, j) of a data matrix D.

The next step is to transform the data into the frequency domain. To avoid interperiod interference in the convolution defined by equation (17.38), we must still zero-pad the measured projection data. As row l of the matrix D holds the K measured projection values at angle θ_l, we must pad that row with zeros to a length of $(2K - 1)_2 = 255$. Keeping in mind that we require a power of two vector length for the FFT algorithm, we affix one more zero to bring the length to $2K = 256 = 2^8$. The zeros are placed in elements $D_{l,-K:-1}$. Repeating this zero padding for each of the $q = 90$ rows converts D into an 90×256 matrix. For ease in programming, the rows in this matrix are then permuted so that the elements $D_{l,-K:-1} = D_{l,-128:-1}$ follow the elements

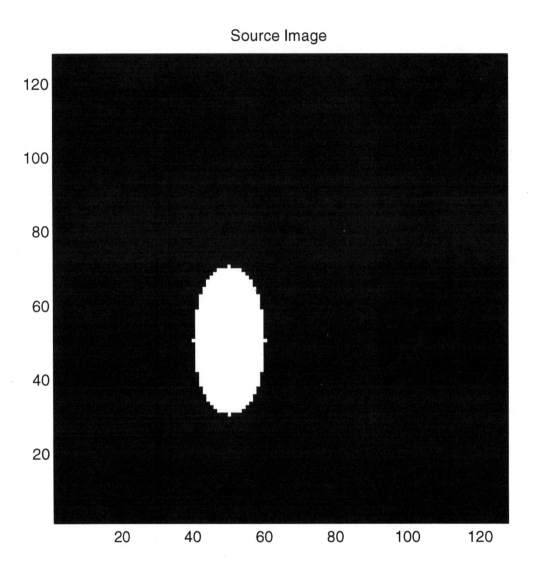

Figure 17.16: The discretized exact image.

$D_{l,0:K-1} = D_{l,0:127}$ for $l = 0, 89$. This removes the need to handle negative matrix indices in the program.

Figure 17.17 shows a gray scale depiction of the matrix D: the pixel block at position (l, j) is colored a shade of gray determined by the value of the projection $D_{l,j}$. A black pixel block corresponds to a zero value. Thus, the right half of the figure is black because of zero padding. If $D_{l,j} = 0$, but $j \leq K - 1 = 127$, the ray at angle θ_l and position t_j does not pass through the object.

The projections for the angles near zero are plotted near the bottom of the figure. The rays at angle $\theta_0 = 0$ pass through the beam section from top to bottom as depicted in figure 17.18. This figure shows fifteen of the $K = 128$ rays in the projection. The first one is at $j = 8$, so every eighth ray is shown.

In figure 17.17, the nonzero (light-colored) projections corresponding to the rays shown in figure 17.18 form a narrow band just to the left of the center of the image. The location of this band corresponds to the location of the beam with respect to the center of the scanner. Rays pass near the center of the projection (ray 64 and those nearby) near the center of the scanner but miss the beam. Only rays 40 through 60 hit the beam. Ray 50 travels through the center of the beam, along the major axis of its cross section, and so is attenuated most. Hence, although there is little variation in color across the band, $D_{0,50}$ shows as a slightly brighter white value. Rays 40 and 60 pass only through the narrow ends of the beam and so are not attenuated as greatly as is ray 50. The corresponding elements of D are thus duller shades of gray.

As the angle value increases, the bright band spreads and swings to the left. The rays no longer cross the beam parallel to the major axis of its cross-section. Therefore, more of the rays in a single parallel projection cross the beam, but none of them travels as great a distance through the beam as does ray 50 at angle $\theta_0 = 0$. The bright band is at its widest and faintest around $\theta_{45} = \frac{\pi}{2}$ where the rays cross the beam as shown in figure 17.19. (The rays in this figure are spaced as in figure 17.18.) At θ_{45}, the rays cross the beam nearly parallel to the minor axis of the ellipse. The sinuous form of the white band in figure 17.17 is typical of such a plot of parallel projections, and so this type of plot is referred to as a *sinogram*.

The rows of the matrix D define the padded projection vectors used in equation (17.38). The next step is to take the FFT of each of these vectors

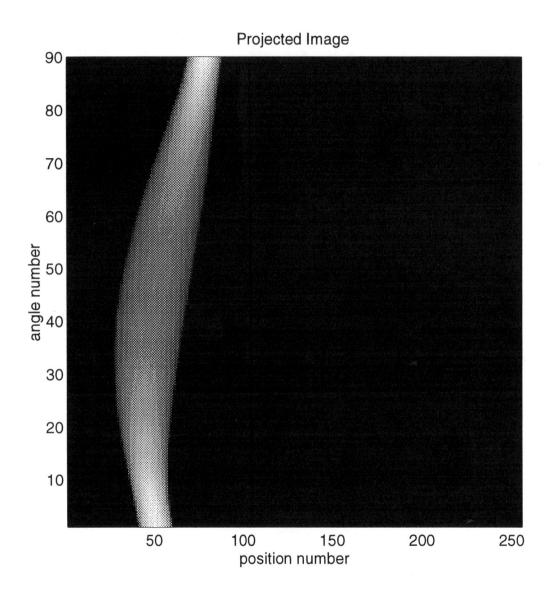

Figure 17.17: The zero padded projected image.

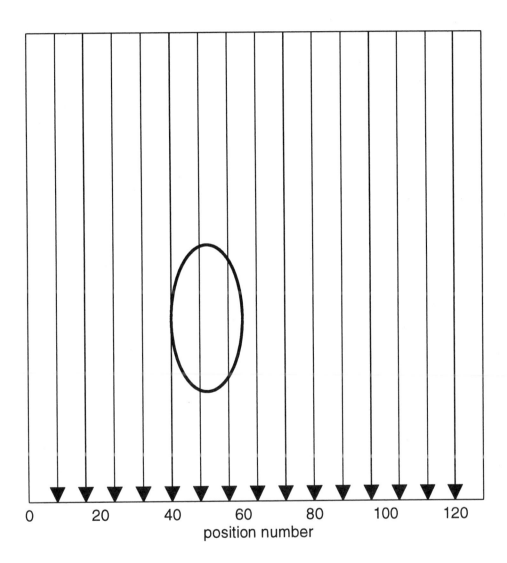

Figure 17.18: The rays at θ_0.

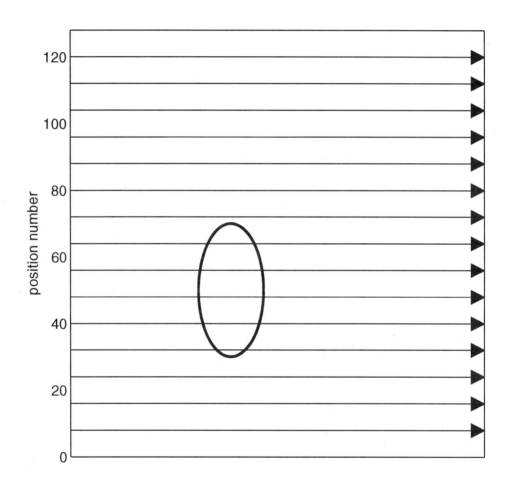

Figure 17.19: The rays at θ_{45}.

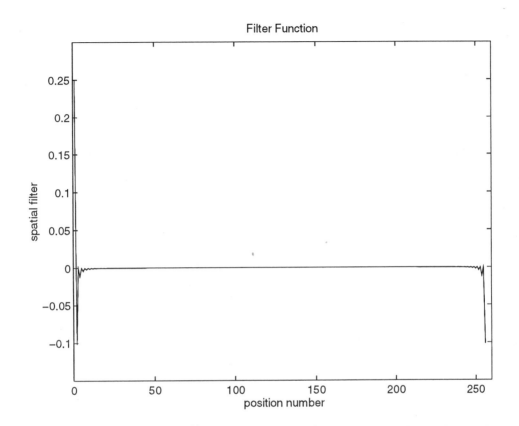

Figure 17.20: The zero padded and swapped filter in the spatial domain.

and compute their elementwise sum with the FFT of the padded spatial filter vector b via equation (17.37). To properly match the reordered vector p, the vector elements $b_{-128:-1}$ must be moved to positions $b_{128:256}$. A plot of this reordered vector is given in figure 17.20. In this figure, the spatial interval Δt is taken to be one so that the filter reaches its maximum of $1/4$ at $j = 0$. Without the element swapping, this peak would appear at the center of the plot, and it would be obvious that the filter function b is symmetric about zero.

The result B of taking the FFT of the padded and swapped vector b is shown in figure 17.21. If the elements $B_{128:256}$ were swapped back into

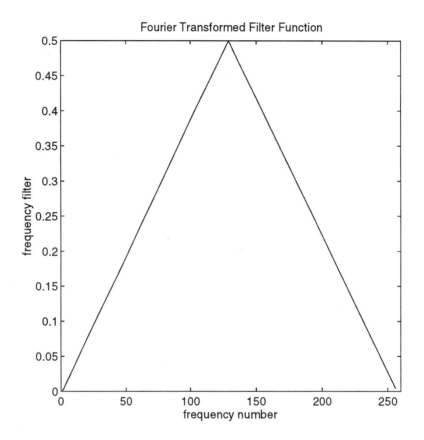

Figure 17.21: The zero padded and swapped filter in the frequency domain.

their original positions $B_{-128:-1}$, this figure would approximate the familiar V shape of the function $|\rho|$ from which it was ultimately derived. It varies from that exact function only near $\rho = 0$, where the computed $B(0)$ curves into a small positive value rather than the exact zero value. (See [Kak & Slaney 88, Rosenfeld & Kak 82].) As $B(256)$ is actually $B(-1)$, neither endpoint of the plotted curve reaches zero.

The next step in the filtered backprojection algorithm is to take the elementwise product of the FFT's of p and b as in equation (17.38). Like the projections, the filtered projections are stored in a matrix with $C(\theta_l, \tilde{t})$ forming row l of that matrix. Figure 17.22 is a gray scale depiction of that matrix. The sinuous form of the sinogram is still evident in the filtered projections, but the zero padding and FFT's have spread the nonzero values out along the whole length of the zero padded vector.

The effects of the filtering are also evident in figure 17.22. The filter B selectively removes low frequencies from the transformed projections. In this way, it serves as a *high-pass* filter. Examination of figure 17.17 shows that the nonzero projections at angles near θ_0 and θ_{90} are fairly uniformly white. That is, except for an abrupt transition from white to black at the edges of the band, the frequency of variation in the projections is low. Thus, the filter causes a noticeable dulling of the band in figure 17.22 for these extreme angles. On the other hand, projections taken at angles near θ_{45} show more and more rapid variation in brightness. The high-pass filter has little effect on the band at these angles, and there is little difference in brightness between figures 17.17 and 17.22. For similar reasons, the filtering accounts for the marked transition from light to dark along the edges of the band in figure 17.22 .

The final step of the filtered backprojection algorithm is to sum the filtered projections $C(\theta_l, \tilde{t})$. In our example, the filtered projection values from the $2Kq = 23,040$ lines $t_j = x \cos \theta_l + y \sin \theta_l$ are mapped onto the 128×128 grid by the summing process described at the end of section 17.7. The result is shown as figure 17.23. Comparing this figure to figure 17.16 shows that the beam's cross-section appears as a bright ellipse of the correct shape in the correct position in the backprojected image. The patterns in the background result from taking a finite number of projection measurements. In particular, they result from using only those frequencies represented by the elements of the vector B in the backprojected image. A detailed study of these *aliasing artifacts* for elliptical and other images is presented in [Kak & Slaney 88].

One final issue to consider in this example is the role of complex numbers in the process. The Fourier transform and inverse Fourier transform clearly move functions back and forth between the real line and the complex plane. Nevertheless, the real figures we have shown in this section give accurate information about the projections, the filters, the filtered projections, and the backprojected image they depict. That is, none of these quantities is complex. The projection data is real because it is measured in the laboratory, and the filter b is real by construction. The symmetry of the filter b ensures that its

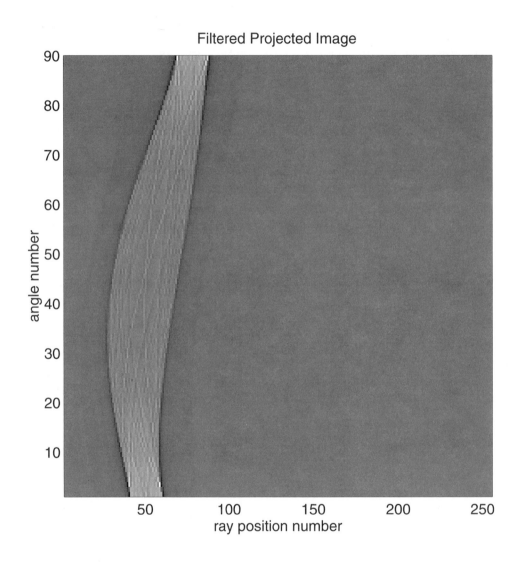

Figure 17.22: The filtered projections.

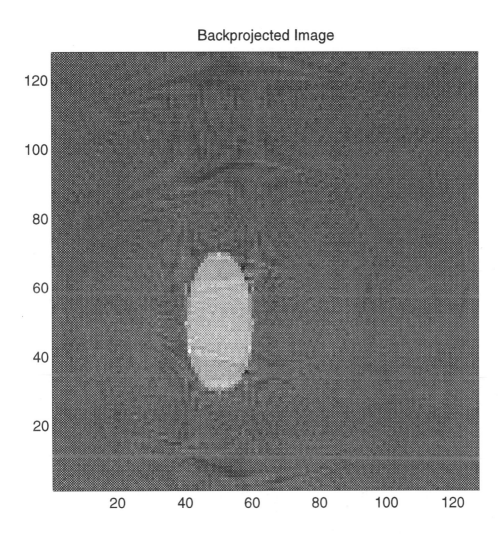

Figure 17.23: The backtransformed image.

exact Fourier transform B is real (in exact arithmetic). The symmetry of B, in turn, ensures that the exact filtered projections are real (again, in exact arithmetic). The real filtered projections are summed to create the real back-projected image. The latter must, of course, be real as it represents a density map of an actual object. In the course of the computation, however, rounding error leads to imaginary components in each of these computed quantities. Fortunately, the stability of the algorithms used at each stage guarantees that the imaginary components are very small so that it is reasonable to ignore them in the figures.

17.8.3   A real life example

In practice, medical CT scans are usually generated in real time in the examination room. Measured projection data is fed into special purpose hardware, typically a custom configuration of digital signal processing chips that perform the FFT's, IFFT's, and sums, and the image is reconstructed directly via special purpose mathematical and imaging software. This process relies on a hardware and software implementation of the filtered backprojection algorithm, supplemented with image refinement techniques such as those presented in [Rosenfeld & Kak 82, Kak & Slaney 88, Russ 92]. The reconstructed image may be viewed as a two-dimensional slice or as a three-dimensional image built up from a series of two-dimensional images.

The efficiency of the process is improved by pipelining the collection of data with reconstruction of the image. To see how this is carried out, we must reexamine equations (17.38) and (17.35). Note that the filtered projection $C(\theta_l, \tilde{t})$ depends only on the single parallel projection p measured at angle θ_l. Thus, as soon as those data are available, computation of the convolution defining $C(\theta_l, \tilde{t})$ can begin. Once the convolution is complete, accumulation of the sum $\mu(x,y) \approx \frac{\pi}{q} \sum_{l=0}^{q-1} C(\theta_l, t)$ can begin. Time to produce an image via this technique varies from several slices per minute to several minutes per slice depending on the amount of data, the resolution required, and the hardware and software used.

In this section, we show the results of the CT scan of a human head. In this scan, $K = 1024$ rays were passed through the head at each of $q = 1200$ angles. The sinogram of this image given in figure 17.24.

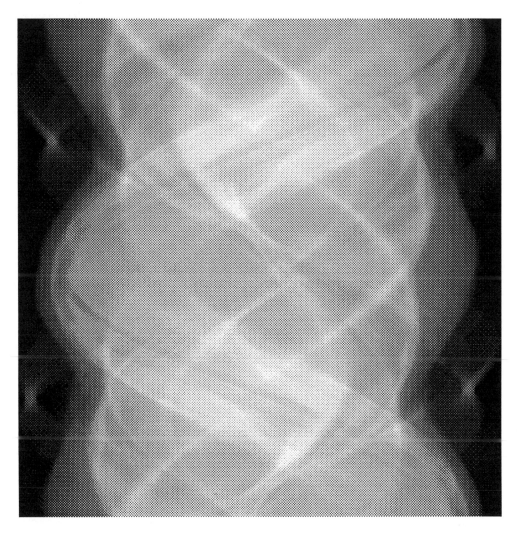

Figure 17.24: The sinogram of the CT scan of a human head (one slice).

The sinogram is not customarily produced in a medical application, but we provide one here just to show how the variations in density through a head complicate the parallel projection data. The sinuous forms of figure 17.17 are still visible, but there are many more such curves of varying brightnesses. Recall that these forms result from varying the projection angle. This sinogram is not zero-padded.

Fourier transforming, filtering and backprojecting the projection data from figure 17.24 gives the two-dimensional axial slice of the head shown in figure 17.25. This is a 512×512 discretized image. The edge of the slice at the patient's face is at the top of the image. The patient's head rests in a carbon fiber frame, visible at the sides and bottom of the image. Rolled towels used to stabilize the head are visible at the sides of the face. Various anatomical features are also visible. This slice was taken at the base of the nose. The palate is visible as an oblong shape at the top center of the image. The dark regions just above it are sinuses. The ears can be seen on the sides of the head, and the ear canals extend inward from them. The bones of the inner ear are visible as bright shapes. The outline of the lobes of the brain and the skull are visible at the bottom of the image. The skull, like all other bone in the image, is bright white. The gray section at the very bottom of the image is the tissue outside of the skull.

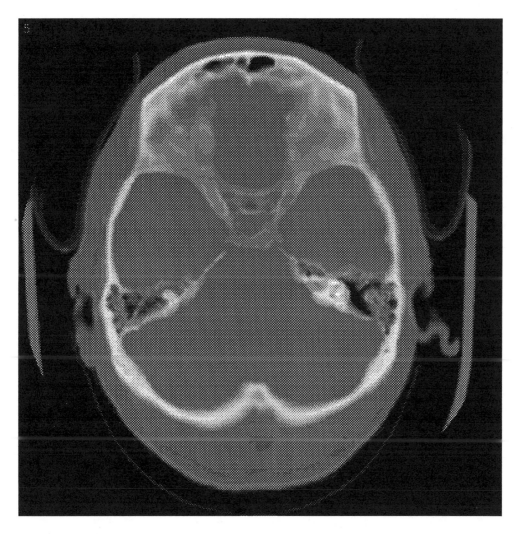

Figure 17.25: A backprojected image of one two-dimensional slice of a human head.

Bibliography

[Adams et al 92] ADAMS, JEANNE C., WALTER S. BRAINERD, JEANNE T. MAR-
TIN, BRIAN T. SMITH, AND JERROLD L. WAGENER. [1992]. *Fortran
90 Handbook*. Intertext Publications. McGraw-Hill Book Company, New
York, NY.

[Alder & Wainright 59] ALDER, B. J. AND T. E. WAINRIGHT. [1959]. Studies in
molecular dynamics I: General method. *Journal of Chemical Physics*,
31:459–466.

[Alder & Wainright 60] ALDER, B. J. AND T. E. WAINRIGHT. [Nov 1960]. Studies
in molecular dynamics II: Behavior of a small number of elastic spheres.
Journal of Chemical Physics, 33(5):1439–1451.

[Alliant 86] Alliant Computer Systems Corporation, Littleton, MA. [Oct 1986].
FX/Series Product Summary.

[Almasi & Gottlieb 89] ALMASI, GEORGE S. AND ALLAN GOTTLIEB. [1989].
Highly Parallel Computing. The Benjamin/Cummings Publishing Company, Inc., Redwood City, CA, 1st edition.

[Almasi & Gottlieb 94] ALMASI, GEORGE S. AND ALLAN GOTTLIEB. [1994].
Highly Parallel Computing. The Benjamin/Cummings Publishing Company, Inc., Redwood City, CA, 2nd edition.

[Anderson et al 92] ANDERSON, E., Z. BAI, C. BISCHOF, J. DEMMEL, J. DON-
GARRA, J. DUCROZ, A. GREENBAUM, AND S. HAMMARLING. [1992].
LAPACK User's Guide. SIAM, Philadelphia, PA.

[Anderson & Dziewonski 84] ANDERSON, D.L. AND A.M. DZIEWONSKI. [Oct.
1984]. Seismic tomography. *Scientific American*, 251:60–68.

[ANSI 91] Americal National Standards Institute, Inc., New York, NY. [May 1991].
Americal National Standard Programming Language Fortran 90. ANSI
X3.198-1991; ISO/IEC 1539:1991.

[Augarten 84] AUGARTEN, STAN. [1984]. *Bit by Bit.* Ticknor & Fields, New York, NY.

[Baer 80] BAER, JEAN-LOUP. [1980]. *Computer Systems Architecture.* Computer Society Press, Rockville, MD.

[Bailey 91] BAILEY, DAVID H. [Oct 1991]. MPFUN: A portable high precision performance multiprecision package. RNR Technical Report RNR-90-022, NASA Ames Research Center.

[Bailey et al 94] BAILEY, DAVID H., ERIC BARSZCZ, LEONARDO DAGUM, AND HORST D. SIMON. [Mar 1994]. NAS Parallel Benchmark results 3-94. RNR Technical Report RNR-94-006, NASA Ames Research Center.

[Barnett et al 91] BARNETT, M., D. G. PAYNE, AND R. VAN DE GEIJN. [1991]. Optimal broadcasting in mesh-connected architectures. Technical Report TR-91-38, Dept. of Computer Science, University of Texas at Austin.

[Barnett et al 94] BARNETT, M., D.G. PAYNE, R. VAN DE GEIJN, AND J. WATTS. [1994]. Broadcasting on meshes with worm-hole routing. Submitted to *Journal of Parallel and Distributed Computing.*

[Baron & Higbie 92] BARON, ROBERT J. AND LEE HIGBIE. [1992]. *Computer Architecture: Case Studies.* Electrical and Computer Engineering. Addison-Wesley Publishing Company, New York, NY.

[Beddow 90] BEDDOW, J. [Oct 1990]. Shape coding of multidimensional data on a microcomputer display. In *Proceedings of Visualization '90*, pages 238–247, Washington, DC. IEEE Computer Society Press.

[Boyle & Thomas 88] BOYLE, R.D. AND R.C. THOMAS. [1988]. *Computer Vision.* Blackwell Scientific Publications, Oxford.

[Brainerd et al 90] BRAINERD, WALTER S., CHARLES GOLDBERG, AND JEANNE C. ADAMS. [1990]. *Programmers Guide to Fortran 90.* McGraw-Hill Book Company, New York, NY.

[Brodlie et al 92] BRODLIE, K.W., L.A. CARPENTER, R.A. EARNSHAW, J.R. GALLOP, R.J. HUBBOLD, A.M. MUMFORD, C.D. OSLAND, AND P. QUARENDON, editors. [1992]. *Scientific Visualization, Techniques and Applications.* Springer-Verlag, New York, NY.

[Burden & Faires 85] BURDEN, RICHARD L. AND J. DOUGLAS FAIRES. [1985]. *Numerical Analysis.* PWS-KENT Publishing Company, Boston, MA, 3rd edition.

[Carrier & Pearson 76] CARRIER, GEORGE F. AND CARL E. PEARSON. [1976]. *Partial Differential Equations: Theory and Practice.* Academic Press, Inc., Orlando, FL.

[Carriero & Gelernter 89] CARRIERO, N. AND D. GELERNTER. [April 1989]. Linda in context. *Communications of the ACM*, 32(4):444–458.

[Chernoff 73] CHERNOFF, H. [1973]. The use of faces to represent points in k-dimensional space graphically. *Journal of the American Statistical Association*, 68(342):361–368.

[Conte & de Boor 80] CONTE, S. D. AND CARL de BOOR. [1980]. *Elementary Numerical Analysis: An Algorithmic Approach.* McGraw-Hill, Inc., New York, NY, 3rd edition.

[Cooley & Tukey 65] COOLEY, J.W. AND J.W. TUKEY. [1965]. An algorithm for the machine calculation of complex Fourier series. *Mathematics of Computation*, 19:297–301.

[Cormack 63] CORMACK, A.M. [1963]. Representation of a function by its line integrals with some radiological applications. *Journal of Applied Physics*, 34:2722–2727.

[Cormack 64] CORMACK, A.M. [1964]. Representation of a function by its line integrals with some radiological applications II. *Journal of Applied Physics*, 35:2908–2913.

[Courant et al 67] COURANT, R., K. FRIEDRICHS, AND H. LEWY. [Mar 1967]. On the partial difference equations of mathematical physics. *IBM Journal of Research and Development*, 108:215–234. This is a translation of the original German article published in *Mathematische Annalen*, 100:32–74 (1928).

[Cray 78] Cray Research, Inc., Mendota Heights, MN. [1978]. *The Cray-1 Computer System: Hardware Reference Manual.* 2240004.

[Cray 89] Cray Research, Inc., Mendota Heights, MN. [1989]. *Fortran (CFT) Reference Manual.* SR-0009 M.

[Cray 91] Cray Research, Inc., Mendota Heights, MN. [1991]. *UNICOS Performance Utilities Reference Manual.* SR-2040.

[Dagum 88] DAGUM, LEONARDO. [1988]. Implementation of a hypersonic rarefied flow particle simulation on the connection machine. RIACS Technical Report 88.46, RIACS, NASA Ames Research Center. Also published as conference paper in 1989.

[Dahlquist & Björck 74] DAHLQUIST, GERMUND AND ÅKE BJÖRCK. [1974]. *Numerical Methods.* Prentice-Hall, Inc., Englewood Cliffs, NJ. Translated by Ned Anderson.

[DEC 90a] Digital Equipment Corporation, Maynard, MA. [Dec 1990]. *DEC AVS Developer's Guide.*

[DEC 90b] Digital Equipment Corporation, Maynard, MA. [Dec 1990]. *DEC AVS User's Guide.*

[DEC 91] Digital Equipment Corporation, Palo Alto, CA 94301. [Dec 1991]. *DECstation 5000 Model 240: Technical Overview.*

[Dewar & Smosna 90] DEWAR, R. AND M. SMOSNA. [1990]. *Microprocessors: A Programmer's Point of View.* McGraw-Hill, Inc., New York, NY.

[Dewdney 90] DEWDNEY, A.K. [Sept. 1990]. Mathematical recreations: How to resurrect a cat from its grin. *Scientific American*, 262:174–177.

[Domik & Gutkauf 94] DOMIK, G. O. AND B. GUTKAUF. [1994]. User modeling for adaptive visualization systems. In *Proceedings of Visualization '94*, pages 117–123, Washington, DC. IEEE Computer Society Press.

[Dongarra & Sorensen 87] DONGARRA, JACK J. AND DANNY C. SORENSEN. [1987]. SCHEDULE: Tools for developing and analyzing parallel Fortran programs. In JAMIESON, LEAH J., DENNIS B. GANNON, AND ROBERT J. DOUGLASS, editors, *The Characteristics of Parallel Algorithms*, pages 363–394. The MIT Press, Cambridge, MA.

[Dongarra 94] DONGARRA, J. J. [1994]. Performance of various computers using standard linear equations software. Technical Report CS-89-85, Oak Ridge National Laboratory, Oak Ridge, TN 37831. `netlib` version as of November 1, 1994.

[Dongarra et al 79] DONGARRA, J. J., C. B. MOLER, J. R. BUNCH, AND G. W. STEWART. [1979]. *LINPACK User's Guide*. SIAM, Philadelphia, PA.

[Dowd 93] DOWD, KEVIN. [1993]. *High Performance Computing*. O´Reilly & Associates, Inc., Sebastopol, CA.

[Dunigan 91] DUNIGAN, T. H. [1991]. Performance of the Intel iPSC/860 and Ncube 6400 hypercubes. *Parallel Computing*, 17(1991):1285–1302.

[Dunigan 92] DUNIGAN, T. H. [1992]. Communication performance of the Intel Touchstone Delta mesh. Technical Report ORNL/TM-11983, Oak Ridge National Laboratory.

[Dunigan 94] DUNIGAN, T. H. [1994]. Early experiences and performance of the Intel Paragon. Technical Report ORNL/TM-12194, Oak Ridge National Laboratory.

[Eberlein & Eastridge 91] EBERLEIN, MARY AND ERIC EASTRIDGE. [Aug 1991]. *A Beginner's Guide to the MasPar MP-2*. Joint Institute for Computational Science (JICS), University of Tennessee, Knoxville, TN.

[Faber 93] FABER, V., [1993]. Los Alamos National Laboratory, Personal Communication.

[Feldman 86] FELDMAN, S. I. [Nov 1986]. Make – a program for maintaining computer programs. In GROUP, COMPUTER SYSTEMS RESEARCH, editor, *Unix Programmer's Manual Supplementary Documents 1*, pages PS1:12–1 to PS1:12–9. USENIX Association, Berkeley, CA.

[Flynn 72] FLYNN, MICHAEL J. [Sep 1972]. Some computer organizations and their effectiveness. *IEEE Transactions on Computers*, C-21(9):948–960.

[Foley et al 90] FOLEY, JAMES D., ANDRIES VAN DAM, STEVEN K. FENIER, AND JOHN F. HUGHES. [1990]. *Computer Graphics: Principles and Practice*. Systems Programming. Addison-Wesley Publishing Company, New York, NY, 2nd edition.

[Fox et al 88] FOX, G. C., M. A. JOHNSON, G. A. LYZENGA, S. W. OTTO, J. K. SALMON, AND D. W. WALKER. [1988]. *Solving Problems on Concurrent Processors*, volume 1. Prentice-Hall, Inc., Englewood Cliffs, NJ.

[Fox et al 94] FOX, GEOFFREY C., ROY D. WILLIAMS, AND PAUL C. MESSINA. [1994]. *Parallel Computing Works!* Morgan Kaufmann Publishers, Inc., San Francisco, CA.

[Friedman 88] FRIEDMAN, A. [1988]. *Mathematics in Industrial Problems.* Springer-Verlag, New York, NY.

[Geist et al 95] GEIST, AL, ADAM BEGUELIN, JACK DONGARRA, WEICHING JIANG, ROBERT MANCHEK, AND VAIDY SUNDERAM. [1995]. *PVM – Parallel Virtual Machine: A Users' Guide and Tutorial for Network Parallel Computing.* Scientific and Engineering Computation. The MIT Press, Cambridge, MA.

[Golub & Van Loan 83] GOLUB, G.H. AND C.F. Van LOAN. [1983]. *Matrix Computations.* The Johns Hopkins University Press, Baltimore, MD, 1st edition.

[Golub & Van Loan 89] GOLUB, G.H. AND C.F. Van LOAN. [1989]. *Matrix Computations.* The Johns Hopkins University Press, Baltimore, MD, 2nd edition.

[Gropp et al 94a] GROPP, W., E. LUSK, AND S. PIEPER. [Oct 1994]. User's guide for the ANL IBM SP-1. Technical Report ANL/MCS-TM-198, Argonne National Laboratory, Argonne, IL.

[Gropp et al 94b] GROPP, WILLIAM, EWING LUSK, AND ANTHONY SKJELLUM. [1994]. *Using MPI: Portable parallel programming with the message passing interface.* Scientific and engineering computation. The MIT Press, Cambridge, MA.

[Haber & McNabb 90] HABER, R. B. AND D.A. MCNABB. [1990]. Visualization idioms: A conceptual model for scientific visualization systems. In NIELSON, G.M., B. SHRIVER, AND L. ROSENBLUM, editors, *Visualization in Scientific Computing*, Tutorial, pages 74–93. IEEE Computer Society Press, Washington, DC.

[Hall 79] HALL, E.L. [1979]. *Computer Image Processing and Recognition.* Academic Press, Inc., San Diego, CA.

[Haltiner & Williams 80] HALTINER, GEORGE J. AND ROGER T. WILLIAMS. [1980]. *Numerical Prediction and Dynamic Meteorology*. John Wiley & Sons, New York, NY, 2nd edition.

[Hannon et al 86] HANNON, L., G. C. LIE, AND E. CLIMENTI. [1986]. Molecular dynamics simulation of flow past a plate. *Journal of Scientific Computing*, 1(2):145–150.

[Heath et al 90] HEATH, M. T., G. A. GEIST, B. W. PEYTON, AND P. H. WORLEY. [1990]. A user's guide to PICL. Technical Report ORNL/TM-11616, Oak Ridge National Laboratory.

[Hennessy & Patterson 90] HENNESSY, JOHN L. AND DAVID A. PATTERSON. [1990]. *Computer Architecture: A Quantitative Approach*. Morgan Kaufmann Publishers, Inc., San Mateo, CA.

[Henning & Volkert 85] HENNING, W. AND J. VOLKERT. [1985]. Programming EGPA systems. In *Proceedings of the Fifth International Conference on Distributed Computing Systems*, pages 552–559, Washington, DC. IEEE Computer Society Press.

[Henrici 62] HENRICI, PETER. [1962]. *Discrete Variable Methods in Ordinary Differential Equations*. John Wiley & Sons, New York, NY.

[Herman 80] HERMAN, G.T. [1980]. *Image Reconstruction for Projections*. Academic Press, Inc., San Diego, CA.

[Higham 93] HIGHAM, NICHOLAS J. [1993]. *Handbook of Writing for the Mathematical Sciences*. SIAM, Philadelphia, PA.

[Hill 90] HILL, F.S. [1990]. *Computer Graphics*. Maxwell Macmillan International Editions, New York, NY.

[Hillis 85] HILLIS, W. DANIEL. [1985]. *The Connection Machine*. The MIT Press, Cambridge, MA.

[Hockney & Jesshope 88] HOCKNEY, R. W. AND C. R. JESSHOPE. [1988]. *Parallel Computers 2*. Adam Hilger, Bristol.

[Hoffmann et al 88] HOFFMANN, G. R., P. N. SWARZTRAUBER, AND R. A. SWEET. [1988]. Aspects of using multiprocessors for meterorological modelling.

In HOFFMAN, G.-R. AND D. F. SNELLING, editors, *Multiprocessing in Meteorological Models*, pages 127–196. Springer-Verlag, New York, NY.

[Hord 90] HORD, R. MICHAEL. [1990]. *Parallel Supercomputing in SIMD Architectures*. CRC Press, Inc., Boston, MA.

[Hoshino 86] HOSHINO, T. [1986]. An invitation to the world of PAX. *IEEE Computer*, 19:68–79.

[Hounsfield 73] HOUNSFIELD, G.N. [1973]. Computerized transverse axial scanning (Tomography): Part I. Description of the system. *British Journal of Radiology*, 16:210–224.

[HPCC 89] Executive Office of the President, Office of Science and Technology, Washington, D.C. [Sep 1989]. *The Federal High Performance Computing Program*. This is often referred to as the Bromley Report. The program is now called the High Performance Computing and Communications (HPCC) program.

[Hwang & Briggs 84] HWANG, KAI AND FAYE A. BRIGGS. [1984]. *Computer Architecture and Parallel Processing*. McGraw-Hill, Inc., New York, NY.

[Hwang 93] HWANG, KAI. [1993]. *Advanced Computer Architecture*. McGraw-Hill, Inc., New York, NY.

[Intel 86] Intel Corporation, Beaverton, Oregon. [Sep 1986]. *iPSC Technical Description*. Order number: 175278-003.

[Intel 93] Intel Corporation, Beaverton, Oregon. [Oct 1993]. *Paragon User's Guide*. Order number: 312489-002.

[Iverson 62] IVERSON, K. E. [1962]. *A Programming Language*. John Wiley & Sons, New York, NY.

[Jessup 95] JESSUP, ELIZABETH R. [1995]. Using the iPSC/2 at CU Boulder. HPSC Course Notes.

[Kahaner et al 77] KAHANER, D., C. MOLER, AND S. NASH. [1977]. *Numerical Methods and Software*. Prentice-Hall, Inc., Englewood Cliffs, NJ.

[Kahaner et al 89] KAHANER, DAVID, CLEVE MOLER, AND STEPHEN NASH. [1989]. *Numerical Methods and Software.* Prentice-Hall, Inc., Englewood Cliffs, NJ. Software on disk with book.

[Kak & Slaney 88] KAK, A.C. AND M. SLANEY. [1988]. *Principles of Computerized Tomographic Images.* IEEE Computer Society Press, Washington, DC.

[Kane 88] KANE, GERRY. [1988]. *MIPS RISC Architecture.* Prentice-Hall, Inc., Englewood Cliffs, NJ.

[Karplus & McCammon 86] KARPLUS, MARTIN AND J. ANDREW MCCAMMON. [Apr 1986]. The dynamics of proteins. *Scientific American*, pages 42–51.

[Kaufman 93] KAUFMAN, L. [1993]. Maximum likelihood, least squares, and penalized least squares for PET. *IEEE Transactions on Medical Imaging*, 12:200–214.

[Kay & Kummerfeld 88] KAY, JUDY AND BOB KUMMERFELD. [1988]. *C Programming in a UNIX Environment.* Addison-Wesley Publishing Company, Reading, MA.

[Keller & Keller 92] KELLER, P. R. AND M. M. KELLER. [1992]. *Visual Cues.* IEEE Computer Society Press, Washington, DC.

[Koelbel et al 94] KOELBEL, CHARLES H., DAVID B. LOVEMAN, ROBERT S. SCHREIBER, JR. GUY L. STEELE, AND MARY E. ZOSEL. [1994]. *The High Performance Fortran Handbook.* Scientific and Engineering Computation. The MIT Press, Cambridge, MA.

[Krol 92] KROL, ED. [1992]. *The Whole Internet: User's Guide and Catalog.* O´Reilly & Associates, Inc, Sebastapol, CA.

[Lawson & Hanson 74] LAWSON, C.L. AND R.J. HANSON. [1974]. *Solving Least Squares Problems.* Prentice-Hall, Inc., Englewood Cliffs, NJ.

[Leighton 92] LEIGHTON, F. THOMSON. [1992]. *Introduction to Parallel Algorithms and Architectures: Arrays, Trees, Hypercubes.* Morgan Kaufmann Publishers, Inc., San Mateo, CA.

[Leiner 94] LEINER, BARRY M. [1994]. Internet technology. *Communications of the ACM*, 37(8):32.

[Leiserson et al 92] LEISERSON, C.E., Z.S. ABUHAMDEH, D.C. DOUGLAS, C.R. FEYNMANN, M.N. GANMUKHI, J.V. HILL, W.D. HILLIS, B.C. KUSZMAUL, M.A. ST.PIERRE, D.S. WELLS, M.C. WONG, S.-W. YANG, AND R. ZAK. [1992]. The network architecture of the Connection Machine CM-5. Technical Report, Thinking Machines Corporation.

[Leithold 76] LEITHOLD, L. [1976]. *The Calculus with Analytic Geometry*. Harper and Row, New York, NY.

[Levkowitz 91] LEVKOWITZ, H. [1991]. Color icons: Merging color and texture perception for integrated visualization of multiple parameters. In *Proceedings of Visualization '91*, pages 164–170, Washington, DC. IEEE Computer Society Press.

[Loan 92] LOAN, C. VAN. [1992]. *Computational Frameworks for the Fast Fourier Transform*. SIAM, Philadelphia, PA.

[Mackinlay 86] MACKINLAY, J. [Apr 1986]. Automating the design of graphical presentations of relational information. *ACM Transactions on Graphics*, 5(2):110–141.

[MasPar 92] MasPar Computer Corporation, Sunnyvale, CA. [Jul 1992]. *MasPar System Overview*. Part Number 9300-0100, Rev. A5.

[MasPar 93a] MasPar Computer Corporation, Sunnyvale, CA. [May 1993]. *MasPar Fortran Reference Manual*. Part Number 9303-0000, Revision A6.

[MasPar 93b] MasPar Computer Corporation, Sunnyvale, CA. [May 1993]. *MasPar Fortran User Guide*. Part Number 9303-0100, Revision A5.

[MathWorks 92a] The MathWorks, Inc., Natick, MA. [Aug 1992]. *MATLAB: Reference Guide*.

[MathWorks 92b] The MathWorks, Inc., Natick, MA. [Aug 1992]. *MATLAB: User's Guide for UNIX Workstations*.

[McCormick et al 87] MCCORMICK, B.H., T.A. DEFANTI, AND M.D. BROWN. [Nov 1987]. Visualization in scientific computing. *Computer Graphics*, 21(6).

[McKim 80] MCKIM, R.H. [1980]. *Experiences in Visual Thinking*. PWS-KENT Publishing Company, Boston, MA, 2nd edition.

[McMahon 86] MCMAHON, F. H. [Dec 1986]. The Livermore Fortran Kernels: A computer test of the numerical performance range. Technical Report UCRL-53745, Lawrence Livermore National Laboratory, Livermore, CA.

[Mesinger & Arakawa 76] MESINGER, F. AND A. ARAKAWA. [1976]. *Numerical Methods Used in Atmospheric Models*, volume 1 of *GARP Publications Series No. 17*. GARP Publications, New York, NY, 2nd edition.

[Metcalf 85] METCALF, MICHAEL, editor. [1985]. *Effective FORTRAN 77*. Oxford University Press, Oxford.

[Miller & Anthes 80] MILLER, ALBERT AND RICHARD A. ANTHES. [1980]. *Meteorology*. Merrill Physical Science. Charles E. Merrill, Columbus, OH, 4th edition.

[Nash 90] NASH, STEPHEN G., editor. [1990]. *A History of Scientific Computing*. ACM Press History Series. Addison-Wesley Publishing Company, Reading, MA.

[nCUBE 94] nCUBE Corporation, Foster City, CA. [Nov 1994]. *nCUBE3: The Teraflops Computing Platform*.

[Olsen 92] OLSEN, D.R. [1992]. *User Interface Management Systems: Models and Algorithms*. Morgan Kaufmann Publishers, Inc., San Mateo, CA.

[Oram & Talbott 91] ORAM, ANDREW AND STEVE TALBOTT. [1991]. *Managing Projects with Make*. Nutshell Handbooks. O´Reilly & Associates, Inc., Sebastopol, CA.

[Page 83] PAGE, REX L., editor. [1983]. *FORTRAN 77 for Humans*. West Publishing Co., St. Paul, MN, 2nd edition.

[Pang 93] PANG, A. [1993]. A syllabus for scientific visualization. In THOMAS, DAVID A., editor, *Scientific Visualization in Mathematics and Science Teaching*. Association for the Advancement of Computing in Education, PO Box 2966, Charlottesville, VA 22902. To be published.

[Patterson & Hennessy 94] PATTERSON, DAVID A. AND JOHN L. HENNESSY. [1994]. *Computer Organization & Design: The Hardware/Software Interface*. Morgan Kaufmann Publishers, Inc., San Mateo, CA.

[Patterson 85] PATTERSON, DAVID A. [Jan 1985]. Reduced instruction set computers. *Communications of the ACM*, 28(1):8–21.

[Pickering 93] PICKERING, R. [1993]. The high-tech science of mummies. *Denver Museum of Natural History Museum Quarterly*, 2:1–5.

[Picket & Grinstein 88] PICKET, R.M. AND G.G. GRINSTEIN. [1988]. Iconographics displays for visualizing multidimensional data. In *Proceedings of the 1988 IEEE Conference on Systems, Man and Cybernetics*, volume I, pages 514–519, Beijing and Shenyang, People's Republic of China.

[Press et al 88] PRESS, WILLIAM H., BRIAN P. FLANNERY, SAUL A. TEUKOLSKY, AND WILLIAM T. VETTERLING. [1988]. *Numerical Recipes in C: The Art of Scientific Computing*. Cambridge University Press, New York, NY.

[Reingold et al 77] REINGOLD, E.M., J. NIEVERGELT, AND N. DEO. [1977]. *Combinatorial Algorithms*. Prentice-Hall, Inc., Englewood Cliffs, NJ.

[Richtmyer & Morton 67] RICHTMYER, ROBERT D. AND K. W. MORTON. [1967]. *Difference Methods for Inital-Value Problems*, volume 4 of *Interscience Tracts in Pure and Applied Mathematics*. John Wiley & Sons, New York, NY, 2nd edition.

[Rosenfeld & Kak 82] ROSENFELD, A. AND A.C. KAK. [1982]. *Digital Picture Processing*, volume 1. Academic Press, Inc., San Diego, CA.

[Rosing et al 91] ROSING, M., R. SCHNABEL, AND R. WEAVER. [1991]. The DINO parallel programming language. *Journal of Parallel and Distributed Computing*, 13:32–40.

[RSI 88] Research Systems, Inc., Boulder, CO. [Sep 1988]. *Introduction to IDL: Interactive Data Language*. Version 1.0.

[RSI 90] Research Systems, Inc., Boulder, CO. [Mar 1990]. *IDL User's Guide: Interactive Data Language*. Version 2.0.

[RSI 92] Research Systems, Inc., Boulder, CO. [Jan 1992]. *IDL Basics: Interactive Data Language*. Version 2.2.

[Russ 92] RUSS, J.C. [1992]. *The Image Processing Handbook*. CRC Press, Inc., Boca Raton, FL.

[Saad & Schultz 89] SAAD, Y. AND M.H. SCHULTZ. [1989]. Data communication in hypercubes. *Journal of Parallel and Distributed Computing*, 6:115–135.

[Sadourney 75] SADOURNEY, ROBERT. [Apr 1975]. The dynamics of finite-difference models of the shallow-water equations. *Journal of Atmospheric Sciences*, 32:680–689.

[Sanz-Serna 92] SANZ-SERNA, J. M. [1992]. Symplectic integrators for hamiltonian problems: an overview. In ISERLES, A., editor, *Acta Numerica 1992*, pages 243–286. Cambridge University Press, Cambridge.

[Scaletti & Craig 91] SCALETTI, C. AND A.B. CRAIG. [1991]. *Using Sound to Extract Meaning from Complex Data*. NCSA. University of Illinois, Champaign IL.

[Sears et al 80] SEARS, F.W., M.W. ZEMANSKY, AND H.D. YOUNG. [1980]. *College Physics*. Addison-Wesley Publishing Company, Reading, MA.

[Sekuler & Blake 85] SEKULER, R. AND R. BLAKE. [1985]. *Perception*. Alfred A. Knopf, New York, NY.

[Sigmon 94] SIGMON, KERMIT. [1994]. *MATLAB Primer*. CRC Press, Inc., Boca Raton, FL, 4th edition.

[Smith 78] SMITH, G. D. [1978]. *Numerical Solution of Partial Differential Equations: Finite Difference Methods*. Oxford Applied Mathematics and Computing Science Series. Oxford University Press, Oxford, 2nd edition.

[Smith 87] SMITH, H.F. [1987]. *Data Structures: Form and Function*. Harcourt Brace Jovanovich, Inc., Orlando, FL.

[Smith et al 76] SMITH, B.T., J.M. BOYLE, J.J. DONGARRA, B.S. GARBOW, Y. IKEBE, V.C. KLEMA, AND C.B. MOLER. [1976]. *Matrix Eigensystem Routines — EISPACK Guide*, volume 6 of *Lecture Notes in Computer Science*. Springer-Verlag, New York, NY, 2nd edition.

[Smith et al 90] SMITH, S., R.D. BERGERON, AND G.G. GRINSTEIN. [1990]. Stereophonic and surface sound generation for exploratory data analysis. In *Proceedings of CHI '90*, pages 125–132.

[Sobell 84] SOBELL, MARK G. [1984]. *A Practical Guide to the UNIX System*. The Benjamin/Cummings Publishing Company, Inc., Redwood City, CA.

[Spencer 89] SPENCER, K.A. [1989]. Computer tomography–an overview. *Journal of Photographic Science*, 37:84–85.

[Spindel & Worcester 90] SPINDEL, R.C. AND P.F. WORCESTER. [Oct. 1990]. Ocean acoustic tomography. *Scientific American*, 263:94–90.

[Stone 87] STONE, HAROLD S., editor. [1987]. *High-Performance Computer Architecture*. Addison-Wesley Publishing Company, Reading, MA, 2nd edition.

[Stone 93] STONE, HAROLD S., editor. [1993]. *High-Performance Computer Architecture*. Addison-Wesley Publishing Company, Reading, MA, 3rd edition.

[Strang 80] STRANG, G. [1980]. *Linear Algebra and Its Applications*. Academic Press, Inc., New York, NY.

[Sun 93] SUN, WEN-YIH. [1993]. Numerical experiments for advection equation. *Journal of Computational Physics*, 108:264–271.

[Tilyou 91] TILYOU, S.M. [1991]. The evolution of positron emission tomography. *Journal of Nuclear Medicine*, 32:15N–26N.

[TMC 88] [May 1988]. Connection Machine Model CM-2 Technical Summary. Technical Report HA87-4, Thinking Machines Corporation, Cambridge, MA.

[TMC 91a] Thinking Machines Corporation, Cambridge, MA. [Jan 1991]. *Connection Machine: Fortran Programming Guide*. Version 1.0.

[TMC 91b] Thinking Machines Corporation, Cambridge, MA. [Jul 1991]. *Connection Machine: Fortran Reference Manual*. Version 1.0 and 1.1.

[TMC 91c] Thinking Machines Corporation, Cambridge, MA. [Jul 1991]. *Connection Machine: Fortran Users's Guide*. Version 1.0 and 1.1.

[TMC 91d] [Jun 1991]. Connection Machine CM-200 Series Technical Summary. Technical report, Thinking Machines Corporation, Cambridge, MA.

[Tondo et al 92] TONDO, CLOVIS L., ANDREW NATHANSON, AND EDEN YOUNT. [1992]. *Mastering Make: A Guide to Building Programs on DOS and UNIX Systems*. Prentice-Hall, Inc., New York, NY.

[van der Steen 94] VAN DER STEEN, AAD J. [Sep 1994]. Overview of recent supercomputers. Technical report, Stichting Nationale Computer Faciliteiten, The Netherlands. Fourth revised edition, `netlib`.

[Washington & Bettge 90] WASHINGTON, WARREN M. AND THOMAS W. BETTGE. [May/Jun 1990]. Computer simulation of the greenhouse effect. *Computers in Physics*, pages 240–246.

[Washington & Parkinson 86] WASHINGTON, WARREN M. AND CLAIRE L. PARKINSON. [1986]. *An Introduction to Three-Dimensional Climate Modeling*. University Science Books, Mill Valley, CA.

[Wehrend & Lewis 90] WEHREND, S. AND C. LEWIS. [1990]. A problem-oriented classification of visualization. In *Proceedings of Visualization '90*, pages 139–143, Washington, DC. IEEE Computer Society Press.

[Weicker 84] WEICKER, REINHOLD P. [Oct 1984]. Dhrystone: A synthetic systems programming benchmark. *Communications of the ACM*, 27(10):1013–1030.

[Williams 85] WILLIAMS, MICHAEL R. [1985]. *A History of Computing Technology*. Computational Mathematics. Prentice-Hall, Inc., Englewood Cliffs, NJ.

[Wilson 94] WILSON, G.V. [Dec 1994]. History of parallel computing. Technical Report TR CSRI-312, Computing Systems Research Institute, University of Toronto.

[Wolfe 89] WOLFE, MICHAEL. [1989]. *Optimizing Supercompilers for Supercomputers*. Research Monographs in Parallel and Distributed Computing. The MIT Press, Cambridge, MA.

[Wyckoff 94] WYCKOFF, R., [1994]. nCUBE Corporation, Personal Communication.

[Zlatev & Waśniewski 92] ZLATEV, ZAHARI AND JERZY WAŚNIEWSKI. [Dec 1992]. Running air pollution models on the connection machine. Technical Report UNIC-92-07, Danish Computing Center for Research and Education, Lyngby, DENMARK.

Index ·

A

advection, 583–664
advection equation
  as total derivative of density, 587
  boundary conditions, 589–592, 658
  characteristic lines, 590–592
  conservation of particles, 588–589
  one-dimensional, 585
  periodic boundary conditions, 592, 594
  periodic boundary equations, 590
  region of solution, 589
  two-dimensional, 658
  verification, 586–587
algorithm
  dissipative, 609
  stability, 609
  von Neumann condition for stability, 610
aliasing artifact, 715
alternate direction exchange (ADE), 472–
  475, 478, 482
Amdahl's Law, 398–402, 459
amplification factor, 603–608
animation, 298, 299, 304, 313, 321, 331
annotation, 314, 320, 331
anonymous ftp, 91–92, 94
arithmetic pipeline, *See* pipeline
arithmetic-logical unit (ALU), 384, 493,
  501
ARPANET, 91
array,
  See also linear array graph; matrix; vec-
    tor
  as data-parallel object, 507–510
  attribute, 508
  constructor

  in CM Fortran, 508–509
  in Fortran 90, 508–509
  in IDL, 234–235, 237–238
  in MATLAB, 188, 189, 204
  section, 193–194, 239, 507, 509
arrow in scientific visualization, 313, 329
assembly code, *See* compiler optimization
automatic compiler vectorization, 348
AVS (Application Visualization System),
  273–291
  AVS Applications, 275–277, 290–291
    Advection application, 291
    Control Panel, 290–291
  AVS network, 277
  color monitor required, 274
  custom modules
    advect_color, 285–286
    advection, 281, 282
    read molecule, 289, 290
  Data Viewer, 273, 275, 277, 286–290
  enter/exit, 275–276
  example of use, 299, 301, 304
  Geometry Viewer, 275, 277, 286–290,
    291
    Cameras Control Panel, 289
    Control Panel, 287–290
    Edit Property, 289
    lights, 289
    normalized image, 287
    Objects stack, 289
    perspective, 289
    Read Object, 287
    SpaceBall, 287
    subdivisions slider, 290
    translucency, 289
  Graph Viewer, 275, 290
  Image Viewer, 275, 290